frederich :

MODERN
SEMICONDUCTOR
DEVICE PHYSICS

MODERN SEMICONDUCTOR DEVICE PHYSICS

Edited by

S. M. Sze

UMC Chair Professor
Department of Electronics Engineering
National Chiao Tung University

A WILEY–INTERSCIENCE PUBLICATION

JOHN WILEY & SONS, INC.

New York • Chichester • Weinheim • Brisbane • Singapore • Toronto

Library of Congress Cataloging in Publication Data:

Modern semiconductor device physics / edited by S. M. Sze.
　　　p.　cm.
　　"A Wiley-Interscience publication."
　　Includes bibliographical references and index.
　　ISBN 0-471-15237-4 (cloth : alk. paper)
　　1. Semiconductors.　I. Sze, S. M., 1936– .
　QC611.M674　1997
　621.3815′2—dc21　　　　　　　　　　　　　　97–4311

Printed in the United States of America

10 9 8 7 6 5 4 3 2 1

CONTENTS

CONTRIBUTORS

P. M. Asbeck, Department of Electrical and Computer Engineering, University of California, San Diego, La Jolla, California, USA

B. J. Baliga, Power Semiconductor Research Center, North Carolina State University, Raleigh, North Carolina, USA

S. Chandrasekhar, Bell Labs, Crawford Hill Laboratory, Lucent Technologies, Holmdel, New Jersey, USA

H. Eisele, Department of Electrical Engineering and Computer Science, University of Michigan, Ann Arbor, Michigan, USA

T. A. Fjeldly, Department of Physical Electronics, Norwegian University of Science and Technology, Trondheim, Norway

M. A. Green, Photovoltaics Special Research Centre, University of New South Wales, Kensington, New South Wales, Australia

G. I. Haddad, Department of Electrical Engineering and Computer Science, University of Michigan, Ann Arbor, Michigan, USA

S. J. Hillenius, Bell Laboratories, Lucent Technologies, Murray Hill, New Jersey, USA

T. P. Lee, Optoelectronic Technology Research, Bell Communications Research (Bellcore), Red Bank, New Jersey, USA

S. Luryi, Department of Electrical Engineering, State University of New York at Stony Brook, Stony Brook, New York, USA

M. S. Shur, Department of Electrical, Computer, and Systems Engineering, Rensselaer Polytechnic Institute, Troy, New York, USA

S. M. Sze, Department of Electronics Engineering, National Chiao Tung University, Hsinchu, Taiwan, ROC

A. Zaslavsky, Division of Engineering, Brown University, Providence, Rhode Island, USA

PREFACE

The electronics industry has enjoyed phenomenal growth in the past 50 years, and has had an unprecedented impact on our society. It is now the largest industry in the world with global sales of over one trillion dollars. The foundation of the electronics industry is the *semiconductor device*. To meet the tremendous demand of the electronics industry, the semiconductor device field has also grown rapidly. Coincident with this growth, the semiconductor device literature has expanded from an annual publication of only a few papers in the 1950s to over 10,000 papers at present.

In 1981, Wiley-Interscience published *Physics of Semiconductor Devices*, Second Edition, which gave a comprehensive introductory account of the physics and operational principles of all classic semiconductor devices. Since that time, more than 120,000 papers on semiconductor devices have been published. For access to this vast literature, we have invited a team of world-renowned experts in the semiconductor device field and prepared *Modern Semiconductor Device Physics (MSDP)*. This book offers detailed coverage of the enhanced performance of classic devices and the physical concepts of novel devices conceived since 1981. It is important to point out that the basic physics of the classic devices remains essentially the same, and about 80% of the contents in the Second Edition remains valid. To keep *MSDP* within reasonable length, we will simply use the results in the Second Edition without derivations, and we will not repeat the material contained there. Therefore, the *MSDP* complements the Second Edition.

Modern Semiconductor Device Physics is intended as a textbook for graduate students in applied physics, electrical and electronics engineering, and materials science. It can also serve as a reference for engineers and scientists actively involved in semiconductor device research and development. It is assumed that the reader has already acquired a basic understanding of device operation as given in *Physics of Semiconductor Devices*, Second Edition (Wiley, 1981). For those who are not familiar with the Second Edition, both books can be studied concurrently. In this case, we assume that the reader has already studied the standard undergraduate textbooks on semiconductor devices such as *Semiconductor Devices: Physics and Technology* (Wiley, 1985).

In the course of writing this text, we had the assistance and support of many people. First we express our appreciation to the management of our academic and industrial institutions, without whose help this book could not have been written. We have benefited from suggestions made by our reviewers: Dr. M. Barnett of the University of Delaware, Prof. R. J. Beresford of Brown University, Dr. R. Bhat and Dr. C. E. Zah of Bell Communications Research, Prof. C. Y. Chang of the National Chiao Tung University, Prof. K. Chang of Texas A&M University, Prof. L. Larson of the University of California, San Diego, Dr. M. Mastrapasqua and Dr. K. K. Ng of Lucent Technologies, Dr. B. J. Moon of Vitesse Semiconductor Corporation, Dr. L. D. Partain of Varian Associates, Prof. P. P. Ruden of the University of Minnesota, and Prof. R. J. Trew of Case Western Reserve University.

We are further indebted to Mr. N. Erdos for technical editing of the manuscript. We wish to thank Mrs. T. W. Sze for preparing the Appendixes, and Ms. P. L. Huang, Ms. C. C. Chang, Ms. L. T. Chou, Ms. H. T. Chua, and Ms. A. Y. W. Wong for preparing the Index. We also thank the Spring Foundation of the National Chiao Tung University for the financial support. I would especially like to thank the United Microelectronics Corporation (UMC), Taiwan, ROC, for the UMC Chair Professorship grant and the hospitality of the Hong Kong University of Science and Technology and the University of Hong Kong that provided the environment to work on this book.

S. M. SZE

Introduction

S. M. SZE

National Chiao Tung University, Hsinchu, Taiwan, ROC

SEMICONDUCTOR DEVICE LITERATURE

Semiconductor devices were studied[1,2] as early as in 1874. However, the major breakthrough came about in 1947 when a research team at Bell Laboratories of Lucent Technologies (formally AT&T Bell Laboratories) invented the bipolar transistor.[3,4] The bipolar transistor and its related semiconductor devices have formed the basis of the modern electronics industry, with global sales second to none.

Coincident with the growth of the electronic industry, semiconductor device literature has also burgeoned and diversified. Figure 1 shows the annual publication of device papers.[5] The INSPECT database for semiconductor devices, the source for Fig. 1, was established in 1969. The total number of device papers published prior to 1969 is estimated to be about 2500. If the device literature exhibits normal life-cycle characteristics (i.e., from incubation to growth, to saturation, and finally to decline), the incubation period was almost three quarters of a century (from 1874 to 1947). After 1947, we moved to the growth period. The initial rapid growth lasted for about 27 years. During this period, the annual number of publications doubled every two years. By 1974, annual publications reached 3000, and the total device papers published in the past 100 years reached 13,000. After 1974, growth has continued but at a slower rate, doubling every 12 years. To date, over 150,000 papers have been published.

We expect that the growth period will continue at least into the early 21st century. If the growth rate remains the same, by the year 2000, the grand total will approach 220,000 papers and by 2005 the grand total would be 350,000! It is indeed overwhelming to realize that we have such a vast literature of semiconductor devices!

Modern Semiconductor Device Physics, Edited by S. M. Sze.
ISBN 0-471-15237-4 © 1998 John Wiley & Sons, Inc.

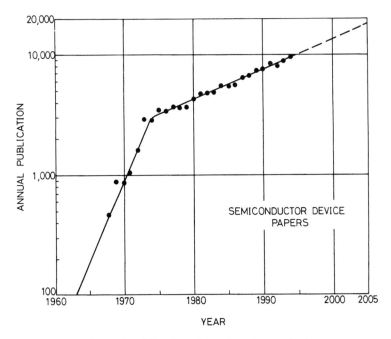

Fig. 1 Annual publication of semiconductor device papers.

DEVICE BUILDING BLOCKS

In the book *Complete Guide to Semiconductor Devices*,[6] a total of 67 major semiconductor devices are identified, together with about 110 device variations that are considered to be related to the former. However, all these devices can be constructed from a small number of device building blocks as shown in Fig. 2.

Figure 2a is the metal–semiconductor interface, which is an intimate contact formed between a metal and a semiconductor. This building block was the first semiconductor device ever studied (in the year 1874). This interface can be used as a rectifying contact, called a Schottky barrier, or as an ohmic contact. We can use this interface to form many useful devices. For example, by using a rectifying contact as the gate, and two ohmic contacts as the source and drain, we can form a MESFET.

The second building block is the *p–n* junction (Fig. 2b), which is formed between a *p*-type and an *n*-type semiconductor. By combining two *p–n* junctions, that is, by adding another *p*-type semiconductor, we form the *p–n–p* bipolar transistor which was invented in 1947 and ushered-in the modern electronics era. If we combine three *p–n* junctions to form a *p–n–p–n* structure, it is a thyristor.

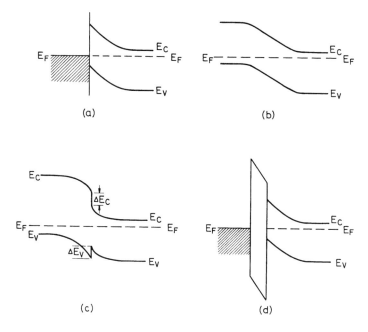

Fig. 2 (a) Metal–semiconductor interface. (b) *p–n* junction. (c) Heterojunction interface. (d) MOS structure.

The third building block (Fig. 2c), is the heterojunction interface, that is, an interface formed between two dissimilar semiconductors. For example, by using $n - Al_xGa_{1-x}As/p - GaAs/p - Al_xGa_xAs$ structure with two heterojunction interfaces, we can form the double-heterostructure laser. Another important example is the resonant-tunneling diode (RTD) which is formed with four heterojunctions, for instance, $GaAs/Al_xGa_{1-x}As/GaAs/Al_xGa_{1-x}As/GaAs$.

Figure 2d shows the metal–oxide–semiconductor (MOS) structure (or MIS structure if the oxide is replaced by an insulator). This structure can be considered as a combination of a metal–oxide interface and an oxide–semiconductor interface. The most important device for ultra-large-scale integration (ULSI) is the MOSFET, which contains a MOS structure as the gate and two *p–n* junctions as the source and drain.

To fabricate these building blocks, we can use the following technologies: (1) a metalization process to form metal–semiconductor contact and metal–oxide interface (which is a part of the MOS structure), (2) ion implantation or impurity diffusion to form the *p–n* junction, (3) an epitaxy process and especially molecular beam epitaxy (MBE) and metalorganic chemical vapor deposition (MOCVD) to form the heterojunction, and (4) an oxidation or thin-film deposition process to form the oxide–semiconductor interface.[7,8]

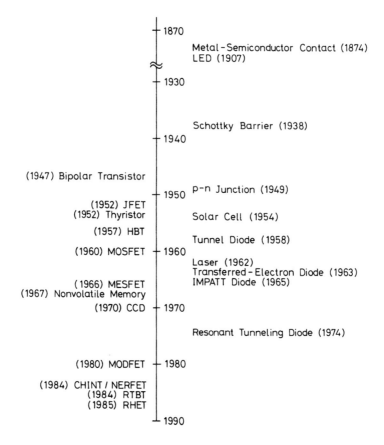

Fig. 3 Chronological chart of major semiconductor devices.

MAJOR MILESTONE

Some major semiconductor devices are shown in Fig. 3 in chronological order; those listed on the right are two-terminal devices, and those on the left are three-terminal and four-terminal devices. The earliest systematic study of semiconductor devices (metal–semiconductor contacts) is generally attributed to Braun, who in 1874 noted the dependence of the resistance on the polarity of the applied voltage and the detailed surface conduction.[1] The electroluminescence phenomenon (for the light-emitting diode) was discovered by Round in 1907, who observed the generation of yellowish light from a crystal of carborundum when he applied a potential between two points on the crystal.[9] In 1938, Schottky suggested that the potential barrier in a metal–semiconductor contact could arise from stable space charges in the semiconductor alone without the presence of a chemical layer; the model is known as the Schottky barrier.[10]

In 1947, the point-contact transistor was invented by Bardeen and Brattain.[3] This was followed by Shockley's classic paper on the $p–n$ junction and bipolar junction transistor[4] in 1949. The transistor had such an unprecedented impact on the electronics industry that the three inventors were awarded the Nobel Prize in 1956. Shockley subsequently proposed the junction field-effect transistor in 1952, the first semiconductor field-effect device ever invented.[11] In the same year Ebers developed the basic model for the thyristor, which is one of the most versatile switching devices.[12] The solar cell, based on the photovoltaic effect, was first developed by Chapin, Fuller, and Pearson in 1954 using a diffused silicon $p–n$ junction and obtained a 6% conversion efficiency.[13] The solar cell is a major candidate to obtain energy from the sun for long-term terrestrial applications.

In 1957, Kroemer proposed the heterojunction bipolar transistor to improve the emitter efficiency;[14] this device is potentially one of fastest semiconductor devices. In 1958, Esaki observed negative resistance characteristics in a heavily doped $p–n$ junction, which led to the discovery of the tunnel diode.[15] The impact of the tunnel diode on the physics of semiconductors has been large, and Esaki was awarded the Nobel Prize in 1973.

The most important device for ultra-large-scale integration (ULSI) is the enhancement-mode MOSFET,* which was reported by Kahng and Atalla[16] in 1960. MOSFET and related integrated circuits now constitute about 90% of the semiconductor device market. Another very important device is the laser diode. In 1962, three groups headed by Hall,[17] Nathan,[18] and Quist[19] announced, almost simultaneously, that they had achieved lasing in semiconductors. Lasers are the key components for a wide range of applications, including optical-fiber communication, laser printing, and atmospheric pollution monitoring.

Three important microwave devices were invented or realized in the next three years. The first device is the transferred-electron diode (TED, also called Gunn diode)[20] by Gunn in 1963. The TED is used extensively in millimeter-wave applications. The second device is the IMPATT diode;[21] its operation was first observed by Johnston, DeLoach, and Cohen in 1965. IMPATT diodes can generate the highest CW (continuous wave) power at millimeter-wave frequencies among all semiconductor devices. The third device is the MESFET,[22] invented by Mead in 1966, which is a very important device for monolithic microwave integrated circuits (MMIC).

The nonvolatile memory[23] was first proposed by Kahng and Sze in 1967. Nonvolatile memory devices have been widely used in integrated circuits such

---

*A similar MOSFET structure was proposed by E. Labate in 1957. The proposal was recorded in *Bell Telephone Laboratories Notebook* No. 57-25647, p. 179. The proposal stated "To control the resistance in a silicon bar by applying a voltage to a metal plate deposited on top of an insulator. This device would control the flow of current from the surface like in a vacuum tube, the resistance can be altered by a voltage applied to the grid of the vacuum tube. This device would be a voltage controlled device." The proposal was witnessed by M. M. Atalla on March 8, 1957.

as the EPROM (erasable programmable read-only memory), EEPROM (electrically-erasable PROM), and the flash memory. Because of its outstanding attributes of high density, electrical rewritability, and nonvolatility, flash memory is projected to surpass DRAM (dynamic random access memory) in system applications and to replace most magnetic memories in the not-too-distant future.

The charge-coupled device (CCD)[24] was invented by Boyle and Smith in 1970. CCD is used extensively in optical sensing applications. The year 1970 represents a line of demarcation for device commercialization. All semiconductor devices that are mass produced, either as discrete devices or as integrated circuits, were invented in or before 1970. Of course, the performances of these devices have been substantially improved over the years with the help of advanced processing technologies, powerful simulation tools, and ingenious modifications of device configurations.

One of the key developments after 1970 was the observation of resonant tunneling in a double-barrier structure[25] by Chang, Esaki, and Tsu in 1974. The resonant-tunneling diode is the basis for most quantum-effect devices, which offer extremely high density, ultra-high speed, and enhanced functionality because they permit a greatly reduced number of devices to perform a given function. Another interesting development is the realization of a high-speed device using modulation doping, the MODFET[26] (modulation-doped FET) by Mimura, Hiyamizu, Fujii, and Nanbu in 1980. With proper selection of heterojunction materials, the MODFET is expected to be the fastest field-effect transistor.

After 1980, most of the new devices were made with heterojunctions.[27] One interesting device, the CHINT (charge injection transistor)[28] which has at least two heterojunctions, was invented by Luryi, Kastalsky, Gossard, and Hendel in 1984. A related device, called NERFET[29] (negative resistance FET) was also invented by the same team. These devices are based on real-space-transfer (RST) mechanism, and they can perform many novel circuit operations and have high functionality. For example, the complex function NORAND can be implemented using only one CHINT. By combining the resonant-tunneling diode with classic devices, many new quantum-effect devices (QED) were proposed. One of them is the RTBT (resonant-tunneling bipolar transistor) which has at least four heterojunctions. This device was proposed by Capasso and Kiehl,[30] and also independently by Ricco and Solomon[31] in 1984. Another QED is the RHET[32] (resonant-tunneling hot-electron transistor) proposed by Yokoyama, Imamura, Muto, Hiyamizu, and Nishi in 1985. These QED and RST devices have significant potential to complement other technologies. For example, the optoelectronic integrated circuit is likely to benefit from the introduction of ultrafast functional elements based on resonant tunneling or real-space-transfer effect.

It should be pointed out that in Fig. 3 we have not included semiconductor sensors (except the photonic devices) that can detect a nonelectrical input

Modern Semiconductor Device Physics

Physics of Semiconductor Devices, 2nd Ed

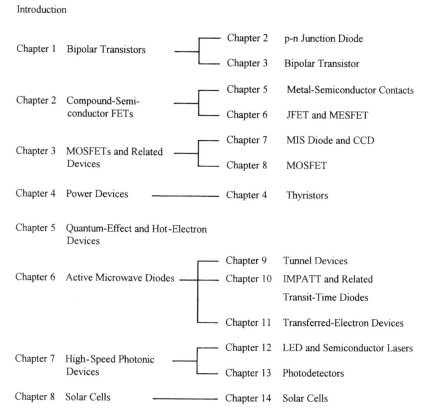

Introduction

Chapter 1 Bipolar Transistors

Chapter 2 p-n Junction Diode

Chapter 3 Bipolar Transistor

Chapter 2 Compound-Semi-
conductor FETs

Chapter 5 Metal-Semiconductor Contacts

Chapter 6 JFET and MESFET

Chapter 3 MOSFETs and Related
Devices

Chapter 7 MIS Diode and CCD

Chapter 8 MOSFET

Chapter 4 Power Devices

Chapter 4 Thyristors

Chapter 5 Quantum-Effect and Hot-Electron
Devices

Chapter 6 Active Microwave Diodes

Chapter 9   Tunnel Devices

Chapter 10 IMPATT and Related
Transit-Time Diodes

Chapter 11 Transferred-Electron Devices

Chapter 7 High-Speed Photonic
Devices

Chapter 12 LED and Semiconductor Lasers

Chapter 13 Photodetectors

Chapter 8 Solar Cells

Chapter 14 Solar Cells

Fig. 4 Arrangement of chapters and relation to *Physics of Semiconductor Devices*, 2nd Ed.

signal and convert it to an appropriate output signal. The semiconductor sensor field is also moving at a rapid pace; its publications now constitute about 20% of all semiconductor device literature (electrical and nonelectrical input signals). Readers interested in semiconductor sensors may consult appropriate sensor textbooks or reference books.[33]

ORGANIZATION OF THE BOOK

Modern Semiconductor Device Physics (MSDP) is written to complement *Physics of Semiconductor Devices*, Second Edition (Wiley, 1981). Therefore, we have adopted the same format and arranged the chapters in a similar sequence as the Second Edition. A comparison of these arrangements is shown in Fig. 4.

MSDP has eight chapters. Chapter 1 deals with the advanced bipolar transistor, heterojunction bipolar transistor, and circuit applications. The corresponding chapters in the Second Edition are Chapters 2 and 3 which provide the background information on bipolar transistors. The metal–semiconductor contacts, MESFET, JFET, and MODFET are considered in Chapter 2 (with corresponding Chapters 5 and 6 in the Second Edition). MOSFETs and related devices such as SOI (silicon-on-insulator), DRAM, and nonvolatile memory are considered in Chapter 3 (Chapters 7 and 8 in the Second Edition). Power devices such as the power MOSFET, insulated gate bipolar transistor, and MOS-gated thyristor are presented in Chapter 4 (Chapter 4 in the Second Edition). The chapter sequence in *MSDP* has been altered here, because many new power devices use combinations of bipolar transistor and MOSFET.

Chapter 5 is concerned with quantum-effect and hot-electron devices. Since most of the devices have been developed after 1981, we do not have a corresponding chapter in the Second Edition. Microwave devices are considered in Chapter 6, which covers tunnel devices, IMPATT diodes, and transferred-electron devices (Chapters 9, 10, and 11 respectively in the Second Edition). High-speed photonic devices, especially quantum-well lasers, distributed feedback lasers, photodetectors, and optoelectronic integrated circuits are considered in Chapter 7 (Chapters 12 and 13 in the Second Edition). Finally, solar cells are presented in Chapter 8 (corresponding to Chapter 14 in the Second Edition). In the Appendixes, we have included the List of Symbols, the International System of Units, Unit Prefixes, the Greek Alphabet, and Physical Constants. Also included are the updated values for Lattice Constants, Properties of Important Element and Binary Semiconductors, Properties of Ternary III–V Semiconductors,[34] and Properties of Insulators.

Semiconductor device literature is presently expanding at a rapid pace, as can be seen from Fig. 1. Many topics, such as ultimate performances of classic devices, novel functionality of quantum-effect devices, and optoelectronic integrated circuits, are still under intensive study. The material presented in this book is intended to serve as a foundation. The references listed at the end of each chapter can supply more information.

REFERENCES

1. F. Braun, "Uber die Stromleitung durch Schwefelmetalle," *Ann. Phys. Chem.* **153**, 556 (1874).
2. Most of the classic device papers are collected in S. M. Sze, Ed., *Semiconductor Devices: Pioneering Papers*, World Scientific, Singapore, 1991.

3. J. Bardeen and W. H. Brattain, "The transistor, a semiconductor triode," *Phys. Rev.* **71**, 230 (1948).

4. W. Shockley, "The theory of *p–n* junction in semiconductors and *p–n* junction transistors," *Bell Syst. Tech. J.* **28**, 435 (1949).

5. From INSPEC database, Bell Laboratories, Lucent Technologies, 1996.

6. K. K. Ng, *Complete Guide to Semiconductor Devices*, McGraw–Hill, New York, 1995.

7. For a reference on basic processing technology, see, for example, S. M. Sze, Ed., *VLSI Technology*, 2nd ed., McGraw–Hill, New York, 1988.

8. For a reference on advanced processing technology, see, for example, C. Y. Chang and S. M. Sze, Eds., *ULSI Technology*, McGraw–Hill, New York, 1996.

9. H. J. Round, "A note on carborundum," *Electr. Wld* **19** (February 9), 309 (1907).

10. W. Schottky, "Halbleitertheorie der Sperrschicht," *Naturwissenschaften* **26**, 843 (1938).

11. W. Shockley, "A unipolar field-effect transistor," *Proc. IRE* **40**, 1365 (1952).

12. J. J. Ebers, "Four terminal of *p–n–p–n* transistors," *Proc. IRE* **40**, 1361 (1952).

13. D. M. Chapin, C. S. Fuller, and G. L. Pearson, "A new silicon *p–n* junction photocell for converting solar radiation into electrical power," *J. Appl. Phys.* **25**, 676 (1954).

14. H. Kroemer, "Theory of a wide-gap emitter for transistors," *Proc. IRE* **45**, 1535 (1957).

15. L. Esaki, "New phenomenon in narrow germanium *p–n* junctions," *Phys. Rev.* **109**, 603 (1958).

16. D. Kahng and M. M. Atalla, "Silicon–silicon dioxide surface device," in *IRE Device Research Conference*, Pittsburgh, 1960. (The paper can be found in Ref. 2.)

17. R. N. Hall, G. E. Fenner, J. D. Kingsley, T. J. Soltys, and R. O. Carlson, "Coherent light emission from GaAs junctions," *Phys. Rev. Lett.* **9**, 366 (1962).

18. M. I. Nathan, W. P. Dumke, G. Burns, F. H. Dill, Jr., and G. Lasher, "Stimulated emission of radiation from GaAs *p–n* junctions," *Appl. Phys. Lett.* **1**, 62 (1962).

19. T. M. Quist, R. H. Rediker, R. J. Keyes, W. E. Krag, B. Lax, A. L. McWhorter, and H. J. Zeigler, "Semiconductor maser of GaAs," *Appl. Phys. Lett.* **1**, 91 (1962).

20. J. B. Gunn, "Microwave oscillations of current in III-V semiconductors," *Solid State Commun.* **1**, 88 (1963).

21. R. L. Johnston, B. C. DeLoach, Jr., and B. G. Cohen, "A silicon diode microwave oscillator," *Bell Syst. Tech. J.* **44**, 369 (1965).

22. C. A. Mead, "Schottky barrier gate field effect transistor," *Proc. IEEE* **54**, 307 (1966).

23. D. Kahng and S. M. Sze, "A floating gate and its application to memory devices,"

Bell Syst. Tech. J. **46**, 1283 (1967).

24. W. S. Boyle and G. E. Smith, "Charge coupled semiconductor devices," *Bell Syst. Tech. J.* **49**, 587 (1970).

25. L. L. Chang, L. Esaki, and R. Tsu, "Resonant tunneling in semiconductor double barriers," *Appl. Phys. Lett.* **24**, 593 (1974).

26. T. Mimura, S. Hiyamizu, T. Fujii, and K. Nanbu, "A new field-effect transistor with selectively doped GaAs/n-Al$_x$Ga$_{1-x}$As Heterojunction," *Jap. J. Appl. Phys.* **19**, L225 (1980).

27. The references for novel device structures invented or developed after 1974 can be found in, for example, S. M. Sze, Ed., *High Speed Semiconductor Devices*, Wiley–Interscience, New York, 1990.

28. S. Luryi, A. Kastalsky, A. C. Gossard, and R. H. Hendel, "Charge injection transistor based on real space hot-electron transfer," IEEE *Trans. Electron Dev.* **ED-31**, 832 (1984).

29. A. Kastalsky, S. Luryi, A. G. Gossard, and R. H. Hendel, "A field-effect transistor with a negative differential resistance," *IEEE Electron Dev. Lett.* **EDL-5**, 57 (1984).

30. F. Capasso and R. A. Kiehl, "Resonant tunneling transistor with quantum well base and high-energy injection: a new negative differential resistance device," *J. Appl. Phys.* **58**, 1366 (1985).

31. B. Ricco and P. M. Solomon, "Tunable resonant tunneling semiconductor emitter structure," *IBM Tech. Disclos. Bull.* **27**, 3053 (1984).

32. N. Yokoyama, K. Imamura, S. Muto, S. Hiyamizu, and H. Nishi, "A new functional resonant tunneling hot electron transistor (RHET)," *Jap. J. Appl. Phys.* **24**, L-853 (1985).

33. For example, S. M. Sze, Ed., *Semiconductor Sensors*, Wiley–Interscience, New York, 1994.

34. O. Madelung, Ed., *Semiconductor-Group IV Elements and III–V Compounds*, Springer-Verlag, Berlin, 1991.

1 Bipolar Transistors

PETER M. ASBECK

University of California, San Diego

1.1 INTRODUCTION

Bipolar transistors were the earliest solid-state devices with useful amplifying properties, and traditionally have been preeminent in high-speed circuits, in analog circuits, and in power devices. In recent years, the leadership of bipolars has been severely challenged by MOSFET technology. Nonetheless, bipolar transistors retain their leading role in these areas. Bipolar technology is also undergoing rapid transformations that promise to maintain or extend this leadership.

Figure 1 shows the schematic cross-section of a bipolar transistor, and the associated band diagram in the direction of electron travel at the center of the emitter. An n–p–n device is pictured, and will continue to be used as an example throughout the chapter, although most of the results apply equally to p–n–p transistors. The bipolar structure provides a number of natural advantages:

1. Electrons travel vertically through the device. The dimensions that control transit time are established by processes with excellent dimensional control, such as diffusion, ion implantation, and epitaxy. It is, therefore, easy to produce structures in which the electron transit time is short, which corresponds to high cutoff frequency, f_T. This can be accomplished without the significant demands on lithography that are typically required with FETs.

2. The entire emitter area conducts current, unlike the situation for FETs, in which only a thin channel conducts current. It is, therefore, possible

Modern Semiconductor Device Physics, Edited by S. M. Sze.
ISBN 0-471-15237-4 © 1998 John Wiley & Sons, Inc.

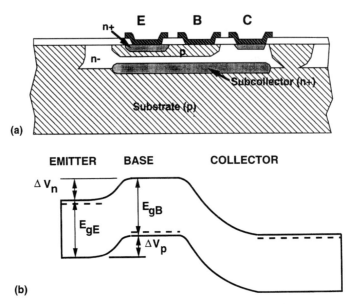

Fig. 1 Schematic *n–p–n* bipolar transistor cross-section (a) and its associated energy-band diagram along the direction of electron travel (b).

to provide large amounts of output current per unit chip area, and maintain high circuit density.

3. The input voltage directly controls the density of the carriers that provide the output current, I_C, leading to I_C variation of the form $\exp(qV/nkT)$ with n close to or equal to 1. This leads, in turn, to transconductance, g_m, of the form $g_m = qI_C/kT$. This is the highest obtainable in any three-terminal device based on voltage modulation of carrier density (a category which encompasses virtually all transistors today). The high transconductance allows circuit operation with low input voltage swings, which is central to low power-delay product in logic applications.

4. The turn-on voltage of bipolar transistors, V_{BE}, is relatively independent of device size and process variations, since it corresponds to the built-in potential of a *p–n* junction. This characteristic minimizes variations in turn-on voltage across a wafer, as needed to minimize offset voltages in analog circuits, and to operate logic circuits with small voltage swings.

5. The input capacitance of bipolar transistors tends to scale with the operating current to the extent that it is dominated by the diffusion capacitance. As a result, bipolar devices can be operated at high or low current, and the input capacitance will adjust accordingly, so that fanout delays do not vary greatly. In bipolar circuitry it is not typically

necessary to tailor device sizes to correspond to the load being driven.

6. For high-voltage, high-current applications, it is possible to obtain conductivity modulation of the lightly doped and resistive collector region, due to double injection of electrons and holes. This effect can provide substantially lower series resistance than obtained with field-effect transistors.

Bipolar transistors also suffer from several disadvantages compared to FETs. Included among these are:

1. Finite DC input (base) current. By contrast, FETs require substantial input current at high frequency, but not at DC.
2. When operated in the saturation regime (forward-biased base–collector junction), base current increases and excess charge is stored. As a result, bipolar transistor switches tend to be slow and are not used in logic circuits. Fast logic families such as emitter-coupled logic employ elaborate techniques to avoid saturation.
3. Turn-on voltage cannot be used as a design variable within a circuit. Only transistors with positive turn-on voltage (enhancement-mode devices) are fabricated at present.

Continuous progress has been made to strengthen the advantages of bipolar transistors, and to minimize their shortcomings. Advanced processes allow smaller device dimensions and minimized parasitics. Heterojunctions add a new degree of freedom in transistor design, enabling higher speeds. Where ultrahigh input impedance is required and pass transistors for signal switching within logic circuits must be implemented, bipolars have been combined with FETs in the BiCMOS technology (refer to Chapter 3).

This chapter discusses the underlying principles of bipolar transistor design and performance. Principles of bipolar transistor operation are first discussed, including primary limits on performance. Advanced silicon homojunction transistor fabrication and heterojunction bipolar transistor technology are then described. The characteristics of heterojunction transistors in various material systems, including silicon, GaAlAs/GaAs, and InGaAs/InAlAs/InP are then presented.

1.2   PRINCIPLES OF BIPOLAR TRANSISTOR OPERATION

1.2.1   Transistor Current-Flow Mechanisms

The application of an input forward-bias voltage to the base–emitter junction (V_{BE}) of a bipolar transistor reduces the conduction-band energy barrier for

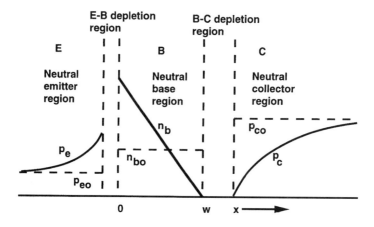

Fig. 2 Minority carrier distributions in a bipolar transistor biased in the forward active region.

electron flow from emitter to base (represented by ΔV_n in Fig. 1). Electrons that surmount this barrier travel across the base by diffusion and, potentially, drift. Those carriers that reach the collector–base junction are swept into the collector by the high fields within the junction region, and provide the desired output collector current. At the same time, the input voltage, V_{BE}, causes holes to flow into the base terminal. Hole charge must build up in the base to balance the charge of the added electrons and maintain quasineutrality. The distribution of electrons and holes within the transistor is shown schematically in Fig. 2 for bias in the active mode (i.e., emitter–base junction forward biased, and base–collector junction reverse biased). Excess carrier recombination occurs between holes and electrons within the base, and holes are injected from base to emitter. To supply the holes, a net current flow is required from the base terminal, which constitutes an undesirable input current. One of the tasks of bipolar transistor design is to maximize the collector current while minimizing the associated base current. For most applications, common-emitter DC current gain on the order of 30–100 is required.

   To quantify the transistor terminal currents, it is convenient to first consider the collector current and estimate its magnitude assuming no recombination between electrons and holes. The base current can then be computed as a perturbation to this current flow.

Collector Current.   If the semiconductor composition and doping are uniform within the base, then the electric fields within the quasineutral base region are small. Minority carriers flow by diffusion only – with no drift component.

The current flow associated with the electrons (collector current in the n–p–n transistor) is then very easily evaluated to be

$$J_c = qD_n \; (dn/dx) = qD_n[n_b(0) - n_b(w)]/w \tag{1}$$

where $n_b(0)$ and $n_b(w)$ are the electron concentrations at the emitter and collector edges of the base and D_n is the electron diffusion coefficient in the base. The electron concentrations can be calculated assuming the injection levels are moderate and recombination is slight, since the carrier quasi–Fermi levels will be constant across the base–emitter and base–collector junctions. Applying Boltzmann statistics,

$$p_b(0)n_b(0) = n_i^2 \; \exp(qV_{BE}/kT) \tag{2}$$
$$p_b(w)n_b(w) = n_i^2 \; \exp(qV_{BC}/kT)$$

where n_i is the intrinsic carrier concentration in the base, and the terminal voltages V_{BE} and V_{BC} correspond to the difference in energy between the carrier quasi–Fermi levels. The electron current is found to be:[1]

$$J_c = qD_n n_i^2 [\exp(qV_{BE}/kT) - \exp(qV_{BC}/kT)]/p_b w \tag{3}$$

where p_b is the hole density in the base, which is approximately equal to the doping level, and w is the quasineutral base thickness.

A variety of corrections to this derivation can be made. Fermi–Dirac statistics may apply in the base. The quasi–Fermi level may not be constant in strongly reverse-biased junctions, although the resultant error is negligible. More importantly, in realistic devices, the composition and doping of the base may be nonuniform. Figure 3 shows the representative doping profile of a modern bipolar transistor and its associated band diagram. Within the base, there are significant acceptor doping gradients, which establish a built-in electric field

$$\mathscr{E} = (kT/q)(dp/dx)/p \tag{4}$$

This electric field insures that the hole current flow normal to the junction vanishes, with the drift component balancing the diffusion associated with the doping gradients. The built-in fields can be appreciable; for example, grading of doping of ×10 across a base whose width is 100 nm results in $\mathscr{E} = 6000$ V/cm. The electrons injected into the base from the emitter, thus, will flow by drift as well as diffusion. In modern transistors, the electric field is intentionally provided to accelerate the electrons towards the collector, increasing the current flow and decreasing the stored electron charge. To calculate the collector current in these transistors, we must consider the overall electron current associated with drift and diffusion:

$$J_c = q\mu n \mathscr{E} + qD_n(dn/dx) \tag{5}$$

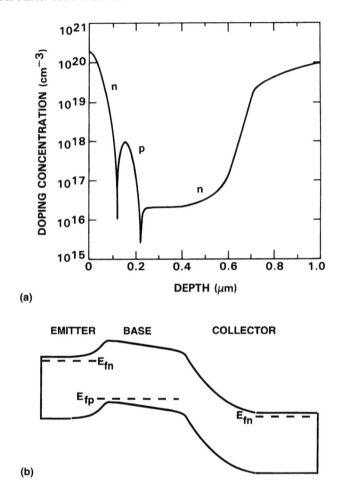

Fig. 3 (a) Representative doping profile for Si homojunction bipolar transistor. (b) Associated band diagram.

When the expression for electric field in Eq. 4 is inserted into Eq. 5 and integrated over the base width, the overall expression for collector current is found to be[2]

$$J_c = qD_n n_i^2 [\exp(qV_{BE}/kT) - \exp(qV_{BC}/kT)]/\int p(x)dx \qquad (6)$$

where the integral is carried out across the quasineutral base. This equation, remarkably, has precisely the same form as Eq. 3—although it is now considerably more general.

A further extension to cover a large class of modern devices can also be carried out. If the material composition and hence bandgap is changed within the base, then the intrinsic carrier concentration will also change. For example, $n_i^2 = N_c N_v \exp(-E_g/kT)$), where N_c, N_v are effective densities of states in conduction and valence bands, and E_g is the bandgap. Under these circumstances, the current flow of carriers proceeds according to a generalization of the drift-diffusion equations, given by[3]

$$J_n = \mu_n n[q\mathscr{E} + (d\Delta E_c/dx] + qD_n[(dn/dx) - (n/N_c)(dN_c/dx)] \qquad (7a)$$

$$J_p = \mu_p p[-q\mathscr{E} + (d\Delta E_v/dx)] + qD_p[(dp/dx) - (p/N_v)(dN_v/dx)] \qquad (7b)$$

where ΔE_c and ΔE_v are conduction and valence band discontinuities. The terms added to $\mathscr{E}$ are associated with composition-dependent quasielectric fields that drive carrier flow. The driving force is the change in the electron or hole binding energy between materials of different composition. The terms added to dn/dx and dp/dx are diffusion-like terms associated with gradients in density of states at the band edges as the material composition changes. As will be described in Section 1.4, it is possible to take advantage of these added terms, particularly the quasielectric fields, to increase the velocity of minority carriers in the base, thus further reducing charge storage and speeding up the transistors. To compute the resulting collector current, it is possible to evaluate the electric field using the relation $J_p = 0$, and then determine the electron current from Eq. 7a. The result is

$$J_c = qD_n[\exp(qV_{BE}/kT) - \exp(qV_{BC}/kT)]/\int[p(x)/n_i^2(x)]dx \qquad (8)$$

where n_i^2 is now a quantity that varies with position within the base, as a result of the changing N_c, N_v, ΔE_c and ΔE_v (or equivalently, changing E_g). This relation, derived by Kroemer,[4] is a straightforward extension of the previous results, remarkable for its simplicity.

Among the consequences of the expression for collector current are the following:

1. There is no dependence of the J_c vs V_{BE} relation on emitter doping, depth, etc.
2. There is no dependence of J_c vs V_{BE} on the detailed distribution of composition or doping in the base, only on the net integrals described above.
3. Transport of electrons across the base in either direction is limited in the same way and has the same voltage dependence; that is, J_c vs V_{BE} has the same form as J_e vs V_{BC} for reverse transistor operation. (Note that the associated hole flows are vastly different for these two conditions, however.)
4. Under different bias conditions, the number of holes in the base can change, and under such circumstances, a change in collector current

is also induced. For example, under very high forward bias, the number of holes can rise above the value of the acceptor doping in the base. This condition is termed high-level injection. Equation 8 demonstrates that there is a drop in collector current. If the injected hole concentration is very much greater than the background acceptor density, then quasineutrality will be established through the relation $p = n = n_i \exp(qV_{BE}/2kT)$. This leads to an expression for J_c with voltage dependence $\exp(qV_{BE}/2kT)$.

If the reverse-bias voltage V_{CB} is increased, then the depletion region within the base will grow, and the integrated hole density will decrease. Through Eq. 8, this must lead to an increase in collector current. The change in J_c is found, for the case of uniformly doped base, to be

$$dJ_c/dV_{CB} = J_c C_{BC}/p_b w \qquad (9)$$

where C_{BC} is the base–collector depletion capacitance per unit area. The resulting expression for J_c as V_{CB} is varied, is given by

$$J_c(V_{CB})/J_c(0) = 1/(1 - V_{CB}/V_A) \approx 1 + (V_{CB}/V_A) \qquad (10)$$

where $V_A = C_{BC}/(qp_b w)$ is known as the Early voltage. The associated output conductance degrades transistor voltage gain, and must be minimized by maintaining a suitably large value of $p_b w$.

The above expressions for J_c correspond to cases in which electron flow is dominated by base transport. This covers most situations encountered in present devices. As described below, additional contributions must be considered in heterojunction bipolar transistors where potential barriers restrict current flow, or produce injected electron distributions far from equilibrium. In the most aggressively scaled devices, further corrections are required, inasmuch as Eq. 7 is not valid if carriers experience few scattering events as they traverse the base.

Base Current.   The hole current which must be supplied from the base terminal has a number of different components, corresponding to recombination of holes with electrons in different regions of the device. As pictured in Fig. 4, there is recombination in the quasineutral base, in the emitter–base space-charge region, at the emitter periphery, in the emitter body, and at the emitter surface. These key components of the base current are evaluated in the following.

Base Recombination.   Recombination of electrons and holes within the base may proceed by direct photon-induced recombination, via deep levels, or by means of the Auger effect. Under most circumstances, a recombination lifetime τ_{rec} can be determined such that the net number of recombinations

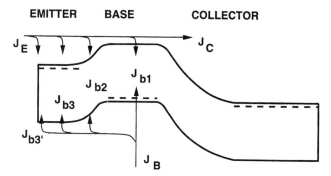

Fig. 4 Schematic components of emitter and base current in n–p–n transistors.

per unit volume in the base is $U = (n - n_{eq})/\tau_{rec}$, where n is the density of electrons injected into the base, and n_{eq} represents the thermal equilibrium minority carrier concentration. The associated base current is the integral of the recombination density over the base:

$$J_{b1} = qN_s/\tau_{rec} \tag{11}$$

where N_s is the integrated density of excess electrons injected into the base. For the simple transistor with uniform base, N_s can be directly evaluated as

$$N_s = [n(0) + n(w)]w/2 = n_i^2[\exp(qV_{BE}/kT) + \exp(qV_{BC}/kT) - 2]w/2p_b \tag{12}$$

In normal operation ($V_{BE} \gg kT/q$, $V_{BC} \ll -kT/q$), only the term involving V_{BE} is important. N_s can be easily related to the collector current J_c for a transistor with uniform doping and composition from Eq. 1 as

$$N_s = J_c w^2/(2D_n q) = J_c \tau_b/q \tag{13}$$
$$\tau_b = w^2/(2D_n)$$

Here τ_b is the base transit time for electrons. Thus the base current J_{b1} and the associated current gain β_1 limited by this mechanism alone are

$$J_{b1} = (\tau_b/\tau_{rec})J_c \tag{14}$$
$$\beta_1 = J_c/J_{b1} = \tau_{rec}/\tau_b$$

The ratio of base transit time τ_b to recombination time τ_{rec} can be viewed as the probability of recombination in the base of a given injected electron. In more complex structures, N_s can also be found to be proportional to J_c, and Eqs. 14–15 continue to hold with an appropriately modified expression for τ_b. In most modern devices, τ_b is of order of picoseconds, whereas τ_{rec}

is of order nanoseconds or microseconds; thus J_{b1} is generally quite small.

Emitter–Base Depletion Region Recombination.   According to the statistics for recombination via deep levels (quantified by Shockley–Read–Hall and others[1]), the effective recombination lifetime for carriers varies strongly with carrier density. The recombination rate reaches a strong maximum under the condition $p = n$, inasmuch as both carriers are relatively plentiful for capture. Within the bipolar transistor, this occurs over a relatively thin region within the emitter–base depletion region, for which $n = p = n_i \exp(qV_{BE}/2kT)$. Integrating the recombination density over the depletion region leads to the net base-current contribution:

$$J_{b2} = q \int U(x)dx = (qn_i/\tau_{eff})(2\pi kT/q\mathscr{E}_p) \exp(qV_{BE}/kT) \qquad (15)$$

where $\mathscr{E}_p$ is the electric field at the plane of maximum recombination. This contribution to base current thus has a voltage dependence of $\exp(qV_{BE}/2kT)$, that is, an ideality factor of 2. Correspondingly, the current gain β_2 limited by this mechanism increases with increasing collector current:

$$\beta_2 = J_c/J_{b2} \propto \exp(qV_{BE}/2kT) \propto J_c^{1/2} \qquad (16)$$

Reverse Injection into Emitter.   When the base–emitter junction is forward-biased, holes flow by diffusion (and potentially drift) into the emitter, recombining with electrons in the emitter body and at the emitter surface. The carrier concentration profile differs significantly among devices, according to the thickness of the emitter relative to the hole diffusion length, and according to the recombination velocity at the surface of the emitter. The carrier profile shown in Fig. 3 corresponds to a situation with a thick emitter, where the diffusion length of holes governs their profile. More frequently, bipolar transistors have thin emitters with metallic contacts (causing high surface recombination velocity), for which the density of holes varies approximately linearly with distance within the emitter. In either case, the hole current density can be calculated taking into account diffusion, drift associated with built-in fields, and varying bandgap, with a method directly analogous to the treatment of electron flow (Eqs. 4–8). This results in hole current J_{b3} given by

$$J_{b3} = qD_p \exp(qV_{BE}/kT)/ \int (n/n_i^2)dx \qquad (17)$$

The DC current gain of the transistor, β_3, as limited by this mechanism, is

$$\beta_3 = J_c/J_{b3} = (D_n/D_p)(\int (n/n_{ie}^2) dx/ \int (p/n_{ib}^2)dx) \qquad (18)$$

where n_{ie} and n_{ib} are the intrinsic carrier concentrations in emitter and base, respectively. For uniform doping of the emitter and base, and a thin emitter with infinite surface recombination velocity, this expression yields

$$\beta_3 = (D_n/D_p)(n_e w_e \exp(E_{ge}/kT)/[p_b w \exp(E_{gb}/kT)] \qquad (19)$$

This equation reveals a number of important factors for transistor design. Current gain is directly proportional to the ratio n_e/p_b (emitter to base doping concentration). To maintain suitably high values of current gain, this ratio is typically chosen to be on the order of 1000. For a silicon homojunction transistor, emitter doping levels of $10^{20}/cm^3$ and base doping levels on the order of 10^{17}–$10^{18}/cm^3$ are typically used. Also seen in Eq. 19 is the fact that the difference in bandgap energy between emitter and base has a major impact on current gain. In homojunction transistors, the principal bandgap difference stems from the bandgap shrinkage resulting from heavy doping on the emitter side of the device. For high doping levels, the effective bandgap energy is reduced because of the effects on carrier energy of the potentials of the donor or acceptor atoms and their associated fluctuations (bandtailing effect), and because of the binding energy of the associated electron and hole gases. The extent of bandgap shrinkage in silicon has been extensively studied theoretically and experimentally.[1,5] As a result of the reduction in emitter bandgap, the current gain is reduced according to the factor $\exp(-\Delta E_{gj}/kT)$, where ΔE_{gj} is the bandgap narrowing at the emitter–base junction edge.

The overall base current is the sum of these components, plus, potentially, the additional currents associated with tunneling at the emitter–base junction and emitter–edge currents. Base current components are often difficult to distinguish in specific circumstances, although in general they differ in their dependence on V_{BE}, on temperature, on base thickness, and on the ratio of emitter periphery to emitter area.

1.2.2 Charge Storage

The AC and transient characteristics of bipolar transistors are dominated by the charge stored within the device that must be increased or decreased when the bias conditions are changed. Within the charge control model, the base current consists of a DC component that is associated with the instantaneous bias voltages at the device terminals and a transient value given by the derivative of the charge stored in the device. For normal device operation, this corresponds to

$$I_B(t) = I_{B0}[V_{BE}(t), V_{BC}(t)] + (dQ_B/dt) \qquad (20)$$
$$I_C(t) = I_{C0}[V_{BE}(t), V_{BC}(t)]$$
$$I_E(t) = I_C(t) + I_B(t)$$

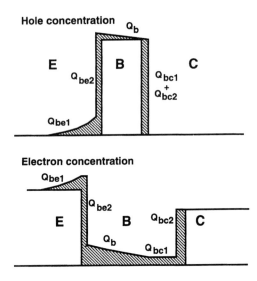

Fig. 5   Excess hole charge and excess electron charge stored in a transistor in forward bias.

Moreover, in the charge control approximation, the value of Q_B is taken to be the value that would apply under steady-state conditions, given the instantaneous values of the terminal voltages. In the spirit of this approximation, from knowledge of the charge distribution in the transistor under steady-state conditions and Eq. 20, the transient and AC performance of the transistor can be calculated.

The charge stored in the transistor, ΔQ_B, in excess of the charge contained in the device at zero bias, Q_{B0}, is pictured in Fig. 5. Here the electron and hole distributions are shown for conditions of zero bias and a representative active bias. The excess charge per unit area ΔQ_B can be computed by integrating over the entire device the contributions corresponding to excess electron charge density, or the contributions corresponding to excess hole charge. The two resulting values are equal in magnitude, since the overall transistor is neutral.

Charge Contributions.   Distinct contributions to the charge, Q_j, may be associated with the different regions of the device (termed Q_{be1}, Q_{be2}, Q_b, Q_{bc1}, and Q_{bc2} in Fig. 5). Each of these contributions is approximately proportional to collector-current density J_c. It is of interest to compute the ratio Q_j/J_c, a quantity with the units of time, which corresponds loosely to the transit time of carriers through the different regions of the device. These are approximately evaluated in the following.

Emitter Region, Q_{be1}.   Under forward-bias of the base–emitter junction, excess holes are injected into the emitter with a distribution that depends on the details of the emitter thickness and surface recombination (as described above). For the simple case of thin emitter with metal contact, the charge stored is

$$Q_{be1} = \int q(p - p_{eq})dx = qw_e n_{ie}^2 \exp(qV_{BE}/kT)/(2n_e)$$
$$\tau_{be1} = Q_{be1}/J_c = w_e wp_b n_{ie}^2/(2D_n n_e n_{ib}^2) \tag{21}$$

τ_{be1} is a significant contribution to the overall delay of silicon bipolar transistors. To minimize it, thin emitters with high doping are desired, while lightly doped, thin bases are also important (as also required to establish high DC current gain). Another effective strategy is to use a wide bandgap emitter, as described later for heterojunction bipolar transistors (HBTs).

Emitter–Base Depletion Region, Q_{be2}.   Charges must be stored at the edges of the depletion region to support the electrostatic fields as the junction voltage is changed. For a change in voltage dV_{BE}, the corresponding change in stored charge is $C_{BE}\,dV_{BE}$. C_{BE} may be approximately evaluated as the capacitance of a *p–n* depletion region. The delay time τ_{be2} associated with this charge is

$$\tau_{be2} = dQ_{be2}/dJ_c = C_{BE}(dJ_c/dV_{BE}) = C_{BE}g_m = C_{BE}qJ_c/kT \tag{22}$$

τ_{be1} and τ_{be2} are frequently lumped together in a contribution τ_e known as the emitter delay.

Base Region, Q_b.   Charge associated with the electrons injected into the base is neutralized with additional holes added to the base. The overall amount of charge, and the associated delay time, have already been discussed for the calculation of base current. For a uniform base, they are given by

$$Q_b = \int q(n - n_{eq})dx = J_c w^2/(2D_n) \tag{23}$$
$$\tau_b = w^2/(2D_n)$$

In modern devices, there are a variety of corrections that must be considered.[6] When significant electrostatic fields are present in the base, electron flow proceeds by drift as well as by diffusion, and τ_b is customarily represented as $\tau_b = w^2/\eta D_n$, where η is a adjustment factor that depends on the magnitude of the electric field present (with $\eta \approx 2[1 + (q\mathscr{E}w/2kT)^{3/2}]$).

The velocity with which electrons may exit from the base is limited to a value of the order of the saturation velocity, v_{sat}. In addition, in thin bases, diffusive flow is not governed by the simple Fick's law expression; rather, it is limited to a velocity known as effusion velocity or thermionic-emission velocity. This velocity corresponds to the situation in which an entire thermal electron population is directed from emitter to collector, and there is no backscattered or returning carrier flow. Under such circumstances, an approximate expression for base transit time is given by $\tau_b = w^2/(2D_n) + w/v_m$, where v_m is the velocity at which the electrons exit the base at the collector edge. v_m is typically given by the effusion or thermionic-emission velocity of electrons, $v_m = (kT/2\pi m^*)^{1/2}$.

Collector Region, $Q_{bc1} + Q_{bc2}$.   As bias conditions are changed, the charge stored in the base–collector depletion region must change, through two mechanisms. In the first mechanism, if V_{BC} changes, then a depletion charge of $Q_{bc1} = C_{BC} \, dV_{BC}$ must be added (or removed) to the base and to the collector edges of the depletion region. Correspondingly, there is delay time τ_{bc1} defined as dQ_{bc1}/dJ_c. By convention, the charge is computed with the collector terminal incrementally short-circuited to the emitter terminal. As a result, by taking into account the external circuit, which may have series resistances, as V_{BE} is changed, the variation of V_{BC} is given by

$$\Delta V_{BC} = \Delta V_{BE} + I_C(R_E + R_C) \tag{24}$$

where R_E and R_C are extrinsic parasitic resistances associated with these terminals. The collector current I_C is given by J_c times the area of the emitter, A_E. The delay time is thus

$$\begin{aligned} \tau_{bc1} &= C_{BC}(dV_{BE}/dJ_c + R_E A_E + R_C A_E) \\ &= C_{BC}(qI_C/kT + R_E A_E + R_C A_E) \end{aligned} \tag{25}$$

A second mechanism causing a variation of the charge in the base (at constant V_{BC}) is associated with the finite velocity of electrons within the depletion region. As J_c increases, the electron density within the collector also increases, and the associated electron charge modifies the space-charge density distributed throughout the collector, from N_D to $N_{Deff} = N_D - J_c/qv_s$ (where N_D is the donor concentration in the collector region, and v_s is the electron velocity, typically at its saturated value). The injected electrons act as acceptor dopants in the depletion region. The resultant change in "doping" changes the amount of charge at the base edge of the depletion region. Within the simple approximation for a one-sided p–n junction, the charge Q_C in the depletion region, and the associated time constant τ_{bc2} are

$$Q_C = [2\varepsilon q N_{Deff}(V_{CB} + V_{bi})]^{1/2} = [2\varepsilon q(N_D - J_c/qv_s)(V_{CB} + V_{bi})]^{1/2} \tag{26}$$
$$\tau_{bc2} = dQ_C/dJ_c = Q_C/(2N_{Deff}qv_s) = w_c/2v_s$$

where w_c is the thickness of the collector depletion region. The result for τ_{bc2} corresponds to one-half the time expected for an electron to traverse the collector depletion region, traveling at the saturated drift velocity. The factor of two accounts for the fact that the charge associated with the electrons distributed through the depletion region is terminated partly at the base depletion region edge, and partly at the collector depletion region edge, under conditions of constant voltage drop across the depletion region. For situations in which the velocity of the electrons varies spatially within the collector, the above simple picture may be modified to give[6]

$$\tau_{bc2} = \int [(1 - x/w_c)/v(x)dx] \tag{27}$$

The overall charge stored in the base is the sum of the contributions defined above. In similar fashion, the overall delay, termed τ_{ec}, the emitter-to-collector delay, is the sum of the contributions described:

$$Q_B = Q_{be1} + Q_{be2} + Q_b + Q_{bc1} + Q_{bc2} \tag{28}$$
$$\tau_{ec} = \tau_{be1} + \tau_{be2} + \tau_b + \tau_{bc1} + \tau_{bc2}$$

Modes of Transistor Operation. The preceding discussion centered on bias conditions in which the emitter–base junction is forward-biased and the base–collector junction is reverse-biased or only moderately forward-biased (called the forward active mode of operation). If the base–emitter junction and base–collector junctions are both reverse-biased, the transistor is in cutoff mode, and no current flows. With the base–collector forward-biased and the base–emitter reverse biased, inverse operation of the transistor is obtained, and emitter current will flow. Under most circumstances, the doping and area ratios employed in the transistor fabrication will lead to a very low value of current gain in this mode of operation, potentially below unity. Finally, if the base–emitter junction and base–collector junctions are both forward-biased, the transistor is said to be in saturation. Representative minority-carrier concentrations in the saturation mode are shown in Fig. 6. The charge stored within the base is considerably greater than what would be stored under normal operation. Also, large amounts of minority-carrier charge are stored in the collector region. The excess charges (holes) must be disposed of during switching operations either by recombination or by extracting them from the base terminal. Typically, they slow the transistor performance dramatically. In saturation, the charge stored is partitioned into contributions (Q_F and Q_R) that are associated with emitter and collector. The base charge of these two portions can be determined from the graphical analysis of Fig. 6. The overall form of the charge control equations, including saturation, then becomes:

$$I_B(t) = I_{B0}[V_{BE}(t), V_{BC}(t)] + (dQ_F/dt) + dQ_R/dt \tag{29a}$$

$$I_C(t) = I_{C0}[V_{BE}(t), V_{BC}(t)] + (dQ_R/dt) \tag{29b}$$

$$I_E(t) = I_{E0}[V_{BE}(t), V_{BC}(t)] + (dQ_F/dt) \tag{29c}$$

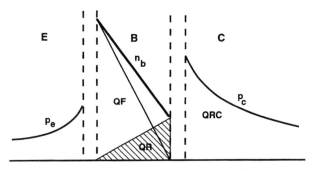

Fig. 6 Minority-carrier distributions and associated stored charge during transistor saturation.

1.2.3 Figures-of-Merit for Transistor Performance

Current Gain Cutoff Frequency (f_T). f_T is the frequency at which the magnitude of the transistor incremental short-circuit current gain, h_{fe}, drops to unity. It is a key estimator of transistor high-speed performance. If we consider AC transistor behavior, assuming small-signal excitation voltages and currents with time dependence $\exp(j\omega t)$, then within the charge control framework, the incremental current gain h_{fe} can be calculated as follows:

$$i_b = i_{b0} + j\omega q_b \qquad (30)$$
$$i_c = i_{c0}$$

$$|h_{fe}| = |i_c/i_b| = |i_{c0}/(i_{b0}+j\omega q_b)| = 1/|[(i_{b0}/i_{c0}) + (j\omega q_b/i_{c0})]| = 1/|(1/\beta + j\omega \tau_{ec})|$$

Here we have introduced notation i_b, i_c, etc. to denote small-signal values of the quantities I_B, I_C, etc., and have noted $q_b/i_{c0} = dQ_B/dI_C = \tau_{ec}$. The current gain h_{fe} has the frequency dependence noted in Fig. 7, reaching a low-frequency value of β, and dropping at a rate of 6 dB/octave ($1/f$ dependence) at frequencies above the beta-rolloff frequency $1/(2\pi\beta\tau_{ec})$. The current gain magnitude drops to unity at a frequency $f_T = 1/(2\pi\tau_{ec})$, an expression arrived at by neglecting $1/\beta$ compared with unity. The delays studied in Section 1.2.2 thus have a key role in governing the AC current gain of the device. A brief summary of the contributions to f_T is frequently written, for normal operation at moderate current densities, as

$$1/(2\pi f_T) = \tau_b + \tau_e' + (w_c/2v_s) + (R_E + R_C)C_{BC}$$
$$+ [(C_{BE} + C_{BC})kT/(qI_C)] \qquad (31)$$

The first three contributions are sometimes referred to as T_F in circuit level models, as described in the following. The last component depends on

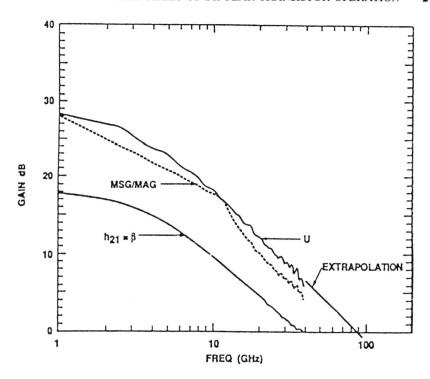

Fig. 7 Representative frequency dependence of incremental short-circuit current gain , h_{fe}, and power gains, MSG, MAG, and U.

collector current as $1/I_C$, and leads to a dramatic slowdown of the transistors at low current levels. To distinguish the contributions, it is customary to plot experimental data of f_T *vs* I_C in the form of $1/(2\pi f_T)$ *vs* $1/I_C$, as shown in Fig. 8; the slope can be identified with $kT/q(C_{BE}+C_{BC})$. As the current I_C reaches very high levels, f_T drops (τ_{ec} rises) due to the base pushout effect described below.

Maximum Frequency of Oscillation (f_{max}).   This is the frequency at which the maximum available power gain of the transistor drops to unity. f_{max} is widely useful to estimate power gain, since over a wide range of frequencies, maximum available power gain, G_p, follows the relation

$$G_p = (f_{max}/f)^2 \tag{32}$$

f_{max} is different from (and typically larger than) f_T, because in addition to current gain, f_{max} takes into account the possibility of voltage gain. A simple analysis based on the simplified hybrid pi equivalent circuit for the bipolar transistor[1] provides the basis for estimating the factors important to f_{max}. The

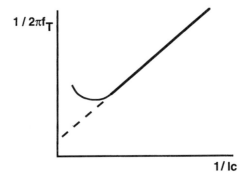

Fig. 8   Schematic dependence of $1/2\pi f_T$ on $1/I_C$ for a bipolar transistor.

maximum available power gain of the transistor is given by

$$G_p = 1/4 \ [\text{Re}(Z_{in})/\text{Re}(Z_{out})]|h_{fe}|^2 \tag{33}$$

where h_{fe} is the current gain defined above, and Z_{in} and Z_{out} are the input and output impedances of the transistor. At high frequency, the real part of Z_{in} is primarily the base resistance R_B. At high frequency, Z_{out} is dominated by the feedback effect of C_{BC}: a voltage source, v_0, at the output will produce a voltage at the input given by $v_0 \, C_{BC}/(C_\pi + C_{BC})$, and an output current, $v_0 g_m C_{BC}/(C_\pi + C_{BC})$ will result. This provides for an output impedance that is real, with

$$\text{Re}[Z_{out}] = 1/(2\pi f_T C_{BC}) \tag{34}$$

where the relationship $f_T = g_m/[2\pi(C_\pi + C_{BC})]$ has been used. The resulting value of power gain, and the corresponding value of f_{max}, are

$$G_p = f_T/(8\pi R_B C_{BC} f^2) \tag{35}$$
$$f_{max} = (f_T/8\pi R_B C_{BC})^{1/2}$$

More exact expressions have been given, where the components of R_B and C_{BC} in different device regions are properly weighted, and phase variations are taken into account.[6]

To realize the power gain inferred from Eq. 35, it is necessary to match input and output impedances of the transistor. Experimental values of f_{max} are derived from S-parameters measured with microwave network analyzers, from which power gain *vs* frequency may be computed. The most important measures of power gain correspond to maximum available gain, MAG, which is the maximum power gain obtainable by conjugate matching of input and output ports, and unilateral or Mason gain, U, which is the maximum gain obtainable by conjugate matching of inputs and outputs, together with use of a lossless feedback network to tune out any internal device feedback. MAG and U both reach unity at the same frequency, f_{max}. For some combination of input and output impedances, many devices are potentially unstable below a characteristic frequency f_1, whereas above f_1, the devices tend to be unconditionally stable. In the regime of potential instability, MAG cannot be defined, and maximum stable gain, MSG, is used as a measure of gain instead.

1.2.4   Limiting Factors in Transistor Performance

Base Punchthrough.   As base regions are made thinner in an attempt to shorten transit time, a limit is reached when at high collector–base biases the collector depletion region meets the emitter depletion region. This condition leads to a reduction in the barrier for current flow from the emitter, and uncontrolled current injection (runaway collector current). The effect may be viewed as a limiting form of the Early effect, or base-width modulation. This effect imposes a constraint on base width and doping (which must be on the order of $10^{18}/cm^3$ or higher for base widths below 50 nm). In small base regions, it is important to consider the statistical nature of the dopant atom distribution. If the number of acceptors within the base in any square of dimension w_c^2 is too small, there will be localized punchthrough.

Base Pushout.   An important limit on collector current density for bipolar transistors is base pushout or Kirk effect. As detailed above, the electron density injected into the collector depletion region $n = J_c/qv_s$ modifies the collector space-charge density. When the injected electron density becomes comparable to the background donor density, the fields in the base–collector depletion region become severely distorted, as shown in Fig. 9. At the current density $J_c = qv_s N_D$, the collector region behaves as if it were undoped, and the depletion region extends all the way to the subcollector. For higher current densities, the collector region behaves as if it were p-doped, causing electric field reduction at the base edge of the depletion region. Finally, at a critical current density J_k, the injected carrier density attains the value of

$$n_c = N_D + 2\varepsilon(V_{CB} + V_{bi})/qw_c^2 \tag{36}$$

and then the electrostatic field at the edge of the base vanishes. Above this current density, holes are no longer confined to the base, and diffuse into the collector, leading to an increased value of base width and reduced collector depletion width. This results in increased base transit time and reduced current gain, together with a slight drop in base–collector transit time

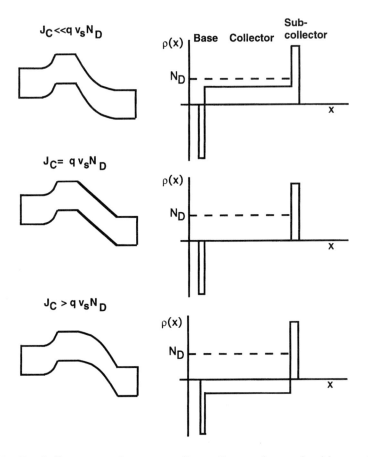

Fig. 9 Band diagrams and corresponding collector charge densities at different collector currents, illustrating base pushout effect.

and increase in capacitance. In silicon-based devices the associated performance degradation is very dramatic, and J_k is typically considered the upper limit for high-speed transistor use. In III–V HBTs, with very small base transit times, the added penalty associated with the induced base is not always excessive and proper operation above this current density is possible.[7]

Avalanche Breakdown. When fields in the base–collector depletion region reach the levels of the semiconductor breakdown field, the passage of current leads to creation of electron-hole pairs via impact ionization. The electrons thus generated flow into the collector, adding to I_C; the holes flow to the base. If the transistor is connected with a voltage source (or short-circuit) at the base terminal, the hole current will flow out the base terminal. However, in many circumstances, a current source (high impedance) is

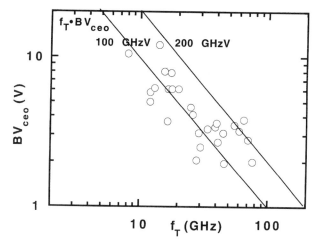

Fig. 10 Relationship between breakdown voltage BV_{ceo} and f_T for various reported bipolar transistors. (After Nakamura and Nakamae, Ref. 8.)

connected to the base terminal. Then the avalanche-generated base current adds to the input current, leading to a further increase in collector current, via the current gain of the transistor. Thus J_c for circuit configurations with the base incrementally open-circuited rises significantly at a voltage (termed BV_{ceo}) much lower than the corresponding voltage for situations with the base incrementally short-circuited (termed BV_{ces}).

As devices are scaled to small dimensions to increase f_T and associated circuit speed, electric fields increase in the collector (for a fixed base–collector voltage). Thus breakdown voltage tends to decrease, and can become a severe limitation to transistor application. Figure 10 illustrates reported values of BV_{ceo} and f_T measured for a wide variety of transistors.[8] The inherent tradeoff of BV and f_T is embodied in the Johnson figure-of-merit, given by the product of f_T and BV_{ceo}. f_T is approximately limited by collector transit delay to a value below $v_s/\pi w_c$; the applied value of V_{CE} (assuming common-base operation) is limited to a value below breakdown voltage $BV = \mathscr{E}_b w_c$, where $\mathscr{E}_b$ is the collector breakdown electric field. Thus the product $BV \cdot f_T$ is limited to a value below $\mathscr{E}_b v_s/\pi$, a constant associated with the material used in the collector, and independent of transistor design.

Parasitic Elements.   To this point, discussion has focused primarily on the intrinsic device, viewed as a one-dimensional structure. Bipolar transistor performance, however, is significantly affected by parasitic elements surrounding the device, as pictured in Fig. 11.

To make contact to the base layer, extrinsic base regions are formed on one or both sides of the emitter. The resistance associated with the base

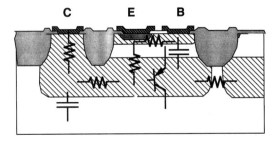

Fig. 11 Schematic cross-section of a bipolar transistor, showing parasitic elements.

contacts and with the regions between the base contacts and the intrinsic device can significantly limit device performance.

To compute the effective base resistance associated with the intrinsic region located under the emitter, and called the "pinch base", a simple approach is to regard the emitter current and associated base current density as uniform across the emitter. The lateral current I_b flowing in the base then varies linearly with position, as shown in Fig. 12a. The base–emitter voltage varies quadratically with position as a result of $I_b R_b$ voltage drops. The average voltage difference across the emitter differs from the voltage at the edge of the emitter by the amount $R_s W/12L$, where R_s is the sheet resistance of the pinch base, and W and L are the width and length of the emitter.

A parasitic element critical for high-frequency performance is the capacitance between the extrinsic base regions and the collector. This extrinsic C_{BC} is typically the dominant part of C_{BC} since the intrinsic base–collector junction area is typically only one-third to one-sixth of the overall junction area. The extrinsic base borders the emitter also and forms sidewalls whose specific capacitance is high because of the high doping levels on both sides of the junction. These are also susceptible to leakage currents and premature breakdown.

Another parasitic element of significance is emitter resistance. The area of the emitter contact is typically equal to the area of the intrinsic device. Values of intrinsic g_m achievable with bipolar transistors are very high (up to 40 mS per μm^2 of emitter), but the extrinsic g_m may be severely limited if R_e is not maintained below $10–50 \, \Omega\text{-}\mu m^2$.

When bipolar transistors are formed into integrated circuits on a silicon substrate, there is ordinarily a parasitic $p–n–p$ transistor formed between base, collector, and substrate. This bipolar transistor can conduct significant currents when the intrinsic transistor is in saturation.

Current Crowding. The AC and DC currents flowing through the base layer produce *IR* drops that tend to reduce the forward bias V_{BE} at the center of the emitter relative to its value at the emitter edge. If the *IR* drops reach

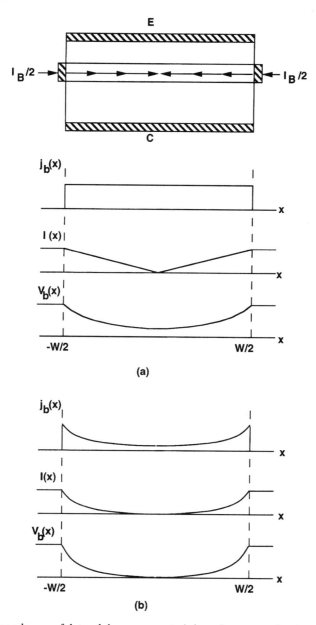

Fig. 12 Dependence of lateral base current, injected current density, and base emitter voltage vs position across the emitter in the simplest approximation (uniform injection into emitter) (a) and in the presence of current crowding (b).

values comparable to kT, they will have a significant impact on the transistor forward current. The center regions of emitter stripes tend to conduct less current during operation at high current levels and high frequencies, reducing the device transconductance. This complicates the theoretical picture of transistor performance at high frequencies.

To quantitatively describe current crowding, a distributed transistor model can be used, as pictured in Fig. 12b. An approximate assessment of the effects can be made by linearizing the transistor behavior about a given bias point. The current density flowing into the base, j_B, is no longer uniform as assumed in Fig. 12a, but varies spatially as a result of the varying voltage across the width of the emitter. Corresponding equations for the incremental junction voltage, v, and lateral base current i_b assuming sinusoidal steady-state AC analysis can be written as

$$dv/dx = -i_b R_s \qquad (37)$$
$$di_b/dx = j_b \sim j_0 + j_0(qv/kT) + j\omega C_{in}v$$

Solution of these equations shows that the voltage varies laterally according to $\exp(\gamma x)$, where

$$\gamma = \{[R_s g_m[(1/\beta) + (jf/f_T)]]\}^{1/2} \qquad (38)$$

The current spreading width decreases at high current operation, at high frequency, for low β, and for high pinch-base resistance. To minimize the impact of the current spreading, emitter widths are made very small (0.25–0.5 μm) for high-speed silicon transistors. These transistors have large pinch-base resistances, $>10 \text{ k}\Omega/\square$, in order to keep base transit time small. This effect is significantly alleviated with HBTs.

An interesting side-consequence of the current-crowding effect is that the base resistance tends to decrease at high current levels or high frequencies. In effect, only areas near the edges of the emitter conduct collector current, and the distance between these regions and the base contacts decreases as current crowding increases.

1.3 SILICON BIPOLAR TRANSISTORS

A representative "conventional" process flow for integrated silicon bipolar transistors is shown in Fig 13. Buried n^+ subcollectors are formed by implantation of arsenic or antimony into a p substrate (a), followed by epitaxial growth of lightly doped silicon to form the collector layer. This is followed by isolation steps based on localized oxidation of the surface (b,c), using silicon nitride masks. Collector contact "plugs" may then be implanted, followed by base formation with boron ion implantation (d). The emitter is formed with the implantation of arsenic or phosphorus; in the process shown,

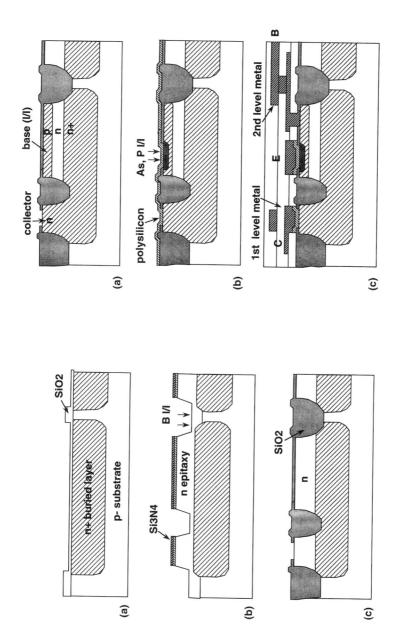

Fig. 13 Representative process flow for "conventional" Si bipolar transistor.

35

the emitter implant is preceded by the deposition of polysilicon, which serves as a diffusion source and contact to the emitter (e). The structure is completed with the deposition of several layers of interconnect metal, separated by interlevel dielectric (f).

In recent years, advances have been made in the fabrication technology of silicon bipolar transistors, which have led to considerable reductions in vertical and lateral device dimensions, and corresponding increases in performance. This section discusses several of the common themes of present technology.

1.3.1   Self-Aligned Emitter and Base Contacts

Significant improvements over conventional technology are possible by employing emitters and extrinsic bases diffused from polysilicon layers, which remain on the wafer forming the respective contacts.[8–11] The polysilicon layers withstand high process temperatures and can be oxidized, so they lend themselves to a variety of sophisticated processing techniques. In particular, the separation between the emitter contact and the base contact can be self-aligned, using a dielectric sidewall spacer whose width can be accurately controlled in the range of several thousand angstroms. The width of the emitter is equal to the width of the opening in the base polysilicon (defined photolithographically) less the sidewall dimensions. A representative device cross-section is shown in Fig. 14a. By repeated application of sidewall spacers, emitter widths down to $0.35\ \mu$m have been made, beginning with $1\ \mu$m lithographic patterns. The emitter polysilicon can extend over a considerably wider region than the area of contact, reducing its series resistance. The base–collector junction area is kept at a minimum, typically three times the emitter area, even for the very narrow emitters. Dramatic savings in device area are made over non-self-aligned devices, which require wide separations between base and emitter to allow for photolithographic alignment tolerances.

The principal limits associated with this technique are related to the region of overlap between the n^+ and p^+ doping distributions produced in the single-crystal silicon from the closely spaced polysilicon areas. If the heavily doped regions overlap too much, then tunneling can occur. Hot electron production can also occur, due to enhancement of the electric field. Carriers can subsequently be injected into the surface oxide, producing surface leakage.

1.3.2   Polysilicon Emitter

The use of polysilicon to form the emitter contact has had a major impact on transistor current gain, and therefore on vertical scaling. The principal role of the polysilicon process is to control the effective recombination velocity, s_0, at the single-crystal emitter surface. With metal-contacted

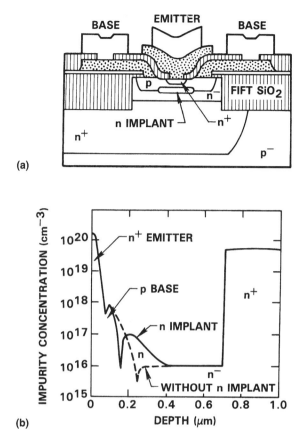

Fig. 14 (a) Cross-section of double polysilicon self-aligned transistor. (b) Doping profile obtained after implantation of phosphorus through emitter opening. (After Konaka et al., Ref. 11.)

emitters, s_0 is very high (on the order of 10^5 cm/s) and, as a result, base current rises with decreasing emitter depth. Lower surface velocities are found experimentally for emitters diffused from polysilicon. In these cases, the distribution of holes is modified, as shown in Fig. 15. The base current is reduced, and for a sufficiently thin single-crystal emitter region, hole storage is also reduced.

The value of s_0 appropriate to polysilicon depends on detailed process conditions, particularly on the amount of SiO_2 at the interface between single crystal and polysilicon. The amount of SiO_2 is typically governed by an HF dip prior to deposition of polysilicon, the subsequent thermal treatment and oxygen exposure of the resultant "hydrogen-passivated" surface, and on heat treatments after polysilicon deposition (which tend to break up the oxide). It is necessary to produce devices with relatively well-controlled values of

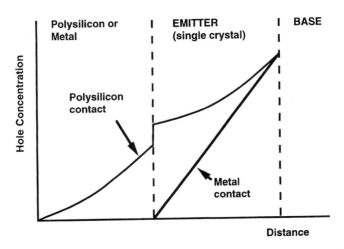

Fig. 15 Schematic minority carrier distribution with polysilicon-contacted emitter.

this thickness; if the oxide is too thick, the series resistance of the emitter contact becomes excessive; if it is too thin, the current gain suffers.

For devices with continuous oxide coverage of the surface, electron and hole currents flow by tunneling. However, the probability for electron tunneling is significantly higher than that for holes, as a result of the energy band lineup. Additional physical effects may be significant in governing current flow at the polysilicon interface, particularly in device with incomplete oxide coverage. Segregation of arsenic impurities has been observed; this could lead to band-bending at the interface in a fashion that presents a barrier for hole transport, thus improving current gain.

With polysilicon, thin emitters can be produced with acceptable current gain. Thin emitters limit hole storage, and allow thin base regions to be formed with adequate control and with higher base doping. An advantage of polysilicon from the processing standpoint stems from the fact that implants of the emitter dopant (As) are made into the polysilicon material, followed by a drive-in anneal. Therefore, implant damage is kept away from the single-crystal material, and by appropriate control of the anneal, extremely shallow emitters can be reproduced. The formation of the base regions is also a critical fabrication step. The conventional method of forming the base layer is by boron implantation prior to emitter deposition. However, boron atoms have a pronounced tendency to channel, giving rise to implant distributions with deep tails. Thus, very low implant energies must be used, with the result that base charge becomes very sensitive to emitter depth. To reproducibly obtain shallow base profiles, boron diffusion is being investigated, using as a source the same polysilicon layer used to form the emitter, which is initially made *p*-type and subsequently arsenic implanted, or using a layer of borosilicate glass deposited prior to the polysilicon and then stripped. Another technique is to use conventional boron implants, and later to implant phosphorous into the collector region underneath the base to suppress the doping associated with the tail of the boron distribution.[11] Figure 14b shows the resultant doping profile. This technique provides several added advantages. The doping in the collector drift region is enhanced to suppress base pushout. This allows high current density to be used, increasing f_T to values above 25 GHz at $J_c = 10^5$ A/cm^2. Another feature of the technique is that the collector implant is patterned by the emitter opening, so that it is formed only in the intrinsic device region. Collector doping in the extrinsic areas remains low to minimize C_{BC}.

1.3.3 Advanced Silicon Bipolar Processes

Sidewall Contact Process. The sidewall base contact structure (SICOS) transistor makes innovative use of polysilicon contacts to achieve consider-able improvement in device performance and to increase the possible circuit uses of the device.[12] A SICOS transistor is shown in Fig. 16. Thick oxide layers separate the base polysilicon from the collector, so that extrinsic base collector capacitance is dramatically reduced. The device structure becomes a nearly perfect embodiment of the one-dimensional model. With ap-propriate doping profiles, a nearly identical geometry can be obtained for current flow downward from emitter to collector and upward from collector to emitter. High-quality upward transistors can be used in circuit applications to simplify layout.

Epitaxial Base. To overcome the problems of base layer formation by ion implantation, and allow thinner base regions with controlled doping profiles, epitaxial growth of the base with *in situ* doping has been explored.[8] The epitaxial process can be configured so that deposition takes place only on exposed single-crystalline regions of the silicon substrate; alternately, deposi-tion can be made all cross the wafer surface, although in regions covered

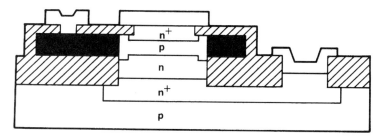

Fig. 16 Cross-section of sidewall contacted structure (SICOS) transistor. (After Nakamura et al., Ref. 12.)

by silicon dioxide, the silicon deposits are polycrystalline. The control achieved during epitaxy permits thinner bases and higher f_T's. It is critical, however, to control the thermal budget of processing subsequent to the base growth, so that the boron dopant does not diffuse excessively. This is a particular concern during the emitter implantation step.

Scaling Issues.   Although polysilicon emitters and advanced processing have allowed shallower emitters and thinner base layers to be produced, there are limits to the vertical scaling achievable.

1. As the base is thinned, base resistance is increased unless doping is also increased. This limit can be counteracted to a considerable extent by making narrow emitters, so that the distance between base contact and center of the emitter stripe is kept at a minimum.
2. It is necessary to avoid base punchthrough at reasonable operating voltages. This requirement roughly corresponds to pinch base resistances on the order of $30\,k\Omega$ or lower, to prevent punchthrough at $V_{CB} = 3\,V$, and requires increasing values of base doping as the structure is scaled.
3. For higher base doping, charge storage in the emitter is increased, decreasing f_T. Current gain drops, due to both increasing hole injection into the emitter, and the appearance of tunneling current at the base–emitter junction (particularly at low bias voltages).
4. As the collector doping is increased to limit collector transit time and allow increasing current density without Kirk effect, the breakdown voltage associated with collector avalanching decreases.

1.4 HETEROJUNCTION BIPOLAR TRANSISTORS

1.4.1 Bandgap Engineering in Bipolar Transistors

The constraints on vertical scaling of bipolar transistors may be overcome if the semiconductor composition can be changed appropriately within the device. Bandgap differences between emitter and base have a critical effect on current gain, as described in Section 1.2.1. In conventional silicon transistors, bandgap shrinkage from heavy emitter doping decreases current gain. Conversely, by intentionally changing the semiconductor composition to produce a bandgap wider in the emitter than in the base, considerable improvement in transistor performance can be obtained. More generally, the importance of bandgap variation stems from the ability to engineer forces on electrons and holes separately. With appropriate changes in the underlying semiconductor material, quasielectric fields are generated from gradients in conduction and valence band energies that drive carrier flow. The ability to tailor these forces provides a powerful new degree of freedom in the design of bipolar devices.[13,14]

The implementation of heterojunction approaches in bipolar devices was delayed by decades because of the technological problem of providing interfaces between dissimilar materials that were free of imperfections, either impurities or structural defects as a result of a mismatch in the lattice constant. Such imperfections typically cause excessive recombination and tunneling currents. Even at the present time, high-performance heterojunction bipolar transistors are limited to relatively few materials systems. The most explored systems are those involving semiconductors that have identical lattice constants (GaAlAs/GaAs and InGaAs/InAlAs/InP) or very nearly identical lattice constants, and use thin layers that adapt to the lattice constant of the substrate (SiGe/Si).

Abrupt and Graded Junctions. At abrupt junctions between different semiconductors, in general there are discontinuities in the energy of the conduction-band minima and valence-band maxima. The change in energy gap ΔE_g between the materials is made up of a conduction-band energy step ΔE_c, and a valence-band energy step, ΔE_v (with $\Delta E_g = \Delta E_c + \Delta E_v$). A representative band diagram for an n–p–n AlGaAs/GaAs HBT with wide-bandgap emitter and an abrupt emitter–base junction is shown in Fig. 17a. An energy barrier appears in the conduction band at the emitter–base junction, which tends to retard the flow of electrons from emitter to base. This decreases the emitter injection efficiency (although it provides benefits for high-speed operation, since the electrons that surmount the energy barrier are injected into the base with high forward velocities). To improve the current gain of the HBTs, the composition of the materials may be graded over distances of several hundred angstroms, and as a result the band diagram of Fig. 17b is obtained. Detailed behavior of the band edges depends on the doping profile and the grading profile used. The conduction-band energy diagram can be readily determined as the sum of the electrostatic potential variation $\phi(x)$ plus the variation of the electron affinity of the material, $\chi(x)$, which varies with material composition, as illustrated in Fig. 18. To avoid the formation of any local maxima in conduction band energy, a quadratic grading of alloy composition with distance from the junction may be used. The variation of $\chi(x)$ will also be quadratic, and it can be cancelled with the variation in electrostatic potential, which is also quadratic for the case of uniform doping. Alternate techniques to reduce the conduction-band barrier include the use of a setback layer in the layer with lower conduction-band energy, or the use of doping dipoles placed judiciously in the junction region, which can create an electrostatic potential change that can counteract the electron affinity variation, as shown in Fig. 19.

Wide-Bandgap Emitter. With wide-bandgap emitters, following the analysis of Eq. 19, the current gain as limited by reverse injection of holes into the

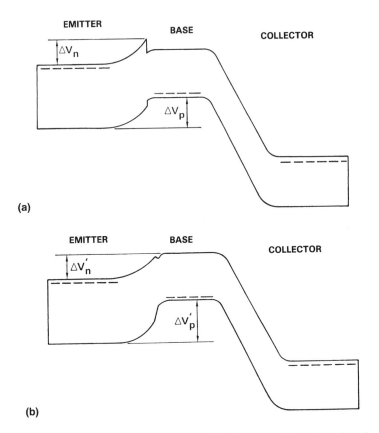

Fig. 17 (a) Energy-band diagram for HBT with abrupt emitter–base junction. (b) Energy-band diagram with graded alloy composition at emitter–base junction.

emitter (emitter injection efficiency) is

$$\beta_3 = (D_n/D_p)n_e w_e \ \exp(\Delta E_g/kT)/(p_b w) \tag{39}$$

The difference in bandgap between emitter and base, ΔE_g, is typically chosen to be greater than 250 meV ($10kT$), providing a factor of 10^4 improvement over the homojunction device. It is therefore possible to insure adequate injection efficiency with very high levels of base doping and very low levels of emitter doping. Representative doping concentrations in present HBTs fabricated in the GaAlAs/GaAs system are shown in Fig. 20, which should be contrasted with those of Fig. 3 for the homojunction case. In HBTs, doping levels up to 10^{20}/cm^3 have been used in the base. As a result, base sheet resistance can be greatly decreased, even with ultranarrow base regions. Transistor f_{max} is thereby greatly increased. The base punchthrough limitation to device scaling disappears as a result of the higher doping. Current-gain

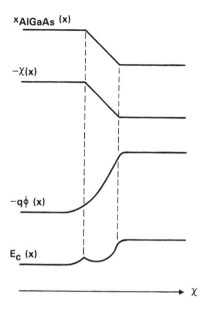

Fig. 18 Relationship between conduction-band diagram, electrostatic potential variation and electron affinity variation at a heterojunction.

reduction from high injection effects at high current density is similarly avoided. The Early voltage of the device increases considerably, because of the fact that base charge is insensitive to output voltage V_{BC}. At the same time, emitter doping levels can be reduced dramatically. This allows the emitter–base space-charge region to broaden considerably on the emitter side of the junction, and leads to a reduction in emitter junction capacitance. The storage of holes in the emitter region approximately vanishes for abrupt heterojunctions, and is greatly minimized for graded heterojunctions, increasing f_T.

Graded Base. With the ability to control semiconductor bandgap, it is worthwhile also to establish a gradual change in bandgap across the base layer from E_{g0} near the emitter to E_{g0}-ΔE_g near the collector, as pictured in Fig. 21a. The high hole conductivity insures that the valence band is effectively flat, and the bandgap shifts establish a conduction-band energy gradient equal to $\Delta E_g/w_b$. This energy gradient is a quasielectric field that drives electrons across the base by drift as well as diffusion. In homojunction devices, drift fields as described in Section 1.2.1 are of the order of 2–6 kV/cm. With HBTs, fields two to five times larger are easily implemented. It is noteworthy that in III–V compounds, electrons can be driven by these high quasielectric fields to velocities in excess of the value predicted by steady-state velocity field curves. Velocity saturation will not occur because the overall voltage drop

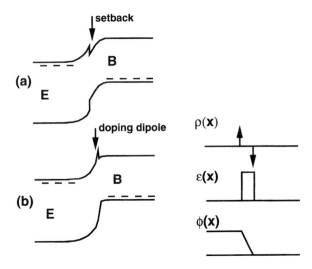

Fig. 19 Band diagrams showing the use of setback layers and doping dipoles to decrease the magnitude of dE_c barriers. Also shown is the charge density, electric field and electrostatic potential variation for the doping dipole.

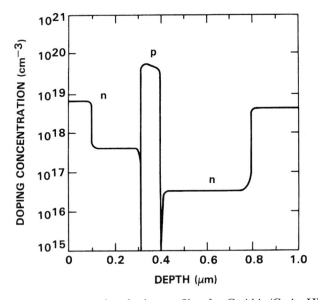

Fig. 20 Representative doping profile of a GaAlAs/GaAs HBT.

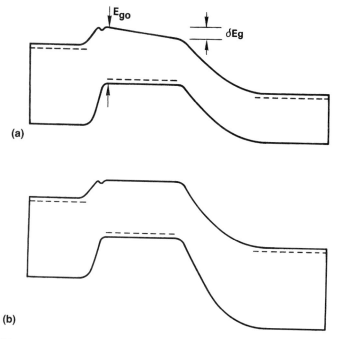

Fig. 21 (a) Band diagram of HBT with graded composition base region. (b) Band diagram of double HBT (with wide-bandgap emitter and base).

across the base is generally limited to values lower than the threshold energy for scattering from the central (Γ) conduction-band minimum to satellite valleys. The result is a considerable improvement in f_T.

Wide-Bandgap Collector. Another possibility offered by bandgap engineering is to increase the bandgap of the collector, as shown in Fig. 21b. This has the benefit of eliminating the injection of holes from the base into the collector when the base–collector becomes forward-biased, analogous to the wide-bandgap emitter effect. This greatly diminishes the saturation stored charge density, and speeds up device turn-off after biasing in the saturation regime. The saturation stored charge, it must be noted, is not fully eliminated, inasmuch as there remains a component of electrons injected into the base from the collector. With double-heterojunction devices, symmetric operation in upward and downward directions can also be established, which provides circuit flexibility.

With double-heterojunction transistors that have wide-bandgap emitters and collectors, it is critical to avoid the formation of a conduction-band barrier at the base–collector junction. As in the case of wide-bandgap emitters, this can be achieved through composition grading, setback layers and doping dipoles.

Another advantage of wide-bandgap collector devices is the increase in breakdown voltage, since impact ionization is reduced in large bandgap materials, and reduction in leakage current. The Johnson figure-of-merit for the collector can be in principle optimized independently from other characteristics of the transistor. It is possible to employ narrow-bandgap base materials, yet not incur the problems of leakage and low voltage range that occur in homojunction devices.

Current Flow in HBTs.   Collector current in HBTs may be limited by (1) transport of electrons across the base as it is in homojunction transistors; (2) transport of carriers across the conduction-band barrier as in the case of abrupt junctions; or (3) supply of electrons from the emitter if the emitter is very lightly doped. For case (1), the collector current J_c has already been derived in Section 1.2.1:

$$J_c = qD_n[\exp(qV_{BE}/kT) - \exp(qV_{BC}/kT)]/\int[p(x)/n_{ib}^2(x)]dx \qquad (40)$$

This equation, the Kroemer–Moll–Ross relation, illustrates that J_c depends solely on the structure of the base, and that the effect of variable base bandgap is simply taken into account by the variation of intrinsic carrier concentration, n_i. The ideality factor associated with J_c is unity.

The more general case where current flow is limited by mechanisms (2) and (3) as well as (1) has been considered.[15] J_c can be expressed in the form

$$J_c = \exp(qV_{BE}/kT)/[(1/J_1) + (1/J_2) + 1/J_3)] \qquad (41)$$

where J_1, J_2, and J_3 are, in general voltage-dependent, current densities governed by the different mechanisms. The turn-on voltage is increased from the value which applies in the base transport limited case. The ideality factor for J_c is somewhat greater than unity because the height of the conduction-band barrier increases somewhat with increasing V_{BE}.

The current gain in HBTs, as limited by emitter injection effects, can be calculated with Eq. 41 for base–emitter junctions with no potential barrier. A noteworthy feature of this expression is that current gain decreases with increasing temperature. This is a natural consequence of the fact that the energy barrier for hole flow is greater than that for electron flow, and hence its thermal activation energy is greater.

1.4.2   Silicon-Based HBTs

To obtain the benefits of heterojunctions combined with standard silicon bipolar technology, considerable efforts have been made to identify a suitable wide-bandgap semiconductor to be used as the emitter or, more recently, a suitable narrow-bandgap semiconductor to be used as the base within silicon devices.

Wide-Bandgap Emitter Approaches.   A variety of materials have been tried as wide-bandgap emitters in silicon-based bipolar transistors, including microcrystalline silicon, amorphous silicon, SIPOS (semi-insulating polysilicon), beta-silicon carbide, and GaP. To date, however, no single material system has emerged as a leading contender for future development. Wide-bandgap emitter approaches have the potential to be easily incorporated into existing silicon bipolar technology. However, several generic problems affect progress.

1. Wide-bandgap materials frequently lead to high emitter resistance, associated either with the material itself, with the emitter–silicon interface, or with contacts to the material.
2. Diffusion of donors into the single-crystal silicon must be suppressed. If the single-crystal region becomes *n*-type, it will become the effective emitter, and have the same stored hole charge as the polysilicon emitter structure.

$Si_{1-x}Ge_x$ HBTs.   ***Band Structure.***   The addition of germanium reduces the bandgap of silicon, leading to an alloy that can be used to form the base region of double-heterojunction bipolar transistors. The lattice constant of the alloy differs considerably from that of silicon, but if the thickness of the epitaxial layers is kept below a critical thickness, then the mismatch between the alloy and the silicon substrate is accommodated elastically and no misfit dislocations form (pseudomorphic growth).[16] Figure 22 shows the critical thickness for $Si_{1-x}Ge_x$ alloys *vs* alloy composition *x*. The SiGe pseudomorphic layers are considerably strained, as indicated by the sizable misfit in lattice constant. Due to the details of the band structure of silicon, this strain provides major benefits for transistor operation. The conduction-band minima in unstrained silicon and SiGe of moderate germanium content correspond to six-fold degenerate valleys located along {100} directions in *k*-space. In the presence of strain, the degeneracy of the valleys is lifted. For growth on (100) substrates, the four valleys oriented normal to the heterojunction are lowered in energy, while the remaining two valleys are raised in energy. As a result, the overall bandgap shrinkage is increased over that of unstrained material with the same germanium content. At the same time, electrons preferentially populate valleys for which the effective mass in the direction of travel is particularly low. In fact, for transport perpendicular to the heterojunction, electron mass corresponds approximately to the transverse mass ($0.19m_0$), which represents a reduction of about 60% from the unstrained case. Similarly, the degeneracy of the top of the valence band of SiGe is lifted. This also contributes to increasing the shift in bandgap energy for a given germanium mole fraction. The energetically favored valence band corresponds to a light hole mass for transport in the plane of the heterojunction, as desired to obtain low base resistance. The effective

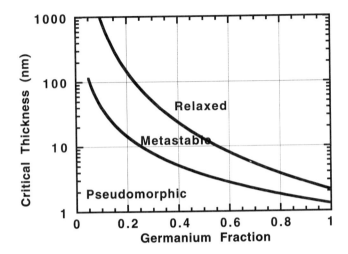

Fig. 22 Critical thickness for pseudomorphic growth of $Si_{1-x}Ge_x$ on Si vs alloy composition.

mass changes from the strain anisotropy more than compensate for the decrease in mobility expected from alloy scattering.

The variation of bandgap of $Si_{1-x}Ge_x$ pseudomorphically grown on silicon is shown in Fig. 23. The bandgap difference between SiGe and silicon appears principally as a valence-band step, a situation of considerable benefit for the formation of n–p–n HBTs since there is no conduction-band discontinuity to retard electron flow.

SiGe Growth and HBT Fabrication.   Techniques for the growth of SiGe include molecular beam epitaxy (MBE), with either solid or gaseous sources; limited reaction processing (LRP), in which epitaxy is carried out from the vapor phase on a controlled basis using heating by high intensity lamps; and ultrahigh vacuum chemical vapor deposition (UHCVD), among others.[16–20] A key feature in the UHCVD technique is the virtual elimination of oxygen, which allows growth to take place at lower temperatures than in ordinary silicon growth. This technique can provide uniform growth over multiple wafers.

Under equilibrium conditions, lattice relaxation of SiGe films grown on silicon will occur if their thickness exceeds the critical thickness. However, films with thickness well in excess of this limit can exist in a metastable state of strain if the growth temperature and temperature of subsequent processing steps are kept low enough. This allows much larger germanium mole fractions than would be possible in equilibrium. Various strategies have evolved for incorporating SiGe and the accompanying process. If the germanium content is limited to amounts consistent with stable strain (to values on the order

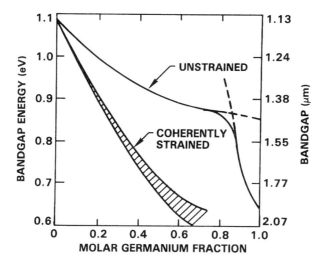

Fig. 23   Variation of energy gap for pseudomorphic $Si_{1-x}Ge_x$ on Si, and unstrained alloy, as a function of alloy composition. (After Iyer et al., Ref. 16.)

of 10% for typical base thicknesses of 50 nm) then the structure can withstand high-temperature processing. On the other hand, if the growth and processing temperature are kept low, then very high germanium fractions, up to more than 35%, can be used.

In the first strategy, SiGe can be incorporated into standard silicon bipolar (and BiCMOS) processing, to make use of the extensive technology already developed. The device designs emphasize the greatest benefit of the SiGe, that of reducing base transit time by providing bandgap grading in the base. A representative layer structure to accomplish this is shown in Fig. 24. The germanium content varies from 0% to 8% from emitter to collector edge of the base, providing a built-in field for electrons of about 14 kV/cm. These structures can achieve very high speed, as noted below. Due to the limited bandgap difference with respect to the emitter, however, the doping level of the base cannot be increased very much. In such transistors, fabrication is generally similar to that of advanced silicon homojunction bipolar transistors (with a subcollector implant, followed by collector epitaxy and localized oxidation isolation). Several possibilities exist for the growth of the SiGe for the base. Conditions can be chosen for selective growth of the SiGe so that it will deposit only in windows opened in silicon dioxide or silicon nitride films; or nonselective growth, in which SiGe is deposited everywhere (although it will be polycrystalline when covering oxide or nitride layers). After the base deposition, the SiGe is covered with polysilicon to form the emitter. Representative device cross-sections are shown in Fig. 25a and b.

If compatibility with standard silicon processing is not required, greater

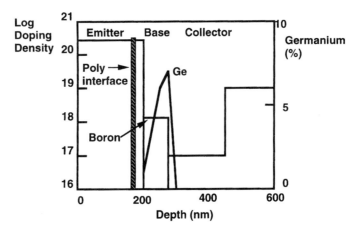

Fig. 24   Schematic doping and Ge concentration in SiGe HBT with graded base. (After Harame et al., Ref. 19.)

amounts of germanium can be incorporated into the base,[20] and the base doping level can be increased considerably, to levels of the order of 10^{19}–10^{20}/cm^3. Transistors of this type make use of a single epitaxial growth for base and emitter (since temperature must be kept too low to deposit or dope the polysilicon). The thin base region can be contacted by low-energy gallium implants, or by selective etching. Wet etches have been developed which are strain sensitive, and will stop when they reach the SiGe base. Low-temperature dielectrics have typically been deposited for passivation. Isolation has been accomplished with mesa etching. A representative device cross-section is shown in Fig. 25c.

Device Characteristics. The decrease in energy gap of the base leads to changes in the I_C–V_{BE} relation towards lower turn-on voltage as expected from Eq. 8. Figure 26 shows experimental I_C results for a series of devices with a constant base Gummel number (2×10^{13}/cm^2) and varying alloy content in the base.[18] The base current is significantly decreased from the homojunction case at a given base doping. The figure also shows the variation of base current with base width, w, at a constant base alloy composition ($x = 0.215$). For w of 300 nm or below, ideality factor is near unity, and base current is independent of w as expected for emitter injection efficiency limited current gain. For w greater than 300 nm, the ideality factor and magnitude of the base current increase, as expected from the formation of misfit dislocations.

Experimental results have confirmed a variety of advantages of SiGe HBTs over transistors based on silicon alone, including:

1.   higher f_T, particularly with graded germanium content in the base.

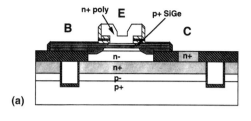

(a)

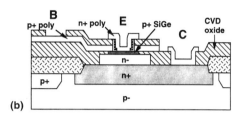

(b)

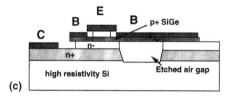

(c)

Fig. 25   Representative cross-sections of SiGe HBTs with different structures.

2. higher base doping for a given current gain, leading to reduced base resistance. This can be translated into higher f_{max}, into reduced Johnson noise, and into reduced current-crowding effects.

3. higher Early voltage.

4. higher base-punchthrough voltage.

The high-speed performance of SiGe HBTs has been excellent. By varying the germanium content across the base to provide a strong built-in field, base transit time could be dramatically reduced.[19] Figure 27 illustrates the advantage obtainable in f_T with respect to standard silicon bipolar devices, with identical processing. By maximizing the built-in field and reducing other delay components, significantly higher f_T, up to 115 GHz, has been demonstrated.

When large amounts of germanium are used in the base, the base thickness can be reduced considerably to achieve high f_T, or the base sheet conductivity can be increased to achieve high f_{max}. Measured gain vs frequency has yielded extrapolated f_{max} values above 150 GHz.

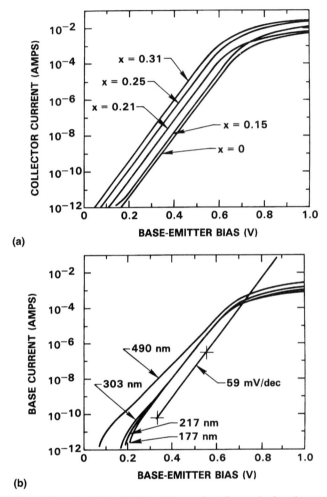

Fig. 26   (a) I_C vs V_{BE} for SiGe HBTs with varying Ge mole fraction x. The HBTs have approximately constant base Gummel number, $2 \times 10^{13}/cm^2$, whereas the homojunction device has Gummel number $10^{12}/cm^2$. (b) I_B vs V_{BE} for SiGe HBTs with fixed Ge mole fraction $x = 0.215$, and varying thickness w. (After Gibbons et al., Ref. 18.)

Early voltage increases with the use of germanium, even if base doping is kept constant, because of the influence of the bandgap on base-transport-limited current flow, as expressed in Eq. 8. In effect, the regions with high germanium content at the collector edge of the base contribute relatively little to the integral of the denominator in Eq. 8 because of the high value of n_{ib}^2 in that region. Thus depletion of the carriers at the collector side of the base does not change the collector current. The product of current gain and Early

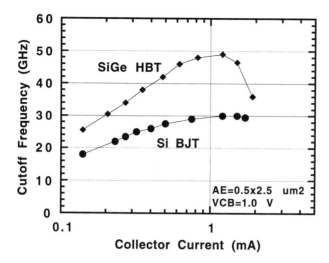

Fig. 27 Measured cutoff frequency, f_T, vs current for SiGe HBT and identically processed Si BJT. (After Harame et al., Ref. 19.)

voltage (βV_A) is an important figure-of-merit for linear (analog) applications. Values of this figure-of-merit are increased by 20–100 times with the addition of germanium.

1.4.3 III–V-Based HBTs

AlGaAs/GaAs HBTs. The $Al_x Ga_{1-x}As/GaAs$ material system provides a variety of advantages for HBTs, including the following:

1. The $Al_x Ga_{1-x}As/GaAs$ material system has an excellent lattice match. The difference in lattice constant between AlAs and GaAs is on the order of 0.14% at room temperature and, due to slight differences in thermal expansion coefficient, is even less at representative crystal-growth temperatures. Matching may therefore be obtained for any choice of aluminum mole fraction, x. This simplifies constraints on composition control during epitaxial growth.

2. Advanced material-growth techniques are available, such as Molecular Beam Epitaxy (MBE) and Metalorganic Chemical Vapor Deposition (MOCVD). These techniques permit the growth of ultrathin device layers with excellent control.

3. Large bandgap differences may be obtained. The bandgap of the alloy system is given by $E_g = 1.424 + 1.247x$ for $x < 0.45$, and $E_g = 1.424 + 1.247 + 1.147(x - 0.45)^2$ for higher x values. Below the critical composition $x = 0.45$, the bandgap is direct. Above this

composition, the conduction-band minima are at X points of the Brillouin zone, while the valence-band maxima remain at Γ. The majority of the bandgap difference (approximately 62%) corresponds to a difference in conduction-band energy, with the remainder (38%) corresponding to valence-band energy difference. As a result of this, abrupt Np heterojunctions have significant conduction band energy barriers.

4. The electron mobility is high in $Al_xGa_{1-x}As$ for $x < 0.45$, as a result of the small effective mass associated with the Γ minimum ($m^* = 0.065$). In pure material, electron mobility reaches 8000 cm²/V-s. The high mobility tends to decrease base transit time, decrease charge storage at the base–emitter junction, and increase conductivity of the undepleted n^- collector.

5. At high electric fields (>3 kV/cm), electrons in GaAs display negative differential mobility under steady-state conditions, and their velocity decreases until it saturates at a value of 0.8–1.0×10^7 cm/s. When electrons enter high field regions, for short times their velocity considerably exceeds this limit. This phenomenon, known as velocity overshoot, is of considerable importance in decreasing the transit time of carriers across the base–collector depletion region.

6. As a result of the wide bandgap of GaAs, the intrinsic carrier concentration is low, $n_i = 2 \times 10^6$/cm³ at room temperature. Intrinsic material has a resistivity on the order of 5×10^8 Ω-cm (frequently described as "semi-insulating"). Substrates can be conveniently produced with resistivities approaching this limit because of the occurrence of easily produced deep levels which can pin the Fermi level near mid-gap. The semi-insulating nature of the substrate simplifies the task of isolating devices and interconnects. The capacitance between devices or interconnects and the substrate, a major factor in silicon-based designs, is reduced to negligible values.

7. GaAlAs/GaAs has been used to fabricate a variety of optoelectronic devices including LEDs, lasers, modulators, and detectors of various types. These devices can be monolithically integrated with circuits based on GaAlAs/GaAs HBTs.

Device Structure.   Figure 28 shows a representative device cross-section, and its associated layer structure. Emitter layers consist of $Ga_{1-x}Al_xAs$ with AlAs mole fraction x chosen to be about 0.25, since with greater values, deep donors known as DX centers begin to appear in n-type GaAlAs. DX centers increase the capacitance of the emitter depletion region, and may contribute to trapping effects. With $x = 0.25$, the bandgap of the emitter is 0.30 eV wider than that of the base, allowing huge increases in injection efficiency. The majority of the energy gap difference corresponds to a conduction-band step (0.2 eV), and in much of the work the alloy composition is graded at the

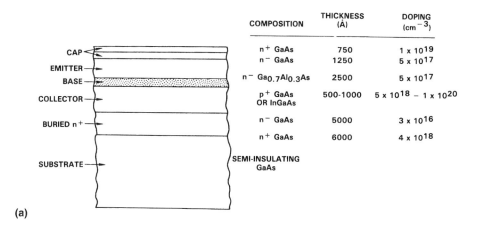

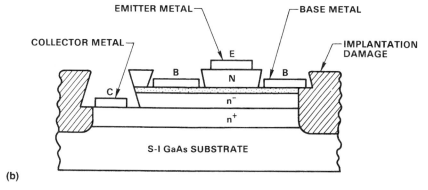

Fig. 28 Representative layer structure (a) and device cross-section (b) for GaAlAs/GaAs HBTs.

base–emitter junction. Base layers are typically made with thicknesses of 0.05–0.1 μm, with values of doping of 5×10^{18}–10^{20}/cm^3. Extremely high values of base doping may be used without incurring hole injection into the emitter, although other sources of base current also are present, and tend to increase with base doping. Resulting values of base sheet resistance are in the range 100–600 $\Omega/\square$. The practical limit to base doping is established by metallurgical considerations. The diffusivity of the most frequently used acceptors (Be in MBE and Zn in MOCVD) increases with concentration. Unless care is exercised during crystal growth, and high temperature processing is eliminated subsequent to growth, the acceptors will penetrate into the lightly doped GaAlAs. If the p–n junction moves into the wide-bandgap material, a potential barrier for electron flow forms, and the barrier to hole flow is reduced, as shown in Fig. 29. This reduces current gain. Structures frequently include layers of InGaAs (not lattice matched to

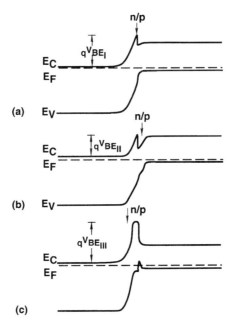

Fig. 29   Band diagrams for GaAlAs/GaAs HBT with different placements of the *p–n* junction (dopant distribution) with respect to the heterojunction.

GaAs) grown on the emitter surface to decrease the emitter contact resistance.

A number of variations in HBT structure have been demonstrated.

1. Grading of the composition of the base has been demonstrated, varying the aluminum content by as much as 10% across the base. InAs can be added to the base layer, producing the alloy GaInAs, whose bandgap is lower than that of GaAs. If the base layer is sufficiently thin, then the layer will grow pseudomorphically.

2. Double heterojunction devices with wide-bandgap collectors have been fabricated and, with appropriate processing have been shown to lead to symmetric device characteristics when operated in upwards and downwards directions.

3. Although most of the work has focused on *n–p–n* devices, *p–n–p* transistors also have been made. Theoretical expectations are that the f_T and f_{max} achievable with *p–n–p*'s should nearly equal those of *n–p–n*'s, although the layer structures should be configured somewhat differently. To maximize f_T in a *p–n–p* device, it is important to reduce the base thickness as much as possible, to avoid large base transit time due to the relatively poor hole mobility. However, the penalty on base

resistance should not be very great, because of the very high electron mobility in GaAs.

4. Collector-up devices, in which the emitter is next to the substrate and the collector layer is uppermost, have been fabricated. Potentially these devices have better RF characteristics than the more conventional emitter-up structures because the extrinsic base–collector capacitance can be effectively eliminated and traded for an extrinsic base–emitter capacitance, which is much less problematic in microwave circuits. In work oriented towards I^2L circuitry, transistors optimized for operation in the upward direction have also been made. In this approach, the base doping is typically produced by ion implantation rather than epitaxy.

5. $Ga_{0.51}In_{0.49}P$ has been used in place of AlGaAs in a number of device structures. $Ga_{0.51}In_{0.49}P$ has the same lattice constant as GaAs, and has a bandgap of 1.89 eV, well above that of GaAs. Potential benefits of this material in place of AlGaAs are: a lower conduction band offset of 0.2 eV, although this is potentially dependent on preparation technique, absence of traps associated with aluminum, and the availability of excellent selective etches to distinguish between GaAs and GaInP.

Fabrication Technology.   Epitaxial layer structures, typically, have been produced with MBE or MOCVD. Devices are then fabricated using a mesa geometry, or by using ion implantation damage to make the epitaxial regions outside the device semi-insulating. Vias are etched to the appropriate layers to contact the base and collector as pictured in Fig. 28, or the material above the layer to be contacted is type-converted by implantation or diffusion. For example, beryllium implantation or zinc diffusion has been used to contact the base. The process of etching down from the surface to the thin base layer is facilitated by the availability of composition-selective etches, which slow considerably when they reach the desired layer. As a result of deep vias to the buried collector, there can be a problem with the step coverage of metalizations. To alleviate the problem, the wafer surface can be planarized, or orientation-selective etched so that etched sidewalls conform to a specific crystallographic orientation that has a tolerable angle with the surface (typically 55°).

A major objective of process development is to produce a self-aligned structure, in which the edge of the emitter, emitter contact, and base contact are produced with the same photolithographic pattern. Various self-alignment techniques have been reported for GaAlAs/GaAs HBTs.[22–25] Figure 30 illustrates a number of the device cross-sections. Central themes of the fabrication approaches are the use of sidewall spacers at the edges of the emitter contact, and the use of refractory emitter contacts that are based on the presence of InAs cap layers. In the dual-liftoff approach, a dielectric layer

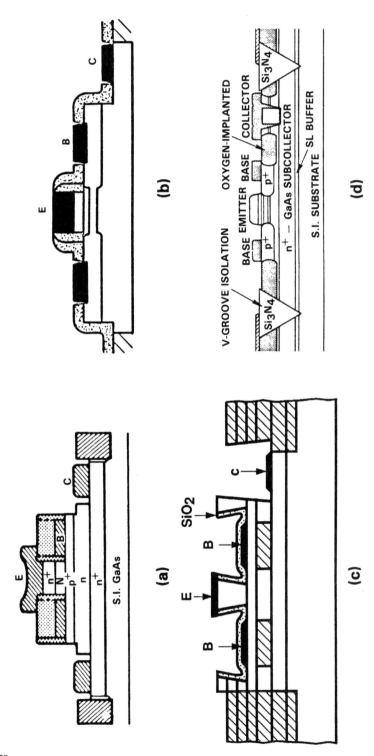

Fig. 30 Various approaches to the fabrication of fully self-aligned GaAlAs/GaAs HBTs.

58

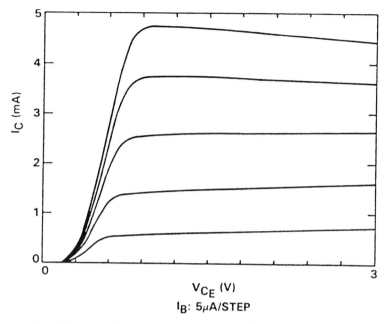

Fig. 31 I_C vs V_{CE} for a representative GaAlAs/GaAs HBT.

is deposited on top of the base contact metal, and both are patterned simultaneously by liftoff. A fabrication technique to minimize extrinsic base–collector capacitance has been developed, based on the use of implants of oxygen or protons into the regions of the n^- collector layer underneath the base contacts. The implants render the material intrinsic, and decrease capacitance per unit area by widening the effective separation between base and collector quasineutral regions. The implants can be carried out through the base without significantly altering its conductivity, because of the high base doping.

Device Characteristics.   Figure 31 shows *I–V* characteristics of a representative single-heterojunction GaAlAs/GaAs bipolar transistor. Distinctive features of the DC characteristics include the output conductance, the offset voltage, and the current gain variation.

Output Conductance.   The curves display high positive output resistance, or even differential negative output resistance. The associated high Early voltage, V_A, in the range 100-500 V, stems from the high doping of the base region, as described in Section 1.2.1. The negative resistance is measurable only on a slow time scale, and is due to the decreasing current gain with increasing temperature.

Offset Voltage.   Figure 31 shows a significant offset voltage, V_{cesat}, which must be applied between collector and emitter before positive I_C will flow. This corresponds to the fact that the turn-on voltage for the base–emitter junction is greater than that for the base–collector junction. For single-heterojunction abrupt HBTs, the difference in turn-on voltage is attributable to the conduction band barrier at the emitter–base junction. For HBTs with graded emitter–base junctions, electron flow for both junctions is equal at comparable bias voltages (as described in Section 1.2.1, corresponding to base-transport-limited current). The offset voltage then results from the difference in hole current between the two junctions. The offset voltage can be decreased substantially by the use of a wide-bandgap collector.

Current Gain.   Current gain varies with collector current, and drops to less than unity for very low I_C values. In most GaAlAs/GaAs HBTs, hole recombination in the quasineutral emitter is entirely suppressed by the emitter bandgap step. The base current arises, in part, from recombination at the base–emitter space-charge region due to deep levels in the GaAlAs and GaAs. Deep-level recombination can be suppressed to some extent by moving the depletion region into the wide-bandgap GaAlAs region, since the Shockley–Read–Hall recombination depends directly on the intrinsic carrier concentration (which is lower in the wider-bandgap material). Another important contribution to base current, dominant in small devices, is recombination associated with the emitter periphery. The surface recombination velocity of GaAs is very high, on the order of 10^6 cm/s. As a result, electrons injected into the base that diffuse to the exposed surface areas of the base near the emitter edges have a high probability of recombining at the base surface. Current gain in such circumstances decreases with the emitter periphery/area ratio, and can be a critical limitation for emitters on the order of 1μm wide. The effect may be suppressed in the following ways:

1. The use of thin base regions with built-in fields from composition grading reduces the number of injected electrons that diffuse to the surface.
2. A device structure can be used which has the surface of the base covered with GaAlAs, so that there is a potential barrier for electrons between the base and the surface. This can be accomplished if the emitter layer is converted to *p*-type by diffusion or implantation in the regions of the base contact or by using a thin layer of *n*-type AlGaAs covering the emitter edges, that is pinched off from surface depletion so that it is effectively an insulating passivation layer.
3. Surface recombination velocity of GaAs can be decreased by chemical treatments such as the deposition of $Na_2S \cdot 9H_2O$.

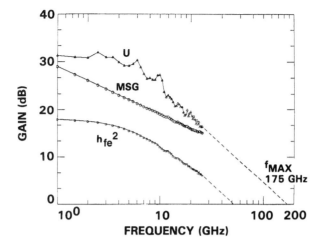

Fig. 32   Gain variation with frequency measured for a GaAlAs/GaAs HBT. H_{fe} is the current gain, MSG the maximum stable power gain, and U is the unilateral gain or Mason gain invariant. (After Chang, Ref. 25.)

RF Characteristics.   Figure 32 shows various measures of the gain of a microwave HBT as a function of frequency. The extrapolated f_T is about 50 GHz, while f_{max} is 175 GHz. Measured f_T values correspond to emitter-to-collector delay times τ_{ec} down to 1–2 ps. The contributions associated with emitter and collector delays decrease with current density J_c, and reach values of the order of 0.4 ps for J_c of 5×10^4 A/cm^2 or above. Typically J_c is limited by base pushout to 0.5–2×10^5 A/cm^2. Hole storage in the emitter does not contribute to τ_e in abrupt base–emitter heterojunction devices, and its contribution is small in graded structures. The base transit time for thin bases is influenced by the thermionic-emission velocity or effusion velocity of electrons (2×10^7 cm/s in GaAs) as well as the diffusion time, and for 600 Å heavily doped bases has a value of about 0.6 ps. The collector space-charge-layer transit time is dependent on collector–base voltage, V_{CB}. As in silicon devices, there is an effect of widening of the space-charge layer with increasing V_{CB}, which increases delay. In addition, there are strong effects on electron velocity in GaAs. The saturation velocity, v_s, is on the order of 10^7 cm/s. However, when electrons enter regions of high field from low-field regions, they are accelerated to velocities much higher than v_s before eventually slowing down due to scattering to satellite valleys of the conduction band. Figure 33 shows calculated results for the velocity of electrons as a function of position near the base–collector junction of a GaAlAs/GaAs HBT, obtained by Monte Carlo techniques.[26,27] Velocities of 8×10^7 cm/s are reached, although the high velocities persist only over distances of about 500 Å. The detailed behavior is dependent on the applied bias V_{CB}. To obtain

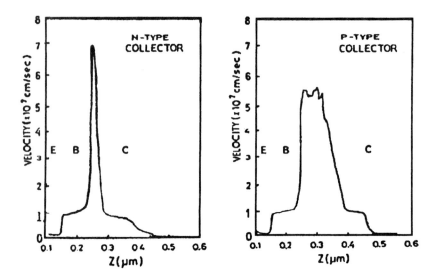

Fig. 33 Calculated electron velocity as a function of position near the base-collector junction of a GaAlAs/GaAs HBT (using the Monte Carlo technique). Values are shown for *n*-type doping in the collector, and *p*-type doping in the collector. (After Katoh and Kurata, Ref. 27.)

an estimate of the distance of significant velocity overshoot, we can calculate the length required for electrons to pick up an energy of 0.36 eV from the electrostatic potential. With this energy they can scatter to the satellite L valleys with the emission of an optical phonon. Novel structures have been devised where the distance of overshoot is increased considerably over conventional devices.[26,28] This can be achieved by decreasing the field at the base–collector depletion edge, by using appropriate acceptor doping of the collector depletion region. Figure 34 shows the band diagram of a structure of this type, the Ballistic Collection Transistor[28] (BCT). With the BCT, an f_T value of 170 GHz has been obtained.

Reliability.   Initial AlGaAs/GaAs HBTs had relatively poor reliability. A critical issue was identified to be the diffusion of beryllium or zirconium dopants in the base, during device operation at or near room temperature. The diffusion of the *p* dopants towards the emitter produced a displacement of the *p–n* junction, and the appearance of a potential barrier, as shown previously in Fig. 29c. The diffusion of the acceptors at such low temperature has been shown to be the result of electron–hole recombination-assisted defect motion. This problem has been overcome by the use of proper growth conditions for the base layer, which provide excess arsenic to avoid the generation of mobile interstitial beryllium species, or by the use of carbon as an acceptor in the base, which has a very low diffusion constant even in

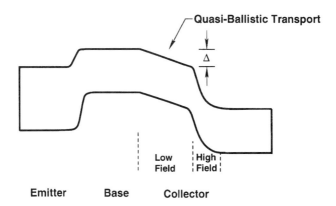

Fig. 34 Band diagram for the ballistic collection transistor, in which the distance of significant velocity overshoot is increased. (After Ishibashi and Yamauchi, Ref. 28.)

the presence of electron–hole recombination. Another problem is associated with increasing amounts of recombination at the edges of the emitter during device operation. This problem has been solved by implementing emitter-edge passivation techniques described above. Currently the reliability of AlGaAs/GaAs HBTs has been shown to be satisfactory[25] for virtually all applications. Extrapolated lifetimes as high as 10^8 hours at room temperature have been quoted.

InGaAs/InP HBTs.   The set of III–V semiconductors that are lattice matched to InP substrates includes $In_{0.53}Ga_{0.47}As$ with energy gap 0.75 eV and $In_{0.52}Al_{0.48}As$ with energy gap 1.5 eV. HBTs configured using $In_{0.53}Ga_{0.47}As$ (abbreviated InGaAs) as the base and InAlAs or InP, whose bandgap is 1.34 eV, as the wide-bandgap emitter have a number of attractive features:[29–31]

1. Well-developed growth technology for the epitaxial layers, based on MBE for InGaAs and InAlAs, or alternatively, MOMBE or MOCVD for InGaAs and InP.

2. High electron mobility in InGaAs, 1.6 times higher than GaAs and nine times higher than silicon in the case of pure material. The extent of transient electron velocity overshoot is also greater in InGaAs, InP and InAlAs than in GaAs, inasmuch as the separation between the Γ conduction-band minimum and satellite valleys is considerably higher than in GaAs. As a result, higher f_T values can be obtained with InGaAs HBTs than with GaAs devices.

3. The bandgap of InGaAs is smaller than that of GaAs or silicon. The resulting V_{BE} turn-on voltage of HBTs is correspondingly smaller for

Layer	Composition	Doping (cm^{-3})	Thickness (μm)
Contact	GaInAs	n=1x10^{19}	0.15
Emitter	AlInAs	n=1x10^{19}	0.1
Emitter	AlInAs	n=5x10^{17}	0.15
Spacer	GaInAs	n=5x10^{17}	0.02
Base	GaInAs	p=5x10^{18}	0.15
Collector	GaInAs	n=1x10^{16}	0.6
Subcollector	GaInAs	n=1x10^{19}	0.7
Substrate	InP	semi-insulating	

(a)

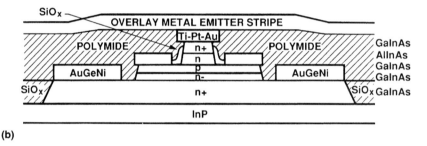

(b)

Fig. 35 Epitaxial layer structure (a) and device cross-section (b) for representative InGaAs HBT. (After Mishra et al., Ref. 34.)

graded-heterojunction bipolar transistors, and as a result, the power supply voltage and power dissipation can be lower in logic circuits, which improves the power–delay product.

4. The recombination velocity at surfaces of InGaAs is much smaller than that for GaAs surfaces (10^3 cm/s vs 10^6 cm/s). As a result, there is less base current caused by emitter periphery recombination, and scaling to small device dimensions is easier.

5. Semi-insulating substrates are available, made of InP doped with iron, which produces a deep acceptor near the mid-bandgap region.

6. Higher substrate thermal conductivity than that of GaAs (0.7 vs 0.46 W-cm/K).

7. Direct compatibility with light sources such as lasers and light-emitting diodes, and photodetectors such as p–i–n diodes, for 1.3 and 1.55 μm radiation, which are important for fiber-optic applications.

Device Structure and Fabrication. Figure 35 illustrates the representative epitaxial layer structure and cross-section of an InGaAs HBT. As noted above, a choice between InAlAs and InP exists for the semiconductor to be used in the emitter (and in the collector of double-heterojunction bipolar

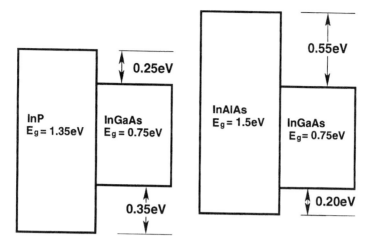

Fig. 36 Band lineups between InP and InGaAs, and between InAlAs and InGaAs.

transistors). The band lineups of these materials with respect to InGaAs are shown in Fig. 36. Abrupt InP junctions exhibit greater valence-band steps than the corresponding InAlAs abrupt junctions. For both systems it is necessary to grade the base–emitter junction to obtain the benefit of low turn-on voltage. But in both cases, if all or part of the conduction-band step is left, then "ballistic launching ramps" which inject electrons into the base with high forward momentum can be formed. This will reduce base transit time considerably, if the base is thin enough. The momentum achievable is much higher in InGaAs than in GaAs because of the low mass, and the fact that the upper limit on energy (i.e., the energy separation between Γ and satellite valleys) is greater.

The low value of bandgap in InGaAs can lead to low values of base–collector breakdown voltage and to relatively high leakage currents at the base–collector junction. To reduce these problems, it is possible to use wide-bandgap collector regions of InAlAs or InP. It is crucial to grade the base–collector junction of such structures, or to use delta doping or setback layers to avoid the formation of a conduction-band barrier. Such a barrier can diminish the collection efficiency for electrons and lead to large base transit times.

High-performance devices have been made with both InP and InAlAs wide-bandgap materials. InP is difficult to grow with conventional MBE because of the difficulty of handling solid phosphorus and its high vapor pressure. The technique of MOMBE conveniently solves this problem by using a gas source (phosphene), as does MOCVD. For graded composition structures, quaternary alloys can be used, which combine indium, gallium, and arsenic with either aluminum or phosphorus. It is necessary to maintain reasonable lattice match to the substrate, which restricts the composition that can be used. The required degree of composition control is relatively difficult to achieve. It is possible, as an alternative, to use "quasiquaternary" alloys, superlattices of short period (10–20 Å) in which the composition is changed abruptly between the endpoints, with a duty cycle that is progressively changed.

Fabrication approaches for InGaAs HBTs are similar to those used for GaAs HBTs. A difference exists in isolation techniques: it is not as straightforward to use implant damage to convert epitaxial layers to semi-insulating material. Therefore, deep mesa or trench isolation is preferred. Self-aligned fabrication techniques have been developed using sidewall dielectric spacers to define separations between base and emitter.

Device Characteristics. The *I–V* characteristics of an InGaAs HBT with an abrupt, single heterojunction structure are shown in Fig. 37a. The high offset voltage is a result of the significant difference in turn-on voltage between base–emitter and base–collector junctions. Breakdown is relatively low, because of the low InGaAs bandgap. Both features can be remedied with the use of wide-bandgap collectors and graded junctions, as shown in Fig. 37b. Current gain in the InGaAs HBTs can be very high, even in structures with narrow emitters, because of the low surface recombination velocity. Devices with emitters as narrow as $0.3 \, \mu m$ have been reported, with gains above 100.

The high-frequency performance of InGaAs HBTs is very impressive.[32,33] f_T above 200 GHz has been demonstrated, the highest of any bipolar transistor. The associated value of τ_{ec} is 0.8 ps. Each of the components of τ_{ec} is minimized in the device. Emitter charging time is reduced due the absence of hole storage at the heterojunction, and by operation at high current density ($2 \times 10^5 \, A/cm^2$). Base transit time is reduced by the injection of electrons with high forward momentum. Collector space-charge-layer transit time is reduced by electron travel at very high velocity, in the overshoot regime. To maintain this condition, narrow depletion regions are used, and the voltage is restricted to a range of V_{CB} bias on the order of 0.6 V, corresponding to the energy separation Γ and L valleys. The ultrahigh speed and scalability of these devices makes them attractive candidates for future digital and microwave circuits.

1.5 BIPOLAR TRANSISTOR MODELING

1.5.1 Physics-Based Modeling

To simulate the characteristics of bipolar transistors, physics-based computations that describe electron and hole distributions, electric fields, currents,

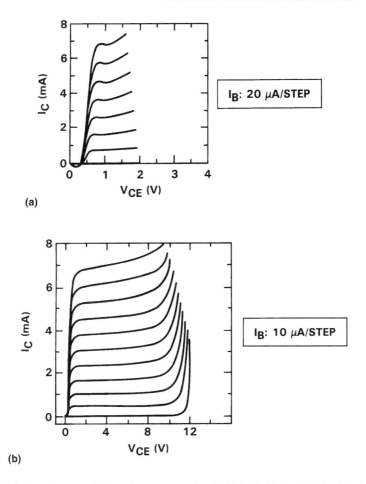

Fig. 37   (a) *I–V* characteristics of representative InAlAs/InGaAs HBT with abrupt base–emitter junction and an InGaAs collector region. (b) *I–V* characteristics of InAlAs/InGaAs HBT with graded base–emitter junction and an InAlAs collector region.

etc. may be made with a variety of simulation tools. One-dimensional codes may be used effectively for many purposes. The characteristics of current crowding, edge leakage, and emitter-periphery capacitance inherently require two- or three-dimensional codes for their analyses. For silicon-based devices a formulation based on drift-diffusion transport is generally adequate. Its accuracy will be impaired for a some situations associated with scaled devices, such as the prediction of impact ionization rates in regions in which the electric field is rapidly changing, and for diffusive flow across very thin bases. For III–V-based HBTs, the accuracy of drift-diffusion simulators is further compromised because of the potentially large energy gradients at abrupt junctions between materials with nonzero ΔE_c, because of the considerable velocity overshoot effects that can be obtained, and because tunneling is a significant transport mechanism in some instances. Monte Carlo simulations are regarded as the most accurate in these circumstances although they are the most expensive in computation time. Hydrodynamic or energy-balance simulations are emerging to provide a less accurate but less expensive approach.

1.5.2   Circuit-Level Modeling

To represent the behavior of bipolar transistors in circuit simulators such as SPICE, the Gummel–Poon model has been used for many years.[2] The schematic for this model is shown in Fig. 38a. Associated with the large-signal model, a small-signal (or "AC" model) is defined, with the schematic shown in Fig. 38b. The schematic diagram corresponds to a pi model, in which the output current generator is connected between collector and emitter. An alternate approach, corresponding to a T model, is shown in Fig. 39, which shows an incremental or "AC" model frequently used with microwave simulators. This model includes a frequency-dependent current source $\alpha(\omega)i$, which is easy to specify for frequency-domain computations. With a proper choice of element values, a one-to-one correspondence between pi and T models can be demonstrated.

In the Gummel–Poon model, the collector current generator corresponds to a current I_C given by Eq. 8. This is represented as

$$I_C = (I_S/K_{qb})[\exp(qV_{BE}/n_f kT) - \exp(qV_{BC}/n_r kT)] \qquad (42)$$

where K_{qb} is a factor to account for the bias dependence of $\int p(x)\,dx$, defined as the ratio of the hole charge in the base at a given bias condition, to that which exists at zero bias, and n_f and n_r are adjustable ideality factors that may be used to fit experimental results if necessary. Expressions are provided within the SPICE model for the dependence of the base-charge factor K_{qb} on V_{BE} and V_{BC} to describe base-width modulation, and on forward current I_C to describe high injection effects, as follows:

$$
\begin{aligned}
K_{qb} = \{&1 + [1 + 4(I_S \exp(qV_{BE}/n_f kT) - 1)/I_K \\
&+ 4(I_S \exp(qV_{BC}/n_r kT) - 1)/I_{KR}]^{1/2}\}/2 \\
&\times 1/[(1 - (V_{BC}/V_A) - (V_{BE}/V_B)]
\end{aligned} \qquad (43)
$$

where V_A, V_B are the forward and reverse Early voltages, and I_K and I_{KR} are forward and reverse knee currents for the onset of high injection. Base current within the model is given by a combination of two sources:

$$I_B = (I_C/\beta_F) + I_{SE} \exp(qV_{BE}/n_e kT) \qquad (44)$$

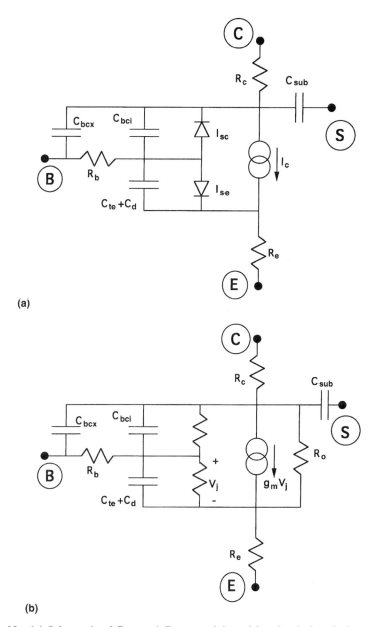

Fig. 38 (a) Schematic of Gummel–Poon model used for circuit description of bipolar transistors. (b) Small-signal equivalent circuit derived from model.

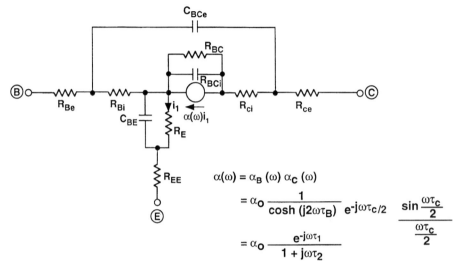

$$\alpha(\omega) = \alpha_B(\omega)\,\alpha_C(\omega)$$

$$= \alpha_0 \frac{1}{\cosh(j2\omega\tau_B)}\, e^{-j\omega\tau_C/2}\; \frac{\sin\frac{\omega\tau_C}{2}}{\frac{\omega\tau_C}{2}}$$

$$= \alpha_0 \frac{e^{-j\omega\tau_1}}{1 + j\omega\tau_2}$$

Fig. 39 Representative T model for small signal description of bipolar transistors.

where β_F is a temperature dependent parameter associated with the ideal base current, and the I_{SE} contribution describes nonideal components such as depletion region recombination. The sum of the two can adequately represent most real situations, where I_B depends on V_{BE} through multiple exponentials with potentially different temperature dependences.

To represent transient and AC characteristics, the Gummel–Poon model utilizes a charge-control approach. The input charge is associated with storage capacitors at the base–emitter or base–collector junctions. The charges Q_{be} and Q_{bc} include depletion region contributions. In addition, Q_{be} includes the stored charge in the base resulting from base transit time and collector transit time contributions, as follows:

$$Q_{be} = Q_{be}(\text{depletion}) + T_F I_S[\exp(qV_{BE}/kT) - 1] \qquad (45)$$

where T_F is defined in Section 1.2.2, as an amalgam of most of (but not all) the contributions to τ_{ec}. To account for bias dependences such as the Kirk effect, T_F is allowed to vary with voltage and current according to

$$T_F = T_{F0}[1 + X_{TF}\exp(V_{BC}/1.44V_{TF})][1 + (I_{TF}/I_S)\exp(qV_{BE}/kT)] \quad (46)$$

where X_{TF}, V_{TF} and I_{TF} are suitable fitting parameters. The small-signal element values for the model are derivable from the large-signal model by differentiating the currents and charges with respect to voltages. The dependence of a large-signal current source on the voltage across its own terminals gives rise to a resistance; the dependence of its current on the voltage across a different terminal pair gives rise to a transconductance. For example, the collector current generator in the large-signal model is responsible for an output resistance, r_o, and a transconductance, g_m, obtained as

$$1/r_o = dI_C/dV_{BC} \qquad (47)$$
$$g_m = dI_C/dV_{BE}$$

Similarly, the charges associated with a given terminal pair are associated with a capacitance (corresponding to the derivative with respect to its own terminal voltage) or a transcapacitance, when the charge depends on the voltage across a different terminal pair. In the Gummel–Poon model, a transcapacitance is introduced through the V_{BC} dependence of the $I_C T_F$ charge at the base–emitter junction.

The Gummel–Poon model has been effective over a period of decades. However, to provide improved accuracy in the description of a variety of modern devices, different models are being pursued. Shortcomings of the Gummel–Poon model include the following:

1. Current crowding effects are not described well. The current crowding is of central concern for silicon bipolar transistors, but can be ignored for most HBTs, which tend to have much greater base doping. To partially account for these effects, an expression is provided in the SPICE model to indicate that the base resistance varies with forward current as described above.

2. The charge storage and the resistance of the collector during saturation is not well modeled. Particularly for transistors which are required to support moderately high voltages, a thick, lightly doped collector region is required. During operation at low V_{BC}, this region is not depleted, and can contribute significant amounts of series resistance. However, if V_{BC} is sufficiently forward biased, holes injected into the collector decrease the resistance. Describing the collector during saturation and "quasisaturation" has been the object of various models.

3. The transient behavior of transistors is described by the charge-control model only to an approximate extent. In particular, the charge distributions on a transient basis differ from what would be obtained by a succession of steady-state distributions as assumed by the charge-control model. To lowest order, the nonstationary charge distributions lead to the result that the output current is delayed in time relative to the input bias voltages. To account for this, a time delay can be incorporated into the collector current source in the model of Fig. 38. The amount of delay necessary to correct the delays already contained in the charge control model is known as excess phase. Such a delay is easily incorporated into frequency-domain simulators by

associating a phase factor $\exp(j\omega t_d)$ with the g_m generator. This strategy is widely used in microwave linear and harmonic balance simulators. Taking accurate account of the extra delay in the time domain is relatively difficult, and can lead to convergence difficulties in transient simulations. One way in which the excess phase can be incorporated is by means of transcapacitors, elements in which charge is stored to an extent proportional to the voltage across a different pair of terminals.

4. To accurately describe the base resistance and base–collector capacitance of a transistor, a distributed resistance–capacitance network should be used. For simplicity, the Gummel–Poon model incorporates only a single base resistance, through which most of the device capacitance must be charged (although a portion of C_{bc} can be placed outside of the resistance, by using the "X_{CJC}" factor). In more accurate descriptions, the base is partitioned into several regions, and various series resistances and associated capacitances are defined.

5. With the high current densities possible with bipolar transistors, there can be appreciable heating of the device with the passage of current. Since various transistor characteristics (such as I_C vs V_{BE}) are strongly temperature dependent, the self-heating effect can modify the measured characteristics. This is particularly important for III–V devices, which have high base conductivity, and as a result large emitter widths. An added complication is that the thermal conductivity is lower than for silicon. To account for self-heating, it is appropriate to consider a thermal subcircuit associated with the transistor. The transistor currents can be computed simultaneously with the device temperature, given a value for the thermal resistance and capacitance.

To provide an improved description of bipolar transistors, several refined models have recently been developed.[34–36]. Figure 40 illustrates the topology for the MEXTRAM model,[35] whereas Fig. 41 shows a model that has been developed particularly for HBTs. Both show the distributed structure of the base resistance and capacitance. Figure 41 also shows the thermal subcircuit to describe self-heating (which appears as an R–C network attached to the basic transistor).

1.6 SUMMARY AND FUTURE TRENDS

Bipolar transistors have long been and, for the foreseeable future, will continue to be the devices of choice for many high-speed applications. Key characteristics in maintaining their advantage are high f_T and high g_m. Both features of bipolar transistors can be achieved with very good control of device characteristics, enabling accurate analog circuits.

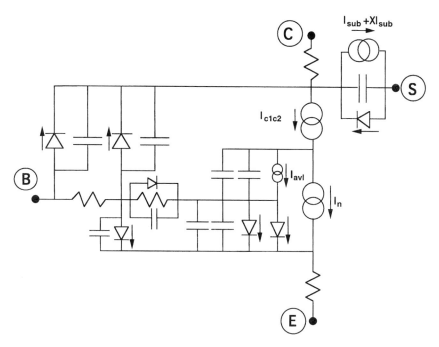

Fig. 40 Schematic of MEXTRAM model for bipolar transistors. (After de Graaf and Klaasen, Ref. 35.)

For digital applications, a major objective has been to increase circuit density. Lateral scaling has been carried out aggressively, with corresponding decreases in device parasitics, interconnect capacitance, and power dissipation per device. This trend will continue as lithography techniques improve. Vertical scaling has a limited scope in homojunction devices. Base doping must be increased due to base punchthrough considerations. However, this tends to degrade current gain, and the emitter stored charge rises, limiting f_T. HBTs provide a way for further improvement of operating speed, by minimizing the emitter stored charge while avoiding base punchthrough. Already f_T's above 200 GHz have been achieved in HBTs.

It is interesting to note that in large, dense digital circuits the importance of lowering base resistivity is not paramount. As emitter width W is scaled, typically collector currents are reduced with logic swings kept constant. Circuit load resistances thus increase as $1/W$, and the tolerable values of R_b also increase. At the same time, the base resistance for a given base sheet resistance (pinch resistance) decreases as W. The permissible base resistivities rise as W^2.

Additional problems for bipolar transistors that will arise from continued dimensional scaling are associated with breakdown, tunnel currents in the base–emitter junctions, and hot-electron effects due to increasing electric

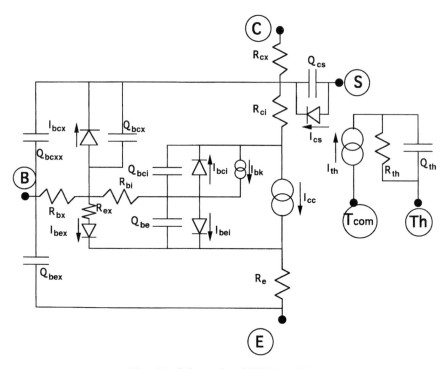

Fig. 41 Schematic of HBT model.

fields. The severity of many of these problems could be reduced if the operating voltages could be scaled down with transistor size. One of the critical problems of bipolar technology is that V_{BE} cannot be scaled with a fixed material system. This makes it difficult to reduce power supply voltage. Inherent problems of the present bipolar logic approaches, such as considerable static power dissipation and saturation charge storage, remain even with scaling.

For analog and microwave applications, the trend to smaller devices is not as clearcut. Power levels used in circuits are dictated by the impedance of free space or of conveniently fabricated transmission lines, together with noise considerations. Therefore currents used in transistors cannot be reduced. When impedance matching is used, there is continued benefit in terms of gain improvement and noise reduction from decreasing R_b down to very low values. Current crowding is a central concern. For analog circuits, matching of devices is frequently of key importance. This is more readily achieved with large lateral device dimensions. Therefore the advantages of HBTs with low base sheet resistances and high gain achievable with large emitter dimensions are more evident.

In addition to these trends based on direct extrapolation of present technology, a variety of new themes will be explored in the future for bipolar transistors. These include increased use of velocity overshoot, new materials, and integration of bipolars with different devices.

1.6.1 Velocity Overshoot Effects

The velocity of electrons in semiconductors, particularly III–Vs, may attain high values, well above 10^7 cm/s, for brief periods after the electrons enter regions with high electric fields. This effect is particularly important in HBTs for electrons traversing the base–collector depletion region, and is emphasized in the BCT. It increases the f_T of the device, by reducing the transit time of electrons across the base–collector depletion region. It also increases the maximum current density, J_{cmax}, that can be carried by an HBT without base pushout, and contributes to reduced base–collector capacitance at a fixed value of f_T by increasing the distance traveled by electrons in a given amount of time. HBTs are very well positioned to take advantage of the velocity overshoot phenomena because the dimensions involved can made small in a reproducible fashion. A major drawback is that the voltage range over which velocity overshoot persists is not very great. To take the fullest advantage of it, logic circuits will have to be configured to keep voltage swings small and the associated values of V_{CE} relatively fixed. For power devices, it is necessary to seek ways to achieve efficient power combining from low-voltage, high-current sources.

1.6.2 New Materials

Exploration of new materials systems is an important avenue to improve devices. Material systems will be tailored to specific applications.

For microwave power generation and amplification, high output power density is desirable, as well as operation at convenient (high) impedance levels. A useful figure of merit for this application is $v_{sat} \mathscr{E}_b^2$, where $\mathscr{E}_b$ is the breakdown electric field. The semiconductors used most extensively for bipolar transistors do not differ significantly in regard to this figure of merit. However, different semiconductors, including SiC, GaN, and diamond, offer the possibility of dramatic increases in this area.

For logic circuits, choice of new materials will be dictated by entirely different considerations. It has been argued that the power–delay product in the limit of very large circuits ("wire-limited chip") is remarkably insensitive to transistor parameters. Key parameters are layout size, which can determine interconnect lengths, and threshold voltage for current flow. One theme of future bipolar evolution will be to minimize V_{BE}. For FET technologies, the threshold voltage for switching (turn-on voltage, V_T) is relatively easily varied by changing channel doping and thickness. FETs, however, tend to have nonuniform V_T and low transconductance, and as a result are difficult to operate with low voltage swings. With bipolar technologies, V_T is fixed by the material system. For this reason, it is relatively uniform across wafers, but it cannot be readily changed to minimize power dissipation. From Eq. 8, V_T ($= V_{BE}$) in HBTs with graded junctions is dependent on the bandgap of the material used in the base, E_{gb}. To minimize dynamic power–delay product, it is important to choose materials providing a low value of E_{gb}. For example, InGaAs can provide an advantage over silicon (while GaAs is worse in this regard because of its large bandgap). InAs base regions may provide further benefit in the future. This trend can be expected to continue until the leakage currents associated with thermal generation of carriers in the small bandgap materials become excessive. The critical leakage of the base–collector depletion region can be minimized by wide-bandgap material in the collector.

1.6.3 Integration of Bipolar Transistors with Different Devices

While bipolar transistors do not have all the characteristics needed for optimal circuits, they can be monolithically combined with other devices that supplement their strengths. One well-known example of this is BiCMOS technology, in which n–p–n bipolar transistors are combined with n-channel and p-channel MOSFETs. BiCMOS logic tends to combine the best features of CMOS and bipolar technologies: low static power dissipation and high drive capability (from high transconductance output transistors).

Related combinations of devices may be expected in the realm of III–V devices. Combinations of HBTs and FETs have been demonstrated, with the objective of combining high-speed and low-power benefits for logic circuits, to achieve combinations of high-power and low-noise characteristics in the microwave domain, and to incorporate devices with high DC input impedance in analog circuits.

A major theme within bipolar technology is the combination of n–p–n and p–n–p devices. This chapter has not covered p–n–p transistors significantly, since p–n–p performance has traditionally lagged that of n–p–n's. Particular synergy occurs, however, with combinations of n–p–n and p–n–p transistors on the same chip. This has long been practiced in analog circuits, where p–n–p's form active loads (current mirrors) in high gain stages, and are used in output drivers (complementary push–pull amplifiers). The p–n–p's most often implemented have been lateral structures with limited f_T. More recently, high-f_T vertical p–n–p's have been developed and incorporated alongside fast n–p–n's, yielding significant performance improvements in circuits. A long-range objective is to develop a logic approach that minimizes static power dissipation, the bipolar analog of CMOS. The most straightforward implementation of such circuits, illustrated in Fig. 42, requires transistors which have adequate current gain when operated in saturation and do not store excess charge under this condition. Such devices have not yet been developed. Alternative approaches, in which logic is carried out by diode switches or by n–p–n's only and the output drivers are implemented

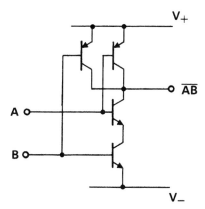

Fig. 42 Circuit diagram for complementary bipolar logic.

with complementary circuits, have also been suggested. On a long-term basis, it can be expected that complementary HBT logic approaches, which will eliminate static power dissipation, operate with low logic swings, and minimize power-supply voltage by appropriate choice of materials, can be developed. By combining such structures with advanced interconnect technology to insure low wiring capacitance, an "optimal" logic approach will have been established.

PROBLEMS

1. Derive Eq. 6 from Eqs. 4 and 5, by integrating over the quasineutral base region, and applying boundary conditions expressed in Eq. 2.

2. Consider a bipolar transistor in which a constant electric field $\mathscr{E}$ is built into the base by a suitable profile of the base doping. Derive an expression for the resultant base transit time, analogous to the equation $\tau_b = w^2/2D$, applicable to base regions without drift fields.

3. Derive the fact that for a bipolar transistor with stripe geometry, in which there is a rectangular emitter and stripe base contacts on either side, the intrinsic base resistance has the value $R_b = R_s W/12L$, where W and L are width and length of the emitter stripe. Derive a similar expression for a bipolar transistor with a circular emitter, of diameter D, surrounded by an annular base contact.

4. Discuss the impact on transistor performance of varying the thickness of the base, w_b. Include the key effects on both AC and DC performance. Derive the value of base thickness that maximizes the value of f_{max}.

5. Discuss the impact on transistor performance (both DC and AC) of varying the thickness of the base–collector depletion region, w_c. What value of w_c leads to the largest value of f_{max}?

6. Discuss the impact on transistor f_T of varying the collector–emitter voltage, V_{CE}. It is frequently observed that for increasing V_{CE}, silicon bipolar transistors have increasing f_T, whereas III–V HBTs have decreasing f_T. Explain this phenomenon, and derive the approximate dependences of f_T on V_{CE}.

7. Discuss the impact on transistor performance of having a large ratio between the base–collector junction area (A_{BC}) and the emitter area (A_E).

8. Consider an HBT in which the bandgap (and composition) of the base is linearly varied between a value E_{g1} at the base–emitter junction and a smaller value E_{g2} at the base–collector junction. For simplicity assume that there is no discontinuity between conduction bands of emitter, base, and collector (such as in the case of Si/SiGe). Show that the transistor current gain is primarily affected by E_{g1}, and the transistor Early voltage is primarily affected by E_{g2}. Derive quantitative relationships for these dependences.

9. Compute the "offset voltage" V_{cesat}, expected for a GaAlAs/GaAs HBT with the doping profile of Fig. 20. Assume that the conduction-band discontinuity between base and emitter is completely graded away, so that electron flow is limited exclusively by base transport. (*Hints*: (1) consider hole flow in the collector; (2) include considerations of Problem 7.)

10. Given a simplified hybrid pi AC model of a bipolar transistor, as illustrated in Fig. 38b, derive an equivalent T model of the transistor. Find the relationship between R_π (between base and emitter in the hybrid pi model) and R_e (between base and emitter in the T model).

REFERENCES

1. S. M. Sze, *Physics of Semiconductor Devices*, 2nd ed., Wiley, New York, 1981.
2. H. K. Gummel and H. C. Poon, "An integral charge control model for bipolar transistors," *Bell Syst. Tech. J.* **49**, 827 (1970).
3. J. E. Sutherland and J. R. Hauser, "A computer analysis of heterojunction and graded composition solar cells," *IEEE Trans. Electron Dev.* **ED-24**, 363 (1977).
4. H. Kroemer, "Two integral relations pertaining to the electron transport through a bipolar transistor with a nonuniform energy gap in the base region," *Solid-State Electron.* **28**, 1101 (1985).

5. D. S. Lee and J. G. Fossum, "Energy band distortion in highly doped silicon," *IEEE Trans. Electron Dev*. **ED-30**, 626 (1983).

6. S. Tiwari, *Compound Semiconductor Device Physics*, Academic Press, New York, 1992.

7. L. H. Camnitz and N. Moll, "An analysis of the cut-off frequency behavior of microwave heterostructure bipolar transistors," in *Heterojunction Transistors*, S. Tiwari, ed., 1994.

8. T. Nakamura and H. Nishizawa, "Recent progress in bipolar transistors," *IEEE Trans. Electron. Dev*. **ED-42**, 390 (1995).

9. D. D. Tang, P. M. Solomon, T. H. Ning, R. D. Isaac, and R. E. Burger, "1.25 μm deep-groove-isolated self-aligned bipolar circuits," *IEEE J. Solid State Circ*. **SC-17**, 925 (1982).

10. T. Yamaguchi, S. Uppilli, G. Kawamoto, J. Lee, and S. Simpkins, "Process and device optimization of a 30 GHz ft submicrometer double-poly-Si bipolar technology," in *Tech. Dig. 1993 Bipolar/BiCMOS Circuits and Technology Meeting*, p. 136.

11. S. Konaka, Y. Amemiya, K. Sakuma, and T. Sakai, "A 20 ps/G Si bipolar IC using advanced SST with collector ion implantation," in *Ext. Abstr. 19th Conf. Solid-State Dev. Matls.*, Tokyo, 1987, p. 331.

12. T. Nakamura, T. Miyazalci, S. Takahashi, T. Kure, T. Okabe, and M. Nagata, "Self-aligned transistor with sidewall base electrodes," *IEEE Trans. Electron Dev*. **ED-29**, 596 (1982).

13. H. Kroemer, "Theory of wide-gap emitter for transistors," *Proc. IRE* **45**, 1535 (1957).

14. H. Kroemer, "Heterostructure bipolar transistors and integrated circuits," *Proc. IEEE* **70**, 13 (1982).

15. A. Marty, G. E. Rey, and J. P. Bailbe, "Electrical behavior of an NPN GaAlAs/GaAs heterojunction transistor," *Solid-State Electron*. **22**, 549 (1979).

16. S. S. Iyer, G. L. Patton, J. M. C. Stork, B. S. Meyerson, and D. L. Harame, "Heterojunction bipolar transistors using Si-Ge alloys," *IEEE Trans. Electron Dev*. **ED-36**, 2043 (1989).

17. H. Temkin, J. C. Bean, A. Antreasyan, and R. Leibenguth, "Ge$_x$Si$_{1-x}$ strained-layer heterostructure bipolar transistors," *Appl. Phys. Lett*. **52**, 1089 (1988).

18. J. F. Gibbons, C. A. King, J. L. Hoyt, D. B. Noble, C. M. Gronet, M. P. Scott, S. J. Rosner, G. Reid, S. Laderman, K. Nauka, J. Turner, and T. I. Kamins, "Si/SiGe heterojunction bipolar transistors fabricated by limited reaction processing," *Tech. Dig. IEDM*, 566 (1988).

19. D. L. Harame, J. H. Comfort, J. D. Cressler, E. F. Crabbe, J. Y.-C. Sun, B. S. Meyerson, and T. Tice, "Si/SiGe epitaxial-base transistors: Part I–Materials, physics and circuits," *IEEE Trans. Electron Dev*. **ED-42**, 455 (1995).

20. A. Gruhle, I. Kibbel, U. Konig, U. Erben, and E. Kasper, "MBE-Grown Si/SiGe HBT's with High Ft and Fmax," *IEEE Electron Dev. Lett*. **EDL-13**, 206 (1992).

21. D. Ankri and L. F. Eastman, "GaAlAs/GaAs ballistic heterojunction bipolar transistor," *Electron. Lett*. **18**, 750 (1982).

22. M. F. Chang, P. M. Asbeck, K. C. Wang, G. J. Sullivan, N. H. Sheng, J. A. Higgins, and D. L. Miller, "AlGaAs/GaAs heterojunction bipolar transistors

fabricated using a self-aligned dual-liftoff process," *IEEE Electron Dev. Lett.* **EDL-8**, 7 (1987).

23. S. Tiwari, "GaAlAs/GaAs heterostructure bipolar transistors: experiment and theory," *Tech. Dig. IEDM*, 262 (1986).

24. H. H. Lin and S. C. Lee, "Super-Gain AlGaAs/GaAs heterojunction bipolar transistors using an emitter edge-thinning design," *Appl. Phys. Lett.* **47**, 839 (1985).

25. M. F. Chang, ed., *Current Trends in Heterojunction Bipolar Transistors*, World Scientific, Singapore, 1996.

26. C. M. Maziar, M. E. Klausmeier-Brown, and M. Lundstrom, "A proposed structure for collector transit-time reduction in AlGaAs/GaAs bipolar transistors," *IEEE Electron Dev. Lett.* **EDL-7**, 483 (1986).

27. R. Katoh and M. Kurata, "Self-consistent particle simulation for AlGaAs/GaAs HBTs under high bias conditions," *IEEE Trans. Electron Dev.* **ED-36**, 2122 (1989).

28. T. Ishibashi and Y. Yamauchi, "A possible near-ballistic collection in an AlGaAs/GaAs HBT with a modified collector structure," *IEEE Trans. Electron Dev.* **ED-35**, 401 (1988).

29. R. J. Malik, J. R. Hayes, F. Capasso, K. Alavi, and A. Y. Cho, "High-gain $Al_{0.48}In_{0.52}As/Ga_{0.47}In_{0.53}As$ transistors grown by molecular beam epitaxy," *IEEE Electron Dev. Lett.* **EDL-4**, 383 (1983).

30. R. N. Nottenburg, H. Temin, B. Panish, R. Bhat, and J. C. Bischoff, "InGaAs/InP double-heterostructure bipolar transistors with near-ideal beta versus Ic characteristics," *IEEE Electron Dev. Lett.* **EDL-7**, 643 (1986).

31. U. K. Mishra, J. F. Jensen, D. B. Rensch, A. S. Brown, M. W. Pierce, L. G. McGray, T. V. Kargodorian, W. S. Hoefer, and R. E. Kastris, "48 GHz AlInAs/GaInAs heterojunction bipolar transistors," *Tech. Dig. IEDM*, 873 (1988).

32. Y. K. Chen, R. N. Nottenburg, M. B. Panish, R. A. Hamm, and D. A. Humphrey, "Subpicosecond InP/InGaAs heterostructure bipolar transistors," *IEEE Electron Dev. Lett.* **EDL-10**, 267 (1989).

33. J.-I. Song, K. B. Chough, C. J. Palmstrom, B. P. Van der Gaag, and W.-P. Hong, "Carbon-doped base InP/InGaAs HBTs with ft = 200 GHz," in *IEEE Device Research Conf.*, 1994.

34. G. M. Kull, L. W. Nagel, S. W. Lee, P. Lloyd, E. J. Prendergast, and H. Dirks, "A unified circuit model for bipolar transistors including quasi-saturation effects," *IEEE Trans. Electron Dev.* **ED-32**, 1103 (1985).

35. H. C. de Graaff and F. M. Klaassen, *Compact Transistor Modeling for Circuit Design*, Springer-Verlag, Berlin/New York, 1990.

36. M. Schroter and H.-M. Rein, "Investigation of very fast and high-current transients in digital bipolar ICs using both a new compact model and a device simulator," *IEEE J. Solid State Circ.* **SC-30**, 551 (1995).

2 Compound-Semiconductor Field-Effect Transistors

MICHAEL S. SHUR

Department of Electrical, Computer, and Systems Engineering,
Rensselaer Polytechnic Institute, Troy, New York

TOR A. FJELDLY

Norwegian University of Science and Technology, Trondheim, Norway

2.1 INTRODUCTION

2.1.1 Principle of FET Operation

The concept of the field-effect transistor (FET) was first proposed as early as the 1930s by Lilienfeld and Heil.[1] But only in the 1950s had semiconductor–material processing technology progressed far enough that Dacey and Ross were able to demonstrate working devices.[2] In the early 1960s, this technology started to displace bipolar junction transistors.

Currently, FET technology plays a dominant role in electronics, and FET devices and integrated circuits are made in a variety of designs and with many different semiconductor materials. Today, most FETs are made from silicon, because of this material's many excellent properties. However, compound-semiconductor FETs, considered in this chapter, occupy respectable niches, especially in high-speed, high-frequency applications, as well as in electronics that has to withstand harsh environments, including high and low temperatures and exposure to high-energy radiation. As a measure of the maturity of compound-semiconductor FET technology, we note that the integration level of GaAs digital circuits has reached more than one million FETs on a chip.

Figure 1a shows the basic structure of a GaAs metal–semiconductor

Modern Semiconductor Device Physics, Edited by S. M. Sze.
ISBN 0-471-15237-4 © 1998 John Wiley & Sons, Inc.

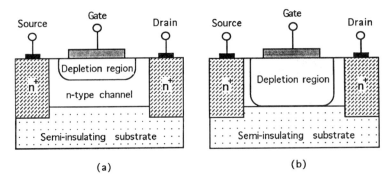

Fig. 1 Schematic MESFET structure at zero drain bias above threshold (a) and below threshold (b).

field-effect transistor (MESFET), which is the most important compound-semiconductor FET. Typically, the GaAs MESFET consists of a semi-insulating or p-type substrate supporting a thin n-type conducting GaAs layer to which three contacts are attached—the ohmic source and drain contacts, and the Schottky gate contact. The conducting layer between source and drain constitutes the FET channel, whose resistance is modulated by the voltage V_{gs} applied between gate and source.* As indicated in Fig. 1b, the rectifying metal–semiconductor gate contact creates a depletion region in the MESFET channel, whose thickness depends on V_{gs}.

When the gate bias is sufficiently negative, the channel is fully depleted (depletion-mode device) and the current between the source and drain becomes very small. The gate–source voltage at which this occurs is called the threshold voltage V_T. Above threshold, the incremental variation of depletion charge per unit area, ΔQ_d, is roughly proportional to the gate voltage variation, ΔV_{gs}:

$$\Delta Q_d = C \Delta V_{gs} \tag{1}$$

where $C = \varepsilon_s/h$ is the differential gate–channel capacitance per unit area, ε_s is the dielectric permittivity of the semiconductor material and h is the thickness of the depletion region. Also, we have $\Delta Q_d = -q \, \Delta n_s$ where n_s is the concentration of conduction electrons per unit area in the channel and q is the unit charge.

This basic principle of the capacitive charge modulation of the conductive channel is common to all FETs. Different FETs use different variations of this principle. They differ by where the channel charge is located, how the isolation between the gate and the channel is achieved, and what materials

---

*V_{gs} is the extrinsic gate–source voltage, which is larger than the intrinsic gate–source voltage V_{GS} to be discussed later.

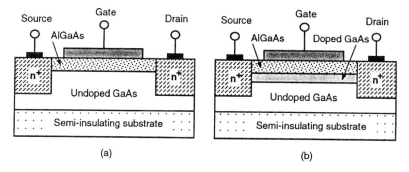

Fig. 2 Schematic structures of (a) HFET and (b) doped-channel HFET.

and doping profiles are used for the gate insulator, the channel, and the substrate.

2.1.2 Types of FETs

In a metal–oxide–semiconductor field-effect transistor (MOSFET), the channel conduction above threshold occurs at the interface of the silicon and the silicon dioxide gate insulator. This device, the most important solid-state device by far, is discussed in Chapter 3.

Heterostructure field-effect transistors (HFETs) are in many ways similar to MOSFETs. In these devices, the gate is separated from the channel by a wide-bandgap semiconductor layer as shown in Fig. 2a, and the channel conduction occurs at the heterointerface.

The MESFET and HFET technologies can be combined in the doped-channel HFET (DCHFET) shown in Fig. 2b. Since compound semiconductors lack good native oxides (such as SiO_2 for Si), MESFETs and HFETs are the devices of choice for compound-semiconductor FETs.

2.1.3 Basic Material Properties

Compound-semiconductor technology is more difficult and is less developed than silicon technology. However, GaAs and several other compound semiconductors have certain advantages over silicon that for many applications outweigh the disadvantages. The advantages include a direct energy gap and, hence, superior optoelectronic properties, a high low-field electron mobility that contributes to smaller parasitic resistances and a higher device speed, a high peak velocity that leads to higher speed and operating frequencies in short-channel devices, and the availability of semi-insulating substrates that make GaAs well suited for microwave and millimeter-wave monolithic integrated circuits. The disadvantages of GaAs compared to

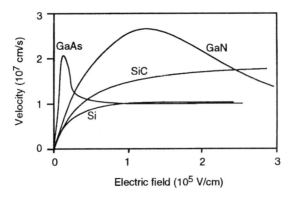

Fig. 3   Velocity vs field dependencies of electrons in Si, GaAs, and the wide-bandgap semiconductors GaN and SiC. (After Shur, Ref. 3.)

silicon include a much lower thermal conductivity and higher material and processing costs.

Figure 3 shows velocity–field dependencies of electrons in Si, GaAs, and the two wide-bandgap semiconductors GaN and SiC.[3] A detailed survey of material properties of compound semiconductors may be found in Ref. 4.

2.2   SCHOTTKY BARRIERS AND OHMIC CONTACTS

The rectifying metal–semiconductor junction used, for example, as the MESFET gate is called a Schottky-barrier junction. The basic equations describing Schottky-barrier contacts are given in Table 1,[1,5] where we use the usual notation and assume fully ionized donors in the depletion region. Figure 4 shows some calculated energy-band diagrams of GaAs Schottky barriers.

The energy-barrier height $q\phi_b$ is a key parameter of the junction, controlling both the width of the depletion region in the semiconductor and the electron current across the interface. It is defined as the energy difference between the semiconductor conduction-band edge at the interface and the Fermi level in the metal (see Fig. 4).

Qualitatively, we can describe the physics of the Schottky-barrier formation as follows. When a metal and an n-type semiconductor are brought in close proximity, an exchange of electrons will take place between the two materials to establish thermal equilibrium, i.e., establish a constant Fermi level throughout the junction. Initially, the barrier height for electron-escape from the metal is usually higher than from the semiconductor. Hence, when thermal equilibrium is reached, a net transfer of electrons from the semiconductor to the metal has taken place, leaving the metal negatively

TABLE 1 Important Equations Describing Schottky Contacts[1,5]

Space-charge density in the depletion layer	$\rho = qN_D$
Field distribution in the depletion layer	$\mathscr{E} = -\dfrac{qN_D(x_n - x)}{\varepsilon_s}$
Potential distribution in the depletion layer	$V = -\dfrac{qN_D(x_n - x)^2}{2\varepsilon_s} = -V_{bi}\left(1 - \dfrac{x}{x_n}\right)^2$
Depletion layer width	$x_n = \sqrt{\dfrac{2\varepsilon_s(V_{bi} - V)}{qN_D}}$
Schottky-diode empirical equation (with series resistance)	$I = I_s\left[\exp\left(\dfrac{V - IR_s}{\eta V_{th}}\right) - 1\right]$
Reverse diode current density (saturation current density)	$J_{ss} = A^*T^2\exp\left(-\dfrac{\phi_b}{V_{th}}\right)$ where
	$A^* = \dfrac{\alpha m^*qk^2}{2\pi^2\hbar^3} \approx 120\alpha\,\dfrac{m^*}{m_0}\left(\dfrac{A}{cm^2 - K^2}\right)$

charged and the semiconductor positively charged. The positive charge in the semiconductor is created by depletion of electrons from a layer near the interface, exposing the positively charged ionized donors. The resulting dipole layer is very similar to that of a p^+-n junction. However, for a given semiconductor material, the Schottky-barrier is usually smaller than the built-in voltage of the p^+-n junction.

The value of $q\phi_b$ depends on the difference between the electron affinities for the metal $q\chi_m$ and the semiconductor $q\chi_s$, that is, $q(\chi_m - \chi_s)$, and on the properties of the metal–semiconductor interface. The termination of any crystalline material naturally gives rise to a large number of interface states, as does the metal–semiconductor interface of a Schottky-barrier junction. If we could ignore the effects of interface states, we would have $\phi_b = \chi_m - \chi_s$, but this is practically never the case.

In the other limiting case of a very high interface-state density, the Fermi level at the semiconductor–metal interface will be pinned at a certain level in the energy gap. The physics of the pinning of the Fermi level can be explained using the following model, illustrated by Fig. 5.

We assume that the interface states are such that they are acceptor-like above and donor-like below a certain energy level $q\phi_0$, the so-called neutral level. Note that the donor-like states are neutral when filled and positively

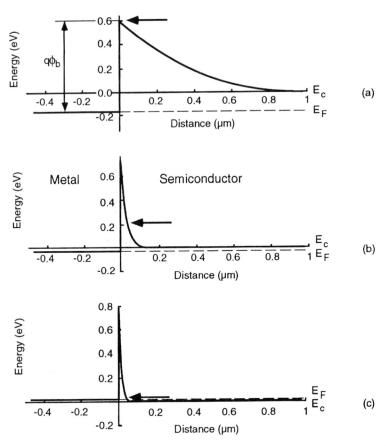

Fig. 4 Band diagrams of Schottky-barrier junctions for GaAs for n-type doping levels $N_D = 10^{15}/cm^3$ (a), $N_D = 10^{17}/cm^3$ (b), and $N_D = 10^{18}/cm^3$ (c). Arrows indicate electron transfer across the junction at forward bias. (After Shur, Ref. 5.)

charged when empty, and the acceptor-like states are neutral when empty and negatively charged when filled. Then the total charge in the interface states is zero if and only if the Fermi level E_F at the interface coincides with the neutral level. Moving E_F above $q\phi_0$ leads to a negative interface charge, $-qN_{sa}(E_F - q\phi_0)$, where N_{sa} is the surface acceptor-like state density, and moving E_F below $q\phi_0$ leads to a positive interface charge, $qN_{sd}(q\phi_0 - E_F)$, where N_{sd} is the surface donor-like state density. Since the position of the Fermi level is determined by the requirement of overall charge neutrality, the Fermi level must coincide with the neutral level in the limit of a very high density of interface states. This conclusion is not dependent on the assumption of a constant density of the interface states. However, an exact calculation of the Schottky-barrier height requires detailed information on the distribution of the interface states and on the other properties of the

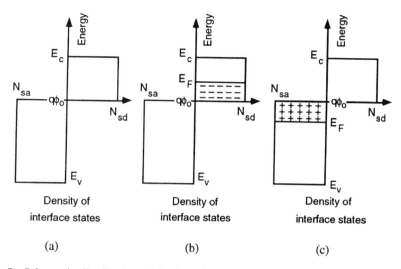

Fig. 5 Schematic distribution of the interface states: (a) neutral, (b) negative, and (c) positive interface charges for different positions of the Fermi level at the interface.

interfacial layer.[6] Since such information is not usually available, the Schottky-barrier height is normally determined from experimental current–voltage and capacitance–voltage characteristics.

As indicated in Fig. 4a, at very low doping levels electrons cross the barrier mainly by passing over the top of the barrier—this is called thermionic emission. At moderate doping levels (Fig. 4b), electrons mainly tunnel through the barrier at some elevated energies where the barrier is sufficiently thin—this is called thermionic–field emission. In highly doped, degenerate semiconductors (Fig. 4c), the depletion region is so thin that even electrons near the Fermi-level tunnel through the barrier—this is called field emission. In the limiting case of a very high doping, the current–voltage characteristic is practically linear, the contact resistance is low, and the metal–semiconductor contact becomes ohmic.

The current–voltage characteristic of a Schottky barrier can be calculated by accounting for the electrons passing over the barrier and those tunneling through the barrier.[1] An exact calculation requires a detailed knowledge of the electron distribution function in a wide range of applied biases, as well as accurate information about scattering mechanisms and doping profiles.[7] Therefore, the analysis of compound-semiconductor FETs is usually based on semi-empirical models. A starting point for such models is the empirical diode equation

$$I = I_s \left[\exp\left(\frac{qV - IR_s}{\eta kT} \right) \right] \tag{2}$$

where I_s is the Schottky-diode saturation current, V is the bias voltage, R_s is the series resistance, η is the ideality factor, k is the Boltzmann constant, and T is the absolute temperature. Ideally, $\eta = 1$, but in practice η is always larger than unity and may be temperature dependent. The saturation current can be expressed as (see Table 1)[1]

$$I_s = A^* T^2 \exp\left(-\frac{q\phi_b}{kT} \right) \qquad (3)$$

where A^* is the Richardson constant ($A^* \approx 8.2$ A/cm^2-K^2 for n-type GaAs and $A^* \approx 74$ A/cm^2-K^2 for p-type GaAs), $q\phi_b$ is the effective barrier height (which accounts for the barrier lowering effects, and, hence, is smaller than the barrier shown in Fig. 4).

Typical metals used for Schottky contacts in GaAs FETs are Al (barrier heights 0.73–0.8 eV), Pt–Al (typical barrier height 0.85 eV), W–Al (barrier heights 0.66–0.71 eV), Ti–Pt–Au, WSi$_2$, and WN$_x$.

As discussed previously, a metal contact to a highly doped semiconductor may function as an ohmic contact. In fact, such contacts to GaAs have achieved fairly low contact resistances, as low as 10^{-7} Ω-cm^2. The key to low contact resistance is to have a very high carrier concentration in the contact region of the semiconductor layer or, better still, to utilize a narrow-gap semiconductor layer between the metal and the GaAs to further reduce the barrier height. An older and less-sophisticated ohmic contact technology involves alloying metals into the semiconductor. Typical alloyed metal contacts are AuGe–Ni, Ag–Sn, or AgIn–Ge for n-type GaAs, and AuZn, Ag–In–Zn, or Ag–Zn for p-type GaAs. When the field emission mechanism is dominant, the contact resistance may be estimated as follows:[5,8]

The contact resistance R_c is taken to be inversely proportional to the tunneling probability in the WKB (Wentzel–Kramers–Brillouin) approximation[1]

$$R_c^{-1} \propto T_t \approx \exp\left(-\frac{2}{\hbar} \int_0^W |p(x)| \, dx \right) \qquad (4)$$

Here $W \approx \sqrt{2\varepsilon_s \phi_b / q N_D}$ is the depletion width at the semiconductor–metal interface, x is the coordinate in the direction perpendicular to the metal–semiconductor interface ($x = 0$ corresponds to the interface), $\hbar$ is the reduced Planck constant, N_D is the donor density, and $p = \sqrt{2m^*[E_c(x) - E_F]}$ is the "missing momentum" needed for a classical penetration to position x, where $E_c(x)$ is the bottom of the conduction band in the depletion region and E_F is the Fermi level. Note that for a degenerate semiconductor, $E_F = E_c(x \to \infty) + \Delta E_F$ where

$$\Delta E_F = \frac{\hbar^2}{2m}\left(3\pi^2 N_D \right)^{2/3}$$

and $E_c(x \to \infty)$ is the bottom of the conduction band far from the metal–semiconductor interface.[5] Hence, Eq. 4 shows that the tunneling probability decreases exponentially with the integral of the "missing momentum" over the tunneling distance.

Since the peak of the barrier gives the largest contribution to T_t, the integrand in Eq. 4 can be linearized with respect to x, resulting in the following estimate of the contact resistance:

$$R_c \propto \exp\left(\frac{q\phi_b}{E_{00}}\right) \tag{5}$$

where

$$E_{00} = \frac{q\hbar}{2}\sqrt{\frac{N_D}{\varepsilon_s m^*}}$$

is the characteristic tunneling energy.

2.3   GaAs MESFETs

2.3.1   MESFET Fundamentals

The GaAs MESFET is an important member of the field-effect transistor family. In any FET, the current flows between two ohmic contacts—the source and the drain. The third contact—the gate—is capacitively coupled to the device channel which connects the source and drain. The gate bias (i.e., the gate voltage between the gate and the channel) determines the concentration of free carriers (electrons or holes) in the FET channel and, hence, controls the drain–source current.

Figure 6 shows schematic diagrams of six different GaAs MESFET designs. The MESFETs are typically fabricated either by a direct implantation into a GaAs semi-insulating substrate (Fig. 6a–e) or by a gate recess as shown in Fig. 6f. Both ion-implanted devices and recessed are used for microwave and millimeter-wave applications. For such applications, the semiconductor layers may also be grown by molecular beam epitaxy to obtain the desired doping and material profiles.

The basic design shown in Fig. 6a uses two implants, one for the channel and another for the highly doped regions, to facilitate the formation of low resistance ohmic contacts. The use of a self-aligned implant (Fig. 6b) reduces the source and drain series resistances. In this case, a refractory Schottky metal gate is used as an ohmic implant mask, since this allows the annealing of the implant without damaging the gate (typically using a rapid thermal anneal (RTA)). The drawbacks of this design are the additional Schottky-gate leakage current caused by implant straggle into the region under the gate,

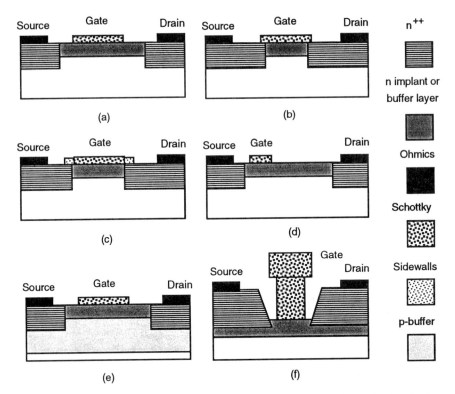

Fig. 6 Schematic diagrams of different GaAs MESFET designs. (a) Basic ion-implanted MESFET. (b) Self-aligned ion-implanted MESFET. (c) Self-aligned ion-implanted MESFET with sidewalls. (d) Ion-implanted MESFET with offset gate. (e) MESFET with *p*-type buffer. (f) Recessed MESFET with T-gate.

indicated as an overlap between the gate metal and the source and drain contacts in Fig. 6b, and a reduced breakdown voltage. This can be remedied by using dielectric sidewalls (typically SiN), as shown in Fig. 6c.

These designs use a symmetrically located gate. However, in power devices, where large drain–source and gate–drain voltages are applied, an offset gate indicated in Fig. 6d decreases the maximum electric field in the region between gate and drain, and thereby increases the breakdown voltage. Alternatively, an additional drain implant may be used.

Figure 6e shows a MESFET structure with a buried *p*-type layer. The additional barrier between the MESFET channel and the substrate, created either by an implant or by using a wide-bandgap semiconductor buffer layer, can improve the MESFET output characteristics. Figure 6f shows a typical short-channel microwave GaAs MESFET which uses a T-gate to minimize the gate series resistance.

To understand the principle of MESFET operation, let us first consider the channel conductance of a device with a uniformly doped active layer at zero drain–source bias. The maximum channel conductance is given by

$$g_0 = \frac{(\text{Conductivity}) \times (\text{Cross-sectional area})}{(\text{Length})} = \frac{qN_D\mu_n Wa}{L} \qquad (6)$$

where N_D is the ionized donor density (which is equal to the electron concentration n), μ_n is the low-field mobility, a is the channel thickness, and W and L are the width and the length of the channel, respectively. The actual channel conductance g_{ds} is smaller than g_0 since the conducting channel thickness b under the gate is reduced by the depletion layer at the interface between the Schottky gate metal and the semiconductor:

$$b = a - h \qquad (7)$$

Here the depletion layer thickness

$$h = \sqrt{\frac{2\varepsilon_s(V_{bi} - V_{GS})}{qN_D}} \qquad (8)$$

depends on the built-in voltage V_{bi} and on the applied gate bias V_{GS} (see Problem 2). The static dielectric permeability ε_s for GaAs is approximately 1.14×10^{-10} F/m. Figure 7 shows the MESFET depletion region schematically.

According to this analysis, ideally, the channel conductance should drop to zero when $b = a$. The corresponding gate bias is called the threshold voltage. From this definition, the threshold voltage (which is probably the most important parameter for a field-effect transistor) becomes

$$V_T = V_{bi} - V_p \qquad (9)$$

where

$$V_p = \frac{qN_D a^2}{2\varepsilon_s} \qquad (10)$$

is called the pinch-off voltage.

Let us now consider what happens when, in addition to the gate bias, a drain–source voltage is applied. This is illustrated in Fig. 7b. The different points in the conducting channel will be at different potentials with respect to the gate contact. If the drain bias is small, then the depletion width as a function of position x can be expressed as

$$h(x) = \sqrt{\frac{2\varepsilon_s[V_{bi} - V_{GS} + V(x)]}{qN_D}} \qquad (11)$$

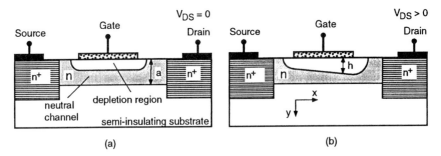

Fig. 7   Schematic shape of the depletion region in a GaAs MESFET with (a) zero drain bias and (b) applied drain bias.

where $V(x)$ is the channel potential relative to the source side. This approach is called the gradual channel approximation (GCA), first used by Shockley in his pioneering work on the theory of FETs.[9] As seen from Eq. 11 and Fig. 7b, the width of the depletion region increases towards the drain. Clearly, this approximation may work only as long as $h(L) < a$. At the drain bias where we have $h(L) = a$, the conducting channel cross-section at the drain side of the gate is pinched off. To maintain a drain current, the velocity of the charge carriers and the longitudinal electric field $\mathscr{E}_x$ at $x = L$ should become infinite. This result is, of course, impossible and forces us to reconsider our analysis.

In fact, Eq. 11 was derived by solving the one-dimensional Poisson equation for the transverse electric field $\mathscr{E}_y$—a procedure that is strictly valid only when $\mathscr{E}_x << \mathscr{E}_y$ (or, more accurately, when $d\mathscr{E}_x/dx << d\mathscr{E}_y/dy$). This analysis clearly shows that this condition is not satisfied near drain when we approach the pinch-off condition (i.e., when $h(L)$ approaches a). To avoid this problem, we may use the fact that the carrier drift velocity tends to saturate when the longitudinal electric field becomes sufficiently large. A realistic $v(\mathscr{E}_x)$ relationship for electrons in GaAs is shown in Fig. 8, together with a simplified piecewise linear model that can be used in the analysis. Here v_s and $\mathscr{E}_s = v_s/\mu_n$ are the saturation velocity and the saturation field of the simple model.

In modern GaAs FETs, the longitudinal channel field very often exceeds the saturation field in normal operation. We then assume that GCA is valid for the entire channel for drain biases $V_{DS} < V_{SAT}$, where the saturation voltage V_{SAT} is the drain–source voltage at which $\mathscr{E}_x(L) = \mathscr{E}_s$. Therefore, below saturation, the drain current I_d is given by the following differential equation, which is obtained by considering the voltage drop dV across a small channel segment of length dx at position x:

$$dV = I_d dR = I_d \frac{dx}{qN_D W\mu_n[a - h(x)]} \tag{12}$$

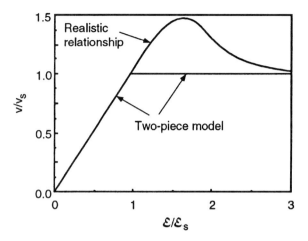

Fig. 8 Simplified piecewise linear model of the dependence of the electron velocity on the electric field, superimposed on a realistic $v(\mathscr{E})$ relationship for GaAs.

Substituting Eq. 11 into Eq. 12 and integrating with respect to x from zero to L, we obtain the fundamental equation of the field-effect transistor (see Problem 4),

$$I_d = g_0 \left(V_{DS} - \frac{2}{3} \frac{[(V_{DS} + V_{bi} - V_{GS})^{3/2} - (V_{bi} - V_{GS})^{3/2}]}{\sqrt{V_p}} \right) \qquad (13)$$

Using this approach, we can also find the electric-field distribution in the channel, from which we can determine the saturation voltage V_{SAT} from the condition $\mathscr{E}_x(L) = \mathscr{E}_s$ (see Problem 6). It can be shown that the Shockley saturation voltage, $V_{SAT} = V_{GT}$, is recovered when $\mathscr{E}_s L \equiv V_L \gg V_p$. In the opposite limit, when $V_L \ll V_p$, corresponding to near velocity saturation in the entire channel, we find $V_{SAT} = V_L$. For intermediate cases, the following interpolation formula can be obtained by combining the results for the two limiting cases:[10]

$$V_{SAT} = \left(\frac{1}{V_L} + \frac{1}{V_{GT}} \right)^{-1} \qquad (14)$$

Here $V_{GT} \equiv V_{GS} - V_T$.

When $V_{DS} = V_{SAT}$, the electron velocity saturates at the drain side of the channel. At higher voltages, the conducting channel can be divided into two regions—the GCA region at the source side of the channel, and the velocity saturation region near the drain. In the velocity saturation region, the GCA is invalid, and $d\mathscr{E}_y/dy$ may be even smaller than $d\mathscr{E}_x/dx$. In this region, the channel depletion-region thickness h_s is nearly constant, and the conducting channel thickness $b_s = a - h_s$ controls the constant drain saturation current I_{sat}:

$$I_{sat} = qN_D W b_s v_s \tag{15}$$

An understanding of the basic physics of the velocity saturation in FETs goes back to the work in Refs. 11–13. Equation 16 approximately expresses the channel saturation current via the MESFET parameters and bias voltages:[10,14]

$$I_{sat} = \frac{\beta V_{GT}^2}{1 + t_c V_{GT}} \tag{16}$$

where t_c is the transconductance compression factor

$$\beta = \frac{2\mathscr{E}_s v_s W}{a(V_p + 3V_L)} \tag{17}$$

is the transconductance parameter.[10] Note that with a nonzero t_c in Eq. 16, I_{sat} vs V_{GT} approaches a linear behavior (constant transconductance) at large V_{GT}, in good accordance with experimental observations.

2.3.2   Advanced MESFET I–V Model

For circuit simulations, a more comprehensive model for MESFET current–voltage (*I–V*) characteristics can be based on a unified description of the channel charge.[15] This model utilizes the universal modeling concept, which has been successfully applied to MOSFETs, GaAs MESFETs, and HFETs[16] (hence the term "universal"). In the extension of this model to GaAs MESFETs, mechanisms such as bias-dependent series source and drain resistances, effects of bulk charge, bias-dependent average low-field mobility, temperature-dependent model parameters, and gate leakage were added. A discussion of the basic properties of this MESFET model follows. For additional details, see Ref. 15.

At drain–source voltages well below saturation, the drain current can be expressed as

$$I_d \approx g_{chi} V_{DS} \approx g_{ch} V_{ds} \tag{18}$$

where g_{chi} and g_{ch} are the intrinsic and extrinsic channel conductances of the linear region, and V_{DS} and V_{ds} are the intrinsic and extrinsic drain–source voltages, respectively, which are related by

$$V_{ds} = V_{DS} + I_d(R_s + R_d) \tag{19}$$

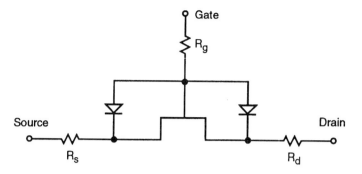

Fig. 9 Equivalent circuit representing MESFET as an intrinsic device in series with the source, the drain and the gate parasitic resistances. In the ideal situation, where the gate leakage current (represented by the diode symbols) is negligibly small, the voltage drop across the gate resistance R_g may be neglected at low frequencies. At high frequencies, this resistance may become very important.

Here R_s and R_d are the series source and drain resistances, respectively (see Fig. 9). In submicron devices, these resistances may become comparable to the intrinsic channel resistance, and therefore play a significant role in the overall performance of compound semiconductor FETs.

The extrinsic gate bias V_{gs} is related to the intrinsic gate–source voltage V_{GS} by

$$V_{gs} = V_{GS} + I_d R_s \tag{20}$$

and $V_{gd} = V_{gs} - V_{ds}$. Note that at small drain bias, we have $V_{ds} \approx V_{DS}$ and $V_{gs} \approx V_{GS}$.

The extrinsic channel conductance in the linear region is related to its intrinsic channel conductance as follows:

$$g_{ch} = \frac{g_{chi}}{1 + g_{chi}(R_s + R_d)} \tag{21}$$

(see Problem 7). Here the intrinsic linear channel conductance is given by

$$g_{chi} = \frac{q n_s W \mu_n}{L} \tag{22}$$

where n_s is the electron density per unit area.

Using Eqs. 7 to 10, we find that the electron sheet density n_{sa} above threshold and at small drain bias (where $V_{GT} \approx V_{gt}$) is given by

$$n_{sa} = N_D a \sqrt{1 - \frac{V_{gt}}{V_p}} \qquad (23)$$

At and below threshold (i.e., for $V_{gs} < V_T$), carrier statistics dictate that the electron sheet density in the channel is exponentially dependent on $E_{Fn} - E_{cmin}$ where E_{Fn} is the electron quasi–Fermi level and E_{cmin} is the conduction band minimum in the channel, located at the boundary between the active layer and the substrate. At zero drain–source bias, E_{Fn} will be very close to the equilibrium Fermi level E_F and becomes constant from source to drain. Moreover, E_{cmin} will be constant along the channel, but its magnitude will depend linearly on the applied gate bias $V_{gt} \equiv V_{gs} - V_T$. Hence, the below-threshold electron sheet density becomes[16]

$$n_{sb} = n_0 \exp\left(\frac{V_{gt}}{\eta V_{th}}\right) \qquad (24)$$

where η is the subthreshold ideality factor, $V_{th} = kT/q$ is the thermal voltage, and

$$n_0 = \frac{\varepsilon_s \eta V_{th}}{qa} \qquad (25)$$

is the value of n_s at threshold.

In the unified charge control model, the subthreshold and above-threshold electron sheet densities are combined as follows:[16]

$$n_s = \frac{n_{sa} n_{sb}}{n_{sa} + n_{sb}} \qquad (26)$$

In order to obtain the correct asymptotic behavior in the subthreshold regime, we replace V_{gt} in Eq. 23 by the effective gate voltage swing[16]

$$V_{gte} = \frac{V_{th}}{2}\left[1 + \frac{V_{gt}}{V_{th}} + \sqrt{\delta^2 + \left(\frac{V_{gt}}{V_{th}} - 1\right)^2}\right] \qquad (27)$$

which asymptotically approaches the thermal voltage V_{th} below threshold and V_{gt} above threshold. The parameter δ determines the width of the transition region.

Combining Eqs. 16 and 20, we derive the expression for the drain current in the saturation regime above threshold (see Problem 8):

$$I_{sat} = \frac{2\beta\zeta V_{gte}^2}{(1 + 2\beta V_{gte} R_s + \sqrt{1 + 4\beta V_{gte} R_s})(1 + t_c V_{gte})} \qquad (28)$$

The empirical parameters t_c and ζ allow us to use this equation for different doping profiles. As an example, for uniformly doped MESFETs with pinch-off voltages $V_p \leqslant 2\,V$, we have $t_c = 0$ and $\zeta = 1$.[10] In general, these parameters have to be extracted either from measurements or from device simulations.

Below threshold, the drain current saturates at $V_{ds} \approx 2V_{th}$ (see, e.g., Ref. 16). This saturation regime is dominated by diffusion current, and therefore

$$I_{sat} = \frac{q n_0 W \mu \eta V_{th}}{L} \exp\left(\frac{V_{gt}}{\eta V_{th}}\right) \tag{29}$$

A unified expression for the saturation current valid both below (I_{satb}) and above (I_{sata}) threshold can be obtained by combining Eqs. 28 and 29 as follows:

$$I_{sat} = \frac{I_{sata} I_{satb}}{I_{sata} + I_{satb}} \tag{30}$$

Based on the results for the various parts of the current–voltage characteristics, a unified description is obtained by introducing an interpolation formula between the linear and the saturation regimes:

$$I_d = \frac{g_{ch} V_{ds}(1 + \lambda V_{ds})}{[1 + (g_{ch} V_{ds}/I_{sat})^\gamma]^{1/\gamma}} \tag{31}$$

The additional empirical factor $(1 + \lambda V_{ds})$ is introduced to account for the finite output conductance in the saturation regime (γ is defined below).

In modern MESFET technologies, such as the buried p-layer LDD (lightly doped drain) MESFET,[17] the bulk charge of the p layer influences the current–voltage characteristics. To model this effect, the use of the following gate-bias-dependent γ in Eq. 31 has been proposed:[15]

$$\gamma = \gamma_0 + \alpha V_{gte} \tag{32}$$

where γ_0 and α are constants.

In modern short-channel FETs, the drain bias affects the device threshold voltage. This effect called the drain-induced barrier lowering (DIBL) (see Fig. 10) is fairly accurately described by the empirical relationship[16]

$$V_T = V_{T0} - \sigma V_{ds} \tag{33}$$

where V_{T0} is the threshold voltage at zero drain–source voltage and σ is a proportionality constant called the DIBL coefficient (see Ref. 16).

Figure 11 compares measured MESFET drain current–voltage characteristics and simulation results based on the universal MESFET model. This

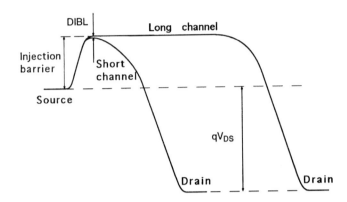

Fig. 10 Qualitative band diagram along a GaAs MESFET channel below threshold for a long-channel and a short-channel device. DIBL is the drain-induced barrier lowering.

model has been implemented in the circuit simulator AIM-Spice.[16] Using a direct extraction method,[16] followed by an optimizing step gives nearly perfect fits to the measured drain current in the subthreshold and the above-threshold regimes of operation, as shown in Figs. 11a and 11b, respectively.[15]

GaAs MESFETs, especially ion-implanted devices operating in a wide temperature range, exhibit a number of nonideal characteristics, which are technology dependent. These nonideal effects include the frequency dependence of the output conductance, back- and side-gating, the kink effect (related to impact ionization and hole trapping in the substrate), and light sensitivity. (The first two of these effects are included into the MESFET AIM-Spice model.[15]) All these effects are very sensitive to the properties of the semi-insulating GaAs substrate. Further studies of semi-insulating GaAs are important for a deeper understanding of the physics of GaAs MESFETs. Controlling these effects would allow us to increase the integration scale and the reliability, and to improve design margins.

2.3.3 MESFET C–V Modeling

For simulation of dynamic events in FET circuits, we also have to account for the variations in the stored charges of the device. In a MESFET, for example, we have stored charges in the gate electrode and in the depletion layer under the gate. Electrically, the variation in these intrinsic charges are expressed in terms of intrinsic device capacitances. In addition, we also have various parasitic capacitances between the device terminals and between the device and its surroundings. Here we consider only the capacitive elements associated with the intrinsic charges of the MESFET.

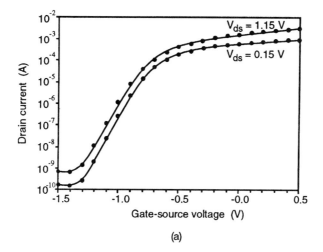

(a)

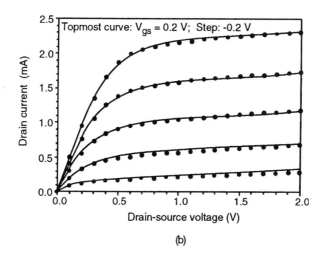

(b)

Fig. 11 Measured (dots) and simulated using AIM-Spice (lines) drain current characteristics of GaAs MESFET operating at room temperature. (a) Sub-threshold characteristics. (b) Above-threshold characteristics. (After Ytterdal et al., Ref. 15.)

As indicated in Fig. 7b, the depletion charge in the MESFET gate region is nonuniformly distributed along the channel when drain–source bias is applied. Hence, the capacitive coupling between the gate electrode and the semiconductor is also distributed, making the channel resemble an *RC* transmission line. In practice, however, because of the short gate lengths and limited bandwidths of FETs, the distributed capacitance of the intrinsic device is usually very well represented in terms of a lumped capacitance model, that is, by capacitive elements connecting the various intrinsic device terminals.

The simplest approach is to model the intrinsic gate–source capacitance C_{GS} and gate–drain capacitance C_{GS} in terms of two separate Schottky-barrier diodes connecting the gate to source and drain, respectively, each occupying half the gate area (see Fig. 12a). If we take the channel potential in each of the two parts of the channel to be constant and equal to the source and the drain potential, respectively, we find using the expressions in Section 2.3.1:[10]

$$C_{GS} = \frac{C_{g0}}{\sqrt{1 - (V_{GS}/V_{bi})}} \tag{34}$$

$$C_{GD} = \frac{C_{g0}}{\sqrt{1 - (V_{GD}/V_{bi})}} \tag{35}$$

where

$$C_{g0} = \frac{WL}{2} \sqrt{\frac{qN_D \varepsilon_s}{2V_{bi}}} \tag{36}$$

These expressions are only valid above threshold. Moreover, they do not account for the charges associated with the depletion zone extensions beyond the gate region, towards the source and drain (see Fig. 7). The capacitances associated with the depletion extensions, together with other parasitic capacitances, will become dominant in the subthreshold regime.

An alternative C–V model for MESFETs is obtained from an adaptation of the Meyer capacitance model for MOSFETs.[16] The analysis by Meyer[18] is based on a simple charge-control model applicable for long-channel MOSFETs, which yields the total gate charge Q_G as a function of the bias voltages. For MESFETs, the modified above-threshold Meyer capacitances for $V_{DS} \leq V_{SAT}$ can be written as[16,18]

$$C_{GS} = \left.\frac{\partial Q_G}{\partial V_{GS}}\right|_{V_{GD}} = \frac{2}{3} C_g \left[1 - \left(\frac{V_{SAT} - V_{DS}}{2V_{SAT} - V_{DS}} \right)^2 \right] \tag{37}$$

$$C_{GD} = \left.\frac{\partial Q_G}{\partial V_{GD}}\right|_{V_{GS}} = \frac{2}{3} C_g \left[1 - \left(\frac{V_{SAT}}{2V_{SAT} - V_{DS}} \right)^2 \right] \tag{38}$$

where the saturation voltage V_{SAT} equals V_{GT} in the long-channel limit, and

$$C_g = \frac{\varepsilon_s WL/a}{\sqrt{1 - (V_{GT}/V_p)}} \tag{39}$$

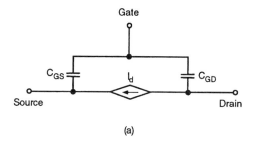

(a)

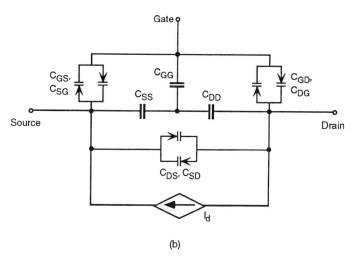

(b)

Fig. 12 MESFET large-signal equivalent circuits according to the two-diode and Meyer models (a), and charge-based modeling (b). (After Nawaz and Fjeldly, Refs. 23 and 25.)

is the total MESFET gate–channel capacitance for $V_{DS} = 0$. In saturation, the capacitances become $C_{GS} = 2C_g/3$ and $C_{GD} = 0$. Figure 13 shows the bias dependencies of the normalized Meyer capacitances C_{GS}/C_g and C_{GD}/C_g vs the normalized bias voltages V_{DS}/V_{SAT} and V_{GT}/V_{DS}.

Accurate modeling of the intrinsic MESFET capacitances requires a careful analysis of the spatial distribution of the depletion charge vs terminal bias voltages. For the MOSFET, such an analysis combined with a proper partitioning of the charge between the various terminals, leads to a set of charge conserving and nonreciprocal capacitances between the terminals.[19,20] Nonreciprocity means that $C_{ij} \neq C_{ji}$, where i and j denote source, drain, and gate (and substrate, when relevant). In fact, it has been shown that the set of Meyer capacitances is incomplete and leads to problems of charge conservation.[19,21] Nonetheless, the resulting error is usually small, except in

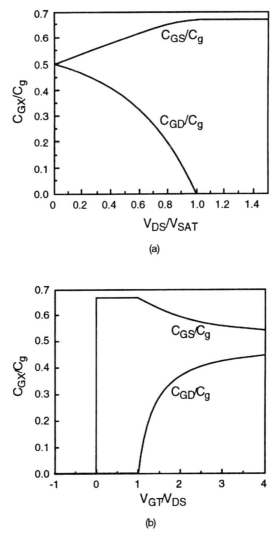

Fig. 13 Normalized above-threshold Meyer capacitances according to Eqs. 37 and 38 vs (a) normalized drain–source bias and (b) normalized gate–source bias.

transient analyses of certain demanding circuits (RAM cells, switched capacitor circuits, charge pumps) where the Meyer model is known to give erroneous results.[21] Similar phenomena have also been reported in MESFETs.[22]

A more accurate model for the intrinsic MESFET capacitances requires a precise analysis of the variation of the charge distribution in the channel vs terminal bias voltages. Moreover, the problem of charge conservation is automatically resolved by the assignment of the channel charge to the source and drain terminals.

Following the procedure by Ward and Dutton, the MESFET depletion charge is divided into a source charge Q_S and a drain charge Q_D, where[23]

$$Q_S = qWN_D \int_0^L \left(1 - \frac{x}{L}\right) h(x)\, dx \tag{40}$$

$$Q_D = qWN_D \int_0^L \frac{x}{L} h(x)\, dx \tag{41}$$

The corresponding gate charge is $Q_G = -(Q_S + Q_D)$. With such an assignment of the charges, charge conservation is automatically assured. Additional charge contributions Q_{Se} and Q_{De} come from the source- and drain-side depletion extensions, respectively. In the linear, above-threshold regime, the depletion extensions can be modeled as quarter circular regions with radii given by the depletion depth at the source and at the drain side of the gate (see Eq. 11). In saturation and in the subthreshold regime, the modeling of these charges is slightly more complicated (see Ref. 10).

Based on this charge assignment, we can define a set of so-called transcapacitances for the MESFET:[23]

$$C_{ij} = \chi_{ij} \frac{\partial Q_i}{\partial V_j} \text{ where } \chi_{ij} \begin{cases} -1 \text{ for } i \neq j \\ 1 \text{ for } i = j \end{cases} \tag{42}$$

where the indexes i and j run over the terminals G, S, D. These are identical to the charge-based nonreciprocal capacitances for the MOSFET.[19,20] The elements C_{ii} are called self-capacitances.

In three-terminal FETs, such as MESFETs, we have a total of nine transcapacitances. This set of nine elements can be organized as follows in a 3×3 matrix (a so-called indefinite admittance matrix):[23]

$$\mathbf{C} = \begin{bmatrix} C_{gg} & C_{gs} & C_{gd} \\ C_{sg} & C_{ss} & C_{sd} \\ C_{dg} & C_{ds} & C_{dd} \end{bmatrix} \tag{43}$$

The elements in each column and each row must sum to zero, owing to the constraints imposed by charge conservation, which is equivalent to obeying Kirchhoff's current law, and for the matrix to be reference independent, respectively.[24] This means that some of the transconductances will be negative, and of the nine elements, only four are independent. In Fig. 12b, we show the large-signal equivalent circuits of a MESFET based on the full set of nine transcapacitances.[23] A complete discussion of the MESFET transcapacitances can be found in Ref. 25.

2.3.4 MESFET Models in SPICE

Some of the models discussed above have been implemented in various versions of the circuit simulator SPICE. The Statz model[23] of Eq. 16, for example, is incorporated in PSpice as model Level 2, and a generalized version of the of this model, known as the TriQuint model, is Level 3 in PSpice. In AIM-Spice,[16] the Statz model is implemented as Level 1, while the more advanced MESFET model according to Ytterdal et al.,[15] including expressions for leakage current, temperature and frequency dependencies of key parameters, and side gating, is Level 2 in AIM-Spice.

Table 2 shows an example of Level 2 MESFET model parameters in AIM-Spice (see Ref. 16 for further details), used for simulating the waveform shown in Fig. 14 of an 11-stage MESFET ring oscillator with ungated load FETs.

2.4   HETEROSTRUCTURE FIELD-EFFECT TRANSISTORS (HFETs)

2.4.1   HFET Fundamentals

In a GaAs MESFET, the channel is highly doped, and ionized impurity scattering reduces the electron mobility from its theoretical limit of close to 9000 cm^2/V-s at room temperature down to 2000–3000 cm^2/V-s. The idea of modulation doping proposed at Bell Laboratories in the late 1970s was to separate dopants from the electrons in the channel to boost the electron mobility in the channel, especially at cryogenic temperatures.[26] In 1980, Mimura et al.[27] demonstrated the first high-electron-mobility transistor (HEMT) and showed a dramatic enhancement in the device current and transconductance in long-channel devices, where the electron mobility plays a dominant role, at cryogenic temperatures. Since that time, many heterostructure transistors have evolved based on the same general principle, and we will use the name heterostructure field-effect transistors (HFETs) for the transistors belonging to this device family.

In an HFET, a wide-bandgap semiconductor layer separates the gate electrode from the channel. Above threshold, a two-dimensional electron gas (2DEG) is formed at the heterointerface between the wide-bandgap semiconductor layer and the narrow-bandgap semiconductor channel. Since, in a 2DEG, electrons confined to a very narrow region near the heterointerface, the electronic motion perpendicular to the heterointerface is quantized, which affects the electron transport properties.

The drain–source current is carried by the 2DEG. In most HFETs, the

TABLE 2 Example of AIM-Spice MESFET Level 2 Model Parameters (SPICE Notation)

Parameter Description	Parameter	Value
Gate length	L	$0.7\,\mu$m
Gate width	W	$20\,\mu$m
Emission coefficient	N	1.44
Drain resistance	RD	$20\,\Omega$
Source resistance	RS	$20\,\Omega$
Saturation velocity	VS	1.9×10^5 m/s
Low-field mobility	MU	0.25 m^2/V-s
Channel thickness	D	10^{-7} m
Threshold voltage	VTO	0.15 V
Knee-shape parameter	M	2
Output conductance parameter	LAMBDA	0.15/V
DIBL parameter 1	SIGMA0	0.02
DIBL parameter 2	VSIGMAT	0.5 V
DIBL parameter 3	VSIGMA	0.1 V

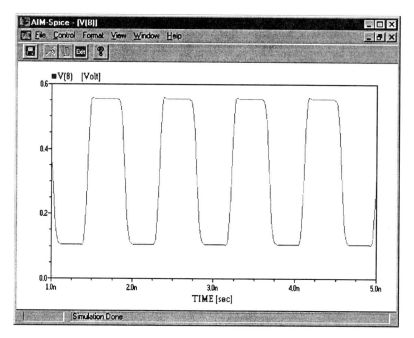

Fig. 14 Output waveform for an 11-stage MESFET ring oscillator simulated in AIM-Spice using the Level 2 MESFET model.

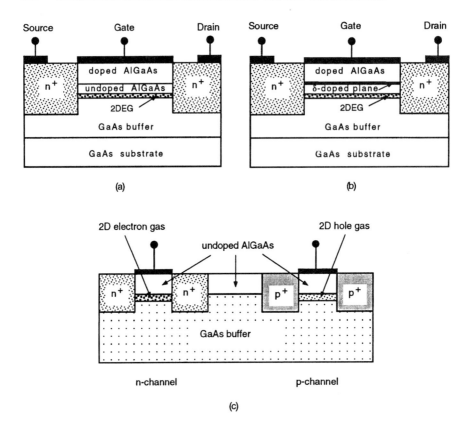

Fig. 15   Schematic HFET structures. (a) Conventional HEMT. (b) Delta-doped HFET. (c) Complementary *n*-channel and *p*-channel HIGFETs. (d) Quantum-well HFET. (e) Inverted HFET. (f) π-HFET. (g) Dipole HFET. Note that in the quantum-well HFET, the 2DEG is localized in a potential well created by two AlGaAs layers. In the inverted HFET, the 2DEG is induced into the GaAs layer under the gate at the GaAs/AlGaAs heterointerface.

dopants in the wide-bandgap semiconductor layer control the device threshold voltage. Typically, the various layers of the HFET are fabricated by molecular beam epitaxy, and the source and drain contacts are ion implanted.

The HFET may either be self-aligned, as shown among the various types of devices in Fig. 15, or non-self-aligned with the gate located similar to that of the MESFET in Fig. 6a. Figure 15a shows a conventional self-aligned HFET. In this device, the threshold voltage is controlled by the dopants in the top AlGaAs layer. However, deep impurities (DX centers) in AlGaAs lead to a variety of problems, such as a time dependence of the device current–voltage characteristics.[10] These problems are somewhat reduced in the δ-doped

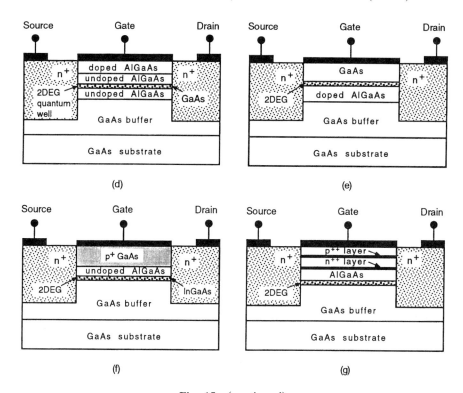

Fig. 15 (continued)

structure shown in Fig. 15b,[28] where all the dopants are located in one plane with a large concentration of donors that controls the threshold voltage.

Complementary heterostructure insulated-gate field-effect transistors (HIGFETs), which combine both n-channel and p-channel devices on the same wafer, are shown in Fig. 15c. This technology, which is similar to Si CMOS, has demonstrated low-power, high-speed operation. Other HFET structures include quantum-well devices, which can have a doped or an undoped channel (Fig. 15d), inverted HFETs (Fig. 15e), π-HFETs (Fig. 15f), and dipole HFETs (Fig. 15g).

In the quantum-well HFET (Fig. 15d), the well consists of a thin GaAs layer between two layers of AlGaAs. The advantage is a better localization of the 2DEG in the channel and higher current-carrying capability as well as a higher output resistance and a smaller leakage current. In inverted HFETs (Fig. 15e), the 2DEG is located close to the gate, which allows an increase of the effective gate capacitance and, hence, of the device transconductance. In addition, the GaAs top layer makes it easier to prepare high-quality ohmic and Schottky contacts.

HFET   structures   can   be   fabricated   from   a   large   variety   of

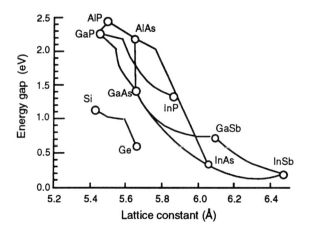

Fig. 16 Energy gaps and lattice constants of semiconductor compounds and solid-state solutions. (After Shur, Ref. 29.)

heterostructure systems, such as AlGaAs/GaAs, AlGaAs/GaInAs/GaAs, AlInAs/GaInAs/InP, or even SiGe/Si. Figure 16[29] shows that some of the heterostructures are lattice matched, such as AlGaAs/GaAs, and others are non-lattice matched, or so-called pseudomorphic structures, such as AlGaAs/GaInAs/GaAs and SiGe/Si. In the latter type, the active narrow-gap layer is made very thin to accommodate the lattice mismatch without causing misfit dislocations.

Ideally, for HFETs with a large conduction-band discontinuity, the above-threshold charge induced into the HFET channel at small drain–source biases is proportional to the gate voltage swing $V_{GT} = V_{GS} - V_T$:

$$n_{sa} = \frac{\epsilon_i V_{GT}}{q(d_i + \Delta d)} \tag{44}$$

where ϵ_i and d_i are the dielectric permeability and the thickness of the wide-bandgap semiconductor, respectively, and Δd can be interpreted as the effective thickness of the 2DEG. Typically, $\Delta d \approx 40$ to 80 Å in AlGaAs/GaAs HFETs. Below threshold, the dependence of n_{sb} on the gate bias is given by the same expression as for GaAs MESFETs (see Section 2.3.2),

$$n_{sb} = n_0 \exp\left(\frac{V_{GT}}{\eta V_{th}}\right) \tag{45}$$

where the HFET sheet density of electrons at threshold is given by

$$n_0 = \frac{\epsilon_i \eta V_{th}}{2q(d_i + \Delta d)} \tag{46}$$

The idealized unified charge control model for HFETs is based on the following equation, which describes both the above-threshold and below-threshold regimes in one continuous expression[30]

$$n_s = 2n_0 \ln\left[1 + \frac{1}{2} \exp\left(\frac{V_{GT}}{\eta V_{th}} \right) \right] \qquad (47)$$

Note that above threshold, when $V_{GT} \gg V_{th}$, this expression reduces to Eq. 44, and below threshold, when $-V_{GT} \gg V_{th}$, it reduces to Eq. 45.

2.4.2 HFET *I–V* Modeling

The above charge-control expressions describe an idealized structure with a large conduction-band discontinuity at the heterointerface, similar to that in a MOSFET. However, for HFETs this discontinuity is, typically, much smaller than in MOSFETs, increasing the statistical probability of finding free electrons inside the wide-bandgap semiconductor. Moreover, with increasing gate bias, a growing fraction of the induced electronic charge in the HFET will reside in the wide-bandgap material, simultaneously limiting the electron sheet concentration that can be induced in the heterointerface channel. In fact, for AlGaAs/GaAs HFETs, the maximum 2DEG carrier density is typically below 2×10^{12} cm^{-2}. Accounting for this effect, the actual HFET 2DEG density n'_s can be approximated by[16,30]

$$n'_s = \frac{n_s}{[1 + (n_s/n_{max})^\gamma]^{1/\gamma}} \qquad (48)$$

where n_{max} is the maximum value of the 2DEG density and γ is a characteristic parameter for the transition to saturation in n'_s.

Once, the unified expression for the surface carrier density in the HFET channel is established, the modeling of the HFET drain current becomes similar to that of the GaAs MESFET. Equation 31, which describes the extrinsic *I–V* characteristics for FETs in both the linear and saturation regimes, still applies. However, the expression for the drain saturation current (see Eq. 28) is somewhat different (see Problem 11):

$$I'_{sat} = \frac{g'_{chi} V_{gte}}{1 + g'_{chi} R_s + \sqrt{1 + 2g'_{chi} R_s + (V_{gte}/V_L)^2}} \qquad (49)$$

Here g'_{chi} is the intrinsic linear channel conductance of Eq. 22, except that n_s is replaced by n'_s of Eq. 48, to incorporate the effect of carrier saturation. V_{gte} is the effective extrinsic gate voltage swing, similar to that of Eq. 27, except that V_{th} is replaced by $2V_{th}$ everywhere, to ensure the correct subthreshold limit of Eq. 49. The corresponding extrinsic saturation voltage

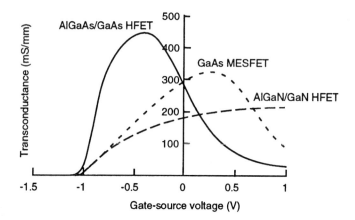

Fig. 17 Computed device transconductance in the saturation region vs gate bias for three devices: 0.5-μm gate GaAs MESFET, 0.5-μm gate AlGaAs/GaAs HFET, and 0.25-μm gate AlGaN/GaN HFET. The source series resistance per unit width for the MESFET and the AlGaAs/GaAs HFET is 0.3 Ω-mm. The source resistance per unit width for the AlGaN/GaN HFET is 2 Ω-mm. (After Shur et al., Ref. 31.)

is given by[16]

$$V_{\text{sat}} = V_{\text{gte}} - I'_{\text{sat}} \left(R_{\text{s}} + \frac{1}{\beta V_{\text{L}}} \right) \tag{50}$$

In order to illustrate the effect of the electron transfer into the wide-bandgap material, Fig. 17[31] compares typical dependencies of the device transconductance, $g_{\text{m}} = \partial I_{\text{sat}}/\partial V_{\text{gs}}$, in the saturation regime for GaAs MES-FET, AlGaAs/GaAs HFET, and AlGaN/GaN HFET. This figure clearly illustrates the drop in the AlGaAs/GaAs HFET transconductance at high gate bias where the saturation of the 2DEG density takes place. In wide-bandgap semiconductors, such as AlGaN/GaN, the conduction-band discontinuity is larger and the effect is drastically reduced. The drop in the transconductance of the GaAs MESFET is related to the increasing gate leakage when the gate Schottky contact is forward biased.

The analytical HFET I–V model discussed in this section accurately reproduces HFET I–V, as shown in Fig. 18.[30,32] The circuit simulator, AIM-Spice,[16] utilizing this model, can adequately simulate HFET integrated circuits.

2.4.3 HFET C–V Modeling

Capacitance models for HFETs can be developed along the same line as for MOSFETs and MESFETs (see Section 2.3.3). Because of the strong

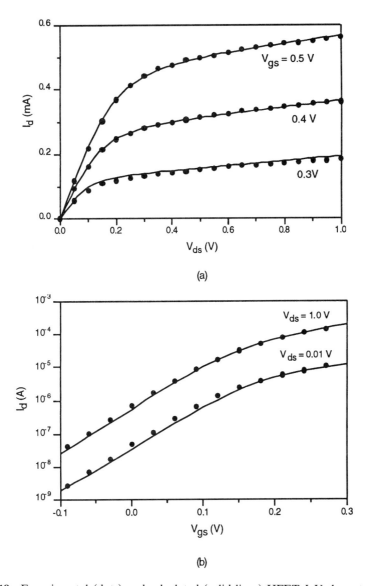

Fig. 18 Experimental (dots) and calculated (solid lines) HFET I–V characteristics. (a) Above-threshold. (b) Subthreshold. (After Fjeldly and Shur, Refs. 30 and 32.)

structural similarity between HFETs and MOSFETs, the Meyer model[18] can readily be adopted for the long-channel HFET case, using the expressions for the above-threshold capacitances C_{GS} and C_{GD} in Eqs. 37 and 38. From Eqs. 47 and 48 for the mobile channel charge, we find the following unified expression for the HFET gate–channel capacitance C_g' at zero drain bias, which replaces C_g in Eqs. 37 and 38:

$$C_g' = WLq\frac{dn_s'}{dV_{GS}} = \frac{C_{gc}}{[1 + (n_s/n_{max})^\gamma]^{1+1/\gamma}} \tag{51}$$

where

$$C_{gc} = WLq\frac{dn_s}{dV_{GT}} = \frac{C_i}{1 + 2\exp\left(-\dfrac{V_{GT}}{\eta V_{th}}\right)} \tag{52}$$

and $C_i = WL\varepsilon_i/(d_i + \Delta d)$ is the above-threshold gate–channel capacitance.

When n_s becomes equal to or larger than n_{max}, C_g' will drop noticeably from its ideal value given by C_{gc}. However, this drop will be compensated by a contribution to the total capacitance from the electron charge residing in the wide-bandgap material.[30]

Using the unified expression for C_g', the Meyer capacitances of Eqs. 37 and 38 become valid both above and below threshold. Moreover, we can replace V_{DS} in these equations by an effective extrinsic drain–source voltage,[16]

$$V_{DSe} = V_{DS}\left[1 + \left(\frac{V_{DS}}{V_{SAT}}\right)^{\gamma_c}\right]^{-1/\gamma_c} \tag{53}$$

such that V_{DSe} approaches V_{DS} when $V_{DS} < V_{SAT}$ and V_{DSe} approaches V_{SAT} when $V_{DS} > V_{SAT}$. Hence, the resulting capacitance expressions will be valid in all regimes of operation, with a smooth transition between the various regimes. The parameter m_c determines the width of the transition region between the linear and the saturation regimes.

As discussed in Section 2.3.3, the Meyer model is a non-charge conserving, long-channel model. However, an improved HFET capacitance model can be established along the same line as for the MESFET. This requires a careful analysis of the distribution of the channel charge, and a partitioning of this charge in a source charge Q_S and a drain charge Q_D according to (see Eqs. 40 and 41)

$$Q_S = qW\int_0^L \left(1 - \frac{x}{L}\right)n_s'(x)\,dx \tag{54}$$

$$Q_D = qW\int_0^L \frac{x}{L}n_s'(x)\,dx \tag{55}$$

The corresponding gate charge is $Q_G = -(Q_S + Q_D)$. The effect of the electrons residing in the wide-bandgap material can be included by replacing n_s' by n_s in these expression. As for the MESFET, we find, using the definition of the transcapacitances in Eq. 42, a set of nine capacitive elements as shown in Eq. 43, and a large-signal equivalent as shown in Fig. 12b.

2.4.4 HFET Models in SPICE

Normally, separate HFET device models are not included in SPICE circuit simulators. Instead, MOSFET models are used for simulating HFET devices and circuits. This approach can be reasonably accurate in some cases, but important effects related to gate leakage and to the transfer of carriers into the wide-bandgap material are, of course, not included in the MOSFET models. For example, as discussed previously, the interlayer transfer of carriers causes a saturation of the carrier density in the conducting channel and a reduction in the current level. This effect is illustrated in Fig. 19 by comparing transfer characteristics (I_d vs V_{gs}) using the conventional Level 3 MOSFET model in SPICE and our universal HFET model, both implemented in AIM-Spice. The results of the simulation clearly indicate the importance of carrier density saturation in the HFET channel, and the shortcomings of the MOSFET model in cases of strong forward gate bias. In addition to the effect demonstrated here, we normally also have a significant gate leakage current at similar gate biases (see Section 2.5).

2.5 GATE LEAKAGE CURRENT

Ideally, the FET gate is capacitively coupled to the channel. But, in fact, the Schottky barrier at the gate–semiconductor interface of MESFETs and HFETs does not provide complete isolation. Strictly speaking, the resulting gate leakage current should be described by analyzing the distributed network of the gate contact. However, as a first approximation, the gate current of a MESFET can be described using a two-diode model as shown in Fig. 9, by assuming that each "diode" represents half the gate area. For an HFET, two pairs of diodes are needed to account for both the heterojunction and the Schottky barrier associated with source and drain.

Here we consider the modeling of a MESFET using the two-diode model of Fig. 9. If we assume that one of the diodes is biased with the gate–source voltage, and the other diode is biased with the gate–drain voltage, the standard diode expression of Eq. 2 (neglecting the series resistances) yields the following gate current:[16]

$$I_g = J_{ss} \frac{LW}{2} \left[\exp \left(\frac{V_{GS}}{\eta_{gs} V_{th}} \right) + \exp \left(\frac{V_{GD}}{\eta_{gd} V_{th}} \right) - 2 \right] \tag{56}$$

In this expression, J_{ss} is the reverse saturation current density, and η_{gs} and η_{gd} are the gate–source and gate–drain ideality factors, respectively. Note

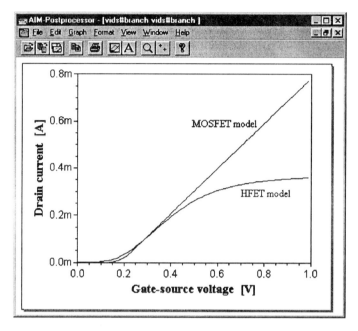

Fig. 19 AIM-Spice simulations of HFET transfer characteristics using the Level 3 MOSFET model and the universal HFET model of AIM-Spice.

that each of the diodes has a cross-section of $WL/2$.

A more accurate description[33,34] introduced effective electron temperatures at the source and drain sides of the channel. The electron temperature at the source side, T_s, is taken to be close to the lattice temperature, and the electron temperature, T_d, at the drain side is assumed to increase with the drain–source voltage to reflect the heating of the electrons in this part of the channel. The resulting gate leakage current can be written as[15]

$$I_g = J_{gs}\frac{LW}{2}\left[\exp\left(\frac{V_{GS}}{\eta_{gs}V_{ths}}\right) - 1\right] + \frac{LW}{2}\left[J_{gd}\exp\left(\frac{V_{GD}}{\eta_{gd}V_{thd}}\right) - J_{gs}\right] \quad (57)$$

where J_{gs} and J_{gd} are the reverse saturation-current densities for the gate–source and gate–drain diodes, respectively, and $V_{ths} = kT_s/q$ and $V_{thd} = kT_d/q$. Note that in the second term in Eq. 57, which accounts for the gate–drain leakage current, the electron transport from the metal to the semiconductor is given by J_{gs}. We use J_{gs} because the effective temperature of the electrons in the metal is assumed to be maintained at the ambient temperature. The net effect is a thermoelectric-current contribution resulting from the electron temperature difference between the gate and the channel near drain.[15]

In forward bias, the reverse saturation-current density is calculated assuming a thermionic-emission mechanism, resulting in

$$J_{\mathrm{gsf}} = A^* T_{\mathrm{s}}^2 \exp\left(-\frac{q\phi_{\mathrm{bs}}}{kT_{\mathrm{s}}}\right) \tag{58}$$

$$J_{\mathrm{gdf}} = A^* T_{\mathrm{d}}^2 \exp\left(-\frac{q\phi_{\mathrm{bd}}}{kT_{\mathrm{d}}}\right) \tag{59}$$

where A^* is the effective Richardson constant and $q\phi_{\mathrm{bs}}$ and $q\phi_{\mathrm{bd}}$ are the effective Schottky-barrier heights at the source and drain sides of the channel, respectively. Note that the subscripts f and r are used to distinguish between the forward- and reverse-bias regimes.

In most GaAs MESFETs, the reverse gate saturation current is dependent on the reverse bias. Dunn[35] proposed to describe this dependence by the following equations:

$$J_{\mathrm{gsr}} = g_{\mathrm{gs}} V_{\mathrm{gs}} \exp\left(-\frac{V_{\mathrm{gs}}\delta_{\mathrm{g}}}{V_{\mathrm{ths}}}\right) \tag{60}$$

$$J_{\mathrm{gdr}} = g_{\mathrm{gd}} V_{\mathrm{gd}} \exp\left(-\frac{V_{\mathrm{gd}}\delta_{\mathrm{g}}}{V_{\mathrm{ths}}}\right) \tag{61}$$

where g_{gs} and g_{gd} are the reverse diode conductances and δ_{g} is called the reverse-bias conductance parameter. However, comparison with experimental data shows that the switching of expressions for J_{gs} at $V_{\mathrm{gs}} = 0$ and J_{gd} at $V_{\mathrm{gd}} = 0$ causes inaccuracies, especially at high temperatures. Furthermore, the approach is not suitable for implementation in circuit simulators because of a discontinuity in the first derivative of the gate current at zero applied voltage. Ytterdal et al.[15] proposed to use the following combination of Eqs. 57 to 61:

$$I_{\mathrm{g}} = \frac{LW}{2}\left\{J_{\mathrm{gsf}}\left[\exp\left(\frac{V_{\mathrm{gs}}}{\eta_{\mathrm{gs}} V_{\mathrm{ths}}}\right) - 1\right] + g_{\mathrm{gs}} V_{\mathrm{gs}} \exp\left(-\frac{V_{\mathrm{gs}}\delta_{\mathrm{g}}}{\eta_{\mathrm{gs}} V_{\mathrm{ths}}}\right)\right. \tag{62}$$

$$\left. + \left[J_{\mathrm{gdf}}\exp\left(\frac{V_{\mathrm{gd}}}{\eta_{\mathrm{gd}} V_{\mathrm{thd}}}\right) - J_{\mathrm{gs}}\right] + g_{\mathrm{gd}} V_{\mathrm{gd}} \exp\left(-\frac{V_{\mathrm{gd}}\delta_{\mathrm{g}}}{\eta_{\mathrm{gd}} V_{\mathrm{ths}}}\right)\right\}$$

This is a unified expression valid for both negative and positive values of V_{gs} and V_{gd}.

Figure 20 shows a comparison of measured and modeled MESFET gate current for different ambient temperatures using the same device as shown in Fig. 11.

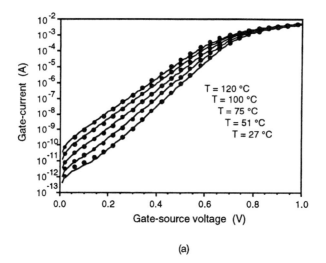

(a)

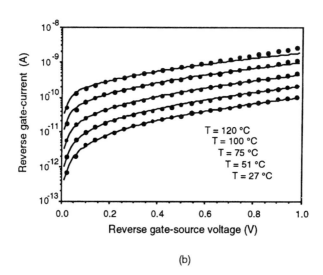

(b)

Fig. 20 Experimental (symbols) and modeled calculated (solid lines) MESFET gate current for (a) positive and (b) negative gate bias for different temperatures; device as in Fig. 11. (After Ytterdal et al., Ref. 15.)

2.6 NOVEL COMPOUND-SEMICONDUCTOR FETs

GaAs MESFETs and AlGaAs/GaAs HFETs represent the mainstream of compound-semiconductor technology. However, several other compound-semiconductor field-effect transistors have been explored and hold promise for future applications. Some of these FETs are briefly reviewed in this section.

2.6.1 Heterodimensional Devices

All semiconductor devices utilize interfaces between different regions—ohmic, p–n junctions, Schottky-barrier junctions, heterointerfaces, and semiconductor–insulator and metal–insulator interfaces. Typically, these interfaces are planes separating different regions. However, recently a new generation of semiconductor devices has emerged. These devices utilize interfaces between semiconductor regions of different dimensions and are called heterodimensional devices.[36] An example of such an interface is the Schottky barrier between a three-dimensional (3D) metal and a two-dimensional electron gas (2DEG). Other possible configurations include the interface between a 2DEG and a 2D Schottky metal, and the interface between a 1DEG and either a 2D or a 3D Schottky metal.

The different heterodimensional Schottky contacts have several features in common—smaller capacitance because of a smaller effective cross-section and a wider depletion region, high carrier mobility related to properties of the 2DEG, smaller electric field, and higher breakdown voltage. The wider depletion region is caused by the perpendicular orientation of the 2DEG plane relative to the Schottky electrode.

Figure 21a shows the layout and the cross-section of such a heterodimensional Schottky diode.[36,37] The 2DEG is created at the interface between a narrow-bandgap substrate (for example, GaAs) and a wide-bandgap barrier layer (for example, AlGaAs) with a narrow region of n-type doping close to the interface. As usual, the electronic charge from the donor impurities will populate the lower energy levels of the energy well at the interface. Contrary to conventional Schottky diodes, the Schottky contact in this case has a vertical orientation, perpendicular to the 2DEG. But, as always, the function of the Schottky contact is to deplete the electron population in the semiconductor adjacent to the metal contact; in this case, in the 2DEG, leaving a sheet charge of uncompensated donors perpendicular to the Schottky contact, as indicated in Fig. 21a. The electric field between the Schottky contact and this depletion charge will clearly be more spread out and weaker than for a conventional 3D depletion region. The reverse-bias capacitance–voltage characteristic of the heterodimensional Schottky diode is shown in Fig. 21b.

The unique characteristics of the new heterodimensional Schottky contacts make them particularly promising for applications in, for example, millimeter-wave electronics and high-speed, ultralow-power integrated circuits. Key devices are novel heterodimensional varactor and mixer diodes, and various transistor designs.

Heterodimensional transistors are very small—typical gate dimensions may be $0.5 \times 0.5 \ \mu m^2$. These small dimensions, the low gate–channel capacitance,

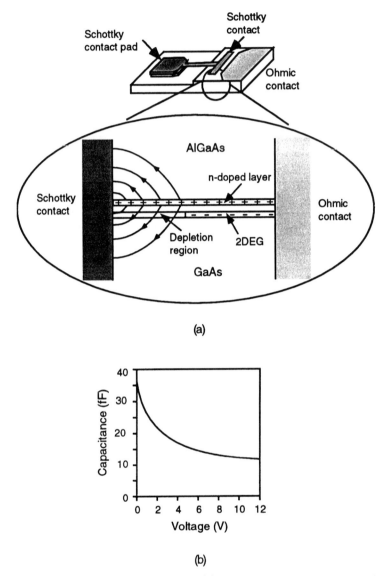

Fig. 21 (a) Schematic structure and (b) measured $C–V$ characteristic of a heterodimensional Schottky diode. (After Peatman et al., Ref. 37.)

and the low parasitic capacitances result in nearly ideal performance with a very small number of electrons in the channel compared to, for example, a typical n-channel MOSFET (see Fig. 22[36]).

The novel two-dimensional metal–semiconductor field-effect transistor (2D-MESFET) utilizes side gates formed by plating gate metal into a trench etched through the plane of the 2DEG, as shown in Fig. 23a. This device

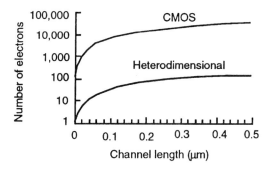

Fig. 22 Number of electrons N vs channel length for $V_{gt} = 0.5$ V. For NMOS, $N = C_i W L V_{gt}/q$, and for heterodimensional transistors, $N = C_h L V_{gt}$. $C_i = \varepsilon_i/a$ is the gate capacitance per unit area and C_h is the gate capacitance per unit length, W, L, and a are the channel width and length and the gate dielectric thickness, respectively, and ε_i is the silicon dioxide permittivity. In this calculation, it was assumed that $a = L/20$, $W = 10L$, $C_h = 10^{-10}$ F/m.

is similar to an HFET, except that the gates are placed perpendicular to and at opposite sides of the conducting layer. This way, the two gates act as Schottky contacts, similar to that shown in Fig. 21a, and are used for modulating the width of the 2DEG channel from opposite sides. The 2DEG conducting layer can, for example, be formed in a pseudomorphic AlGaAs/InGaAs heterostructure. The gates are formed by etching through the plane of the conducting layer and by electroplating Pt/Au onto the walls using resist as the mask. The metal thickness is easily varied by adjusting the plating parameters. Otherwise, conventional HFET processing techniques are used. Further details of the fabrication are described in Ref. 38.

A related device is the coaxial MESFET, schematically shown in Fig. 23b.[39] In this device, the electron gas is controlled from three sides—by the top Schottky gate and by the two side Schottky gates, which are all connected. This "coaxial" design should permit very precise control of the electron density from a 2D gas to a 1D gas, to only a few electrons in the channel.

The width of the conventional FET cannot be made too small, since the parasitic edge capacitances will diminish the gate control, leading to an increase in the threshold voltage and to a decrease in the transconductance, as well as to a spread in the device parameters. These phenomena are generally referred to as the narrow-channel effect.

The 2D-MESFETs and, especially, the coaxial MESFET are particularly promising for low-power, high-speed integrated-circuit applications, because they practically eliminate the narrow-channel effect. They also may achieve high speed because of reduced parasitic capacitances C_p. In a conventional FET, the gate capacitance C_g can be reduced by either decreasing the device

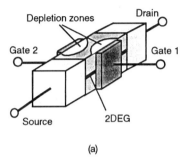

(a)

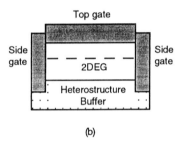

(b)

Fig. 23 Schematic diagram of (a) a 2D-MESFET and (b) a coaxial MESFET. The 2DEG in (a) is in the (horizontal) plane perpendicular to the side gates and to the source and drain. (After Peatman et al., Ref. 39.)

area or by increasing the gate–channel separation. Both changes in device dimensions lead to a substantial increase of C_p/C_g since the parasitic fringing capacitance decreases roughly in proportion to the gate periphery and C_g decreases roughly in proportion to the gate area.

Once the power-delay product becomes limited by $C_p \Delta V^2$, where ΔV is the voltage swing, a further decrease in C_g only leads to a deterioration in speed without any further reduction of power consumption. Hence, we conclude that a decrease in both parasitic and gate capacitances is needed in order to achieve lower power. A very sharp pinch-off and an extremely small leakage current are also expected in a coaxial MESFET.

Since the 2D-MESFET allows us to reduce the parasitic capacitances, we expect improvements not only in the power–delay product but also in the device speed at a given power level.

Figure 24 compares qualitative distributions of the electric field lines in a conventional HFET, a 2D-MESFET, and a coaxial MESFET. As can be seen from this figure, most of the streamlines in a 2D-MESFET, and even more so in a coaxial MESFET, terminate on the gate electrode. Hence, the

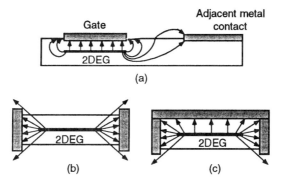

Fig. 24 Gate and fringing electric field streamlines from 2DEG in (a) a conventional HFET, (b) a 2D-MESFET, and (c) a coaxial MESFET. (After Shur et al., Ref. 41.)

parasitic capacitance of a 2D-MESFET and the coaxial MESFET is smaller than that of conventional FETs. This shows that the heterodimensional devices greatly reduce the detrimental narrow-channel effect and permit the use of narrower and lower-power devices, since they allow a reduction in the gate capacitance without a commensurate increase in the relative importance of the parasitic capacitance. Of course, this does not solve the problem of driving the interconnects. The circuit layout for low-power electronics must have short interconnects, except for a few long interconnects which have to be driven by special drivers. If the number of such long interconnects is not large, then the share of the drivers of the total power budget can be small or, at least, manageable.

Peatman et al.[40,41] have demonstrated a $1 \times 1\ \mu m^2$ AlGaAs/InGaAs 2D-MESFET with a peak drain current of 210 mA/mm, a transconductance of 210 mS/mm, and a subthreshold slope of 75 mV/decade corresponding to an ideality factor of 1.3. This performance is comparable to that of state-of-the-art 10-μm-wide HFETs. For this device, the estimated cutoff frequency was about 21 GHz which is comparable to that of the best 1-μm-long HFETs.

By eliminating the narrow-channel effect and reducing parasitic capacitances, this new technology enables the gate width to be scaled to submicrometer dimensions, allowing a large reduction in power consumption without the loss of speed. For an enhancement device with a 0.5-μm-wide channel, the threshold voltage was zero while the knee (*on*) voltage was about 0.2 V. This device should operate at a drain–source voltage of less than 1 V. Based on a charge-control model of the 2D-MESFET, a power–delay product of 0.1 fJ was estimated, which is an order of magnitude lower than that of existing state-of-the-art technologies. Peatman et al.[40,41] also observed a minimal threshold voltage shift with temperature and an almost total absence of DIBL and other short-channel effects in these devices. Figure 25 shows

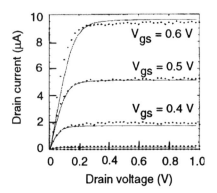

Fig. 25 Measured and simulated current–voltage characteristics of a 0.5-μm 2D-MESFET.

typical 2D-MESFET I–V characteristics with a very high output conductance in saturation.

The 2D-MESFETs described above were fabricated using AlGaAs/GaInAs/GaAs pseudomorphic heterostructures, grown by molecular beam epitaxy. More recently, ion-implanted GaAs 2D-MESFETs were made, which demonstrated performance characteristics only slightly inferior to the MBE-grown devices,[42] and new logic element, which uses 2D-MESFETs with multiple gates as well as resonant-tunneling diode loads.[43,44] These recent results clearly show the potential of this technology for low-power, high-speed applications, as illustrated in Fig. 26.[45]

The same principle has been applied to a resonant-tunneling transistor (see Fig. 27a).[43,44] This device is based on a resonant-tunneling structure, shown schematically by the two thick solid lines, which controls the current flow between source and drain. The side Schottky gates modulate the effective cross-section of the device and consequently the current flow (see Fig. 27b).

This Schottky-gated resonant-tunneling transistor (SG-RTT) has demonstrated high transconductance at room temperature. It may find applications as a load and switching device in ultralow-power electronic circuits.

2.6.2 Wide-Bandgap Semiconductor FETs

As we discussed in Section 2.3, the drain current in FETs below threshold is controlled by an energy barrier between source and drain, with an exponential decrease in the subthreshold current with increasing barrier height. The maximum barrier height is typically about half the semiconductor energy gap. This means that FETs made from wide-bandgap semiconductors will have a large *on*-to-*off* current ratio, low static-power consumption in

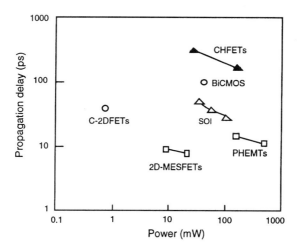

Fig. 26 Power–delay chart for different low-power technologies. (After Peatman et al., Ref. 45.)

integrated circuits, and tolerable leakage current up to high temperatures. All this may lead to applications such as nonvolatile solid-state memories (to replace mechanical hard drives), and high-temperature electronics. Wide-bandgap semiconductor materials are also interesting for optical devices operating as emitters and detectors in the blue and even in the ultraviolet range of frequencies.

All these advantages have created a great interest in wide-bandgap semiconductor devices in recent years. The research in these materials is primarily concentrated on four material systems: diamond, ZnSe and other related II–VI compounds, different modifications of SiC (called polytypes), and GaN and related materials. Diamond does not show much promise for applications in field-effect transistors, at least not in the foreseeable future (except for applications in heatsinks utilizing its superior thermal conductivity). II–VI compounds have primarily been explored for applications in green and blue semiconductor lasers. The materials of interest for use in electronic devices, including FETs, are primarily those based on SiC and GaN.

As early as 1907, Round[46] reported on the semiconductor properties of SiC. Material and device scientists first became interested in GaN in the late 1920s and early 1930s because of its excellent mechanical properties. In the 1970s, Pankove[47] demonstrated GaN light-emitting diodes. More recently, the superb electronic and optoelectronic properties of this material have attracted the attention of many research groups worldwide. GaN is a direct, wide-bandgap semiconductor that is suitable for applications in visible- and ultraviolet-light emitters and detectors. Its good electronic transport properties, its ability to form high-quality heterostructures with AlGaN, and

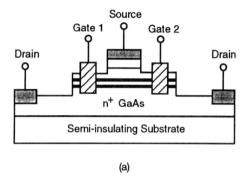

(a)

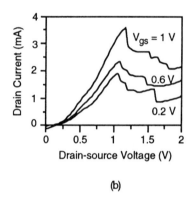

(b)

Fig. 27 (a) Schematic structure of a Schottky-gated resonant-tunneling transistor (SG-RTT). (b) Its current–voltage characteristics at $T = 300$ K. (After Robertson et al., Ref. 43.)

its chemical stability, make the GaN-based material system very attractive for FETs, especially for devices operating at high temperatures in harsh environments. In fact, within the last few years, research in GaN and AlGaN/GaN FETs has resulted in tremendous progress, with the best results challenging the GaAs-based material system.

The device structure of an AlGaN/GaN HFET is shown in Fig. 28.[48] The device epilayer structure was deposited on a basal-plane sapphire substrate using low-pressure MOCVD. It consists of a 0.5 μm highly insulating GaN layer followed by a 10-nm-thick conducting channel and a 10-nm-thick $Al_{0.1}Ga_{0.9}N$ layer. The unintentional doping in these layers is estimated to be around 5×10^{17} cm^{-3}, which results in a channel depletion at zero gate bias.

HFET structures with varying gate lengths, gate widths, and source–drain spacings are fabricated on isolated mesas. These mesas are formed by reactive

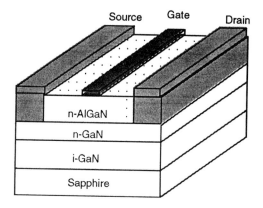

Fig. 28 Device structure of AlGaN/GaN HFETs. (After Khan et al., Ref. 48.)

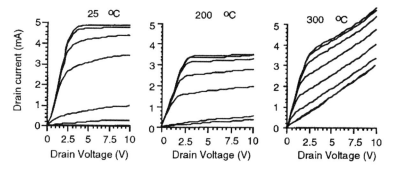

Fig. 29 Current–voltage characteristics of AlGaN/GaN HFETs at different temperatures. Top curve: $V_{gs} = 1$ V; step: -0.5 V. (After Khan et al., Ref. 49.)

ion etching in a CCl_4 plasma using photoresist as an etch mask. Ti/Al is used as ohmic source and drain metal, and titanium is the Schottky-barrier metal for the gate. These devices operate at temperatures up to 300°C (see Fig. 29[49]).

Recently reported experimental data on the microwave operation of GaN/AlGaN HFETs indicate a fairly high maximum frequency of oscillations and cutoff frequency ($f_{max} > 97$ GHz, $f_T > 36$ GHz at room temperature).[48,50] However, their performance still falls far short of the theoretically predicted performance. It can be dramatically improved by reducing the source series resistance and by optimizing the device design. Analyses show that transconductances over 1000 mS/mm should be achievable in submicrometer AlN/GaN HFETs, with a relatively small drop at elevated temperatures.[31]

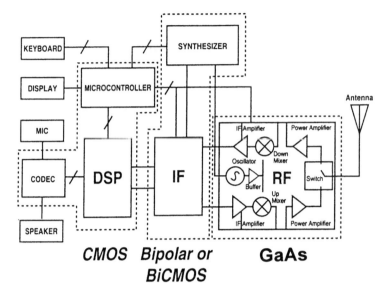

CMOS   Bipolar or   GaAs
BiCMOS

Fig. 30 Schematic diagram of "three-chip phone". MIC: microphone; CODEC: coder/decoder; DSP: digital signal processor; IF: intermediate frequency module; RF: radio frequency module (GaAs). (After Ref. 51.)

2.7 SUMMARY AND FUTURE TRENDS

Traditional compound-semiconductor field-effect transistors have nearly reached the end of the scaling curve with gate lengths of 0.1 μm and below. They have reached cutoff frequencies in excess of 300 GHz, picosecond switching speeds, and integration scales of more than a million transistors on a chip. Their applications range from consumer electronics to communications and defense. Technically, they are a success story. In terms of market share, achievements are much more modest. However, continuing research and development efforts hold promise of creating many new enabling technologies based on compound-semiconductor FETs.

The trends in GaAs-based microwave and millimeter-wave technology point toward reduced costs, improved yields, and integration with more conventional silicon-based electronics. As an example, we show in Fig. 30 the schematics of a future "three-chip" wireless handset phone envisioned by TriQuint Semiconductor, Inc.[51] According to the company, such a combination of silicon and GaAs-based technology should provide the lowest cost and the minimum battery requirements.

Another new GaAs-based technology integrates HFETs and HBTs (heterostructure bipolar transistors). Figure 31 shows layout, compositions and doping layers for a low-noise receiver chip described by TRW, Inc.[52]

A definite trend is the re-emergence of competitive GaAs-based digital

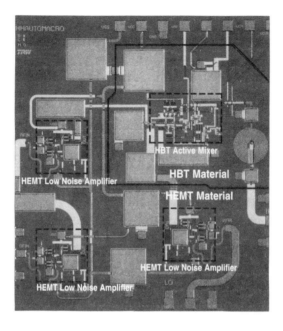

(a)

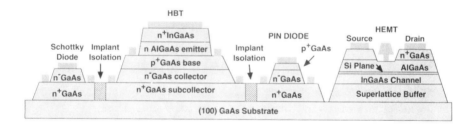

(b)

Fig. 31   Integrated HEMT/HBT circuits from TRW, Inc. (a) Low noise receiver chip. (b) Schematic of composition and doping layers. (After Ref. 52.)

technology. This technology relies both on n-channel GaAs MESFETs (used, for example, in the chips manufactured by Vitesse, Inc.) and on complementary GaAs technology analogous to silicon CMOS. Most applications of digital GaAs integrated circuits are in A/D converters and in high-speed communications. It is expected that GaAs digital technology will move towards 150-mm wafers by the year 2000. Although this goal is modest by silicon technology standards, its realization will lead to a dramatic increase in the production of GaAs-based digital ICs.

$Ga_{0.47}In_{0.53}As$ is lattice-matched to InP and can be grown on InP substrates. The transport properties of this material are superior to those of GaAs because the electron effective mass in $Ga_{0.47}In_{0.53}As$ is approximately one-half of that in GaAs. The advantages of InP-based devices utilizing GaInAs lattice matched to InP include higher mobility and peak velocity, related to the smaller effective mass in this material, a higher conduction-band discontinuity at the AlInAs/GaInAs heterointerface (approximately 0.5 eV compared to 0.3 eV for the AlGaAs/GaAs heterointerface), with a commensurate increase in the maximum density of the two-dimensional electron gas and in the maximum HFET current, and a higher thermal conductivity of InP compared to GaAs (although still much smaller than that of Si).

At frequencies above 100 GHz, InP-based HFETs using $Ga_{0.47}In_{0.53}As$ are superior to HFETs grown on GaAs substrates. These devices have reached maximum frequencies of operation close to 600 GHz! It is also important that this technology is compatible with optoelectronic InP-based devices. At present, the limiting factors for InP-based HFET technology are the very high cost of InP substrates and difficulties related to handling of the fragile InP wafers.

At the low end of compound-semiconductor technology, the competition to GaAs-based FETs is expected to come from SiGe technology.[53] Much of the potential of this technology, which uses the heterojunctions between silicon and SiGe, is linked to SiGe HBTs, which may compete with GaAs MESFETs for applications in RF systems. However, SiGe HFETs and complementary SiGe HFETs have been demonstrated as well. A big advantage of this technology is its compatibility with conventional silicon processes. The difficulties are related to the large lattice mismatch between silicon and germanium (approximately 4%). Recent developments of the SiGeC materials system hold promise of nearly perfectly matched, stress-free Si/SiGeC heterostructures.[54]

Recent breakthroughs in wide-bandgap semiconductor technology, including the development of GaN-based HFETs, make this technology one of the hot emerging areas. SiC and GaN FETs are expected to dominate in high-power and high-temperature applications.

All in all, future developments in compound-FET technology seem to be not so much in further scaling of the device dimensions but in exploring and integrating different materials systems, exploring new device ideas, integrating FET, bipolar, and optoelectronic technology, in reducing costs, and in improving manufacturability.

PROBLEMS

1. Calculate and plot current–voltage characteristics of a GaAs Schottky diode at room temperature with a cross-section of $1 \times 10 \ \mu m^2$ using the Richardson constant $A^* = 8.2 \ A/cm^2\text{-}K^2$ and the ideality factor $\eta = 1.1$.

Compare the results for the smallest and largest values of the Schottky-barrier height typical for such diodes (0.6 eV and 0.9 eV).

2. Using the solution of the one-dimensional Poisson equation for the electric field $\mathscr{E}$ in the direction y perpendicular to the metal–semiconductor interface

$$\frac{d\mathscr{E}}{dy} = \frac{\rho(y)}{\epsilon_s}$$

a. Calculate and plot the dependence of the depletion width d_d on the bias voltage V between the Schottky metal and the semiconductor for the doping profile shown in Fig. P1 Here $\rho(y) = qN_D$ is the charge density of the ionized donors in the depletion region, and $\epsilon_s = 1.14 \times 10^{-12}$ F/cm is the dielectric permittivity. The built-in voltage is $V_{bi} = 0.6$ V.

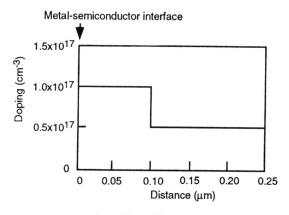

Fig. P1.

b. Calculate the threshold voltage of a MESFET using semiconductor layers with this doping profile, assuming that the total active-layer thickness is 0.25 μm.

3. Estimate the built-in voltage and calculate the GaAs MESFET channel conductivity per unit area at zero drain–source voltage vs the gate–source voltage V_{GS} at room temperature for the doping profiles shown in Fig. P2. Active-layer thickness $a = 0.25$ μm, $\mu_n = 3000$ cm^2/V-s, $\epsilon_s = 1.14 \times 10^{-12}$ F/cm, Schottky-barrier height $q\phi_b = 0.8$ eV, effective density of states $N_c = 4.7 \times 10^{17}$ cm^{-3}. Assume that the semiconductor is nondegenerate.

4. Derive the fundamental equation of the field-effect transistor, Eq. 13.

5. Estimate the maximum lateral electric field $\mathscr{E}_{xmax}$ of a GaAs MESFET

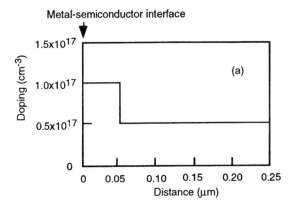

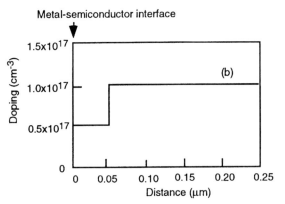

Fig. P2.

channel using a procedure similar to that used for deriving the fundamental FET equation. Assume the following parameters:

$$V_{bi} = 0.6 \text{ V}$$

$$V_p = 2 \text{ V}$$

$$\mu_n = 0.3 \text{ m}^2/\text{V-s}$$

$$a = 0.1 \ \mu\text{m}$$

$$W = 20 \ \mu\text{m}$$

$$L = 1 \ \mu\text{m}$$

$$v_s = 10^5 \text{ m/s}$$

$$V_{DS} = 0.5 \mathscr{E}_s L$$

$$V_{GS} = 0 \text{ V}$$

Calculate $d\mathscr{E}_x/dx$, compare with $d\mathscr{E}_y/dy$, and comment on the validity of the GCA.

6. For the same device as in Problem 5, find an implicit expression for the drain voltage V_{SAT} at which $\mathscr{E}_x(L) = \mathscr{E}_s$ at the drain contact. Determine the numeric value of V_{SAT} for $V_{GS} = 0\,V$.

7. Consider an FET with source and drain series resistances as indicated in Fig. P3. Derive the equation linking the intrinsic and extrinsic transconductances.

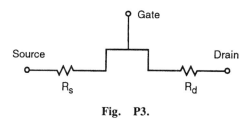

Fig.   P3.

8. Use the following expression for the above-threshold saturation current in terms of the intrinsic gate–source voltage:

$$I_{sat} = \beta(V_{GS} - V_T)^2$$

and Eq. 20, relating the intrinsic and extrinsic gate–source voltages, to derive the equation for the saturation current in a GaAs MESFET, accounting for the source series resistance. Compare the result with Eq. 28 for $t_c = 0$ and $\zeta = 1$.

9. Design a GaAs MESFET with a maximum device transconductance of at least 200 mS/mm and a drain saturation current of 200 mA/mm at zero gate–source bias.

10. Calculate and plot the dependence of the channel capacitance of GaAs MESFETs on the gate bias with the doping profiles shown in Fig. P3. Assume that the built-in voltage is 0.7 V for both profiles. Comment on the expected linearity of the device transconductance for these doping profiles.

11. Use Eq. 44 for the 2DEG carrier density and the piecewise linear velocity field relation shown in Fig. 8 to find an intrinsic expression for the above-threshold saturation current of an HFET. From this, derive the following extrinsic expression, where the effect of the source series resistance is included (see Eqs. 19 and 20):

$$I_{SAT} = \frac{g_{chi}V_{gt}}{1 + g_{chi}R_s + \sqrt{1 + 2g_{chi}R_s + (V_{gt}/V_L)^2}}$$

and compare it with Eq. 49. (*Hint*: calculate the electric field distribution in the channel and use the condition $v(L) = v_s$ where v_s is the electron saturation velocity.)

12. Design an AlGaAs/GaAs HFET with a maximum device transconductance of at least 400 mS/mm and a drain saturation current of 600 mA/mm at zero gate-source bias. Assume that the wide-bandgap semiconductor is uniformly doped, in which case the threshold voltage is given by

$$V_T \approx \phi_b - \frac{q N_D d_i^2}{2\varepsilon_i} - \frac{\Delta E_c}{q}$$

where $q\phi_b = 0.8$ eV is the barrier height between the gate and AlGaAs, $\varepsilon_i = 1.06 \times 10^{-10}$ F/m for AlGaAs, and $\Delta E_c = 0.15$ eV is the conduction-band discontinuity at the heterojunction.

13. Estimate the maximum transconductance of an AlGaN/GaN HFET (in mS/mm) assuming the following parameters: breakdown field = 7000 kV/cm for the barrier layer, maximum carrier density in the channel = $3 \times 10^{17}/m^2$, minimum channel length = 0.07 μm, velocity saturation = 2×10^5 m/s, low-field mobility = 0.15 m^2/V-s, and dielectric permeability = 9×10^{-11} F/m.

14. Comment on the expected differences between the capacitance–voltage characteristics of a GaAs MESFET and an AlGaAs/GaAs HFET.

15. Explain how the conduction-band discontinuity between the narrow-bandgap semiconductor (where the device channel resides) and the wide-bandgap semiconductor (the barrier layer) affects the maximum 2DEG density in an HFET channel. Illustrate your answer by sketching the band diagrams of the device for the direction perpendicular to the heterointerface for different gate–channel voltages.

16. What is the most important factor limiting the maximum HFET transconductance?

   a. The finite conduction-band discontinuity.
   b. The gate leakage current.
   c. The barrier-layer breakdown.

   What does the answer to this question depend on?

REFERENCES

1. S. M. Sze, *Physics of Semiconductor Devices*, 2nd ed., Wiley, New York, 1981.
2. G. C. Dacey and I. M. Ross, "Unipolar field-effect transistor," *Proc. IRE* **41**, 970 (1953).

3. M. S. Shur, "Wide band gap semiconductors. Good results and great expectations," in *Proc. NATO Advanced Research Workshop*, Ile de Bendor, France, July 1995, S. Luryi, ed., Kluwer Academic Publishers.

4. M. E. Levinshtein, S. Rumyantsev, and M. Shur, eds., *Handbook of Semiconductor Material Parameters*, Vol. 1, World Scientific, Singapore, 1996.

5. M. S. Shur, *Introduction to Electronic Devices*, Wiley, New York, 1996.

6. E. H. Rhoderick and R. H. Williams, *Metal-Semiconductor Contacts*, 2nd ed., Oxford University Press (Clarendon), London/New York, 1988.

7. U. Bhapkar and R. J. Mattauch, "Numerical simulation of the current-voltage characteristics of heteroepitaxial Schottky-barrier diodes," *IEEE Trans. Electron Dev.* **ED-40**(6), 1038 (1993).

8. C. Y. Chang, Y. K. Fang, and S. M. Sze, "Specific contact resistance of metal-semiconductors barriers," *Solid State Electron.* **14**, 541 (1971).

9. W. Shockley, "A unipolar field-effect transistor," *Proc. IRE* **40**, 1365 (1952).

10. M. S. Shur, *GaAs Devices and Circuits*, Plenum, New York, 1987.

11. A. B. Grebene and S. K. Ghandi, *Solid State Electron.* **12**, 573 (1969).

12. R. A. Pucel, H. A. Haus, and H. Statz, "Signal and noise properties of gallium arsenide microwave field-effect transistors," in *Advances in Electronics and Electron Physics*, Vol. 38, Academic Press, New York, p. 195, 1975.

13. R. E. Williams and D. W. Shaw, "Graded channel FET's improved linearity and noise figure," *IEEE Trans. Electron Dev.* **ED-25**, 600 (1978).

14. H. Statz, P. Newman, I. W. Smith, R. A. Pucel, and H. A. Haus, *IEEE Trans. Electron Dev.* **ED-34**, 160 (1987).

15. T. Ytterdal, B.-J. Moon, T. A. Fjeldly, and M. S. Shur, "Enhanced GaAs MESFET model for a wide range of temperatures," *IEEE Trans. Electron Dev.* **ED-42**(10), 1724 (1995).

16. K. Lee, M. S. Shur, T. A. Fjeldly, and T. Ytterdal, *Semiconductor Device Modeling for VLSI*, Prentice–Hall, Englewood Cliffs, N.J., 1993.

17. M. Noda, K. Hosogi, T. Oku, K. Nishitani, and M. Otsubo, "A high-speed and highly uniform submicrometer gate BPLDD GaAs MESFET for GaAs LSI's," *IEEE Trans. Electron Dev.* **ED-39**(4), 757 (1992).

18. J. E Meyer, "MOS models and circuit simulation," *RCA Rev.* **32**, 42 (1971).

19. D. E. Ward, *Charge Based Modeling of Capacitance in MOS Transistors*, Ph.D. thesis, Stanford University, 1981.

20. D. E. Ward and R. W. Dutton, "A charge-oriented model for MOS transistors," *IEEE J. Solid-State Circ.* **SC-13**, 703 (1978).

21. O. G. Johannessen, T. A. Fjeldly, and T. Ytterdal, "Unified capacitance modeling of MOSFETs," *Phys. Scripta* **T54**, 128 (1994).

22. D. Divekar, "Comments on 'GaAs FET device and circuit simulation in SPICE'," *IEEE Trans. Electron Dev.* **ED-34**(12), 2564 (1978).

23. M. Nawaz and T. A. Fjeldly, "A charge conserving capacitance model for GaAs MESFETs for CAD applications," *Phys. Scripta* **T69**, 142 (1997).

24. N. Arora, *MOSFET Models for VLSI Circuit Simulation*, Springer-Verlag, Berlin/New York, 1993.

25. M. Nawaz and T. A. Fjeldly, "A new charge conserving capacitance model for GaAs MESFETs," *IEEE Trans. Electron Dev.* (accepted for publication).

26. R. Dingle, H. L. Stormer, A. C. Gossard, and W. Wiegman, *Appl. Phys. Lett.* **37**, 805 (1978).

27. T. Mimura, S. Hiyamizu, T. Fujii, and K. Nambu, "A new field effect transistor with selectively doped GaAs/n-Al$_x$Ga$_{1-x}$As heterostructures," *Jpn. J. Appl. Phys.* **19**, L225 (1980).

28. N. C. Cirillo, A. Fraasch, H. Lee, L. F. Eastman, M. S. Shur, and S. Baier, "Novel multilayer modulation doped (Al,Ga)As/GaAs structures for self-aligned gate FETs," *Electron. Lett.* **20**(21), 854 (1984).

29. M. S. Shur, "Introduction," in *Compound Semiconductor Technology. The Age of Maturity*, M. S. Shur, Ed., World Scientific, Singapore, 1996.

30. T. A. Fjeldly and M. Shur, "Unified CAD models for HFETs and MESFETs," (invited paper), *Workshop Proceedings of 21st European Microwave Conference*, Microwave Exhibitions and Publishers, Stuttgart, 1991, 198.

31. M. S. Shur, A. Khan, B. Gelmont, R. J. Trew, and M. W. Shin, "GaN/AlGaN field effect transistors for high temperature applications," *Inst. Phys. Conf. Series*, No 141: Ch. 4, 419 (1995). (Invited paper presented at Int. Symp. Compound Semicond., San Diego, CA, Aug. 18–22 1994.)

32. T. A. Fjeldly and M. Shur, "Simulation and modeling of compound semiconductor devices," in *Compound Semiconductor Technology. The Age of Maturity*, M. S. Shur, Ed., World Scientific, Singapore, 1996.

33. M. Berroth, M. Shur, and W. Haydl, "Experimental studies of hot electron effects in GaAs MESFETs," in *Extended Abstracts of the 20th Intern. Conf. on Solid State Devices and Materials (SSDM-88)*, Tokyo, Aug. 1988, p. 255.

34. K. Y. Lee, B. Lund, T. Ytterdal, P. Robertson, E. Martinez, J. Robertson, and M. Shur, "Enhanced CAD model for gate leakage current in heterostructure field effect transistors," *IEEE Trans. Electron Dev.* **43**(6), 845 (1996).

35. C. Dunn, *Microwave Semiconductor Devices and Their Circuit Applications*, H. A. Watson, Ed., McGraw-Hill, New York, 1969.

36. M. S. Shur, W. C. B. Peatman, M. Hurt, R. Tsai, T. Ytterdal, and H. Park, "Heterodimensional technology for ultra low power electronics," in *Proc. NATO Advanced Research Workshop*, Ile de Bendor, France, July 1995, S. Luryi, Ed., Kluwer Academic Publishers.

37. W. C. B. Peatman, T. W. Crowe, and M. S. Shur, "A novel Schottky/2-DEG diode for millimeter and submillimeter wave multiplier applications," *IEEE Electron Dev. Lett.* **13**(1), 11 (1992).

38. W. C. B. Peatman, H. Park, and M. Shur, "Two-dimensional metal-semiconductor field effect transistor for ultra low power circuit applications," *IEEE Electron Dev. Lett.* **15**(7), 245 (1994).

39. W. C. B. Peatman, H. Park, B. Gelmont, M. S. Shur, P. Maki, E. R. Brown, and M. J. Rooks, "Novel metal/2-DEG junction transistors," in *Proc. 1993 IEEE/Cornell Conf.*, Cornell,Univ. Press, Ithaca, N.Y., 1993, p. 314.

40. W. C. B. Peatman, R. Tsai, T. Ytterdal, M. Hurt, H. Park, J. Gonzales, and M. S. Shur, "Sub-half-micron width 2-D MESFET," *IEEE Electron Dev. Lett.* **17**(2), 40 (1996).

41. M. Shur, W. C. B. Peatman, H. Park, W. Grimm, and M. Hurt, "Novel heterodimensional diodes and transistors," *Solid State Electron.* **38**(9), 1727 (1995).

42. M. Hurt, M. S. Shur, W. C. B. Peatman, and P. B. Rabkin, "Quasi-three-dimensional modeling of a novel 2-D MESFET," *IEEE Trans. Electron Dev.* **ED-43**(2), 358 (1996).

43. J. Robertson, T. Ytterdal, W. C. B. Peatman, R. Tsai, E. Brown, and M. S. Shur, "2-D MESFET/RTD logic elements for compact, ultra low-power electronics," in *Proc. of Int. Semicond. Device Res. Symp.*, Charlottesville, VA, Dec. (1995), p. 365.

44. J. Robertson, T. Ytterdal, W. C. B. Peatman, R. Tsai, E. Brown, and M. S. Shur, "RTD/2-D MESFET/RTD logic elements for compact, ultra low-power electronics," *IEEE Trans. Electron Dev.* **44**(7), (1997).

45. W. C. B. Peatman, M. Hurt, H. Park, R. Tsai, and M. S. Shur, "Narrow channel 2-D MESFET for low power electronics," *IEEE Trans. Electron Dev.* **ED-42**(9), 1569 (1995).

46. H. J. Round, *Electr. Wld* **19**, 308 (1907).

47. J. I. Pankove, *J. Lumin.* **7**, 114 (1973).

48. M. A. Khan, Q. Chen, M. S. Shur, B. T. Dermott, J. A. Higgins, J. Burm, W. Schaff, and L. F. Eastman, "Short channel GaN/AlGaN doped channel heterostructure field effect transistors with 36.1 GHz cutoff frequency," *Electron. Lett.* **32**(4), 357 (1996).

49. M. A. Khan, M. S. Shur, J. N. Kuznia, J. Burm, and W. Schaff, "Temperature activated conductance in GaN/AlGaN heterostructure field effect transistors operating at temperatures up to 300°C," *Appl. Phys. Lett.* **66**, 1083 (1995).

50. M. A. Khan, Q. Chen, M. S. Shur, J. N. Kuznia, J. Burm, B. T. Dermott, J. A. Higgins, J. Burm, W. Schaff, and L. F. Eastman, "High temperature performance of doped channel AlGaN/GaN heterostructure field effect transistors," in *Proceedings of High Temperature Conference*, Sandia, Albuquerque, NM, June (1996).

51. "TriQuint enters RF power amp market, forecasts a three chip phone," *Compound Semicond.* **2**(13), Jan./Feb. (1996).

52. "Device feature.W new integrated HEMT/HBT circuits," *Compound Semicond.* **1**(4), Sept./Oct. (1995).

53. R. People, "Physics and applications of GeSi/Si heterostructures," *IEEE J. Quant. Electron.* **22**, 1696 (1986).

54. H. G. Grimmeiss and J. Olajos, "Physics and applications of GeSi/Si heterostructures," *Phys. Scripta.* **T69**, 52 (1997).

3 MOSFETs and Related Devices

S. J. HILLENIUS

Bell Laboratories, Lucent Technologies, Murray Hill, New Jersey

3.1 INTRODUCTION

This chapter reviews recent advances in the design of MOS devices and some of the basic applications of these devices. To establish the foundation for the descriptions to follow, a basic description of the operation of a MOS device will first be presented. We assume that readers are familiar with the materials in *Physics of Semiconductor Devices*.[1]

The MOS transistor consists of four parts that are also the electrical designations of the device, that is, the source, drain, gate, and substrate. The transistor is a four-terminal device with these four parts determining the electrical characteristics. The source and drain regions of the transistor are sometimes referred to as the diffusion regions and are formed to be self-aligned to the gate region. The schematic in Fig. 1 shows basic elements of this structure. The transistor is formed wherever the gate feature crosses a diffusion feature. The basic operation of the MOS transistor uses the charge on the gate of the device to control the movement of charge between the source and drain through the channel under the gate.

To discuss the most recent advances in the basic CMOS device structure requires that we first look at the individual parts of the MOS transistors. We must examine how the shrinking of the dimensions and the lowering of the supply voltages has affected the way we look at the device characteristics and how we characterize the transistors. The traditional way of looking at the MOS device can be summarized in Eqs. 1–4. These equations are presented as the starting point of this discussion with no derivation. (Ref. 1 gives a detailed derivation of these equations.) The basic premises are also summarized here. The intent is to start with these premises and to explain

Modern Semiconductor Device Physics, Edited by S. M. Sze.
ISBN 0-471-15237-4 © 1998 John Wiley & Sons, Inc.

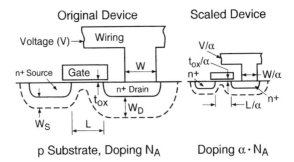

Fig. 1 Schematic diagram of an *n*-MOS transistor showing the features and scaled features.

the recent developments that will modify or change the interpretation of these equations.

The formula for the drain current in the linear region (for small V_D) is given as

$$I_D = (Z/L)\mu C_{ox} \left[(V_G - V_T)\, V_D - \left(\frac{1}{2} + \frac{\sqrt{\varepsilon_s q N_n / \psi_B}}{4 C_{ox}} \right) V_D^2 \right] \tag{1}$$

The threshold voltage is defined as

$$V_T = 2\Psi_B + \left(\frac{\sqrt{2\varepsilon_s q N_n (2\Psi_B)}}{C_{ox}} \right) \tag{2}$$

The drain current in the saturation region is given as

$$I_D = (Z/L)\mu C_{ox}(V_G - V_T)V_D \qquad \text{(for large } V_D) \tag{3}$$

and the transconductance is given as

$$g_m = (dI_D/dV_G) = (Z/L)\mu C_{ox}V_D \tag{4}$$

where Z is the width of the transistor, L is the length, μ is electron mobility, Ψ_B is the surface potential at flat band voltage and C_{ox} is the gate capacitance.

These characteristics are all derived with basic solutions to the Poisson equation determining the charge distribution in the gradual channel approximation. We also assume that there are no fixed oxide charges. These equations have been used extensively to predict device and circuit performance. The limitations in these equations become evident as the devices are scaled and modified to optimize the circuit density, performance, reliability,

Table 1. Comparison of Constant-Field and Generalized Scaling Rules

Parameter	Constant Electric Field	Generalized Scaling
Lateral dimension	$1/\alpha$	$1/\alpha$
Electric field	1	$\mathscr{E}$
Voltage	$1/\alpha$	$\mathscr{E}/\alpha$
Substrate doping	α	$\mathscr{E}\alpha$

and the power consumption. Corrections to the basic assumptions over the last few years will be presented here. These corrections must be made because many of the effects that were considered to be negligible in the past have now become important factors in device design.

The history of MOS devices has evolved to the point where virtually all of the MOS devices used in modern electronics are created in a complementary metal oxide semiconductor (CMOS) technology that uses both n-type (n-MOS) and p-type (p-MOS) devices manufactured simultaneously on the same silicon substrate. CMOS technology permits both the inherent low power of CMOS-based logic as well as very dense integrated circuits. Therefore, the topics in this chapter will emphasize the basic device developments that affect the operation and improvement in the CMOS technologies.

3.2   SCALED MOSFETs

The phenomenal success of the MOS transistor has been partially due to the capability of the MOS transistor to take advantage of the lateral scaling improvements in the technologies. Lateral scaling results in simultaneous improvements in both the performance and the packing density of the devices. The fundamental logic behind the scaling of the MOS devices is to reduce the lateral dimensions by some fundamental scaling factor α. There are two major device scaling methods which have been discussed over the years.[2] The first is "constant electric field scaling," which allows the field across the gate oxide of the transistor to remain constant. The second, more generalized, scaling rules allow the voltage to remain constant and are the more realistic approach to applications that require constant supply voltage constant. A comparison of the constant field and generalized scaling requirements is shown in Table 1 and Fig. 1.[3] This figure shows the shrinking of the device by the parameter α for the t_{ox}, junction depth, contact size, gate length, and channel doping.

Although generalized scaling has served well for the last few decades, many of the technology advances that allow the devices to continue improving

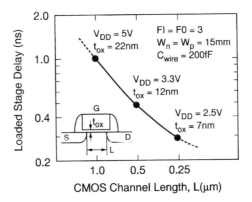

Fig. 2 Circuit performance of a loaded stage delay circuit, showing the effect of scaling on the performance. (After Davari et al., Ref. 3. Copyright ©1995 IEEE.)

the performance and the packing density are approaching fundamental physical limitations. Further improvements will be achieved by considerations other than simply device size reductions. Future device improvements will require the devices to be either optimized for voltage reduction, high performance, or reliability. These three parameters cannot be optimized simultaneously, so there needs to be tradeoffs among these three constraints. The performance scaling with voltage is best illustrated by the stage delay vs voltage plot shown in Fig. 2 for a loaded ring oscillator consisting of transistors with the width of the n-MOS (W_n) and p-MOS (W_p) devices equal to 15 μm and the fan-in (FI) and fan-out (FO) of the oscillators equal to 3. This simply means that each stage of the oscillator is driving the load of three stages. This figure illustrates a high-performance scaling of a CMOS technology in which the gate oxide thickness is scaled nearly linearly with channel length. The voltage levels are reduced by about the square root of L. The scaling is shown as the loaded stage delay of a test circuit where both the device width and wiring capacitance (the load) are kept constant.

An improvement in the generalized scaling rules was presented using empirical observations resulting from simulations.[4] The resulting scaling rules are derived from an improvement of the generalized design guide:

$$L_{min} = A \ (x_j \ t_{ox} \ W^2_{SD})_{1/3} \tag{5}$$

where W_{SD} is the sum of the depletion widths of the source and drain which is a function of the channel doping N. x_j is the junction depth and t_{ox} is the gate oxide thickness, and L_{min} is the minimum channel length (L_c) allowed for the device. We can see that this generalized design guide is incorrect, since it indicates that the L_{min} will tend to zero when either X_j or t_{ox} are scaled to zero. The relation is also derived from the assumption that a fixed

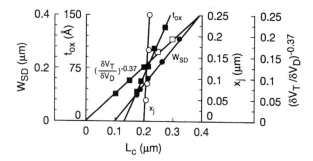

Fig. 3   Plots that were used to determine the functions $f_1 (\delta V_T/\delta V_D), f_2(t_{ox}), f_3 (W_{sd})$, and $f_4 (x_j)$ and the constants C, D, and E from Eq. 6. (After Ng et al., Ref. 5. Copyright © 1993 IEEE.)

amount of subthreshold current is allowed, which is too stringent a requirement. The improved guidelines use an equation of the following form:[5]

$$L_c = Bf_1(\delta V_T/\delta V_D)[f_2(t_{ox}) + C][f_3(W_{SD}) + D][f_4(x_j) + E)] \qquad (6)$$

where the functions f_{1-4} were determined by fitting this equation to a set of over 100 device simulations in which the parameters L_c, t_{ox}, N, and x_j were varied. The constants C, D, and E are used to account for the nonzero L_c when t_{ox}, W_{SD}, and x_j approach zero. This was fitted to a series of simulated devices and the parameters were extracted from a fit shown in Fig. 3. Each parameter was varied individually, with the other parameters kept constant. The fit of each parameter is quite good, with the derived formula having the values shown in Eq. 7:

$$L_c = (2.2 \ \mu m^{-2})(\delta V_T/\delta V_D)^{-0.37}(t_{ox} + 0.012 \ \mu m)[(W_{SD} \\ + 0.15 \ \mu m)(x_j + 2.9 \ \mu m)] \qquad (7)$$

An extension of this analysis was performed[6] in which the scaling relationships between the effective channel length (L_{eff}), the device speed (g_m/WC_{ox}), and the drain-induced barrier lowering ($\delta V_T/\delta V_{DS}$) were the parameters of interest, and the L_c, t_{ox}, V_T, and V_{DD} were kept constant. The other variables were allowed to vary with the tradeoff between g_m/WC_{ox} and $\delta V_T/\delta V_D$ and L_{eff} optimized. The relationship was then empirically fitted to a power-law relationship by using measured data and simulations. The results showed that the tradeoffs were dominated by the source and drain junction parameters. In particular it is the junction depth, R_{SD}, and the abruptness of the junction that are the dominant characteristics in deciding these tradeoffs. The power-law relationship of $(\delta V_T/\delta V_D)^{-0.37}$ (Eq. 7) was shown to reproduce with a value of $(\delta V_T/\delta V_D)^{-0.39}$.

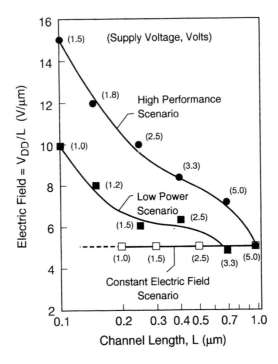

Fig. 4 Electric field in the channel of the *n*-MOS transistor for channel lengths down to 0.1 μm using three different scaling scenarios. (After Davari et al., Ref. 3. Copyright © 1995 IEEE.)

Although this empirical approach is not as elegant or satisfying as the physically derived scaling rules of Table 1, it is more accurate. The complexity is due to the fact that the device operation becomes much more three-dimensional when the changes in the lateral dimensions affect the other dimensions in a complicated interaction with the vertical features. Scaling the devices below feature sizes of 0.1 μm is difficult because of the fundamental limitations of the silicon device parameters. The limitations of these 0.1-μm devices involve the limitations coming from the problems of making very thin gate oxide, ultrashallow junctions for the sources and drain, and working with the very low operating voltages necessary for these very small devices.

The conventional scaling approach described previously has given way to more specialized scaling scenarios where the devices are optimized for either high-performance or for low-power applications.[7] This approach is ex-emplified in Fig. 4, which shows the electric field as a function of channel length for these two approaches plus the constant-electric-field scaling. The high-performance scenario optimizes the power supply voltage for maximum speed while maintaining the reliability constraints. The low-power scenario arbitrarily allows the performance to degrade by a factor of 1.5 and then to minimize the power within that constraint. The third curve in the figure is for constant electric-field scaling.

Virtually all of the advances in the MOS devices and the improvements in the ULSI CMOS technologies have been focused on the individual aspects of the scaling issues. Therefore, most of the remaining parts of this chapter will be devoted to the individual parts of the MOS transistors and to the physical limitations that affect the choices of device dimensions and materials.

3.3   CMOS/BiCMOS

The discussion of advances in MOS devices is basically the discussion of the improvements in the CMOS structures. This section will discuss the improvements in the individual parts of the MOS devices. Bipolar devices in conjunction with CMOS devices are used for certain applications. This technology is referred to as BiCMOS. Although this technology can be important for devices that need powerful drivers for high-speed applications, the basic physics of the MOS devices is identical to that of the CMOS devices. Bipolar devices are covered in Chapter 1 of this book, so this section applies to both the CMOS devices and the MOS part of a BiCMOS structure.

3.3.1   Source and Drain Structures

An important aspect of most modern scaled CMOS devices is the contribution to the resistance of the device made by the source and drain regions. This was evaluated by splitting the components of the resistance into several pieces[8] (see Fig. 5), which were described as contact resistance (R_{co}), sheet resistance (R_{sh}), and a combined spreading (R_{sp}) and accumulation resistance (R_{ac}). The contact resistance is given by

$$R_{co} = \rho_{sh}S/w \tag{8}$$

where

$$\rho_{sh} = \rho/x_j \tag{9}$$

The parameter ρ_{sh} is the sheet resistance of the source–drain region ($\Omega/\square$), w is the width of the device, S is the distance from the edge of the channel to the edge of the contact window, ρ is the bulk resistivity in the heavily doped region of the source and drain, and x_j is the junction depth. Although R_{co} and R_{sh} can be considered to be constant as a function of the gate and drain voltages for the operating range of the devices, R_{sp} and R_{ac} are dependent on the gate voltage and will vary across the range of operating voltages. The results of this study showed that R_{sp} and R_{ac} must be considered simultaneously and that these parameters will both be affected by the gate voltage.

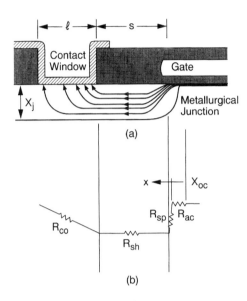

Fig. 5 Contact for an MOS transistor, showing the components of the series resistance.

The precise dependence on the gate voltage depends on the geometry of the drain doping profile. These values do not affect the scaled devices significantly, and it is seen that the most dramatic effect on the series resistance of the devices as the devices are scaled is from the specific contact resistance, which gives a value of

$$R_{co} = (\sqrt{\rho_{sh}\rho_c/r_j})/w \tag{10}$$

where w is the contact width and the value of the accumulation and spreading resistance will dominate the scaled devices. The value of $R_{ac} + R_{sp}$ is

$$R_{ac} + R_{sp} \approx (4\rho/\pi W)[\Delta(\delta\rho_{ch}/\rho) + \ln(r_j/2r_{ch})] \tag{11}$$

where Δ is the additional resistance due to the spatial dependence of the junction profile under the gate and r_{ch} is the surface channel depth. The conclusion drawn from this analysis on the relative influence of the various components of the series resistance on the performance of the MOS devices is that the $(R_{ac} + R_{sp})$ term will dominate the series resistance of the device, with the abruptness of the junction being a critical factor in this term.

Another consideration for the design of the drain is the reduction of gate induced drain leakage (GIDL). This is caused by the effects of the high field region under the gate in the region of the drain overlap. This region is shown schematically in Fig. 6a, where the high field region of the drain becomes

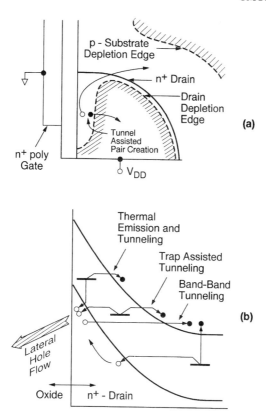

Fig. 6 Mechanism for the gate-induced drain leakage. (b) Band diagram showing the various mechanisms that can create the resulting hole current caused by gate-induced drain leakage. (After Brews, Ref. 9.)

a nonequilibrium deep depletion region. In this region, pair creation can occur, which is indicated in the figure. There are several possible mechanisms to create this current,[9] which are shown in Fig. 6b. These include thermal emission and tunneling, trap-assisted tunneling, and band-to-band tunneling. It is the band-to-band tunneling that has the most relevance at the voltages and structures of modern MOS devices. Minimizing tunneling also dictates the use of abrupt junctions to minimize the amount of deep depletion that occurs in the overlap region.

One interesting new method that has been proposed to get around the need for a shallow junction is to use a structure that raises the silicon in the source–drain region to allow the benefits of the shallow junction to be realized but still allow the contacts to be made to a relatively thick region of silicon. This structure is obtained by selectively forming silicon in the source and drain region of the MOS device after the gate is formed. The resulting structure can be used to reduce the contact and sheet resistance of the transistor without affecting the advantages of a shallow junction.

3.3.2 Channel Structures

The channel profile of the MOS devices is engineered to minimize the short-channel effects, maximize the current drive, and to insure that the device turns off. The extremes of the channel profiling allow the device characteristics to be maintained for the shortest channel lengths. The scaling rules described above assume that the doping concentration in the channel stays constant. However, there is significant benefit to having a nonuniform channel doping. The most important advantages to having a nonuniform channel doping is to optimize the device characteristics with respect to short, channel effects and device performance. The two most important characteristics for the operation of digital CMOS devices are that the device produces as much drive current as possible for the given operating voltage and that the current that the device draws while it is off is as small as possible. This can be summarized by saying that we need to simultaneously minimize I_{off} while maximizing I_{on}. Both of these effects are determined to a large extent by the channel doping profile.

One way to minimize the short-channel effects and still maintain the benefits of a relatively lightly doped channel region is to use very abrupt doping profiles in the channel region. This allows the region close to the surface to have relatively light doping, so that the mobility is not degraded as much, and the region deeper into the channel to have doping heavy enough to reduce the short-channel effects. The doping under the source and drain regions must also be relatively light to minimize the amount of parasitic junction capacitance of the source and drain. One structure[10] that satisfies these constraints is the pulse-shaped doping structure shown in Fig. 7. This device has a lightly doped region close to the surface with a more heavily doped region deeper in the silicon. The lightly doped region allows the field close to the surface to be low enough so that the mobility is not degraded compared to the field produced if the channel region was doped heavily enough to reduce the short-channel effects. The pulse-doped region reduces the short-channel effects, yet does not extend deep enough into the silicon to contribute to excessive junction capacitance for the source and drain. This particular structure will be discussed further in this chapter when we can compare it to the silicon-on-insulator device.

The well-known short-channel effects cause a decrease in the threshold voltage as the channel length is decreased. This is due to the charge sharing in the channel that cause the gate voltage to have less control of the channel charge as the channel length gets smaller. An entirely different effect causes the threshold voltage to increase as the channel length gets smaller. This effect is due to redistribution of the dopant in the channel region, which causes the channel doping at the surface of the channel to increase, which

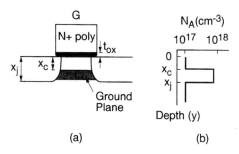

Fig. 7 (a) Cross-section of the pulse-doped structure. (b) Doping density of the channel region. (After Yan et al., Ref. 10. Copyright © 1992 IEEE.)

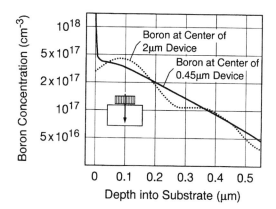

Fig. 8 Channel profiles of boron concentration for with devices with identical processing but with different channel lengths, showing the redistribution of the boron due to enhanced diffusion from the source–drain regions. After Rafferty et al., Ref. 11. Copyright © 1994 IEEE.)

causes the threshold voltage to increase. This redistribution is caused by the influence of the source–drain implants and the introduction of enhanced diffusion of dopants caused by point defects.[11] The redistribution of the dopant is shown in Fig. 8. The implications of this effect have dramatic consequences on the channel-length dependence on the transistor characteristics, since the channel doping of the transistor now has a channel-length dependence, which cannot be easily described analytically. The doping for the 0.45 μm transistor in this figure shows the boron concentration with a value of about $10^{18}/cm^{-3}$ at the surface whereas the 2 μm device has a value of about $3 \times 10^{17}/cm^{-3}$ and a different profile. It is this redistribution of the boron that causes an increase in the threshold voltage.

One other major device parameter that is essential for high performance

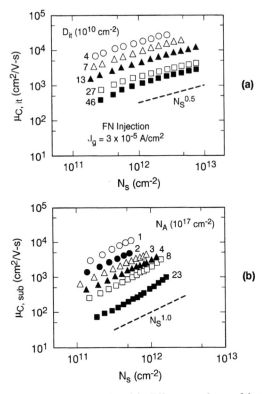

Fig. 9 Mobility vs channel charge for (a) different values of interface charge and (b) different channel doping, showing the effect of surface scattering and impurity scattering. (After Koga et al., Ref. 12. Copyright © 1994 IEEE.)

is the inversion layer mobility. The mobility has been described through the universal relationship between effective mobility, μ^{eff} and the effective normal field, $\mathscr{E}_{\text{eff}}$ in the high-field regions, where the influence of scattering is minimal. The effects of surface scattering and the Coulomb scattering within the channel are becoming more important in scaled devices. High mobility at low surface-carrier density is essential for high performance at low voltages. A recent study[12] of the Coulomb-scattering effects and interface-scattering effects gives a correction to the universal mobility curve resulting from Coulomb scattering from two separate sources, interface traps and substrate doping. In Fig. 9a the mobility $\mu_{\text{c,it}}$ is plotted as a function of N_s, which is the surface carrier density. This mobility was determined experimentally by changing the trap density through injecting charge into the gate oxide and monitoring the mobility. Mobility was measured for several values of time during which the gate oxide was injected with charge with a current of 3×10^{-5} A/cm^2. These times (ranging from 20 to 600 s) produced

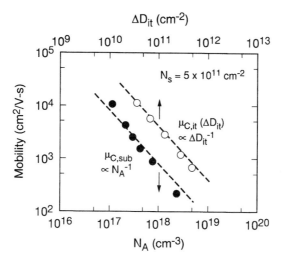

Fig. 10 Dependence of $\mu_{c,sub}$ and D_{it} on N_A, showing that the mobility is inversely proportional to the total number of Coulomb scattering centers in each case. After Koga et al., Ref. 12. Copyright © 1994 IEEE.)

increasing levels of interface traps. The interface-trap effect on mobility was found to be

$$\mu_{c,it}(D_{it}) \propto \sqrt{N_s}/D_{it} \tag{12}$$

where D_{it} is the density of interface traps and N_s is the surface carrier density.

The effect of the substrate concentration was also measured by varying the substrate concentration and the mobility component of the substrate was found to be

$$\mu_{c,sub} \; \alpha \; N_s/N_A \tag{13}$$

In Fig. 9b the effect of the channel doping on the mobility is shown and indicates that $\mu_{c,sub}$ is proportional to N_s, whereas $\mu_{c,it}$ is proportional to $\sqrt{N_s}$. The weaker dependence of $\mu_{c,it}$ can be explained physically by considering the electron distribution in the MOS inversion layer. At higher N_s, electrons are scattered more frequently by the interface charges because the electron distribution shifts towards the surface. This reduces the mobility enhancement caused by the screening effect. These two results are summarized in Fig. 10, where the mobility is seen to be inversely proportional to the total number of scattering centers in both cases.

3.3.3 Gate Structure

The gate structure of the CMOS devices has not changed dramatically in the past several years. The most common choice of gate materials for modern devices is a layered structure of polysilicon with a metal silicide on top of the gate. This produces a silicon gate that determines the work function of the MOS devices combined with a low-resistance metal silicide to form a relatively high-conductivity interconnect between individual devices. The issues over the last several years concern dimensional effects that complicate the understanding of the field produced by the gate and exactly how to model the MOS device. The gate can no longer be treated as if it were a metal with a work function that is equal to that of a degenerate n-type or p-type layer of silicon. Also, the channel can no longer be treated as if it is a classic semiconductor. The electrical properties of the gate must be considered in light of the quantum-mechanical phenomena that cause modifications in the assumption of what the charge distribution will be.

One clear influence on the characterization of the device is the degree that quantum effects affect the calculation of the threshold voltage. The threshold voltage of the MOS can no longer be calculated by assuming that there is a charge sheet formed right at the interface of the silicon and SiO_2. There are corrections that need to be made, which are due to depletion effects that occur in the gate and channel and to the fact that there is a quantum-mechanical limitation to how close the charge sheet can exist to the interface.[13,14] This has the effect of causing the measured capacitance of the gate to have, effectively, three components: the oxide capacitance, C_{ox}, the substrate capacitance, C_s, and the gate electrode capacitance, C_p. The relationship among these terms is that of three capacitors in series

$$C_{gate} = [(1/C_{ox}) + (1/C_p) + (1/C_s)]^{-1} \qquad (14)$$

$$C_{ox} = \varepsilon ox/t_{ox} \qquad (15)$$

When C_p and C_s are large, this equation says that the gate capacitance and the oxide capacitance are the same. However, for gate oxides below about 10 nm, the other terms become very important.

The gate electrode was modeled as either a perfect conductor, or as a heavily doped single-crystal silicon. The free-carrier charge densities in silicon were described using either Maxwell–Boltzmann (MB) or Fermi–Dirac (FD) statistics. Figure 11a shows the simulated gate capacitance for the structures where each of these effects is included. The quantum-mechanical effects take into account that the high field in the silicon surface causes the electrons to be confined to distances that are comparable to their de Broglie wavelengths. This is the dominant effect on the deviation of the gate capacitance from the standard model. The effect of each of these components on the calculated C–V curves is shown in Fig. 11a, which shows

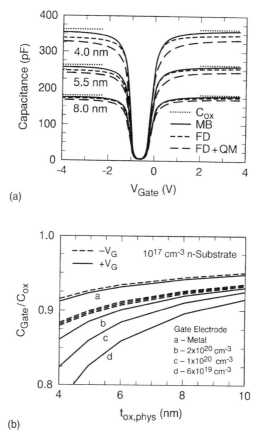

Fig. 11 (a) Gate oxide capacitance simulations including Maxwell–Boltzmann (MB) or Fermi–Dirac (FD) statistics and quantum-mechanical effects in the channel. (b) Attenuation in the gate oxide capacitance for different gate oxide thickness. (After Krisch et al., Ref. 14. Copyright © 1995 IEEE.)

the increased impact of these effects as the gate oxide becomes thinner. In Fig. 11b the effects are plotted as a function of gate oxide thickness. The calculated C_{gate} is expressed as a ratio to the C_{ox}.

3.3.4   Gate Dielectric Properties

The limitation on how thin the gate oxide can be is determined by many factors. This section will present some of the recent device implications of the very thin gate dielectric. Figure 12 shows a cross-section of a 2.5-nm gate oxide (GOX) taken using transmission electron microscopy (TEM) to illustrate the scale of the thin gate oxides in modern devices. The TEM picture also shows the individual silicon atoms in the single crystal region,

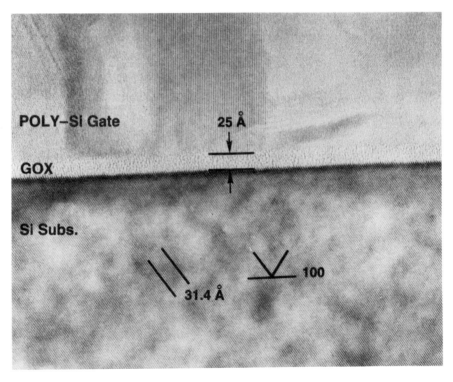

Fig. 12 TEM cross-section of a 2.5 nm (25 Å) gate oxide. The lattice spacing of the silicon substrate is used to accurately determine the oxide thickness. (After Liu et al., Ref. 15. Copyright © 1996 IEEE.)

which allows a very accurate measurement of the gate oxide thickness by using the silicon lattice spacing as the distance calibration.[15]

One of the more serious physical limitations in the design of the scaled MOSFETs is the onset of significant tunneling currents through the gate oxide in very thin oxides. This causes the devices to draw more power and may also result in reliability problems resulting from damage caused by the currents flowing through the oxides. The improvement in device performance is inversely related to the oxide thickness. However, at high-enough fields (7–8 MV/cm) the electron tunneling from the gate to the channel of the device can cause significant leakage current. This effect can result either from Fowler–Nordheim current or from direct tunneling through the oxide if the oxide is thin enough. The effect of the current on device operation and on the limitations to the gate oxide dimensions will be reviewed here. The reliability effects resulting from the gate currents will be discussed later.

The Fowler–Nordheim currents follow the expression

$$J = A\mathscr{E}_{ox}^2 \exp(-B/\mathscr{E}_{ox}) \tag{16}$$

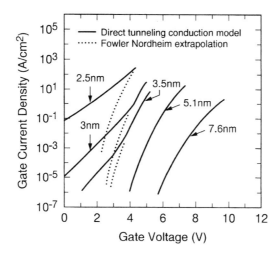

Fig. 13 Components of the gate current as a function of gate voltage for different oxide thicknesses, showing the onset of direct tunneling. (After Hu et al., Ref. 7.)

where A and B are constants ($A = 1.25 \times 10^{-6}$ A/V^2 and $B = 233.5$ MV/cm for thermally grown oxide), J is the current density in A/cm^2 and $\mathscr{E}_{ox}$ is the oxide field in V/cm. The direct tunneling through the gate oxide becomes a more important consideration as oxides are thinned to the point that the direct tunneling currents are appreciable. Figure 13 shows the gate-tunneling currents for various gate oxide thicknesses.[7] It can be seen that the direct tunneling current is appreciable even at 3 V for a gate oxide thickness of about 3.5 nm. For gate oxides less than 3.5 nm, the tunneling current becomes a significant part of the total current drawn by the devices. One interesting approach to this problem is to see how much current can flow through the gate oxide before the device characteristics are no longer useful.

The device characteristics of transistors with a 1.5-nm gate oxide thickness were evaluated.[16] The results show that although the gate current is an issue for the transistors with long gate lengths, the short-gate-length devices show good device characteristics, and record I_{on} currents were demonstrated using these devices.[17] These results are summarized in Fig. 14, where the transistor I–V curves for both the long-gate-length devices and the short-gate-length devices are shown. There are two contributions to the drain current, one from the source and the other from the gate. The gate current component is detrimental to the device characteristics but is proportional to the gate length, whereas the contribution from the source increases as the gate length decreases. For very-short-channel transistors the gate current is not a significant component of the drain current, so the transistors show good device characteristics. These n-channel transistors reported[17] with

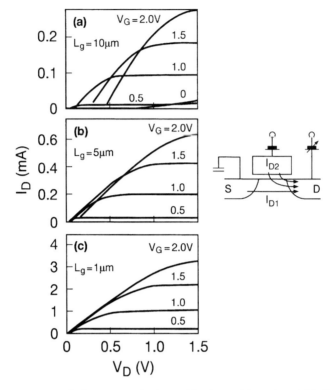

Fig. 14 Gate-current influence on devices with channel lengths of (a) $L_g = 10\,\mu m$, (b) $L_g = 5\,\mu m$, and (c) $L_g = 1.0\,\mu m$. The gate oxide is 1.5 nm thick. The two components of drain current are shown on the right. (After Momosa et al., Ref. 17. Copyright © 1996 IEEE.)

$L_c = 0.1\,\mu m$ show that a record room-temperature transconductance of over 1000 mS/mm can be achieved (see Fig. 15).

3.4   RELIABILITY

3.4.1   Hot Carriers

n-MOS Devices.   Hot-carrier effects are a source of serious device degradation because of the charging of the gate oxide in the MOS transistor. This charging can occur in both the *n*-MOS and *p*-MOS devices, but is a more serious concern in *n*-MOS devices. The basic mechanism of hot-carrier aging is shown in Fig. 16. Channel electrons gain energy in the high-field region of the drain and are accelerated towards the gate oxide. This causes the

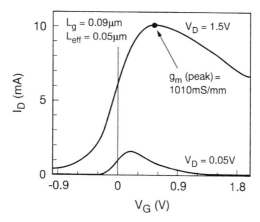

Fig. 15 Peak transconductance of a transistor with oxide 1.5 nm thick and $L_g = 0.09\ \mu$m. (After Momosa et al., Ref. 17. Copyright © 1996 IEEE.)

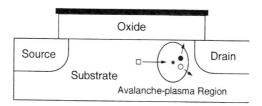

□ Channel hot electron
● Electron created by impact ionization
○ Hole created by impact ionization
* Impact ionization event

Fig. 16 Hot-electron generation at the drain of the MOSFET, resulting from impact ionization at high fields.

charge to be injected into the gate oxide and creates a fixed charge in the oxide. The fixed charge increases the threshold voltage of the device and reduces the drive. The effect of the fixed charge is usually determined by evaluating the time (t) dependence of the device. The gate voltage is set at one- half of the drain voltage, and the lifetime is extrapolated through the formula

$$dV_T = At^n \tag{17}$$

where n has a value of around 0.5–0.7, dV_T is the change in the threshold voltage due to device aging, and A is a constant that can be associated with the device process fabrication steps.[18]

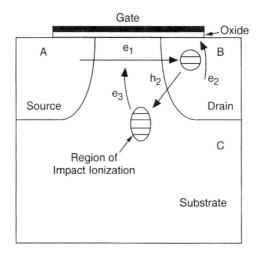

Fig. 17 Hot-electron degradation mechanisms illustrated by the channel electron e_1 causing secondary holes h_2 to form in the high field region of the drain, in turn causing secondary electrons e_2 to be created through impact ionization. (After Bude, Ref. 20. Copyright © 1995 IEEE.)

For future devices, hot carriers will be somewhat reduced by lowering V_{DD}. The actual voltage reduction will depend on particular application, but reliability will still be a factor in considering the voltage. It was assumed that as the voltage was reduced to around 1 V or less, hot carriers would no longer be a concern. Recent results have shown that impact ionization can occur at voltages that are significantly less than 1 V. This means that hot-carrier degradation is still valid for these voltages.[19] Charge damage occurs at these voltages through a three-step mechanism that causes an energy enhancement resulting from a feedback mechanism, namely impact-ionization feedback through the drain–bulk junction.[20]

The impact-ionization feedback effect is shown schematically in Fig. 17. Channel electrons (e_1) are injected into the drain where they gain enough energy to impact ionize, forming low-energy electron–hole pairs in region B. The secondary electrons (e_2) formed in this region leave through the drain, but the holes (h_2) are then accelerated back into region C where they can once more gain enough energy to impact ionize forming more electron–hole pairs. The electrons formed in region C can then fall back through the potential drop and gain more energy. This effect cannot be observed directly, but can be simulated by using statistical distribution functions of the electron energies and allowing the impact ionization to occur. The carriers can only contribute to the gate current if their energies are above the effective Si–SiO$_2$ conduction-band discontinuity, which is about 3.2 eV. The results of these simulations show an energy tail in the electron distribution shown in Fig. 18,

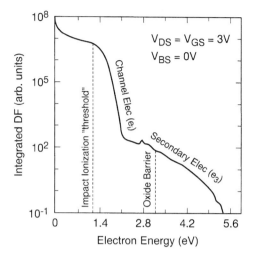

Fig. 18 Monte Carlo calculation of the hot-carrier distribution function (DF) for the case where the impact ionization causes a secondary electron contribution out to higher energies. (After Bude, Ref. 20. Copyright © 1995 IEEE.)

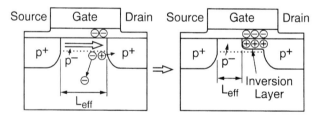

Fig. 19 Schematic illustration of the mechanism causing the hot-electron-induced punchthrough effect. (After Woltjer et al., Ref. 22. Copyright © 1995 IEEE.)

which indicates a considerable population of electrons with an energy above 3.2 eV.

***p*-MOS Devices**   The degradation of *p*-MOS devices is due to a different mechanism than that of the *n*-MOS devices. The mechanism is triggered by the electrons in the inversion region of the channel that are injected into the gate oxide. The trapped electrons cause the surface of the channel to invert, which effectively extends the p^+ region of the drain into the channel. Figure 19 shows this mechanism and indicates the charged area effectively extending the drain. Extending the drain reduces the effective channel length of the transistor and actually increases the transconductance as a function of the time the device is operated at a high voltage. The mechanism is called hot-electron-induced punchthrough (HEIP).[21] This is the dominant cause of p-MOS device degradation. A second mechanism that can also contribute to p-MOS degradation is one in which the holes produce interface traps that reduce the transconductance. As the transistors are scaled down into the deep-submicron region, a third mechanism becomes important—the injection of holes into the gate oxide, causing a positive oxide charge. This mechanism is most important for deep-submicron p-MOSFETS with a nitrided gate oxide, and results in a decrease in the transconductance.[22] The impact of these effects can be described by the following relationship in which a combination of negative charge and interface traps are contributing:

$$\Delta g_m/g_m = 0.1 \times 5 t_{ox}/8 L_{eff}[\log_{10}(1 + 10^{\sqrt{8L_{eff}/5t_{ox}}} \, t/\tau_{gm,-})]^2 \qquad (18)$$
$$- 0.1(t/\tau_{gm,it})^{0.45}$$

Whereas the degradation mechanism for a combination of positive charge and interface states yields

$$\Delta g_m/g_m = 0.1 \times 5 t_{ox}/8 L_{eff}[\log_{10}(1 + 10^{\sqrt{8L_{eff}/5t_{ox}}} \, t/\tau_{gm,+})]^2 \qquad (19)$$
$$- 0.1(t/\tau_{gm,it})^{0.45}$$

where $\tau_{gm,it}$, $\tau_{gm,-}$, and $\tau_g{}^*{}_{m,+}$ are the times required to meet 10% $\Delta g_m/g_m$ for the surface state, negative charge generation, and positive charge generation, respectively.

These three mechanisms can all contribute to the degradation of the p-MOS devices and may also have an impact on the noise margins of very small devices. Figure 20 shows three different aging conditions for $\tau_{gm,ss}$, $\tau_{gm,-}$, and $\tau_{gm,+}$ as a function of $1/V_D$. The $\tau_{gm,it}$ is due to interface traps generated at $V_G = V_D/2$, $\tau_{gm,-}$ is due to negative charge generated at $V_G = V_T$, and $\tau_{gm,+}$ is due to positive charge generated at $V_G = V_D$.

Although the mechanisms involved with the aging of p-MOS devices are becoming better understood, the aging of the p-MOS devices will be of less concern in general than the aging of the n-MOS devices.

3.4.2   Dielectric Wearout

The wearout mechanisms for thin dielectrics are due to hole or electron trapping in the thin oxide, which leads to breakdown of the oxide and a current path through the insulator. DRAM and EPROM applications, as well as the scaled MOS devices, have caused the electric fields across the thin oxides to be increased as the technologies have improved. The result of these increased fields in the oxides has been an improved understanding of the mechanisms of dielectric breakdown as well as improvements in the reliability of the thin dielectrics that are currently being used in devices. Charge trapping has been the most important wearout mechanism for the MOS

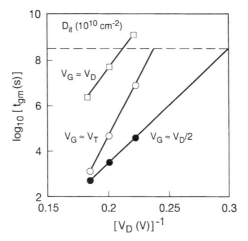

Fig. 20 p-MOS lifetime extrapolations for 10% change in transconductance vs $1/V_D$ for different degradation mechanisms. (After Woltjer et al., Ref. 22. Copyright © 1995 IEEE.)

devices and is the factor that most limits the maximum electric field in the device. A charge-to-breakdown value Q_{bd}, which is commonly used for evaluating oxide quality has been in common practice for some time. The most recent advance in the understanding of the dielectric wearout mechanisms is to identify the charging of the thin oxides when they are in the direct-tunneling regime.[23] These new results may indicate that the thinner dielectrics have a different charging mechanism and Q_{bd} is no longer a good parameter to use. The time-to-breakdown data (see Fig. 21) shows a projection for the very thin dielectrics that extend well into the sub-3-nm range with useable supply voltages. Figure 22 is a projection for the minimum gate oxide if the limitation is either tunneling current or time-dependent dielectric breakdown (TDDB). If we can tolerate the gate current in the devices described above, then we will be limited only by the TDDB at gate oxides that are less than 2.5 nm. Unfortunately, the wearout mechanism of the very thin oxides is still not well characterized or understood.

3.5 SOI AND 3D STRUCTURES

The significant improvements of scaled MOS devices that have been described have fundamental limitations and cannot be improved further by changing the dimensions. The structures can then be improved only by changing the basic operation of the transistor, which allows the device to be extended into the silicon. One important area of development over the last several years has been to use silicon-on-insulator (SOI) devices to improve the performance.[24]

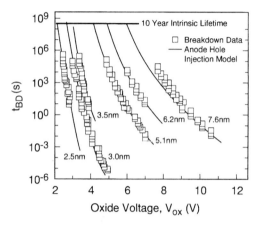

Fig. 21 Time-to-breakdown measurements as a function of oxide voltage for thin gate oxides. (After Schuegraf and Hu, Ref. 23.)

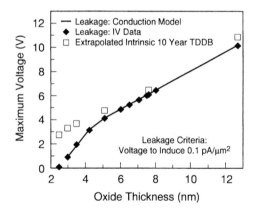

Fig. 22 Plot of maximum acceptable oxide thickness for two different failure determinations, leakage and TDDB. (After Schuegraf and Hu, Ref. 23.)

The advantages of using SOI structures are that the parasitic capacitance can be significantly reduced as well as some unique properties of SOI that allow low-power and low-voltage operations to be improved. SOI devices also have a significantly better resistance to radiation damage in some applications and have been used for these purposes for many years.

3.5.1 Partially Depleted SOI MOSFETs

The most common device to be developed using SOI materials is called partially depleted SOI. This name comes from the fact that the devices are

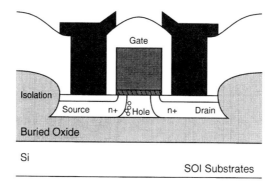

Fig. 23 SOI cross-section showing the device structure and the collection of holes in the channel region causing the floating-substrate effect.

constructed so that the channel region is thick enough that the depletion width of the channel is smaller than the silicon thickness of the channel. This structure is shown in Fig. 23. The advantage of this kind of structure is that the device operation and device design is very close to the operation and design of a bulk CMOS device. There are several significant differences that will be discussed here.

The floating-substrate effect is the phenomenon of the accumulation of excess holes in the channel region, which causes the MOS device to conduct both in the channel region under the gate and at the silicon–SiO$_2$ interface of the buried oxide. This is sometimes called the kink effect since it results in a kink in the transistor I–V characteristics as the device turns on and the channel region charges. It can also cause a low drain breakdown voltage. One solution to this problem is to form a contact within the channel region, which allows the charge to be removed from the channel to eliminate the back biasing. This can also be accomplished by reducing the lifetime of the majority carriers to reduce the amount of stored charge. However, the techniques of reducing lifetimes can also increase the source-to-drain leakage.

A more direct approach to resolving the floating-substrate problem is to prevent the charge from forming by creating a direct contact on the substrate to the source contact of the transistor. This eliminates the floating-substrate charging but it complicates the layout of the device. Applications for partially depleted SOI devices may be found when the supply voltages drop low enough to eliminate the floating-substrate effects.

Another recent technique to reduce the floating-substrate problem is to bandgap engineer the silicon layer to narrow the bandgap at the source to increase the hole flow to the source.[24] Implanting germanium into the silicon layer, forming a $Si_{1-x}Ge_x$ region, creates band narrowing by some factor ΔE that will increase the hole current by $\exp(\Delta E / K_B T)$. This is shown in Fig.

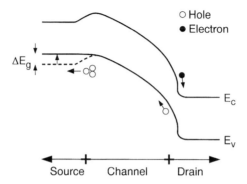

Fig. 24 Band structure diagram showing the effect of the Ge implant on the band structure at the source end of the transistor allowing the sinking of the holes generated in the channel. (After Terauchi et al., Ref. 24. Copyright © 1995 IEEE.)

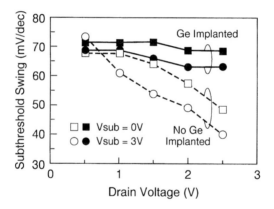

Fig. 25 Drain-voltage dependence of the subthreshold swing for the cases of having and not having a Ge implant. (After Terauchi et al., Ref. 24. Copyright © 1995 IEEE.)

24, and the experimental results showing the impact on the subthreshold swing, S, are shown in Fig. 25. This figure shows that S is about 70 mV/dec with the germanium implant. Without the germanium implant the measured swing is much less than this, and is as low as 40 mV/dec for a 2.5 V drain voltage. This is not the subthreshold swing resulting from (kT/q) term in the formula for drain current but is an indication of the modification of the V_T of the transistor resulting from the floating-substrate effect. The V_T decreases as the gate voltage is increased due to the substrate charging and shows an apparent reduction of the $\log I_D - V_G$ swing.

Structure	Conventional	Gate-All-Around	Ground Plane
Schematic	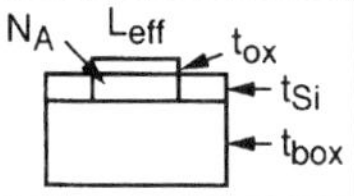	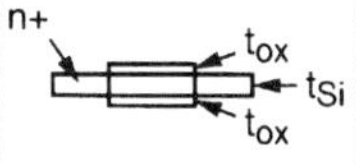	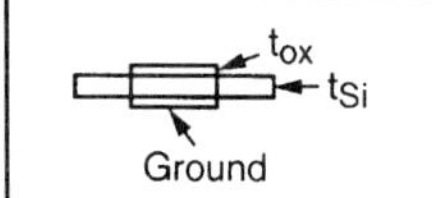
Scaling	$\lambda = \sqrt{\dfrac{\varepsilon_{Si}}{\varepsilon_{ox}}}\, t_{Si}\, t_{ox}$	$\lambda = \sqrt{\dfrac{\varepsilon_{Si}}{2\varepsilon_{ox}}}\, t_{Si}\, t_{ox}$	$\lambda = \sqrt{\dfrac{\varepsilon_{Si}}{2\varepsilon_{ox}}}\; \dfrac{t_{Si}\, t_{ox}}{\left(1 + \dfrac{\varepsilon_{Si}\, t_{ox}}{\varepsilon_{ox}\, t_{Si}}\right)}$

(a) (b) (c)

Fig. 26 Scaling relations for the conventional SOI MOSFET, the gate-all-around configurations and the ground-plane structure. (After Yan et al., Ref. 10. Copyright © 1992 IEEE.)

3.5.2 Fully Depleted SOI MOSFETs

Fully depleted SOI MOSFETs are devices made on very thin silicon layers on top of the insulating oxide. The silicon layer is thin enough to allow the channel region of the transistors to be fully depleted, which allows the fields to be lowered in the channel region. Lower fields can reduce the hot-carrier effects, reduce short-channel effects, and increase the drive. In addition to these advantages, the capacitance of the source–drain region is reduced because the junctions can now be surrounded by oxide, which has a lower dielectric constant than silicon. These effects have been quantified, and a generalized treatment of the scaling advantages of the fully depleted SOI devices was generated using a scaling parameter λ, where

$$\lambda = (\sqrt{\varepsilon_{Si}/\varepsilon_{ox}})\, t_{Si} t_{ox} \tag{20}$$

for the fully depleted SOI device.[10] Figure 26 shows a comparison of the characteristic scaling dimensions along with the device structures.

Another kind of device that can further improve the performance of the SOI transistors is one in which the gate completely surrounds the channel of the silicon. This structure[25] can be most simply described as doubling the effective field in the channel, so that the scaling parameter becomes

$$\lambda = (\sqrt{\varepsilon_{Si}/2\varepsilon_{ox}})\, t_{Si} t_{ox} \tag{21}$$

This scaling parameter can be thought of as the characteristic length of the distance between the highest field point in the drain to the lowest field point in the channel. For effects such as short-channel effects, this is a good figure of comparison for the relative scaling dimensions of these SOI devices. This configuration can also be made with a more complicated structure by creating another polysilicon layer in the substrate, which allows the channel region to be biased using a back gate potential to allow the V_t of the front channel to be modified by changing the voltage on the back channel.[26]

One interesting alternative approach to getting the advantages of the fully depleted SOI device is a similar structure that can achieve the same kind of reduction in the length of the characteristic field variation by creating a heavily doped region under the channel, which produces a thin active region with a ground layer underneath. This structure has the characteristic length

$$\lambda = \sqrt{\varepsilon_{Si}/2\varepsilon_{ox}}\,(t_{Si}t_{ox})/[1 + (\varepsilon_{Si}t_{ox}/\varepsilon_{ox}t_{Si})] \tag{22}$$

In this structure, the ground-plane region defines the thickness of the channel region but does not extend below the drain junction, so it does not add any unnecessary capacitance to the device. This is the pulse-doped channel device described in Section 3.3.1. It also has the important advantage of not requiring the buried oxide structure that adds considerable complexity to the manufacture.

3.6   MEMORY STRUCTURES

3.6.1   DRAM

The most important feature of the DRAM structure that separates this technology from the other applications is the capacitor. The single-transistor DRAM cell requires a capacitor that can store sufficient charge to allow the cell state to stay true between refresh cycles. The most significant development in the DRAM devices has been the advance in the capacitor design, which has allowed the cell charge retention to keep up with (and exceed) the needs of the DRAM circuits. The DRAM capacitors have been improved in two ways: increasing the surface area and increasing the capacitor dielectric constant. Since the capacitance of the DRAM capacitor is governed by the simple equation

$$C = \varepsilon_i\,A/t_{ox} \tag{23}$$

where ε_i is the permittivity of the dielectric, A is the area of the capacitor, and t_{ox} is the dielectric thickness; for a minimum t_{ox}, the only two parameters that are adjustable are, ε_i and A.

One area of progress in making a capacitor with a larger surface area is to form the capacitor as a trench. This technique can produce a trench capacitor extending nearly $7\,\mu$m into the silicon, and has the advantage of allowing the transistors to be formed on a nearly planar surface with the trench extending below the active device area. Figure 27 shows a cross-section

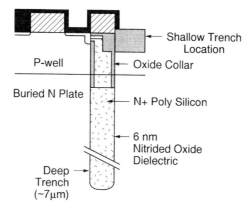

Fig. 27   Cross-section of the DRAM cell using a deep-trench storage capacitor. After Bronner et al., Ref. 27. Copyright © 1995 IEEE.)

of a trench capacitor cell.[27] The effect of the large surface area and the ability to thin the gate dielectric by improving the reliability of the thin oxides has resulted in a DRAM that can store more charge in smaller surface areas. The cell in the figure uses a 6 nm nitrided oxide as the dielectric.

Another area of tremendous progress in DRAM technology goes in the opposite direction and forms the capacitors in geometries that extend above the silicon to create a large surface area. The large surface areas can be created using a large planar capacitor shaped as a dome or a crown, as shown in Fig. 28. These structures also take advantage of higher-dielectric-constant materials for the interlevel dielectric for the capacitor, such as Ta_2O_5, (Ba, Sr)TiO_3, or $SrTiO_3$. Dielectric constants[28] of up to 200 have been reported for (Ba, Sr)TiO_3. A cross-section of the crown-type capacitor structure incorporating a Ta_2O_5 dielectric is shown in Fig. 28, illustrating how the high-dielectric-constant materials are included in the DRAM cell. Figure 29 shows a comparison of all of the DRAM capacitor geometries with the different kinds of dielectric materials used. This figure shows three different kinds of capacitor geometry: plane, pedestal, and crown. The crown geometry is the one shown in Fig. 28. The pedestal geometry is similar to the crown, except that the top surface is flat. The plane geometry is simply the flat capacitor structure, similar to the gate of an MOS device.

3.6.2 SRAM

The specific physics and device needs of the SRAM have not changed significantly in the last several years except for the constant need to shrink the size of the SRAM cell beyond the scaling allowed by simply shrinking the device dimensions. The most significant structural change that has occurred to accomplish this is the extensive use of thin-film transistors (TFT)

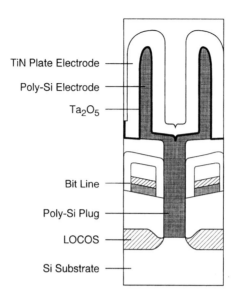

Fig. 28 Cross-section of the DRAM cell of a device with a high-dielectric-constant capacitor material using a crown type of capacitor structure. After Ohji et al., Ref. 28. Copyright © 1995 IEEE.)

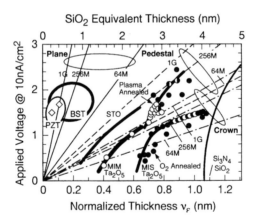

Fig. 29 Plot of the voltage needed for the application of materials and structures showing three generations of DRAM as a function of normalized thickness when high dielectric constant capacitor dielectrics are used. (After Ohji et al., Ref. 28. Copyright © 1995 IEEE.)

SRAM Cell Trend

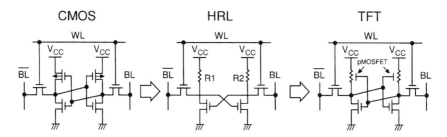

Fig. 30 Circuit schematic of three different SRAM cells. (a) CMOS six-transistor cell. (b) High-resistance load (HRL) four-transistor cell. (c) TFT six-transistor configuration.

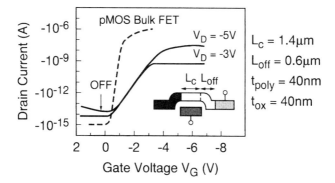

Fig. 31 Drain current for the *p*-MOS load device, showing the effect of the drain extension for improving the off current of the TFT transistor.

to allow the *p*-channel load transistor to be put on top of the *n*-channel transistor to produce a cell stacked in the third dimension. The evolution of the SRAM cell from a full six-transistor CMOS cell to a four-transistor cell using resistors for the load and finally to a TFT structure, is shown in Fig. 30. The characteristics of the *p*-MOS TFT are critical for the successful implementation of this cell structure. This is the most common structure for a high-performance SRAM cell, and some significant improvements in the TFT structure have been achieved.[29] The most important of these was to improve the off current and the current drive of the device by lightly doping a region of the drain; this reduces the drain leakage resulting from the high field of the drain, the cause of band-to-band tunneling. This structure is shown in Fig. 31, where a comparison of the off currents for the TFT and bulk *p*-MOS devices is illustrated.

A fundamentally new way of looking at storing information is to store the

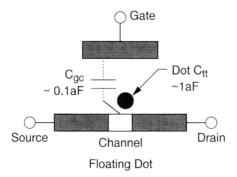

Floating Dot

Fig. 32 Schematic showing the geometry of the storage dot and the TFT transistor. (After Yano et al., Ref. 30. Copyright © 1995 IEEE.)

information in the form of a single electron. Several ways of creating this kind of storage element have been proposed, but they all require the formation of an island of silicon in a dielectric to store a single electron. Most of these structures require the device to be cooled to low temperatures to insure that the electron can be detected. Recently, however, a room-temperature version was demonstrated.[30] This single-electron memory is a structure that uses the charge from a single electron to store a single bit of information. A single-transistor serves as the sensing element of the charge. The single-electron sensing element and the storage dot are included in the same structure to maximize the sensitivity. Figure 32 shows the structure and the position of the silicon dot with respect to the channel of the TFT. The gate is located above the dot. The basic operation is to have a single electron enter the potential well created by the silicon particle. The electron charge will then cause the TFT to shift the threshold voltage enough to be a viable state for the memory. Figure 33 shows the threshold shift after the dot has been written. The erase-to-write cycle is shown, with the arrows indicating the drain current used to trap an electron in the dot. Figure 34 shows the energy diagram of the device before and after the cell is written.

3.6.3 Nonvolatile Memory

Nonvolatile memory is distinct from the DRAM and SRAM in that the information stored is maintained after the power supply is removed. The nonvolatile memories come in several types that are differentiated by the ease of programming. The PROM (programmable read-only memory) is programmed by either a mask-determined logic state or a fusible link that allows the user to customize the memory once for a specific application. The basic transistor cell for the nonvolatile memory is the floating-gate transistor. This basic structure is usually as shown in Fig. 35, where the floating-gate and the control gate are indicated. In the EPROM (electrically programmable

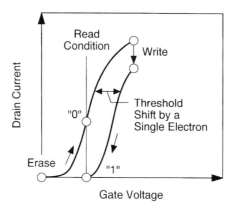

Fig. 33 Operation of the single-electron memory cell, showing the threshold shift caused by the effect of the stored electron. (After Yano et al., Ref. 30. Copyright © 1995 IEEE.)

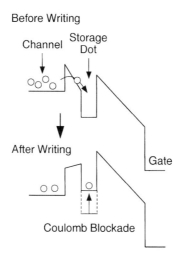

Fig. 34 Energy-band structure showing the effect of trapping of the electron in the storage dot. The Coulomb blockade indicates the effect of the electron reducing the potential in the well and blocks the transfer of another electron. (After Yano et al., Ref. 30. Copyright © 1995 IEEE.)

read-only memory), the programming is done by hot-electron injection or tunneling to a floating gate, as shown in Fig. 35. An EEPROM is an electrically erasable PROM, where each cell can be individually selected and erased. This requires a second transistor to be used in each memory cell. A flash EPROM (or flash EEPROM) is an EPROM that can be electrically erased, but all of the bits must be erased globally. This allows a one-transistor

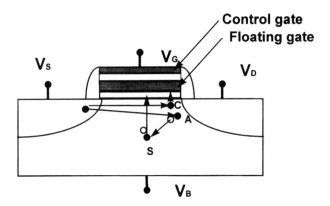

Fig. 35 Cross-section of EEPROM cell showing the hot-electron injection mechanisms. The electrons (solid circles) flow from the source to the drain and inject charge through CHEI (C), DAHC (A), and SCIHE (S). (After Hu et al., Ref. 31. Copyright © 1995 IEEE.)

cell to be maintained, but it loses the bit selectivity for the erase cycle. The floating gates of the EEPROM devices are usually written by applying a sufficient voltage across the gate to cause Fowler–Nordheim (FN) tunneling to occur.

The most recent advance in EEPROMS is in the way that the devices are programmed. The channel hot-electron injection technique (CHEI) uses the hot electrons in the channel to charge the gate (see Fig. 35). Drain-avalanche hot-carrier (DAHC) injection uses the effect of the electrons created in the high-field region to charge the gate. Both of these techniques require relatively high voltages and power consumption. The substrate-current induced hot-electron (SCIHE) technique allows for smaller voltages in the programming cycle.[31] This technique uses the substrate current generated by channel impact ionization as opposed to the high drain fields needed for the DAHC technique. This is consistent with the trend towards lower supply voltages necessary for the scaling of the devices. The ability to lower the voltage of the write cycles of the device allows the EEPROM to be more closely coupled with the CMOS logic and even embedded into the same silicon device. This technique and channel-initiated secondary electron-injection (CISEI)[32] are low-voltage options to program the floating gate without having to rely on higher voltage necessary for FN tunneling current. Each of these techniques uses electrons in the channel that achieve sufficient energy to travel over the barrier of the oxide.

The lowest voltage programming method has been achieved by using the CISEI method, where programming voltages down to 2.5 V have been achieved. The mechanism, shown in Fig. 36, is the same as the hot-electron degradation current shown in Fig. 17. The resulting gate

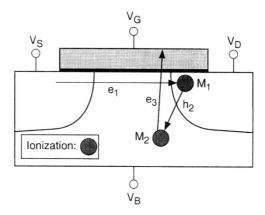

Fig. 36 Schematic of the CISEI process in an n-MOSFET transistor illustrating the channel electrons, e_1, impact ionizing and current multiplying to create holes, h_2. Current multiplication of the holes with multiplication M_2 then creates electrons, e_3, that enter into the gate with probability T. (After Bude et al., Ref. 32. Copyright © 1995 IEEE.)

charging is given by

$$I_G = I_{DS} M_1 M_2 T \tag{24}$$

where the gate current I_G is determined by the multiplication M_1, creating holes that then ionize in the substrate with multiplication M_2, and then reach the gate with probability T. By creating a structure as shown in Fig. 37, where a p-type halo is created around the drain region of the transistor, a higher field is created in the drain region, which increases the multiplication factor M_2. The resulting devices show an increase in gate current as a result of changing the substrate bias (see Fig. 38).

3.7 LOW-VOLTAGE/LOW-POWER DEVICES

Low-voltage operation and low power consumption of devices is a driving force behind the advances in modern CMOS technologies. The scaling rules and the device operations that have been outlined all contribute to the reduction of the power consumption. The power used by the transistors driving a capacitive load will be proportional to V_{DD}^2, so reducing the supply voltage will significantly reduce the power used. To go beyond the simple scaling rules that will reduce the power, some special considerations must be made in order to continue the power reduction into the region where the fundamental limitations of the device scaling prevent the further reduction of power consumption.

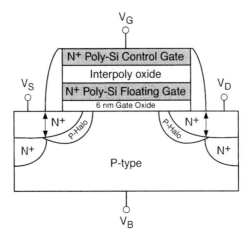

Fig. 37 Cross-section of CISEI memory cell showing the *p*-halo region used to increase the generation of channel-induced substrate electrons. (After Bude et al., Ref. 32. Copyright © 1995 IEEE.)

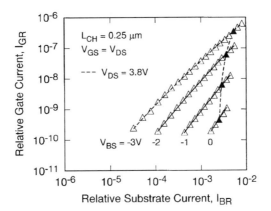

Fig. 38 Gate current of the CISEI memory cell as a function of substrate current, for different back-bias voltages. This effect of the back bias on the gate current is an indication of the CISEI as the source of the gate current. (After Bude et al., Ref. 32. Copyright © 1995 IEEE.)

3.7.1 Low-Threshold-Voltage Devices

Optimizing transistors for low-voltage operation requires an allowance for the higher off currents of the devices. This higher off current causes the device to draw more current while it is idle, but can be balanced with higher performance. The implementation of these devices would be in a circuit

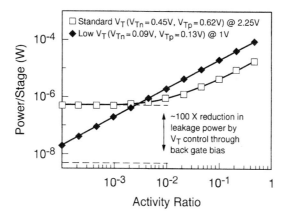

Fig. 39 Plot of the simulated power/stage for a circuit as a function of the activity ratio, showing the effect of the back gate bias on the power consumed. (After Chen et al., Ref. 33. Copyright © 1995 IEEE.)

configuration that allows parts of the circuit to be selectively turned off by back biasing the transistors.

The ability to lower the threshold voltage and still maintain the off current and the drive current has become a focus of the new device structures. One method to accomplish this is to allow the threshold voltage to vary or be adjusted by changing the back bias of the transistor. This technique can be implemented quite simply by designing devices with very low threshold voltages that are limited simply by how much power is used in the circuit in the active mode. This can be defined as an activity ratio α, which will change depending on the type of circuit. The threshold can be reduced until the off current is high enough to be an appreciable part of the total operating power.[33] Devices can then be designed to have this low threshold voltage and the performance optimized without appreciably increasing the power consumption in the active mode. When the circuit is idle the standby current can be decreased tremendously by raising the threshold voltage to a higher value by back biasing the circuit. Typically the ratio α can be quite small for many circuit applications. Figure 39 shows the impact of this kind of approach on a series of simulated circuits. These devices were designed for high performance at low voltage by providing low threshold voltages at the cost of allowing considerable off currents. The standby power could still be maintained quite low for the devices by including a back-gate-bias threshold-voltage control for the idle circuits.

Another method of achieving low off currents but maintaining the performance of devices is to use variable threshold devices.[34] These devices use the technique of tying the gate potential to the substrate potential to create a lateral bipolar action in the devices. This can be looked at as reducing the V_T of the MOS device because the back biasing of the transistors occurs as the device is turned on. The low-power advantages are not truly realized unless the devices are constructed on an SOI substrate, where the channel region can be contacted and the individual transistors are isolated from the rest of the device.[35]

3.7.2   Noise Effects for Low-Voltage Devices.

As the supply voltage gets smaller and the threshold voltage of the devices is reduced, the size of the signal available for the circuit designer becomes smaller. Consequently, applications become very sensitive to fluctuations in the signal. This sensitivity to these fluctuations, or noise, makes understanding the sources of noise a very important aspect of the design of the transistors. There have been a number of improvements that have helped in understanding the sources of noise and in the design of devices that to have lower noise figures. The $1/f$ noise in the MOS transistor is dominated by the effects of the interface. The noise in the MOSFET is due to the fluctuations of the current in the channel. This can be caused by fluctuations in the number of free charge carriers or by fluctuations in the mobility of the carriers. The most current view is that the noise in the n-MOS device is due to the fluctuations in the number of electrons, whereas the noise in the p-MOS device is due to the variation of the mobility of the holes in the channel.[36] The noise is described with a spectral density S_G, where

$$S_G/G^2 = \alpha/Nf \qquad (25)$$

where G is the conductance, α is the $1/f$ noise parameter, which has a value between 10^{-7} and 10^{-4}, N is the total number of free carriers, and f is the frequency. The number of carriers is for a homogeneous sample with perfect contacts or one with contacts where the reduced number is well defined for a device with nonuniform fields. The n-MOS noise is due to the channel electrons populating and depopulating traps close to the oxide interface.

Noise in the p-MOS transistor is more problematic, but the prevailing opinion is that the dominant component of noise for the p-MOS device is the fluctuation of the mobility due to surface scattering. This opinion is supported by the fact that the surface-channel p-MOS device has a significantly higher noise factor than the buried-channel transistors, where the channel is further from the interface.

3.8   SUMMARY AND FUTURE TRENDS

The last decade has been very important in the development of the very-large-scale integrated circuits and in the application of these devices. This is the period in which we have seen various technologies become more integrated into a single chip and more and more functionality incorporated into these "systems on a chip" architectures. The future of these systems will be in the increased integration of these different structures with an ever-increasing dependence on the three-dimensional nature of these devices. This includes the memory structures and the interconnect as well as the individual transistor designs.

One trend that we have observed recently is that the understanding and design of the current and future devices will rely more and more on detailed simulation. Simulations and calculations using statistical techniques and three-dimensional simulation tools will be required to model three-dimensional effects in very small devices. This is evident in some of the empirical scaling rules that were reviewed and in the distribution functions calculated to explain gate current generation. These results are somewhat less satisfying from an intuitive viewpoint, but it is clear that the trend will continue as the complexity of the devices increases.

We have also seen some of the "ultimate barriers" to the device dimension get pushed back as we get closer to the limits. The gate oxide thickness was thought to be ultimately limited to about 3 nm until devices with 1.5 nm gate oxides were shown to have potentially useful characteristics.

Another direction of development that may have very interesting consequences for the future is the recent progress made in the area of organic field-effect transistors.[37,38] These devices take the application out of the realm of silicon-based structures and into the realm of carbon-based devices. These devices are basically thin-film transistors that use organic polymers in place of silicon as the active layer. The potential advantages of these materials is that they can be easily deposited as a thin film by sublimation. They are compatible with materials such as flexible plastics, and the processing temperature is well below that normally required for silicon technology. One promising material is the thiophene oligomer α-hexathieylene. This material has produced devices of the configuration shown in Fig. 40 with on-to-off current ratios of 10^6. These devices work in the enhancement mode, and have

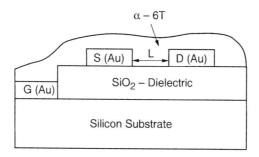

Fig. 40 Schematic of an organic TFT with α-6T active layers. (After Torsi et al., Ref. 38.)

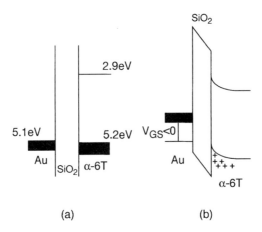

Fig. 41   Energy-band diagram of the heterojunction transistor in the off (a) and on (b) states. (After Torsi et al., Ref. 38.)

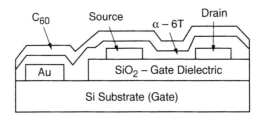

Fig. 42   Schematic of a heterojunction organic TFT with α-6T and C_{60} active layers. (After Dodabalapur et al., Ref. 37.)

been shown to be surface-acting devices, meaning that the conduction is independent of the thickness of α-6T. The cross-section in Fig. 40 shows the polymer deposited on top of an SiO_2 dielectric that is on top of a silicon substrate. The silicon substrate acts as the gate for this FET. The operation of the transistor can be seen schematically in Fig. 41, where the band structure is shown with the gate at 0 V and at a negative bias.

A further refinement of the organic FET is to have a complementary set of devices (*n*-channel and *p*-channel) to be able to create a CMOS type of circuit. This has been achieved[37] using a combination of α-6T and the fullerene compound C_{60}. This allows a heterojunction of the two materials to be formed, which enables the single device to conduct as an *n*-channel enhancement device or a *p*-channel enhancement device. This device is shown in Fig. 42; it has the same configuration as the α-6T device except for the additional layer of C_{60}. The energy diagram looks quite different,

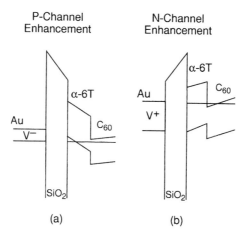

P-Channel
Enhancement

N-Channel
Enhancement

(a)   (b)

Fig. 43 Energy-band diagram of the heterojunction transistor in the *n*-channel and (a) *p*-channel (b) modes of operation. (After Dodabalapur et al., Ref. 37.)

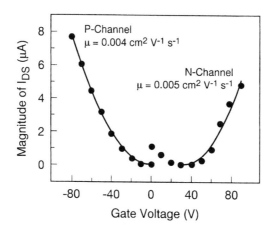

Fig. 44 Magnitude of the drain current as a function of operation ($V_{DS} = 40$ V). The solid lines are the data and the points are a parabolic fit from which the field effect mobilities were extracted. (After Dodabalapur et al., Ref. 37.)

however, with the heterojunction allowing an accumulation layer to be formed with both polarities of the gate voltage. Figure 43 is the energy-band diagram of this device, showing that the current will flow for either polarity of gate voltage. Figure 44 shows the magnitude of the drain current as a function of operation ($V_{DS} = 40$ V). This indicates the current flow for both the *p*-channel and the *n*-channel devices. The voltages are quite high and

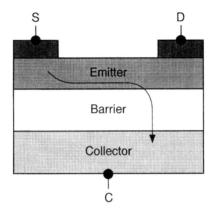

Fig. 45 Schematic diagram of a charge-injection transistor (CHINT). Carriers in the emitter layer are heated by an electric field applied between source and drain. They then undergo a transfer into the collector region after acquiring enough energy to surmount the barrier. (After Mastrapasqua et al., Ref. 39. Copyright © 1996 IEEE.)

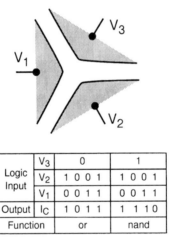

Logic Input		V_3	0	1
	V_2		1 0 0 1	1 0 0 1
	V_1		0 0 1 1	0 0 1 1
Output	I_C		1 0 1 1	1 1 1 0
Function			or	nand

Fig. 46 Truth table showing the function of a three-input CHINT structure. (After Mastrapasqua et al., Ref. 39. Copyright © 1996 IEEE.)

the currents are very low compared to a silicon device. However, the appeal of a structure like this is the possibility of applications that would require the flexibility of new materials.

Another important future direction for CMOS-based logic is to generate structures that will allow more complicated circuit functions to be formed with single transistor structures. One example of this trend is the charge-injection transistor (CHINT),[39] which is shown in Fig. 45. The basic function of a CHINT device is that carriers in the emitter layer are heated by an electric field between the source and drain. These carriers can then achieve enough energy to spill over the barrier and end up in the collector area. Devices have been made using Si/SiGe heterostructures to create multiple- input CHINT, where the combination of the three terminals allows the input voltages to determine the output voltage in ways that allow more complicated logic to be achieved. Figure 46 shows an example of the principle of a multiterminal CHINT device, with a logic table showing the function for the various combinations of voltages on the inputs. The only point to be made from this example is that the single device peforms a specific logic function that would take several transistors to accomplish using a normal CMOS design. This trend will be important as the technology evolves to a point where the transistor dimensions approach practical limits, and the only way to achieve more functionality is to design around the need for more transistors.

PROBLEMS

1. Calculate the scaled I_D of two n-MOS devices, one with constant voltage scaling and one with constant field scaling. The initial device parameters are $L = 1\,\mu m$, $I_D = 500\,\mu A/\mu m$ and $V_D = 5\,V$. The initial gate oxide thickness is 10 nm.

2. Calculate the change in the mobility of a transistor after injecting gate current causes the density of interface traps to increase by $10^{11}/cm^2$.

3. What would be the required change in substrate doping to cause a change in mobility equal to the effect of the interface traps of Problem 3.

4. Calculate the scaled λ for the device of Problem 1 when it is a pulse-doped device with a channel thickness of 100 nm.

5. Calculate the scaled λ for an SOI device and a gate-all-around device as the device of Problem 4 with the thickness of the silicon film in the channel region is 100 nm.

6. For a gate oxide thickness of 4 nm, the capacitance of the gate resulting from the quantum-mechanical effect is about 10% of the C_{ox}. Calculate the change in gate current in saturation.

7. Calculate the gate current of an $L = 0.5\,\mu m$ square transistor that has a gate oxide thickness of 5 nm. Assume that $V_{DD} = 4V$.

8. What is the capacitance of a DRAM capacitor if it is planar, $1\,\mu m^2$ with an oxide thickness of 10 nm? Now calculate the capacitance if the same surface area is used for a trench that is $7\,\mu m$ deep and with the same oxide thickness.

REFERENCES

1. S. M. Sze, *Physics of Semiconductor Devices*, 2nd ed., Wiley, New York, 1981.
2. R. H. Dennard, F. H. Gaensslen, H. N. Yu, V. L. Rideout, E. Bassous, and A. R. LeBlanc, "Design of ion implanted MOSFETs with very small physical dimensions," *IEEE J. Solid-State Circ.* **SC-9**(5), 256 (1974).
3. B. Davari, R. Dennard, and G. Shahidi, "CMOS scaling for high performance and low power—next ten years," *Proc. IEEE* **83**(4), 595, (1995).
4. J. Brews, W. Fichtner, E. Nicollian, and S. Sze, "Generalized guide for MOSFET miniaturization," *IEEE Electron Dev. Lett.* **EDL-1**, 2 (1980).
5. K. Ng, S. Eshraghi, and T. Stanik, "An improved generalized guide for MOSFET Scaling," *IEEE Trans. Electron Dev.* **ED-40**(10), 1895 (1993).
6. H. Hu, J. Jacobs, L. Su, and D. Antoniadis, "A study of deep-submicron MOSFET scaling based on experiment and simulation," *IEEE Trans. Electron Dev.* **ED-43**(4), 669 (1996).
7. C. Hu, "MOSFET scaling in the next decade and beyond," *Semicond. Int.* June, 105 (1994).
8. K. Ng and W. T. Lynch, "Analysis of the gate-voltage dependence of series resistance of MOSFETS," *IEEE Trans. Electron Dev.* **ED-33**, 965 (1986).
9. J. Brews, "Submicron MOSFET," in *High-Speed Semiconductor Devices*, S. M. Sze, Ed., Wiley, New York, 156 (1990).
10. R. H. Yan, A. Ourmazd, and K. F. Lee, "Scaling the Si MOSFET: from bulk to SOI TO bulk," *IEEE Trans. Electron Dev.* **ED-39**(7), 965 (1992).
11. C. Rafferty, H. Vuong, S. Eshraghi, M. Giles, M. Pinto, and S. Hillenius, "Explanation of reverse short channel effect by defect gradients," in *IEDM Tech. Dig.*, 809 (1994).
12. J. Koga, S. Takagi, and A. Toriumi, "A comprehensive study of MOSFET electron mobility in both weak and strong inversion regimes," in *IEDM Tech. Dig.*, 475 (1994).
13. B. Ricco, R. Versari, and D. Esseni, "Characterization of polysilicon-gate depletion in MOS structures", *IEEE Electron Dev. Lett.* **EDL-17**(3), 103 (1996).
14. K. Krisch, J. Bude, and L. Manchanda "Gate capacitance attenuation in MOS devices with thin gate dielectrics," *IEEE Electron. Dev. Lett.* **EDL-17**(11), 521 (1996).
15. C. Liu, Y. Ma, K. Cheung, L. Fritzinger, J. Becerro, H. Luftman, H. Vaidya, J. Colonell, A. Kamgar, J. Minor, R. Murray, W. Lai, C. Pai, and S. Hillenius, "25A gate oxide without boron penetration for $0.25\,\mu m$ and $0.3\,\mu m$ PMOS-FETS," in *VLSI Sym. Dig. Tech. Papers*, 18 (1996).
16. H. Momosa, M. Ono, T. Yoshitomi, T. Ohguro, S. Nakamura, M. Saito, and H. Iwai, "Tunneling gate oxide approach to ultra-high current drive in small-geometry MOSFETS," in *IEDM Tech. Dig.*, 593 (1994).
17. H. Momosa, M. Ono, T. Yoshitomi, T. Ohguro, S. Nakamura, M. Saito, and H. Iwai, "1.5 nm direct tunneling gate oxide Si MOSFETS," *IEEE Trans. Electron Dev.* **ED-43**(8), 1233 (1996).
18. B. Doyle, M. Bourcerie, J. Marchetaux, and A. Boudou, "Interface state creation and charge trapping in the medium to high gate voltage range (Vd/2 > Vg > vd)

during hot-carrier stressing of n-MOS transistors," *IEEE Trans. Electron Dev.* **ED-37**(3), 744 (1990).

19. L. Manchanda, R. Storz, R. Yan, K. Lee, and E. Westerweick, "Clear observation of sub-band gap impact ionization at room temperature and below in 0.1 μm Si MOSFETS," in *IEDM Tech. Dig.*, 994 (1992).

20. J. Bude, "Gate current by impact ionization feedback in sub-micron MOSFET technologies," in *VLSI Sym. Dig. Tech. Papers*, 101 (1995).

21. M. Koyanagi, A. Lewis, R. Martin, T. Huang, and J. Chen, "Hot-electron-induced punchthrough (HEIP) effect in submicrometer PMOSFET's," *IEEE Trans. Electron Dev.* **ED-34**(4), 839 (1987).

22. R. Woltjer, G. Paulzen, H. Pomp, H. Lifka, and P. Woerlee, "Three hot-carrier degradation mechanisms in deep submicron PMOSFETS," *IEEE Trans. Electron Dev.* **ED-42**(1), 109 (1995).

23. K. Schuegraf and C. Hu, "Reliability of thin SiO_2" *Semicond. Sci. Technol.* **9**, 969 (1994).

24. M. Terauchi, A. Yoshimi, Murakoshi, and Y. Ushiku, "Suppression of the floating-body effects in SOI MOSFETs by bandgap engineering," in *VLSI Sym. Dig. Tech. Papers*, 35 (1995).

25. J. Colinge, Gao, A. Ramano-Rodriguez, H. Maes, and C. Claes, "Silicon-on insulator gate-all-around devices," in *IEDM Tech. Dig.*, 595 (1990).

26. I Yang, C. Vieri, A. Chnadrakasan, and D. Antoniadis, "Back gated CMOS on SOIAS for dynamic threshold voltage control," in *IEDM Tech. Dig.*, 877 (1995).

27. G. Bronner, H. Aochi, M. Gall, J. Gambino, S. Gernhardt, E. Hammer, H. Ho, J. Iba, H. Ishiuchi, M. Jaso, R. Kleinhenz, T. Mii, M. Narita, L. Nesbit, W. Neumueller, A. Nitayama, T. Ohiwa, S. Parke, J. Ryan, T. Sato, H. Takato, and S. Yoshikawa, "A fully planarized 0.25 μm CMOS Technology for 256 Mbit DRAM and beyond," in *VLSI Sym. Dig. Tech. Papers*, 15 (1995).

28. Y. Ohji Y. Matsui, T. Itoga, M. Hirayama, Y. Sugawara, K. Torii, H. Miki, M. Nakata, I. Asano, S. Iijima, and Y. Kawamoto, "TaO_5 capacitors, dielectric material for giga-bit DRAMS," in *IEDM Tech. Dig.*, 111 (1995).

29. T. McNelly, J. Hayden, A. Perera, J. Pfister, C. Subramanian, M. Blackwell, B. James, S. Ajuria, W. Feil, Y. Ku, T. Lii, J. Lin, F. Nkansah, C. Philbin, C. Sun, M. Thompson, and M. Woo, "High performance 0.25 μm SRAM technology with tungsten interpoly plug," in *IEDM Tech. Dig.*, 927 (1995).

30. K. Yano, T. Ishii, T. Hashimoto, T. Kobayashi, F. Murai, and K. Seki, "Room temperature single electron memory," *IEEE Trans. Electron Dev.* **ED-41**(9), 1628 (1994).

31. C, Hu, D. Kencke, S. Banerjee, R. Richart, B. Bandyopadhyay, B. Moore, E. Ibok, and S. Garg, "Substrate-current-induced hot electron (SCIHE) injection: a new convergence scheme for flash memory," in *IEDM Tech. Dig.*, 283 (1995).

32. J. Bude, A. Frommer, M. Pinto, and G. Weber, "EEPROM/flash sub 3.0V drain-source bias hot carrier writing," in *IEDM Tech. Dig.*, 990 (1995).

33. Z. Chen, J. Burr, J. Shott, and J. Plummer, "Optimization of quarter micron MOSFETS for low voltage/low applications," in *IEDM Tech. Dig.*, 63 (1995).

34. S. Verdonckt-Vanderbroek, S. Wong, and J. Woo, "High-gain lateral bipolar action in a MOSFET structure," *IEEE Trans. Electron Dev.* **ED-38**(11), 2487 (1991).

35. F. Asserderaghi, D. Sinitsky, S. Parke, J. Bokor, P. Ko, and C. Hu, "A dynamic threshold voltage MOSFET (DTMOS) for ultra-low voltage operation," in *IEDM Tech. Dig.*, 809 (1994).

36. L. Vandamme, X. Li, and D. Rigaud, "1/f noise in MOS devices, mobility or number fluctuations?" *IEEE Trans. Electron Dev.* **ED-41**(11), 1937 (1994).

37. A. Dodabalapur, H. E. Katz, L. Torsi, and R. C. Haddon, "Organic field-effect transistors," *Science* **269**, 1560 (1995).

38. L. Torsi, A. Dodabalapur, and H. E. Katz "An Analytic model for short channel organic thin-film transistors" *J. Appl. Phys.* **78**(2), 1088 (1995).

39. M. Mastrapasqua, C. King, P. Smith, and M. Pinto, "Functional devices based on real space transfer in Si/SiGe structure", *IEEE Trans. Electron Dev.* **ED-43**(10), 1671 (1996).

4 Power Devices

B. JAYANT BALIGA

Power Semiconductor Research Center, North Carolina
State University, Raleigh

4.1 INTRODUCTION

Among the power-electronics community, power-semiconductor devices are regarded as critical components whose characteristics dictate improvements in the performance of systems. The transition to semiconductor-device-based power electronics occurred with the introduction of the power bipolar transistor and thyristor in the 1950s. Improvements in the power-handling capability and switching speed of these devices over the next two decades were crucial to the reduction in the cost and size of power electronic systems, leading to an increase in the number of applications. However, these current-controlled bipolar devices required significant input power. The complexity of the control circuits, which must be implemented using discrete components, hindered further reductions in the cost and size of systems.

With the advent of MOS technology for CMOS integrated circuits, a new class of power devices became possible in the 1970s—the power MOSFETs.[1] Since the MOSFET is a voltage-controlled device that can be gated with insignificant steady-state input current, it is possible to integrate its control circuit, leading to a large reduction in the size and complexity of the power electronic systems. In addition, the MOSFET can be switched at a much faster rate than the bipolar transistor because it operates in the unipolar mode, which eliminates the delays associated with stored charges within bipolar devices. The higher switching speed of the MOSFET allowed major improvements in the performance of power electronic systems, such as switch-mode power supplies for computers. The power MOSFET was initially anticipated to replace the bipolar transistor in all its applications. However,

Modern Semiconductor Device Physics, Edited by S. M. Sze.
ISBN 0-471-15237-4 © 1998 John Wiley & Sons, Inc.

the on-state voltage drop of the power MOSFET was significantly larger than that of the bipolar transistor when designed to operate at above 300 V, resulting in higher power losses in applications. Consequently, the power MOSFET has successfully replaced the bipolar transistor only in systems operating at voltages below 200 V.

In the 1980s, a new class of power devices called MOS–bipolar structures were created to provide power switches suitable for high-voltage applications such as motor control.[1] The insulated-gate bipolar transistor (IGBT) first reported in 1982 has now replaced the bipolar transistor in all high voltage applications.[2] In the IGBT, a high input impedance is achieved by an MOS-gate structure, and low on-state voltage drop is achieved by bipolar current conduction. Its current saturation capability and excellent safe operating area make the IGBT well suited for compact smart power applications. Although IGBTs with voltage ratings of 4.5 kV are now under development to replace the gate turn-off (GTO) thyristor in high-power motor controls for traction (electric street-cars and locomotives), MOS-gated thyristors are alternate candidate devices that offer a lower on-state voltage drop.

Recently, there has been considerable interest in the development of power devices based on silicon carbide. This interest has been created by a fundamental analysis[3] that projected a 200-times reduction in the specific on-resistance of the drift region when silicon carbide replaces silicon. On this basis, it has been theoretically demonstrated that unipolar power rectifiers and switches can be developed in silicon carbide with breakdown voltages up to 5 kV, which possess superior on-state characteristics when compared with even bipolar silicon devices. In addition, these devices would offer superior switching behavior and higher operating-temperature capability. High-performance Schottky power rectifiers with breakdown voltages up to 1000 V have been successfully fabricated by using an edge termination that produces nearly ideal breakdown voltage.[4] SiC power-switch structures are now under development that could replace silicon devices in the twenty-first century.

This chapter discusses the physics and electrical characteristics of power devices that have gained prominence in the last decade. To keep pace with the rapid increase in the operating frequency of power switches, there have been advances in power rectifiers that are discussed in the next section. This is followed by sections on the power MOSFET and IGBT. The last two sections deal with MOS-gated thyristors and silicon carbide power devices. A more detailed treatment of power devices, including the crucially important edge termination design, can be found in a comprehensive textbook *Power Semiconductor Devices* by Baliga.[5]

4.2 POWER RECTIFIERS

In this section, recent advances in power rectifiers will be discussed. For low (<100 V) operating voltages, the Schottky-barrier rectifier has been the device of choice for power-electronics applications because of its relatively low on-state voltage drop and high switching speed. It is extensively used in switch-mode power supplies. Recently, two structural innovations—the junction-barrier-controlled Schottky (JBS) rectifier and the trench-MOS-barrier Schottky (TMBS) rectifier—have enabled significant improvements in its performance. For high (>200 V) operating voltages, the p–i–n rectifier has been the device of choice. However, due to its poor reverse recovery switching behavior, several alternate structures have been proposed to reduce the stored charge during the on-state. This section provides an overview of these developments.

4.2.1 Schottky-Barrier Rectifier

The basic physics of the metal–semiconductor (or Schottky) contact has been discussed in many textbooks[5,6] and reviewed in Chapter 2. The structure of the Schottky power rectifier consists of a lightly doped drift region in series with the Schottky contact, as shown in Fig. 1. The n-drift region is designed to support the required reverse blocking voltage. If ideal breakdown is assumed at the edges of the device, the doping concentration (N_D) of the n-drift region can be related to the desired breakdown voltage (V_B) by[6]

$$N_D = 2 \times 10^{18} V_B^{-4/3} \quad (\text{cm}^{-3}) \tag{1}$$

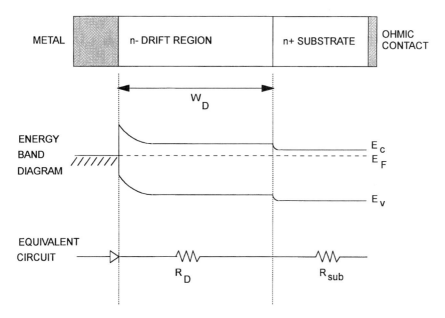

Fig. 1 Power Schottky-barrier rectifier structure.

where N_D is in cm^{-3} and V_B is in volts. The thickness of the drift region (i.e., the depletion region at breakdown) is given by

$$W_D = 2.67 \times 10^{10} N_D^{-7/8} \quad \text{(cm)} \tag{2}$$

Using these relationships, the series resistance per unit area (called the specific on-resistance) of the drift region is given by the product of the resistivity and W_D:

$$R_{\text{on,sp}} = \frac{W_D}{q \mu_n N_D} = 5.93 \times 10^{-9} V_B^{2.5} \quad (\Omega\text{-cm}^2) \tag{3}$$

For a Schottky power rectifier with a reverse breakdown voltage of 50 V, the drift region has a specific on-resistance of about 1×10^{-4} Ω-cm^2, leading to a voltage drop of only 10 mV across this region at a current density of 100 A/cm^2. If the other series resistance components shown in Fig. 1 (contact and substrate) are neglected, then the voltage drop across the Schottky power diode is determined by the barrier height of the metal–semiconductor contact. The on-state voltage drop is then given by thermionic emission theory (for negligible series resistance):[6]

$$V_F = \phi_B + \frac{kT}{q} \ln\left(\frac{J_F}{A^* T^2}\right) \tag{4}$$

where ϕ_B is the Schottky-barrier height, k is Boltzmann's constant, T is the absolute temperature, J_F is the on-state current density, and A^* is Richardson's constant. For a Schottky barrier height of 0.8 V, the on-state voltage drop is found to be 0.5 V at a current density of 100 A/cm^2. However, if the breakdown voltage is increased, the series resistance due to the drift region increases very rapidly in accordance with Eq. 3, leading to higher on-state voltage drop ($V_F + J_F R_{\text{on,sp}}$). This can be seen in Fig. 2 for the case of rectifiers with breakdown voltages above 200 V. Although the on-state voltage drop can be reduced by reducing the Schottky-barrier height, this leads to a severe increase in the reverse leakage current.

At small reverse-bias voltages, the leakage current of the Schottky diode is given by the saturation current of the contact. However, in the case of the high voltages that power devices must support, Schottky-barrier lowering must be accounted for. When the image force analysis[5,6] of Schottky-barrier lowering is taken into account, the reverse leakage current density is given by:

$$J_R(V_R) = -A^* T^2 e^{-q(\phi_B - \Delta\phi_B)/kT} \tag{5}$$

with

$$\Delta\phi_B = \sqrt{\frac{q \mathscr{E}_m}{4\pi\varepsilon_s}} \tag{6}$$

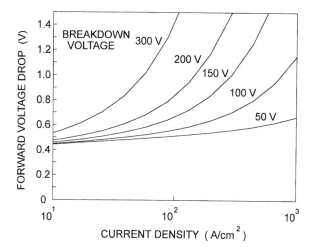

Fig. 2 On-state characteristics of silicon Schottky rectifiers with breakdown voltages ranging from 50 to 300 V.

where the maximum electric field ($\mathscr{E}_{\mathrm{m}}$) is related to the applied reverse-bias by

$$\mathscr{E}_{\mathrm{m}} = \sqrt{\frac{2qN_{\mathrm{D}}}{\varepsilon_{\mathrm{s}}}(V_{\mathrm{R}} + V_{\mathrm{bi}})} \qquad (7)$$

The calculated leakage currents of a Schottky rectifier with and without the Schottky-barrier-lowering effect are compared in Fig. 3. Note the considerable increase in the reverse leakage current resulting from the Schottky-barrier lowering at high reverse voltages—in fact, the actual leakage current of Schottky rectifiers is even worse than that predicted by the Schottky-barrier lowering.

In order to account for this larger reverse leakage current at high reverse-bias voltages, it is necessary to take into account the pre-avalanche multiplication of carriers at the high electric fields when the applied reverse-bias approaches close to the breakdown voltage.[7] The optimization of the characteristics of the Schottky power rectifier requires a trade off between forward voltage drop and reverse leakage current. As the Schottky-barrier height (ϕ_{B}) is reduced, the forward voltage drop decreases but the leakage current increases. A tradeoff curve that can be useful during device design is the relationship between the forward voltage drop and the reverse leakage current:

$$J_{\mathrm{R}} = J_{\mathrm{F}}e^{-(qV_{\mathrm{F}}/kT)} \qquad (8)$$

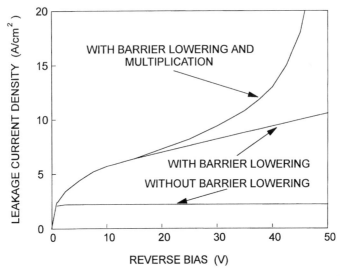

Fig. 3   Reverse blocking characteristics of a silicon Schottky rectifier.

The calculated tradeoff curves for Schottky rectifiers at several ambient temperatures are provided in Fig. 4. Note that the tradeoff curves are not dependent on the semiconductor material used for device fabrication (except for small changes in the Richardson constant A^*).

The ultimate limiting factor that determines the choice of the Schottky-barrier height is the power dissipation in the rectifier. The power dissipation during forward conduction depends upon the forward conduction current, forward voltage drop, and duty cycle. The power dissipation during reverse blocking depends upon the leakage current, the reverse bias voltage, and the duty cycle. In choosing the Schottky-barrier height, it is important to calculate the total power dissipation as a function of temperature using

$$P_D = J_F V_F \frac{t_{on}}{T} + J_R V_R \frac{T - t_{on}}{T} \tag{9}$$

where t_{on} is the time period during which the diode is in its on-state, T is the total period, J_F and J_R are the forward (on-state) and reverse leakage current densities, respectively, and V_F and V_R are the forward (on-state) and reverse-bias voltages, respectively. The switching losses have been neglected in this equation. Figure 5 shows the calculated power dissipation as a function of temperature for four Schottky-barrier heights. These curves were calculated using a forward current density of 100 A/cm^2, a reverse blocking voltage of 20 V and a duty cycle of 0.5. The figure shows that the total power dissipation can be reduced by lowering the Schottky-barrier height but this

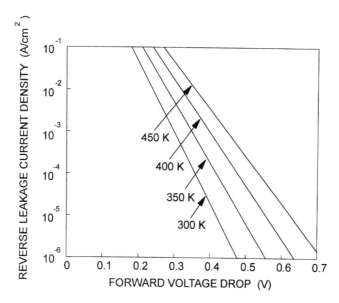

Fig. 4 The tradeoff curve between on-state voltage drop and reverse leakage current for low breakdown voltage silicon Schottky rectifiers.

is accompanied by a reduction in the maximum operating temperature as indicated by the arrows. Although the arrows are shown at the minimum power dissipation temperature, the actual maximum operating temperature is slightly higher and is determined by the thermal resistance of the heatsink.

4.2.2 Junction-Barrier-Controlled Schottky Rectifier

The junction-barrier-controlled Schottky (JBS) rectifier is a Schottky-rectifier structure with a p–n junction grid integrated into its drift region.[8–10] A cross-section of the device structure is shown in Fig. 6. The junction grid is designed so that its depletion layers do not pinch off under zero bias conditions. With this design, the device contains conductive channels under the Schottky barrier through which current can flow during forward-biased operation. When a positive bias is applied to the n^+ substrate, the p–n junctions and the Schottky barrier become reverse-biased. The depletion layers formed at the p–n junctions spread into the channel. In the JBS rectifier, the junction grid is designed so that the depletion layers will intersect under the Schottky barrier when the reverse bias exceeds a few volts. After depletion-layer pinch-off, a potential barrier is formed in the channel. Once the potential barrier is formed, any increase in applied voltage is supported by it with the depletion layer extending toward the n^+ substrate. Therefore, the potential barrier shields the Schottky barrier from the applied voltage.

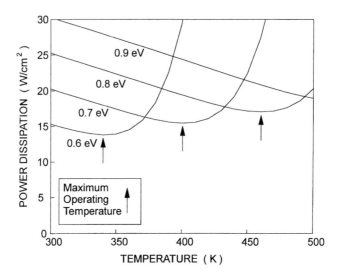

Fig. 5 Power dissipation in silicon Schottky rectifiers as a function of temperature.

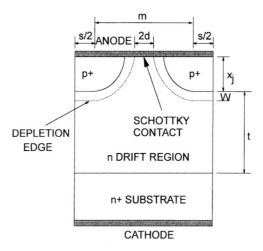

Fig. 6 Cross-section of the JBS rectifier structure.

This shielding prevents the Schottky barrier-lowering phenomenon and eliminates the large increase in leakage current observed for conventional Schottky rectifiers. Once the pinch-off condition is established, the leakage current remains relatively constant with increasing reverse-bias up to the point of avalanche breakdown.

The forward conduction characteristics of the JBS rectifier can be analyzed using the same methods used for the Schottky rectifier by allowing for the increase in the series resistance of the drift region due to current constriction in the channel structure. Consider the case of a p–n junction grid with stripe geometry and cross-sectional dimensions defined in Fig. 6. In this structure, the junction grid is formed by planar diffusion through a diffusion window of width s with a masked region of width m. The lateral diffusion of the junction is assumed to be 85% of the vertical depth (x_j). The current density across the Schottky barrier (J_{FS}) is

$$J_{FS} = \frac{m+s}{2d} J_{FC} \tag{10}$$

where J_{FC} is the cell current density obtained by dividing the total JBS rectifier current by the cell area. Using this expression

$$V_{FS} = \phi_B + \frac{kT}{q} \ln\left(\frac{m+s}{2d} \frac{J_{FC}}{A^*T^2}\right) \tag{11}$$

In addition to the voltage drop across the Schottky barrier, the voltage drop across the drift region must also be included. On the basis of current spreading from a cross-section at the top with a width of $2d$ to a cross-section at the bottom with a width of $m+s$, the drift region resistance is[9]

$$R_D = \rho \frac{(x_j+t)(m+s)}{m+s-2d} \ln\left(\frac{m+s}{2d}\right) \tag{12}$$

where ρ and t are the resistivity and thickness of the drift region required to obtain the desired breakdown voltage. Combining the voltage drop across the Schottky-barrier and the drift region, the forward voltage drop of the JBS rectifier is obtained:

$$V_{FS} = \phi_B + \frac{kT}{q} \ln\left(\frac{m+s}{2d} \frac{J_{FC}}{AT^2}\right) + \rho \frac{(x_j+t)(m+s)}{(m+s-2d)} \ln\left(\frac{m+s}{2d}\right) J_{FC} \tag{13}$$

As in the Schottky rectifier, the reverse leakage current of the JBS rectifier is determined by thermionic emission, including the effect of barrier lowering:

$$J_L = \left(\frac{2d}{m+s}\right) A^* T^2 \exp\left[-\left(\frac{q\phi_B}{kT}\right)\right] \exp\left(\frac{q}{kT}\sqrt{\frac{q\mathscr{E}}{4\pi\varepsilon_s}}\right) \tag{14}$$

where $\mathscr{E}$ is the electric field at the metal–semiconductor interface. An important feature of the JBS rectifier is that the electric field at the Schottky barrier remains approximately constant (independent of the reverse bias) once the channel potential barrier forms. Its value then corresponds to the voltage at which the channel pinch-off occurs:

$$\mathscr{E} = \sqrt{\frac{2qN_D}{\varepsilon_s}(V_P + V_{bi})} \tag{15}$$

where V_P is the channel pinch-off voltage given by[9]

$$V_P = \frac{qN_D}{8\varepsilon_s}(m - 1.7x_j)^2 - V_{bi} \tag{16}$$

To suppress the barrier-height-lowering phenomenon further, a device structure with the p–n junction formed on the sidewalls of a trench etched in the top surface has been explored.[11] The vertical sidewalls of the trench produce a superior barrier in the channel between the p regions, as demonstrated in JFETs,[12] leading to a lower leakage current.

The lowest on-state voltage drop can be obtained by making the width (s) of the junction diffusion window as small as possible because this minimizes the dead space below the junctions where the current does not flow. The best JBS-rectifier characteristics can, therefore, be expected when submicron lithography is used to pattern the diffusion windows for the p^+ diffusions. Experimental results have been obtained on JBS rectifiers capable of supporting 30 V in the reverse direction, fabricated using 0.5 μm diffusion windows.[13] A much better tradeoff curve between the on-state voltage drop and the reverse leakage current was obtained in this case when compared with 2 μm diffusion windows. The JBS rectifier fabricated using 0.5 μm design rules has been shown to have superior characteristics to the device with the trench structure because its leakage current is lower by a factor of 45 for the same on-state voltage drop of 0.3 V. On the basis of this reduction in leakage current, the maximum operating temperature of the JBS rectifier can be extended from 100°C, for the conventional Schottky rectifier to 175°C while reducing the total power dissipation (P_D) by a factor of 2.

4.2.3   Trench-MOS-Barrier Schottky Rectifier

The trench-MOS-barrier Schottky (TMBS) rectifier is a Schottky rectifier containing a trench region with an oxide layer on its sidewall and bottom surfaces. A cross-section of its structure is shown in Fig. 7. The Schottky metal is deposited on the sidewalls and the top of the trench surface. The mesa region between the trench sidewalls has a doping concentration and width (m) chosen to produce a redistribution of the electric field under the metal–semiconductor contact. This redistribution occurs due to the coupling of the charge in the mesa region with not only the Schottky metal on the top surface but also with the metal on the sidewalls. The depletion layer then extends from the top surface (as in the conventional Schottky rectifier) and the sidewalls of the trench. Using two-dimensional numerical simulations,

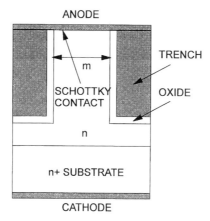

Fig. 7   Cross-section of the TMBS rectifier structure.

it has been found[14] that if the charge in the mesa region (product of doping concentration and thickness) is about 5×10^{12}/cm^2, the electric field no longer peaks at the surface below the metal but takes the form shown in Fig. 8. In this figure, the well-known triangular electric-field profile observed in the conventional Schottky rectifier is also shown for comparison. To obtain a reverse blocking voltage of 30 V, the doping concentration for the conventional Schottky rectifier was chosen as 3×10^{16}/cm^{-3}. In contrast, the doping concentration in the mesa region for the TMBS rectifier can be much greater $(1 \times 10^{17}$/cm$^{-3})$. For this high doping concentration, the breakdown voltage for a parallel-plane junction is calculated to be only 9.5 V. However, it is possible to sustain 27 V across the TMBS rectifier structure because of the change in the electric-field distribution, which produces a high electric-field zone near the bottom of the trench. The high doping concentration in the TMBS rectifier reduces the series resistance during current flow in the on-state, allowing the device to operate at a low on-state voltage drop.

Another important observation is that the electric field under the Schottky contact is much smaller in the TMBS rectifier compared with that in the conventional Schottky rectifier. For a trench depth of over 1 μm, more than two-fold reduction in electric field is observed. This greatly reduces the Schottky-barrier-lowering phenomenon because the increase in leakage current is exponentially dependent upon the electric field at the metal–semiconductor interface. The voltage supported across the TMBS rectifier can be calculated by integration of the electric-field distribution shown in Fig. 8. A trench depth of 2 μm is optimum for obtaining a low series resistance within the TMBS structure, while obtaining a breakdown voltage of 30 V.

The greatly superior electric-field profiles in the TMBS rectifier, compared with the conventional Schottky and the JBS rectifier, mean that the trade-off curve between on-state voltage drop and reverse leakage current is greatly

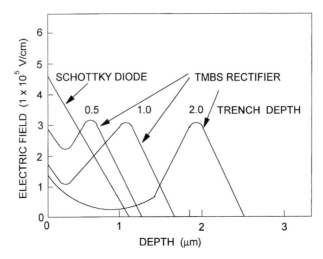

Fig. 8 Electric-field distribution in the TMBS rectifier structure.

improved. Figure 9 shows a comparison between the performance of the TMBS rectifier and the JBS rectifier fabricated with submicron technology. For identical operating conditions (on-state current density 60 A/cm², reverse bias 10 V), the leakage current for the TMBS rectifier is three orders of magnitude smaller than for the JBS rectifier. The lower power dissipation permits smaller heatsinks in applications such as switch-mode power supplies.

4.2.4 *p–i–n* Rectifier

The *p–i–n* rectifier is widely used in high-voltage power circuits. In contrast with the Schottky rectifier, the *n*-drift region (which is referred to as the intrinsic region or *i* region) in the *p–i–n* rectifier is flooded with minority carriers during forward conduction. Consequently, the resistance of the *i* region becomes very small during current flow allowing these diodes to carry a high current density during forward conduction. However, the high concentration of minority carriers in the *i* region creates problems when the *p–i–n* rectifier is switched to the reverse blocking mode because this stored charge must be removed before the device can support voltage.

Consider the cross-section of a *p–i–n* rectifier, shown in Fig. 10, with an *n*-drift region. To support large reverse blocking voltages, it is necessary to use a low doping concentration (N_B) and a large thickness ($2d$) for the drift region. During on-state current flow, as the current density increases, the injected carrier density increases and ultimately exceeds the relatively low background doping of the *n*-drift region. This condition is called high- level injection. When the injected hole density becomes much greater than the

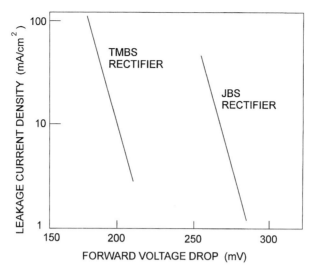

Fig. 9 Comparison of the tradeoff curve between on-state voltage drop and leakage current for the TMBS and JBS rectifiers.

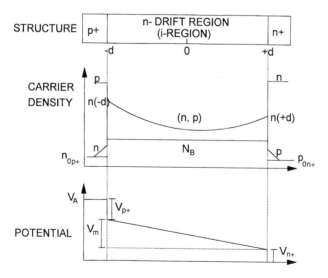

Fig. 10 Carrier and potential distributions within the *p–i–n* rectifier in the on-state.

background doping, charge neutrality in the *n*-drift region requires that the concentrations of holes and electrons become equal:

$$n(x) = p(x) \qquad (17)$$

These concentrations can become far greater than the background doping level resulting in a large decrease in the resistance of the *i* region. This phenomenon, called conductivity modulation, is an extremely important effect that allows transport of a high current density through the *p–i–n* rectifier with low on-state voltage drop.

Under steady-state conditions, current flow in the *p–i–n* rectifier can be accounted for by the recombination of holes and electrons in the *n* base region and the anode/cathode end regions. First, consider the case where the recombination in the end regions is negligible, that is, the end regions have unity injection efficiency. The current density is then determined by recombination in the *n*-drift region:

$$J = \int_{-d}^{+d} qR \, dx \tag{18}$$

where R is the recombination rate given by

$$R = \frac{n(x)}{\tau_{HL}} \tag{19}$$

If the high-level lifetime (τ_{HL}) is assumed to be independent of the carrier density, then

$$J = \frac{2qn_a d}{\tau_{HL}} \tag{20}$$

where n_a is the average carrier density and d is half the *n*-drift-region width. From this equation, we see that the free-carrier density in the drift region increases in proportion to the current density. Since this increase in carrier density results in a proportional increase in the conductivity of the drift region, it can be concluded that the voltage drop in the drift region will be independent of the current density. This is important for maintaining a low on-state voltage drop even at high operating current densities in *p–i–n* rectifiers.

To solve for the actual free-carrier distribution in the drift region, we must use the continuity equation

$$\frac{dn}{dt} = 0 = -\frac{n}{\tau_{HL}} + D_a \frac{d^2 n}{dx^2} \tag{21}$$

where D_a is the ambipolar diffusion coefficient. Defining an ambipolar diffusion length as

$$L_a = \sqrt{D_a \tau_{HL}} \tag{22}$$

the preceding equation can be written as

$$\frac{d^2 n}{dx^2} - \frac{n}{L_a^2} = 0 \tag{23}$$

The boundary conditions needed to solve this equation are obtained by the current transport occurring at the p^+ and n^+ ends of the diode. At the n^+ boundary, the current transport is due to electrons and the hole current goes to zero. Thus

$$J = J_n(+d) = 2qD_n \left(\frac{dn}{dx}\right)_{x=+d} \tag{24}$$

Analysis of the boundary condition at the opposite end of the device gives

$$J = J_p(-d) = -2qD_p \left(\frac{dn}{dx}\right)_{x=-d} \tag{25}$$

The solution of Eq. 23 with these boundary conditions is given by[5]

$$n = p = \frac{\tau_{HL}J}{2qL_a} \left(\frac{\cosh(x/L_a)}{\sinh(d/L_a)} - \frac{\sinh(x/L_a)}{2\cosh(d/L_a)}\right) \tag{26}$$

This catenary carrier distribution is illustrated in Fig. 10. The hole and electron concentrations are the highest at the p^+–$n(-d)$ and n–$n^+(+d)$ junctions with the minimum closer to the cathode side because of the difference in the mobility of electrons and holes. The extent of the drop in the carrier concentration away from the junctions is dependent upon the ambipolar diffusion length. At medium current densities, this diffusion length is controlled by the high-level lifetime.

To determine the voltage drop across the rectifier, we must first obtain the electric-field distribution. The voltage drop across the i region (V_m) can then be obtained by integrating the electric-field distribution. The following approximations can be used to calculate the voltage drop in the middle region:[5]

$$V_m = \frac{3kT}{q} \left(\frac{d}{L_a}\right)^2 \qquad \text{for } d \leq L_a \tag{27}$$

and

$$V_m = \frac{3\pi kT}{q} e^{d/L_a} \qquad \text{for } d \geq L_a \tag{28}$$

It is important to note that the voltage drop in the middle region is independent of the current density because the free-carrier concentration increases in proportion to the current density. The voltage drop across the end regions is given by

$$V_{p+} + V_{n+} = \frac{kT}{q} \ln\left(\frac{n(+d)n(-d)}{n_i^2}\right) \tag{29}$$

Combining Eq. 29 with Eq. 26, the current density of a forward-biased diode at high-injection levels, in the absence of recombination in the end regions, is given by

$$J = \frac{2qD_a n_i}{d} F\left(\frac{d}{L_a}\right) e^{qV_a/(2kT)} \tag{30}$$

where

$$F\left(\frac{d}{L_a}\right) = \frac{(d/L_a)\tanh(d/L_a)}{\sqrt{1 - 0.25\tanh^4(d/L_a)}} e^{-qV_m/(2kT)} \tag{31}$$

From Eq. 30, we can conclude that a low forward drop occurs when the function F is large, which occurs at (d/L_a) values close to unity.

A major limitation to the performance of $p\text{--}i\text{--}n$ rectifiers at high frequencies is the power loss during switching from the on-state to the off-state. This process of switching from the on-state to the off-state is called reverse recovery. As illustrated in Fig. 11, a large reverse transient current occurs in $p\text{--}i\text{--}n$ rectifiers during reverse recovery. Since the voltage across the rectifier is also large during the second portion of the recovery following the peak in the reverse current, the rectifier dissipates a large amount of power. In addition, the peak reverse current adds to the average current flowing through the transistors that control the current flow in the power circuit. This not only produces an increase in the power dissipation in the transistors but creates a high internal stress that can cause second-breakdown induced failure. Consider the linearized turn-off current waveform shown in Fig. 11 with a constant di/dt during turn-off from an initial current density J_F. The reverse-recovery time t_{rr} can be determined by relating the charge removed during the reverse-recovery process to the initial stored charge within the middle region. Since the charge removed during the reverse-recovery process is the area under the reverse-recovery current waveform and the initial stored charge is the product of the average carrier concentration n^* in the middle region and its thickness $(2d)$:

$$\tfrac{1}{2}J_{PR}t_{rr} = Q_s = qn^*2d = J_F\tau_{HL} \tag{32}$$

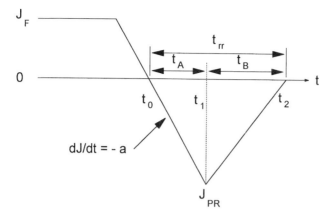

Fig. 11   Reverse-recovery waveform for the *p–i–n* rectifier.

Using this equation

$$t_{rr} = 2\tau_{HL}\frac{J_F}{J_{PR}} \tag{33}$$

From this expression, we can conclude that a smaller reverse-recovery time can be obtained by reducing the high-level lifetime. Lifetime can be reduced by introducing recombination centers into the *i* region. Among the many possible approaches, the diffusion of gold and platinum, and the use of high-energy electron irradiation have been most carefully studied.[15–17]

4.2.5   Merged *p–i–n*/Schottky Rectifier

The current flow in high-voltage *p–i–n* rectifiers during the reverse-recovery transient is a significant source of power loss in power-electronic circuits. The reverse-recovery current can be reduced by the use of lifetime control methods. However, this produces an increase in the on-state voltage drop, resulting in a tradeoff between on-state losses and switching losses. The merged *p–i–n* Schottky (MPS) rectifier is an alternate approach to reducing the switching losses in high-voltage power rectifiers without increasing the on-state voltage drop.[18] The device structure, shown in Fig. 12, is similar to that of the JBS rectifier. However, its physical operation is quite different. In the JBS rectifier, the *p–n* junction is used exclusively to reduce the leakage current by preventing Schottky-barrier lowering during reverse blocking. This feature is also used in the MPS rectifier to obtain a high breakdown voltage in spite of the presence of the Schottky region. However, in the MPS rectifier, the *p–n* junction becomes forward biased in the on-state, unlike in the JBS rectifier, since the drift region has a very high resistance because it is designed

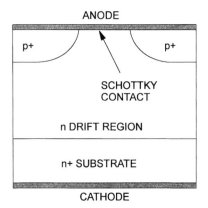

Fig. 12   Cross-section of the MPS rectifier structure.

to support high voltages during reverse blocking. The forward bias on the *p–n* junction produces the injection of holes into the *n*-drift region. This results in conductivity modulation of the drift region similar to that in the *p–i–n* rectifier, which drastically reduces its resistance to current flow and allows large current flow through the Schottky region. The injection level required to reduce the resistance in series with the Schottky region is not as large as that observed in the *p–i–n* rectifier. Consequently, the stored charge in the MPS rectifier is much smaller than that in the *p–i–n* rectifier.

The on-state characteristics of the MPS rectifier have been analyzed by two-dimensional numerical simulation and compared with those for the *p–i–n* rectifier and Schottky rectifier with the same high voltage drift region parameters designed to support 900 V. These characteristics are compared in Fig. 13. Although the current flow in the Schottky rectifier begins at relatively low forward-bias voltages across the barrier, the current becomes limited by the high series resistance of the unmodulated drift region. This results in a low on-state current density with high forward voltage drop for the Schottky rectifier. In contrast, for the *p–i–n* rectifier, very little current flow occurs at forward-bias voltages below 0.5 V. However, at larger forward-bias voltages, the *p–n* junction begins to inject holes into the drift region and reduces its series resistance. This results in a low on-state voltage drop of about 1 V at high on-state current densities. Under typical on-state conditions, the injected carrier concentration is about $1 \times 10^{17}/\text{cm}^3$, which results in a high stored charge in the drift region.

In the case of the MPS rectifier, current flow begins at low forward-bias voltages as in Schottky rectifiers. At on-state biases below 0.5 V, the current flow occurs primarily across the Schottky region. Since the area of the Schottky region is about half the total cell area, the on-state current density in the MPS rectifier is about half that for the Schottky rectifier at these low

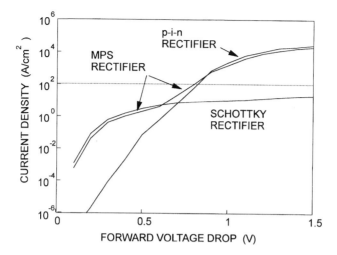

Fig. 13   On-state characteristics of the MPS, *p–i–n*, and Schottky rectifiers.

forward-bias values. However, when the forward bias is increased to above 0.6 V, the *p–n* junction in the MPS rectifier begins to inject holes into the drift region. This produces conductivity modulation of the drift region, resulting in a low resistance in series with the Schottky region. As a consequence, the on-state current density in the MPS rectifier becomes even larger than that for the *p–i–n* rectifier for forward bias voltages between 0.6 and 0.9 V. For a typical on-state current density of 100 A/cm^2, indicated by the dotted line, the on-state voltage drop of the MPS rectifier is even lower than that for the *p–i–n* rectifier. The on-state losses for the MPS rectifier are, therefore, even lower than those for the *p–i–n* rectifier. Although it is interesting to note that the on-state voltage drop of the MPS rectifier is the lowest among all the high-breakdown-voltage rectifiers, the difference between the MPS rectifier and the *p–i–n* rectifier is less than 0.1 V.

Figure 13 shows that, when the forward bias voltage is increased beyond 0.9 V, the curves for the MPS and *p–i–n* rectifiers cross each other. This takes place because at very high current densities typical of surge conditions, the current flow occurs primarily via the *p–n* junction within the MPS rectifier structure. Since the area of the *p–n* junction in the MPS rectifier is about half that for the *p–i–n* rectifier, the current density within the MPS rectifier also becomes half that for the *p–i–n* rectifier. The relatively small difference in the on-state voltage drop means that the surge-current handling capability of the MPS rectifier is essentially the same as that for the *p–i–n* rectifier.

The stored charge in the MPS rectifier can be shown to be smaller than in the *p–i–n* rectifier by solving the continuity equation using the boundary conditions for the MPS rectifier in the Schottky region. This region acts as an interface with infinite recombination rate making the minority carrier concentration go to zero in this region. On this basis, the boundary condition at $x = -d$ for the Schottky portion of the MPS rectifier can be written as

$$p(-d) = 0 \qquad (34)$$

On the basis of high-level injection conditions with charge neutrality applied to the boundary at $x = +d$,

$$\left(\frac{dp}{dx}\right)_{x=+d} = \left(\frac{dn}{dx}\right)_{x=+d} = \frac{J_F}{2qD_n} \qquad (35)$$

where J_F is the on-state current density, and d is half the thickness of the n^- drift region. The solution for the carrier distribution is[5]

$$p(x) = \frac{J_F L_a}{2qD_n} \frac{\sinh[(x + d)/L_a]}{\cosh(2d/L_a)} \qquad (36)$$

This carrier profile is compared with that within a p–i–n rectifier in Fig. 14. In this figure, the carrier profile is shown for the MPS rectifier along a line through the Schottky region and the p–n junction region. Equation 36 describes the profile through the Schottky region. It can be seen that the carrier concentration at the p–n junction interface is also smaller than that for the p–i–n rectifier. From these carrier profiles, we can conclude that the stored charge in the MPS rectifier is substantially smaller than that in the p–i–n rectifier. An estimate for the stored charge in the MPS rectifier can be obtained by integration of the carrier profile given by Eq. 36:

$$Q_s = q \int_{-d}^{+d} p(x)\,dx = \frac{J_F L_a^2}{2D_n}\left[1 - \frac{1}{\cosh(2d/L_a)}\right] \qquad (37)$$

The stored charge in the Schottky region predicted by this equation is approximately one-fifth that of the p–i–n rectifier operating at the same on-state current density. The reverse recovery behavior of the MPS rectifiers has been observed to be significantly superior to those for p–i–n rectifiers because of the reduced stored charge.

4.2.6 Static Shielding Diode

The static shielding diode (SSD) is a p–i–n rectifier structure containing two p-type regions with different injection efficiencies.[19] The first p-type region is highly doped to produce a high injection efficiency as in the conventional p–i–n rectifier. The second p-type region is a shallow, lightly doped region that has a low injection efficiency. These regions are integrated, as shown in the cross-section in Fig. 15. The characteristics of the SSD structure can

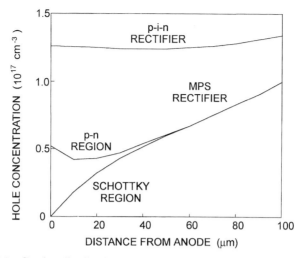

Fig. 14   Carrier distribution within the MPS rectifier in the on-state.

be adjusted between that of the *p–i–n* rectifier (when the second *p*-type region is made deeper and more heavily doped) to that of the MPS rectifier (when the second *p*-type region is made very shallow and lightly doped).

The effect of changing the surface doping concentration of the second *p*-type region has been studied by numerical simulations.[20] As shown in Fig. 16, when the surface concentration of the second region is reduced to $1 \times 10^{15}/cm^{-3}$, the stored charge is reduced by 35%, whereas the on-state voltage drop increases from 1.2 to 1.43 V. This occurs because the injected carrier density at the second *p*-type region is reduced as a result of its poor injection efficiency. As in the case of the MPS rectifier, under reverse-bias conditions, the leakage current in the SSD is suppressed by the potential barrier formed under the second *p*-type region. This structure has a trade-off curve between on-state voltage drop and stored charge similar to that of the MPS rectifier. However, it requires the additional process step of ion implantation and drive-in for the second *p*-type region.

4.3   POWER MOSFETs

Three types of discrete vertical-channel power-MOSFET structures have been explored by the industry. Although the VMOSFET was the first commercial structure, it was superseded by the DMOSFET because of stability problems during manufacturing and a high local electric field at the tip of the V groove. A cross-section of the DMOSFET structure is shown in Fig. 17. This DMOS structure is fabricated by using planar diffusion technology with a refractory gate, such as polysilicon, as a mask with the

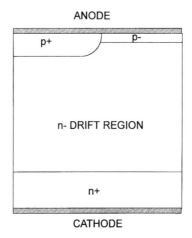

Fig. 15   Cross-section of the SSD rectifier structure.

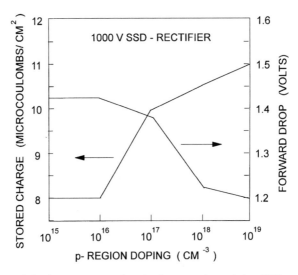

Fig. 16   Effect of doping concentration in the *p*-region of the SSD rectifier on the on-state voltage drop and stored charge.

p-base region and the n^+ source regions defined by the edge of the polysilicon gate. The name for this device is derived from this double-diffusion process. The difference in the lateral diffusion between the *p*-base and n^+ source regions defines the surface channel region. The third power-MOSFET structure that has been explored is the UMOSFET structure shown in Fig. 18. The name for this structure is derived from the U-shaped groove formed in the gate region by reactive ion etching. The U-groove structure has a higher

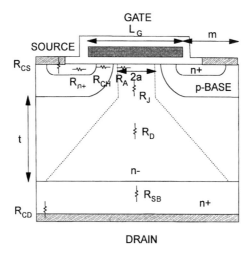

Fig. 17 Cross-section of DMOSFET structure illustrating its internal resistances.

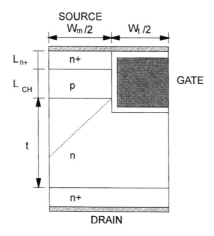

Fig. 18 Cross-section of UMOSFET structure.

channel density (channel width per cm^2 of active area) than either the VMOS or DMOS structures, which allows significant reduction in the on-resistance of the device. The technology for the fabrication of this structure was derived from the trench-etching techniques developed for the storage capacitor in memories. In both structures, note that the p base region is shorted to the n^+ source region.

With the gate shorted to the source in all power MOSFET structures, a positive drain voltage reverse-biases the p-base–n-drift region junction. The depletion layer extends primarily into the n-drift region because of the higher doping level of the p-base region. Its doping concentration and width must be based on the criteria established for avalanche breakdown of p–n junctions. A higher drain blocking-voltage capability requires lower drift-region doping and larger width, producing a higher resistance to current flow in the on-state. To carry current from drain to source, a positive bias is applied to the gate electrode. The gate bias creates an inversion layer in the channel region resulting from the strong electric field created normal to the semiconductor surface through the oxide layer, as discussed in Chapter 3. This inversion layer provides a conductive path between the n^+ source regions and the drift region. A positive drain voltage now results in current flow between drain and source (via the n-drift region and the channel), which is limited by the resistances shown in Fig. 17. At low drain voltages, the current flow is essentially resistive, with the on-resistance determined by a combination of the channel and drift-region resistances. The channel resistance decreases with increasing gate bias, whereas the drift-region resistance remains invariant, resulting in the total resistance decreasing with increasing gate bias until it approaches a constant value, indicated by the dotted line in Fig. 19 as the on-resistance (R_{on}). The on-resistance is an important power-MOSFET parameter because it is a measure of the current-handling capability of the device. At high drain voltages, the current saturates, as shown in Fig. 19. The current saturation in power MOSFETs can be used to provide a current-limiting function in power circuits as long as the power dissipation in the devices is kept within reasonable limits. Note that the current flow in power MOSFETs occurs solely by transport of majority carriers (electrons for n-channel devices).

To switch the power MOSFET into the off-state, the gate bias voltage must be reduced to zero by shorting the gate electrode to the source electrode externally. When the gate voltage is removed, the electrons are no longer attracted to the channel and the conductive path from drain to source is broken. The power MOSFET then switches rapidly from the on-state to the off-state without the delays associated with minority-carrier storage and recombination observed in bipolar devices. The turn-off time is controlled by the rate of removal of the charge on the gate electrode because this charge determines the conductivity of the channel. Turn-off times of under 100 ns can be achieved with a moderate gate drive current flow required to discharge of the input gate capacitance of the device.

An important device parameter for power MOSFETs is the transconductance (g_m), defined as the rate of change of drain current with change in gate voltage. A large transconductance is desirable to obtain a high current-handling capability with low gate drive voltage, and for achieving high-frequency response. In the saturated-current region of operation, the output characteristics are controlled by the gate-induced channel characteristics. The transconductance is, therefore, determined by the design of the channel and gate structure.

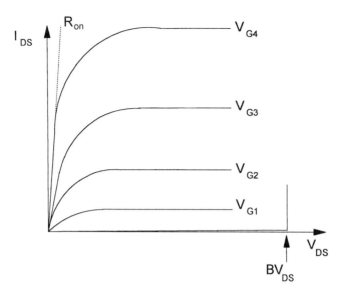

Fig. 19 The output characteristics of a power MOSFET.

4.3.1 Forward Blocking Capability

In the forward blocking mode, the gate electrode of the power MOSFET is externally shorted to the source. Under these conditions, no surface channel forms under the gate at the surface of the p base region. When a positive drain voltage is applied, the p-base–n-drift-layer junction becomes reverse-biased and supports the drain voltage. The breakdown voltage of this junction cannot be predicted by the simple parallel-plane analysis for several reasons.[21,22] The power MOSFET structure consists of an n^+–p–n transistor in which the p base region is shorted at selected points to the n^+ emitter by the source metallization. This structure will conduct current as soon as the depletion layer in the p base reaches through to the n^+ emitter because the emitter–base junction then becomes forward biased and the emitter injects electrons into the p-base region. The reach-through breakdown condition is determined by the p-base doping profile. The surface concentration (N_{SP}) of the p-base diffusion and the n^+-emitter depth combine to determine the peak doping (n_{AP}) in the p base. The peak p-base doping is an important parameter for the doping profile because it controls the threshold voltage of the power MOSFET, i.e., the minimum gate voltage required to induce a surface channel. For a typical threshold voltage of 2 to 3 V for an n-channel power MOSFET, the peak base concentration (n_{AP}) is about $10^{17}/cm^3$.

The channel length is another important design parameter in power

MOSFETs. It has a strong influence on the on-resistance and the transconductance. The channel length is determined by the difference in the depths of the p-base and n^+-emitter diffusions, i.e., $x_p - x_{n^+}$. Despite the shorting of the n^+ emitter to the p base by the source metal, a parasitic n^+-p-n bipolar transistor exists in the power MOSFET. When the p-base–n-drift-layer junction is reverse-biased, the depletion layer in the p base can extend to the n^+-emitter–p-base junction and cause premature reach-through breakdown. It is important to design the p-base diffusion profile so that sufficient charge is resident in the p base to prevent reach-through of the depletion layer to the n^+ emitter. When the p-base surface concentration is below $10^{18}/\text{cm}^3$, the depletion width on the diffused side can extend well over 1 μm especially for low drift-layer concentrations. Consequently, the fabrication of devices with channel lengths of less than 1 μm is not possible unless the device is required to support only low voltages.

The breakdown voltage of the power MOSFET can be determined by either breakdown at the edges of the device or by breakdown within the MOS cell structure. The edge breakdown is determined by the edge termination. Since the on-resistance of the power MOSFET increases very rapidly with increasing breakdown voltage, as shown later in this section, it is important to use an edge termination that approaches the ideal breakdown voltage. With such terminations, the breakdown voltage can shift to the MOS cell structure. The effect of the MOS cell structure on the breakdown voltage has been discussed in detail in Ref. 5.

4.3.2 On-State Characteristics

Forward current flows in an n-channel power MOSFET when a positive gate bias is applied to create a conductive path across the p base region underneath the gate. The current flow is limited by the total resistance between the source and drain. This resistance consists of many components, as illustrated in the DMOS cross-section in Fig. 17, which combine to determine the on-state voltage drop. The resistances of the n^+ emitter (R_{n^+}) and substrate (R_S) regions are generally negligible for high-voltage power MOSFETs that have high drift-region resistance. They become quite important when the drift and channel resistance become small as in the case of low (<100 V) breakdown voltage devices. The channel (R_{CH}) and accumulation layer (R_A) resistances are determined by the conductivity of the thin surface layer induced by the gate bias. These resistances are a function of the charge in the surface layer and the electron mobility near the surface. In addition to these resistances, the drift layer contributes two more components to the total on-resistance. The portion of the drift region that comes to the upper surface between the cells contributes a resistance R_J that is enhanced at higher drain voltages due to the pinch-off action of the depletion layers extending from adjacent p base regions. This phenomenon has been termed the JFET action. Finally, the main body of the drift region contributes a large series resistance (R_D)

especially in high-voltage devices. The analysis of each of these components of the on-resistance is provided in this section.

Threshold Voltage. The voltage on the gate electrode at which strong inversion occurs in the MOS structure is an important design parameter for power MOSFETs because it determines the minimum gate bias required to induce an *n*-type conductance in the channel. This voltage is called the threshold voltage. For proper device operation, its value cannot be too large or too small. If the threshold voltage is large, a high gate bias voltage will be needed to turn on the power MOSFET. This imposes problems with the design of the gate drive circuitry. It is also important that the threshold voltage not be too low because the device can then be triggered into conduction inadvertently either by noise signals at the gate terminal or by the gate voltage being pulled up during high-speed switching. Typical power MOSFETs are designed for threshold voltages that range between 2 and 3 V.

In the absence of any difference in the work function of the metal and the semiconductor, the threshold voltage is given by[5,6]

$$V_T = \frac{Q_s}{C_{ox}} + 2\psi_B \tag{38}$$

and

$$Q_s = \sqrt{4\varepsilon_s kT n_{AP} \ln(n_{AP}/n_i)} \tag{39}$$

$$C_{ox} = \frac{\varepsilon_{ox}}{t_{ox}} \tag{40}$$

$$\psi_B = \frac{kT}{q} \ln\left(\frac{n_{AP}}{n_i}\right) \tag{41}$$

where Q_s is the charge in the depletion layer per unit area, C_{ox} is the gate oxide capacitance per unit area, ψ_B is the difference between the Fermi level and the intrinsic level in the bulk of the semiconductor, n_i, is the intrinsic carrier concentration, ε_s and ε_{ox} are the permittivities of the semiconductor and the oxide, and t_{ox} is the gate oxide thickness. This expression shows that the threshold voltage will increase linearly with gate oxide thickness and approximately as the square root of the semiconductor doping concentration (n_{AP}). In actual metal–oxide–semiconductor structures, the threshold voltage is altered as a result of several factors.

1. An unequal work function for the metal and the semiconductor. If the barrier height between silicon dioxide and metal is ϕ_B, the difference

between the metal and the semiconductor work function can be obtained as[5,6]

$$q\phi_{ms} = q\phi_B + q\chi_{ox} - \left(q\chi_s + \frac{E_g}{2} + q\psi_B\right) \qquad (42)$$

where χ_{ox} and χ_s are the electron affinity of the oxide and semiconductor, respectively.

2. The presence of fixed oxide charge (Q_f) at the oxide–silicon interface.
3. The presence of mobile ions in the oxide with charge Q_m.
4. The presence of trapped charges at the oxide–silicon interface with charge Q_{it}.

All of these charges cause a shift in the threshold voltage:[5,6]

$$V_T = \phi_{ms} + \frac{Q_s}{C_{ox}} + 2\psi_B - \left(\frac{Q_f + Q_m + Q_{it}}{C_{ox}}\right) \qquad (43)$$

which must be taken into account during device design.

Channel Resistance. The channel resistance is determined by the electron charge available for transport in the surface inversion layer as well as the surface mobility of these electrons.[23,24] When the surface potential (ψ_S) exceeds twice the bulk potential (ψ_B), a strong inversion layer begins to form. Since the band bending is small beyond this point, the inversion layer charge available for current conduction is given by

$$Q_n = C_{ox}(V_G - V_T) \qquad (44)$$

Thus, the channel resistance at low drain voltages, where the voltage drop along the channel is negligible, is given by

$$R_{ch} = \frac{L}{Z\mu_{ns} C_{ox}(V_G - V_T)} \qquad (45)$$

where Z and L are the width and length of the channel and μ_{ns} is the surface mobility of electrons. As the drain current increases, the voltage drop along the channel between drain and source becomes significant. The positive drain potential opposes the gate bias voltage and reduces the surface potential in the channel. The channel charge near the drain is then reduced by the voltage drop along the channel. When the drain voltage becomes equal to $V_G - V_T$, the charge in the channel at the drain becomes zero. This condition is called channel pinch-off. At this point, the drain current saturates. Further increase in drain voltage results in no further increase in drain current. The drain voltage is now supported across an extension of the depletion layer under the gate.

The channel *I–V* characteristics can be derived as a function of the gate and drain voltages under the gradual channel approximation[5,6]

$$I_D = \frac{\mu_{ns} C_{ox} Z}{2L} [2(V_G - V_T)V_D - V_D^2] \qquad (46)$$

As the drain voltage and current increase, the second term becomes increasingly important and causes the drain current to saturate. Physically, this corresponds to the reduction in the channel inversion-layer charge near the drain with increasing drain voltage. Ultimately, the inversion layer charge at the drain end of the channel becomes zero and the drain current saturates with a value

$$I_{DS} = \frac{\mu_{ns} C_{ox} Z}{2L} (V_G - V_T)^2 \qquad (47)$$

The saturated drain current is an important parameter because it determines the maximum current that the channel will support. The transconductance of the device in the saturated-current region of operation can be obtained by differentiating Eq. 47 with respect to the gate voltage:

$$g_{ms} = \frac{dI_D}{dV_G} = \mu_{ns} C_{ox} \frac{Z}{L} (V_G - V_T) \qquad (48)$$

This device parameter is useful for defining the gate-drive requirements for the power MOSFET.

DMOSFET Specific On-Resistance. The specific on-resistance of the power DMOSFET is determined by the resistance components illustrated in Fig. 17:

$$R_{on} = R_{n^+} + R_{CH} + R_A + R_J + R_D + R_S \qquad (49)$$

where R_{n^+} is the contribution from the n^+-source diffusion, R_{CH} is the channel resistance, R_A is the accumulation-layer resistance, R_J is the contribution from the drift region between the *p* base regions, R_D is the drift-region resistance, and R_S is the substrate resistance. Additional resistances can arise from a non-ideal contact between the source–drain metal and the n^+ semiconductor regions, indicated by R_{CS} and R_{CD} in Fig. 17, as well as from the leads used to connect the device to the package. The resistance contributions from the n^+ source region and the contacts is usually negligible in modern power devices.

Consider the ideal case where the resistances of the n^+ emitter, n^+ substrate, n-channel region, accumulation region, and JFET region are negligible. The specific on-resistance of the power MOSFET will then be determined by the drift region alone. In addition, if we assume that the current flows uniformly through the drift region without current-spreading effects, the resistance of the drift region is referred to as the ideal specific on-resistance of the power MOSFET. This is the resistance of a drift region with the doping concentration and thickness required to support the desired breakdown voltage. Using equations relating the doping concentration (n_D) and thickness (W_D) of the drift region to the breakdown voltage,[5] the ideal specific on-resistance is given by the expression in Eq. 3 for n-channel devices.

The contribution from the substrate can be neglected for high-voltage power MOSFETs. However, in devices with breakdown voltages below 50 V, it can contribute significantly to the on-resistance. This is especially true because the substrate must be sufficiently thick to impart adequate strength to the wafers during device fabrication. It can be assumed that the current density is uniform within the substrate because of rapid current spreading at the drift region interface. The specific resistance contributed by the substrate is then given by

$$R_{SB,sp} = \rho_{SB} t_{SB} \tag{50}$$

where ρ_{SB} is the resistivity of the substrate and t_{SB} is its thickness. In the case of a typical antimony-doped substrate with a thickness of 0.05 cm and a resistivity of 0.01 Ω-cm, the substrate resistance per unit area is 5×10^{-4} Ω-cm^2. This value is comparable to the ideal specific resistance of the drift region for a 50 V device. The substrate resistance can be reduced by using 0.001 Ω-cm arsenic-doped substrates and by lapping the substrate to reduce its thickness after fabrication of the device structure on the top surface.

On the basis of the channel resistance analysis, the contribution from the channel can be minimized by making the channel length (L_{CH}) small and keeping its width (Z) large. The channel resistance per square centimeter for the linear cell structure is given by[5]

$$R_{CH,sp} = \frac{L_{CH}(L_G + 2m)}{2\mu_{ns} C_{ox}(V_G - V_T)} \tag{51}$$

Note that the channel resistance decreases when the cell pitch ($L_G + 2m$) is reduced. This occurs because the channel density (channel width per cm^2 of active cell area) increases. The channel resistance can also be reduced by decreasing the gate-oxide thickness while maintaining the same gate-drive voltage.

The resistance of the accumulation layer (R_A) accounts for the current spreading from the channel into the JFET region. The accumulation-layer resistance depends upon the charge in the accumulation layer and the mobility (μ_{nA}) of free carriers at the accumulated surface. For the linear cell geometry, the accumulation layer resistance per square centimeter is given by[5,25,26]

$$R_A = \frac{K(L_G - 2x_P)(L_G + 2m)}{2\mu_{nA}C_{ox}(V_G - V_T)} \tag{52}$$

where the factor K is introduced to account for the two-dimensional nature of the current flow from the channel into the JFET region via the accumulation layer. Good agreement with experimental results has been observed for $K = 0.6$, which implies that the effective resistance to the drain current flow is 60% of the total accumulation-region resistance. The accumulation-layer resistance can be reduced by decreasing the length (L_G) of the gate electrode between the cells. However, this has an adverse effect upon the JFET resistance (R_J).

The resistance of the drift region between the p-base diffusions is referred to as the JFET resistance because the current flow resembles that in a junction field-effect transistor with the p base regions acting as the gate regions. This resistance contribution can be calculated easily if the effect of the voltage drop along the vertical direction on the depletion region is neglected. Under the assumption that the current flows uniformly from the accumulation layer into the JFET region, the resistance of the JFET region becomes that of a semiconductor region with a cross-sectional area

$$A_{JFET} = aZ = \left(\frac{L_G}{2} - x_P\right)Z \tag{53}$$

where Z is the width of the cell orthogonal to the cross-section. The JFET contribution to the specific resistance is then given by[5]

$$R_J = \frac{\rho_D(L_G + 2m)(x_P + W_0)}{L_G - 2x_P - 2W_0} \tag{54}$$

where ρ_D is the resistivity of the JFET region. In high-voltage power MOSFETs, the drift-region doping must be small to obtain the desired breakdown voltage. The depletion-layer extension (W_0) can then be a significant fraction of the gate length (L_G) leading to a large resistance contribution from the JFET region. This problem can be solved by increasing the gate length. However, this leads to a poor channel density and to reduced cell-breakdown voltage. It is, therefore, preferable to increase the doping concentration in the JFET region while maintaining a lower doping concentration in the drift region to obtain the desired breakdown voltage. The maximum doping concentration in the JFET region must be kept below about $5 \times 10^{16}/\text{cm}^3$ to avoid a high local electric field and to prevent significant alteration of the channel doping.

The drift region is assumed to begin below the bottom of the p-base diffusion. The current spreads from the JFET region into the drift region as shown by the dotted lines in the Fig. 17. One model[5] that allows a reasonably accurate estimation of the drift-region spreading resistance is based on the current spreading from a cross-section of $a = L_G - 2x_P$ at a 45° angle. The specific resistance contribution for the drift region is then given by

$$R_{D,\text{sp}} = \frac{\rho_D(L_G + 2m)}{2} \ln\left(\frac{a + t}{a}\right) \tag{55}$$

It is important to note that even if ideal breakdown voltage is assumed at the device edge termination and within the DMOS cell structure, the specific resistance of the drift region is not equal to the ideal specific on-resistance because of the effect of the current spreading from the JFET region into the drift region. The deviation of the drift-region specific resistance from the ideal value becomes worse as the cell size becomes large when compared with the drift region thickness. The reason is that the drain current tends to flow only under the gate region, creating a significant dead space under the p base region.

The contributions to the on-resistance from various terms are dependent upon the device geometrical design parameters. When the gate length (L_G) is small, the JFET and drift region resistances become large because of the small width (a) through which the current must flow into the channel. At the same time, the accumulation layer resistance becomes small because of the shorter path along the surface, and the channel resistances become small because of a reduction in the cell pitch, which is equivalent to an increase in the channel density. The opposite trends occur when the gate length is increased. This implies that there is an optimum gate length at which the total specific on-resistance has a minimum value.[5,26] Therefore, it is necessary to calculate the specific on-resistance as a function of the polysilicon gate length during the optimization of the design of a power DMOSFET structure.

The impact of increasing the polysilicon gate length (L_G) upon the various components of the on-resistance for a 50-V DMOSFET is shown in Fig. 20 together with the total specific on-resistance. The channel and accumulation layer resistances increase as the gate length (L_G) increases. Concurrently, the resistances of the JFET and drift regions decrease because the cross-sectional area for the current flow increases. A minimum in the total on-resistance occurs at an optimum gate length of 12 μm. The minimum specific on-resistance for this example is found to be $3\,\text{m}\Omega\text{-cm}^2$. In comparison, the ideal specific on-resistance for a breakdown voltage of 50 V is $0.1\,\text{m}\Omega\text{-cm}^2$. Thus, the device specific on-resistance deviates from the ideal

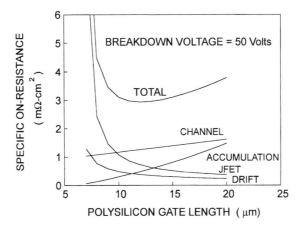

Fig. 20   On-resistance components of a power DMOSFET with a breakdown voltage of 50 V.

value by a large factor (30). It is worth pointing out that the channel resistance at the optimum gate length is significantly larger than all the other components.[27] This indicates that improvements in the performance of the low-breakdown-voltage power DMOSFETs can be obtained by reducing the cell pitch, the channel length, and the gate oxide thickness. One method for obtaining a smaller cell window is to eliminate the need for opening a contact window. This can be done, with the structure shown in Fig. 21, by forming an oxide spacer.[28] The spacer is created by conformal deposition of an oxide layer after patterning the polysilicon layer followed by reactive-ion-etching. With this method, devices with breakdown voltage of 50 V can be fabricated with specific on-resistance below $1\,m\Omega\text{-}cm^2$.

UMOSFET Specific On-Resistance.   The specific on-resistance of silicon power DMOSFETs is significantly higher than the ideal specific on-resistance. This deviation is caused by the resistance contributions from the channel region, the accumulation layer, and the JFET regions in the DMOS structure. The UMOSFET structure was proposed[29] to reduce the resistance contributions from these regions. It can be seen from Fig. 18 that the UMOSFET structure does not have a JFET region. Instead, the current spreads out from an accumulation layer formed on the surface of the U groove. The UMOSFET resistance then consists mainly of the channel resistance and the drift region resistance per unit area. For the UMOSFET structure, the channel resistance is given by[5]

$$R_{CH,sp} = \frac{L_{CH}(W_m + W_t)}{2\mu_{ns} C_{ox}(V_G - V_T)} \tag{56}$$

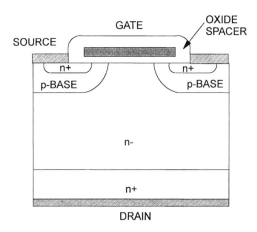

Fig. 21   Cross-section of DMOSFET structure with an oxide spacer.

The UMOS structure can be fabricated with narrow mesa and trench regions because of the absence of the JFET region. In combination with good lithographic tools, it is possible to reduce the UMOS cell size to less than 6 μm, resulting in a much higher channel density than in the DMOS structure. For a typical gate oxide thickness of 1000 Å and a channel length of 1 μm, the specific on-resistance of the channel is calculated to be less than 0.2 mΩ-cm^2.

The drift region spreading resistance can be derived[5,29]

$$R_D = \rho_D \left\{ \left[\left(\frac{W_m + W_t}{2} \right) \ln \left(\frac{W_m + W_t}{W_t} \right) \right] + \left(t_D - \frac{W_m}{2} \right) \right\} \tag{57}$$

where the first term is associated with the portion of the drift region where the current spreads at 45°, and the second term is associated with the portion of the drift region where the area of cross-section is equal to the cell area. Unlike the DMOS structure, an overlap of the current spreading occurs even for low breakdown voltage designs for the UMOS structure because of the very small half-width for the mesa region.

Unlike the DMOS structure, there is no optimum design for the UMOS cell. In the UMOSFET structure, it is beneficial to reduce the mesa and trench widths as much as possible. As these dimensions become smaller, the contribution of the channel resistance becomes smaller the channel density increases. In addition, the spreading resistance of the drift region approaches the ideal specific on-resistance as the mesa width (W_m) becomes smaller. For a UMOSFET with a breakdown voltage of 50 V fabricated using mesa and trench widths of 3 μm, the total specific on-resistance is calculated to be

0.35 mΩ-cm^2 vs 0.1 mΩ-cm^2 for the ideal case. Thus, the UMOSFET structures allow specific on-resistances approaching the ideal value. Trench-gate devices with self-aligned contacts have been reported with specific on-resistances of 0.5 to 1 mΩ-cm^2 for a device with breakdown voltage of 50 V.[30,31]

A further reduction in the specific on-resistance can be obtained by extending the trench until it penetrates the n^+ substrate, as shown in Fig. 22.[32,33] In this structure, the drain current flows not only via the n-drift region but also along the sidewalls of the trench through the formation of an accumulation layer. The specific on-resistance is then no longer limited by the resistance of the drift layer and can become even lower than the ideal value when the cell pitch is reduced sufficiently.[33] Although this structure has a very low specific on-resistance of 0.20 mΩ-cm^2, its breakdown voltage is limited to 25 V because the entire drain voltage must be supported across the gate oxide. This problem can be solved by using a double-gate structure with a thin gateoxide layer up to the bottom of the p base region, and a thick oxide in the lower portion of the extended trench.[34] A breakdown voltage of 70 V and a specific on-resistance of 0.7 mΩ-cm^2 was experimentally obtained for this structure. The higher specific on-resistance arises from the poor conductivity of the accumulation layer for the portion of the trench with the thick oxide layer.

4.3.3   Switching Characteristics

The switching speed of the power MOSFET is limited by the charging and discharging of its input (gate) capacitance. In addition to the gate-to-source capacitance (C_{GS}), a significant gate-to-drain capacitance (C_{GD}) must be included in the analysis due to the overlap of the gate electrode over the drift

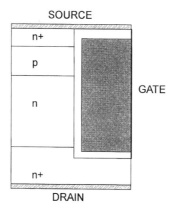

Fig. 22   Cross-section of UMOSFET structure with an extended trench gate.

region. This capacitance is amplified by the Miller effect into an equivalent input gate capacitance[5]

$$C_M = (1 + g_m R_L) C_{GD} \tag{58}$$

where g_m is the transconductance and R_L is the load resistance. The total input capacitance is

$$C_{INPUT} = C_{GS} + C_M \tag{59}$$

The input gate-to-source capacitance for this structure contains several components: (a) the capacitance C_{n^+} arising from the overlap of the gate electrode over the n^+ source region, (b) the capacitance C_p arising from the MOS structure created by the gate electrode over the p base region, and (c) the capacitance C_o arising from running the source metal over the gate electrode. The gate-to-drain capacitance varies with gate and drain voltage with a high value during the on-state because of the accumulation region at the surface of the drift region. As the drain voltage increases and the device supports high voltages, the gate-to-drain capacitance decreases. This capacitance can severely reduce the frequency response because of amplification by the Miller effect. It is therefore important to reduce its magnitude. The gate-to-drain overlap capacitance can be drastically reduced by eliminating the gate overlap over the drift region and restricting the source electrode to the cell diffusion window, allowing a significant improvement in the switching performance of power MOSFETs.[35,36] However, it is important to note that eliminating the gate-to-drain overlap creates a high electric field at the edge of the gate that can reduce the cell breakdown voltage to below that for the edge termination. In addition, it adversely impacts the on-resistance by increasing the resistance between the channel and the drift region because an accumulation layer is no longer formed over a portion of the surface between the base regions. The cell breakdown voltage can be improved with a reduced drain–gate capacitance by incorporating a shallow p-type diffusion in the portion where the gate electrode has been interrupted.[36,37] This p-type region behaves like a guard ring for the electrode, reducing the electric-field crowding at its edge. However, it also enhances the JFET action within the cell during current flow. Thus, the capacitance reduction is achieved at the drawback of an increase in on-resistance.

Due to the inherent high-speed turn-on and turn-off capability of power MOSFETs, they are often used as power switches in high-frequency power circuits. For a clamped inductive load with a steady-state current I_L flowing through it, the turn-on time can be shown to be given by[5,38]

$$t_{on} = \frac{(V_S - V_F) R_G C_{GD}}{[V_G - (V_T + I_L/g_m)]} \tag{60}$$

where V_S is the supply voltage, V_F is the on-state voltage drop across the MOSFET, R_G is the resistance in series with the gate drive voltage source, and V_G is the gate drive voltage. The turn-off time for the power MOSFET is given by[5,38]

$$t_{off} = R_G(C_{GS} + C_{GD}) \ln\left(\frac{I_L}{g_m V_T} + 1\right) \tag{61}$$

The power loss can be reduced by keeping these time intervals short by reduction of the gate series resistance (R_G) and the drain–gate capacitance (C_{GD}). However, reducing the series resistance of the gate drive circuit has the disadvantage of increasing the cost of the drive circuit. It is, therefore, important to use the methods discussed in the earlier section to reduce the drain–gate capacitance.

4.3.4 Power MOSFET Safe Operating Area

The safe operating area defines I–V boundaries that allow operation of the device without destructive failure. The maximum current at low drain voltages is limited by power dissipation if the leads are sufficiently thick to prevent fusing. The maximum voltage at low drain currents is determined by the avalanche breakdown phenomena. However, under the simultaneous application of high current and high voltage, the device may be susceptible to destructive failure even if the duration of the transient is small enough to prevent excessive power dissipation. This failure mode has been referred to as second breakdown. The term second breakdown refers to a sudden reduction in the blocking-voltage capability when the drain current increases. This phenomenon originates from the parasitic bipolar transistor in the power MOSFET structure. When the drain voltage is increased to near the avalanche breakdown voltage, current flows into the p base region in addition to the normal current flow within the channel inversion layer. The avalanche current collected within the p base region flows laterally along the p base region to its contact. The voltage drop along the p base region forward biases the edge of the n^+ emitter furthest from the base contact. When the forward bias on the emitter exceeds 0.7 V, it begins to inject carriers. The parasitic bipolar transistor is then no longer capable of supporting the p-base–n-drift-layer breakdown voltage (BV_{CBO}) and its breakdown voltage is reduced to BV_{CEO}, which is typically 60% of BV_{CBO}. Consequently, it is desirable to include a deep p^+ diffusion in the center of the DMOS cell. Commercial power MOSFETs now have excellent safe operating area because of the very efficient shorting of the emitter–base junction of the parasitic bipolar transistor.

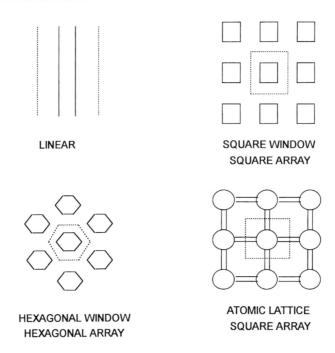

LINEAR

SQUARE WINDOW
SQUARE ARRAY

HEXAGONAL WINDOW
HEXAGONAL ARRAY

ATOMIC LATTICE
SQUARE ARRAY

Fig. 23   Cell layout topologies used for power DMOSFETs.

4.3.5   Cell Topology

In previous discussions of power MOSFETs in this section, cross-sections of devices were shown without considering the cell layout surface topology. The DMOS and UMOS structures allow any conceivable cell topology as long as it meets all technological constraints, such as alignment tolerances. The cell windows that are commonly used for the design of power MOSFETs have linear, square, circular, hexagonal, or the atomic-lattice-layout (A-L-L) shapes, as illustrated in Fig. 23. These windows can be located in either a square or hexagonal cell pattern. The impact of the layout of these cells upon the resistance has been analyzed under the assumption that they all have the same drift-region doping concentration.[39] This analysis demonstrates that the on-resistances of all cellular designs are equal if the size of the cell window and the ratio of the area of the cell window to the total cell area are the same. However, it has been demonstrated that the drift-region doping concentration must be adjusted on the basis of the electric-field crowding effect within the cell.[36,40] When this is taken into account, the A-L-L design is superior to the other designs. The A-L-L design also has a lower drain-overlap capacitance, which is favorable for high-frequency operation.[36]

4.4 INSULATED-GATE BIPOLAR TRANSISTORS

As discussed in the previous section, the on-resistance of the power MOSFET increases very rapidly when its breakdown voltage increases. This makes the on-state power losses unacceptable for medium and high-power applications where high DC supply voltages are used. Since bipolar current conduction allows operation at high on-state current densities with a low on-state voltage drop, it was advantageous to develop a new category of power semiconductor devices in which bipolar current transport is controlled via an MOS-gate structure.[1] Among these devices, the insulated-gate bipolar transistor (IGBT) has become the most commercially successful power switch because of its superior on-state characteristics, reasonable switching speed, and excellent safe operating area. These devices have replaced bipolar power transistors in medium-power applications, where the blocking voltages are between 300 and 2000 V.

Figure 24 shows a cross-section of the DMOS IGBT structure.[41] In this structure, current cannot flow when a negative voltage is applied to the collector with respect to the emitter because the lower junction (J_1) becomes reverse-biased. Thus, the IGBT structure exhibits high reverse blocking capability. In this mode of operation, the depletion region extends into the n-drift region. When a positive voltage is applied to the collector with the gate shorted to the emitter , the upper junction (J_2) becomes reverse-biased and the device operates in its forward blocking mode. In this mode of operation, the voltage is supported by a depletion region formed in the n-drift region. The forward and reverse blocking capabilities shown in Fig. 25 are approximately equal because they are determined by the thickness and resistivity of a common n-drift layer.

If a positive gate bias is applied of sufficient magnitude to invert the surface of the p base region under the gate when the device is in its forward blocking mode, the IGBT can be switched to its on-state. In the forward conducting state, electrons flow from the n^+ emitter to the n-drift region providing the base drive current for the vertical p–n–p transistor in the IGBT structure. Since the emitter junction (J_1) for this bipolar transistor is forward biased, the p^+ region injects holes into the n base region. When the positive bias on the collector terminal of the IGBT is increased, the injected-hole concentration increases until it exceeds the background doping level of the n-drift region. In this regime, the device characteristics are similar to those of a forward-biased p–i–n diode. Consequently, these devices can be operated at high current densities even when designed to support high blocking voltages.

If the inversion-layer conductivity is reduced by a gate bias close to the threshold voltage, a significant voltage drop occurs across this region because of the electron current flow similar to that observed in conventional MOSFETs. When this voltage drop becomes comparable to the difference between the gate bias and the threshold voltage, the channel becomes

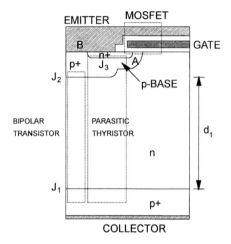

Fig. 24 Cross-section of the DMOS IGBT structure.

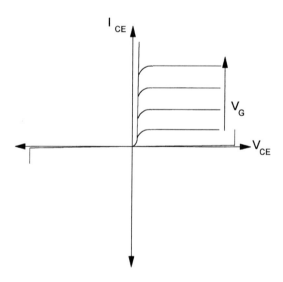

Fig. 25 The output characteristics of an IGBT with symmetric blocking capability.

pinched off. At this point, the electron current saturates. Since this limits the base drive current for the *p–n–p* transistor, the hole current flowing through this path is also limited. Consequently, the device operates with current saturation in its active region with a gate-controlled output current, as shown in Fig. 25.

To switch the IGBT from its on-state to the off-state, it is necessary to discharge the gate by shorting it to the emitter. In the absence of a gate voltage, the inversion region at the surface of the p base under the gate cannot be sustained. Removing the gate bias cuts off the supply of electrons to the n-drift region, which initiates the turn-off process. The turn-off does not occur abruptly because there is a high concentration of minority carriers injected into the n-drift region during forward conduction. At first, the anode current is abruptly reduced because the electron current through the channel is terminated. The collector current then decays gradually with a characteristic time constant determined by the minority-carrier lifetime.

The IGBT structure contains a parasitic p–n–p–n thyristor between the collector and the emitter terminals, as indicated in Fig. 24. If this thyristor latches up, the current can no longer be controlled by the MOS gate. Latch-up can be suppressed by preventing the injection of electrons from the n^+ emitter region into the p base during device operation by shorting these regions using the emitter metal. The equivalent circuit for the IGBT then consists of a MOSFET driving the base of the p–n–p transistor in a Darlington configuration.[5]

4.4.1   Reverse Blocking Operation

When a negative collector voltage is applied, a large voltage can be supported by the IGBT structure shown in Fig. 24 because junction J_1 becomes reverse-biased. When junction J_1 becomes reverse-biased, its depletion layer extends primarily into the lightly doped n-drift region. The breakdown voltage during reverse blocking is determined by an open-base transistor formed between the p^+ collector, n-drift region, and the p base region. This structure is prone to punch-through breakdown if the n-drift region is too lightly doped or it is too thin. To obtain the desired reverse blocking capability, it is essential to design the resistivity and thickness of the n-drift region optimally. The design is also affected by the minority-carrier diffusion length. As a general guideline, the width of the n-drift region is chosen so that its thickness is equal to the depletion width at the maximum operating voltage plus one diffusion length. Since the forward voltage drop increases with increasing n-drift region width, it is important to obtain the desired breakdown voltage with a minimum width for the n-drift region. When the blocking voltage requirement increases, the n-drift-region width (d_1) must be correspondingly increased:

$$d_1 = \sqrt{\frac{2\varepsilon_s V_m}{qN_D}} + L_p \tag{62}$$

where V_m is the maximum blocking voltage and L_p is the minority-carrier diffusion length. At large blocking voltages, the depletion-layer width becomes much larger than the diffusion length. The n-drift-region width then increases approximately as the square root of the blocking voltage.

4.4.2 Forward Blocking Operation

To operate the IGBT in the forward blocking mode, the gate must be shorted to the emitter. This prevents the formation of the surface inversion layer under the gate. When a positive collector bias is applied, the IGBT can then support a large voltage because the p-base–n-drift-region junction (J_2) becomes reverse-biased. A depletion layer extends from this junction on both sides at the top and merges at relatively small voltages due to the low doping concentration in the drift region. The forward-blocking capability can be severely degraded by reach-through of the depletion layer of junction J_2 to the lower junction (J_1). In symmetric devices (i.e., devices with equal forward and reverse blocking capability that are used in AC circuits), the n-drift-region width must be chosen using Eq. 62 to design the reverse blocking capability. In DC circuit applications, the IGBT is not required to support reverse voltage. This offers the opportunity to reconfigure the device structure to optimize the forward conduction characteristics for a given forward blocking capability without considering the reverse blocking capability. In the asymmetric blocking IGBT structure,[42] the uniformly doped n-drift region of the symmetric IGBT is replaced by a two-layer n-drift region which contains a highly doped buffer layer at junction J_1. This alters the electric-field distribution from a triangular form in the symmetric device to a rectangular form in the asymmetric device. Under these circumstances, the same forward blocking capability can be obtained in the asymmetric device with half the thickness of the drift region of the symmetric device. This results in a superior on-state characteristics for the asymmetric structure.

4.4.3 On-State Characteristics

The IGBT can be operated in its forward-conduction mode by applying a positive gate bias to create an inversion layer under the MOS gate. This layer forms a conducting channel that connects the n^+ emitter to the n-drift region. As in a power MOSFET, the gate voltage must be sufficiently above the threshold voltage to make the channel resistance small during current flow. Once the channel is formed, forward current flows by the injection of minority carriers across the for ward-biased collector junction (J_1). Over most of the n-drift region, the injected carrier density is, typically, 100 times greater than the n-drift region doping level, resulting in a drastic reduction of its series resistance. This feature allows operation of the IGBT at very high current densities during forward conduction.

We can analyze the forward conduction characteristics by regarding the IGBT as a p–i–n rectifier in series with a MOSFET. The current density in the p–i–n rectifier can be assumed to be approximately equal to the collector current density (J_C) because the current spreads from the bottom of the drift region and is uniformly distributed across the cross-section of the device cell over most of the distance between the collector and p base region. The voltage drop across the p–i–n rectifier ($V_{F,pin}$) is then related to its forward-conduction current density ($J_{F,pin}$) by[5]

$$V_{F,pin} = \frac{2kT}{q} \ln\left(\frac{J_C d}{2qD_a n_i F(d/L_a)} \right) \tag{63}$$

where $d = d_1/2$. The various terms in this equation are the same as those defined in Section 4.2.4. Since the p–i–n rectifier current flows through the MOSFET channel, the MOSFET current is

$$I_{MOSFET} = J_C WZ \tag{64}$$

and the voltage drop across the MOSFET is related to the current flowing through it and the gate bias voltage by[5]

$$I_{MOSFET} = \frac{\mu_{ns} C_{ox} Z}{2L_{CH}} [2(V_G - V_T)V_{F,MOS} - V_{F,MOS}^2] \tag{65}$$

where the term $V_{F,MOS}$ is the voltage drop across the MOSFET portion of the IGBT, and L_{CH} is the MOSFET channel length. In the forward conduction mode, sufficient gate voltage is applied such that the forward voltage drop across the device is low. Under these conditions, the MOSFET section of the IGBT operates in its linear region. The voltage drop across the MOSFET section is then given by[5]

$$V_{F,MOS} = \frac{I_C L_{CH}}{\mu_{ns} C_{ox} Z(V_G - V_T)} \tag{66}$$

The forward voltage drop across the IGBT is the sum of the voltage drop across the MOSFET and the p–i–n rectifier:

$$V_F = \frac{2kT}{q} \ln\left(\frac{I_C d}{2qWZD_a n_i F(d/L_a)} \right) + \frac{I_C L_{CH}}{\mu_{ns} C_{ox} Z(V_G - V_T)} \tag{67}$$

On the basis of this model, the IGBT forward-conduction current density will rise exponentially with the forward bias voltage as in a p–i–n rectifier. This behavior has been experimentally observed in 600-V devices, as shown in Fig. 26. The figure also shows the forward conduction characteristics of a bipolar transistor operating at a current gain of 10 and a power MOSFET for comparison. When the forward drop exceeds 1 V, the current density of

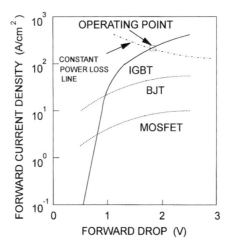

Fig. 26 On-state characteristics of a 600 V IGBT compared with those for the power MOSFET and bipolar transistor.

the IGBT surpasses that of the power MOSFET and bipolar transistor. At a typical operating forward voltage drop of 2 to 3 V for devices with breakdown voltages of 600 V, the IGBT current density is 20 times that of the power MOSFET and five times that of the bipolar transistor. The on-state voltage drop at which a device can be operated is usually determined by the ability to extract the heat generated by the current flow through the device without exceeding a junction temperature of about 200°C. For a fixed thermal impedance, this implies that the operating point on the on-state characteristics can be obtained by drawing a constant power loss line, as shown in Fig. 26. The intersection of this line with the on-state I–V characteristic of a device defines its operating point. On the basis of this criterion, the on-state voltage drop of the IGBT is lower than that for the other devices, whereas its on-state current density is substantially greater than that for the other devices, permitting a smaller chip size.

Using the p–i–n rectifier–MOSFET model, it is also possible to derive the IGBT characteristics under current saturation. When the gate bias is reduced to a value close to the threshold voltage such that the voltage drop across the MOSFET channel becomes significant, the MOSFET limits the current. The transconductance for the IGBT is larger than that for the MOSFET with the same cell size and channel length because of the additional current from the n–p–n bipolar transistor within the IGBT structure. When this current is included, the transconductance is given by[5]

$$g_{ms} = \frac{1}{1 - \alpha_{pnp}} \frac{\mu_{ns} C_{ox} Z}{L_{CH}} (V_G - V_T) \tag{68}$$

Since the current gain (α_{pnp}) of the p–n–p transistor is typically about 0.5, the transconductance of the IGBT can be two times larger than that of a MOSFET with the same channel aspect ratio.

Although the on-state characteristics of the IGBT resemble those for a p–i–n rectifier, the carrier distribution within the n-drift region is not the same as that in a p–i–n rectifier. This difference is due to the reverse biased junction J_2 within the IGBT structure. Since this junction is reverse-biased, the free-carrier density must become zero at this boundary. Thus, although the carrier distribution follows the behavior in a p–i–n rectifier due to high-level injection conditions, the boundary conditions for the IGBT are not identical to those for the p–i–n rectifier. We can derive the free-carrier distribution in the IGBT along a line extending from junction J_1 to junction J_2 by solving the one-dimensional continuity equation under steady-state conditions, with the boundary conditions

$$p(d_1) = 0 \tag{69}$$

and

$$J_p(x) = J \quad \text{at } x = 0 \tag{70}$$

$$J_n(x) = 0 \quad \text{at } x = 0 \tag{71}$$

where J is the collector current density. The carrier distribution in the n^- drift region is given by[5]

$$p(x) = \frac{JL_a}{2qD_p} \frac{\sinh[(d_1 - x)/L_a]}{\cosh(d_1/L_a)} \tag{72}$$

Figure 27 illustrates this charge distribution together with the catenary carrier distribution profile observed in the p–i–n rectifier. Comparing these profiles shows that the conductivity modulation of the n-drift region is quite similar near the collector junction J_1. However, there is a significant difference between them at junction J_2, where the conductivity modulation in the IGBT is much less than that in the p–i–n rectifier. From this difference, it can be concluded that, although the on-state characteristics of the IGBT resemble those for a p–i–n rectifier, the on-state voltage drop for the IGBT will be larger than that for a p–i–n rectifier due to a substantial voltage drop in the JFET region.[5]

4.4.4   Switching Behavior

In the IGBT, the collector current flow can be interrupted by reducing the gate voltage to zero. When the gate voltage falls below the threshold voltage

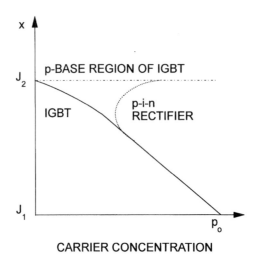

Fig. 27   Carrier distribution with the IGBT in its on-state.

of the MOS gate structure, the channel inversion layer can no longer exist. At this point, the electron current (I_e) flowing through the channel ceases. If the gate turn-off is performed using a low external resistance in the gate drive circuit, so as to abruptly reduce the gate voltage to zero, the collector current will drop abruptly, as shown in Fig. 28, because the channel current is suddenly discontinued. However, the collector current continues to flow because the hole current (I_h) does not cease abruptly. The high concentration of minority carriers stored in the n-drift region during the on-state supports the hole current flow. The decay of the minority carriers due to recombination leads to a gradual reduction in the collector current. This current flow is sometimes referred to as a current tail.

It is customary to define the turn-off time as the time required for the collector current to decrease from its on-state value (I_{CO}) to 10% of this value. To determine this time interval, it is necessary to obtain the magnitude of the initial abrupt fall in the collector current. Since the abrupt drop in the collector current (I_{CD}) is due to the cessation of the electron current provided during the on-state via the MOSFET channel, its magnitude is determined by the current gain of the p–n–p transistor:

$$I_{CD} = I_e = (1 - \alpha_{pnp})I_{CO} \tag{73}$$

After this abrupt drop, the collector current is sustained by the hole current flow from the stored charge in the n-drift region. Initially, the hole current flow is equal to its value during on-state conduction just prior to turn-off.

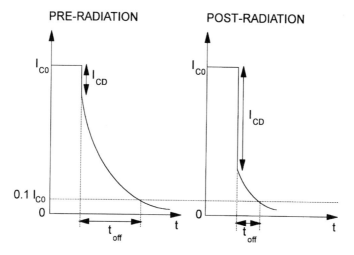

Fig. 28 Turn-off waveforms for the IGBT before and after electron irradiation.

Its magnitude (I_1) is given by

$$I_1 = I_{CO} - I_{CD} = I_h = \alpha_{pnp} I_{CO} \tag{74}$$

After this, the collector current decays exponentially at a rate determined by the minority-carrier lifetime. Since significant current flow occurs during this time with a large density of free carriers in the n-drift region, it is appropriate to use the high-level lifetime to characterize the rate of decay of the current. Thus

$$I_C(t) = I_1 e^{-t/\tau_{HL}} = \alpha_{pnp} I_{CO} e^{-t/\tau_{HL}} \tag{75}$$

The turn-off time (t_{off}) is then given by[5]

$$t_{off} = \tau_{HL} \ln(10\alpha_{pnp}) \tag{76}$$

Electron irradiation of IGBTs has been found to be the most widely accepted method for controlling lifetime in the drift region.[43] When the IGBT is irradiated, the turn-off time decreases not only because of a reduction in the lifetime (τ_{HL}), but also because of a decrease in the current gain (α_{pnp}). The changes in the turn-off waveform for the collector current after electron irradiation can be seen in Fig. 28 where waveforms for pre- and post electron-irradiated devices are compared. After electron irradiation, the magnitude of the abrupt drop of the collector current (I_{CD}) increases and the time for the current tail to decay to zero decreases. Since the turn-off time is dominated by the recombination tail, the turn-off time will decrease in proportion to the minority-carrier lifetime. The minority carrier lifetime after irradiation (τ_f) depends upon the pre-irradiation lifetime (τ_i) and the electron irradiation dose (ϕ):

$$\frac{1}{\tau_f} = \frac{1}{\tau_i} + K\phi \tag{77}$$

where K is the radiation-damage coefficient. At high doses, the first term in this equation becomes negligible if the initial lifetime is large, and the lifetime after irradiation decreases inversely proportional to the radiation dose. The turn-off time can be reduced from over 20 μs to less than 200 ns by a radiation dose of 16 Mrad using 3 MeV electron irradiation.

In all bipolar power devices, an improvement in switching speed is accompanied by a degradation in current conduction capability. When the diffusion length is reduced by the electron irradiation, the conductivity modulation of the drift region is also reduced and the IGBT on-state voltage drop increases similar to that observed with p–i–n rectifiers. The increase in switching speed of the IGBTs is accompanied by a loss in current-handling capability. The IGBT performance is still superior to that of a power MOSFET with the same 600-V forward blocking capability. Since a short turn-off time is desirable to reduce switching losses and a low forward voltage drop is desirable to reduce the conduction losses, it becomes necessary to trade off between these characteristics. This can be done conveniently by using a plot of the forward voltage drop vs the turn-off time as shown in Fig. 29 for the symmetric and asymmetric IGBT structures. Depending upon the application, the appropriate device characteristics can be selected by choosing the irradiation dose. In circuits operating at low frequencies with large duty cycles, where the conduction losses dominate the switching losses, IGBTs with turn-off times in the range 5–20 μs are the best. An example of this is line-operated phase control circuits. For higher-frequency circuits with shorter duty cycles, where the switching losses are comparable to the conduction losses, IGBTs with turn-off times in the range 0.5–2.0 μs are appropriate. An example of this type of application is in AC motor drives operating at frequencies ranging from 1 to 10 kHz. For high-frequency circuits, the switching losses dominate. Therefore, IGBTs with gate turn-off times ranging from 100 to 500 ns are necessary. An example of such circuits is in uninterruptible power supplies (UPS) operating at between 20 and 100 kHz. Thus, the IGBTs can be tailored to match the power-switching requirements of a broad range of applications. Figure 29 shows that the power losses are smaller for the asymmetric IGBT structure when compared with the symmetric IGBT structure.

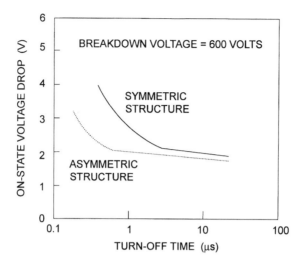

Fig. 29 Tradeoff curves of on-state voltage drop and turn-off time for the symmetric and asymmetric IGBT structures.

4.4.5 Latch-Up of Parasitic Thyristor

The maximum operating current for the IGBT is limited by the parasitic thyristor within the structure. As shown in Fig. 24, this thyristor is formed between the n^+ source region of the MOSFET acting as the cathode of the thyristor, the p base region, the n-drift region, and the p^+ collector region. With the IGBT in the on-state, electrons are supplied via the MOSFET channel and holes injected from junction J_1 are collected at junction J_2. The hole current flows to the emitter terminal through the resistance of the p base region. This current flow produces a voltage drop that forward biases the junction J_3. At normal operating current levels, this voltage drop can be made much smaller than that for a forward-biased diode (V_{bi}) by making the shunting resistance small. Under these conditions, the current gain of the n–p–n transistor is very small and the thyristor cannot latch up. However, when the on-state current density is increased, the forward bias on junction J_3 can become large enough to increase the current gain of the n–p–n transistor. If the sum of the current gains of the n–p–n and p–n–p transistors exceeds unity, the thyristor can latch up. The collector current can now flow directly to the emitter terminal, bypassing the MOSFET channel so that the IGBT current is no longer controlled by the gate bias.

It is essential to suppress the turn-on of the parasitic thyristor to maintain gate control over the collector current by reducing the gain of either the

n–p–n transistor or the p–n–p transistor. There are two basic techniques for reducing the gain of the p–n–p transistor

1. By reducing the base transport factor, which can be achieved by electron irradiation.[43]
2. By decreasing the injection efficiency of junction J_1 by incorporating a buffer layer.[42]

   Reduction of the gain of the n–p–n transistor can be achieved by

1. Adding a deep, highly doped p^+ region within the IGBT cell structure.[44]
2. Forming a shallow p^+ region that is self-aligned to the edge of the polysilicon gate located deeper than the n^+ emitter.[45]
3. Reducing the doping concentration of the n^+ emitter region.[5]
4. Reducing the hole current component flowing in the p base region using an IGBT cell referred to as the minority carrier bypass design.[46]
5. Increasing the surface concentration for the p-base diffusion while obtaining an acceptable threshold voltage using an additional n-type ion implantation in the channel region to compensate for the higher boron concentration.[47]
6. Reducing the gate oxide thickness.[48]
7. Using the atomic-lattice-layout (A-L-L) cell topology.[49]
8. Adding a diverter region in the IGBT cell.[50]

A detailed systematic treatment of latch-up suppression can be found in Ref. 5. Since the on-state current flow in the IGBT occurs partially through the p–n–p transistor, a reduction in its gain produces an increase in the on-state voltage drop. Consequently, it is preferable to reduce the gain of the n–p–n transistor to suppress thyristor latch-up. Although the latch-up of the parasitic thyristor was a serious limitation to IGBT operation when it was first commercially introduced, recent progress in latch-up suppression using the above techniques has eliminated this problem for most applications.

4.4.6   IGBT Safe Operating Area

The safe operating area (SOA) is defined as the area within the output characteristics of the IGBT where it can be operated without destructive failure if the power dissipation is kept within the thermal constraints of the device package. In general, there are three distinct boundaries for the SOA. With high applied voltages at low current levels, the maximum voltage that can be supported is limited by the breakdown voltage of the edge termination.

At high current levels with small collector voltages, the maximum collector current for the IGBT is limited by the onset of latch-up of the parasitic thyristor. Latch-up is observed at high gate bias voltages, especially when the device is operating at higher temperatures. This phenomenon is sometimes referred to as current induced latch-up because it occurs when the collector current exceeds a certain current level irrespective of the collector bias, as long as the collector voltage is relatively small.

In addition to these boundaries for the SOA there is a boundary at which the current and voltage simultaneously become large. Because of the high power dissipation within the device under these conditions, one limitation to the maximum current–voltage product is the temperature rise in the device. This thermal limit is determined by the die mount-down, the package, and the heatsink. If the time duration during which the device is subjected to the simultaneous high current and voltage stress is short, then the power dissipation is no longer the limiting factor. The SOA is then dictated by a phenomenon referred to as avalanche-induced second breakdown. This phenomenon can occur during two modes of IGBT operation.

The forward-biased safe operating area (FBSOA) of the IGBT is defined by the maximum voltage that the device can withstand without destructive failure while the collector current is saturated. During this mode of operation, electrons and holes are transported through the drift region while it is supporting a high voltage. The electric field in the drift region is sufficiently large to result in velocity saturation for the carriers. Consequently, the electron and hole concentrations in the drift region are related to the corresponding current densities by[5]

$$n = \frac{J_{\text{n}}}{q v_{sn}} \tag{78}$$

and

$$p = \frac{J_p}{q v_{sp}} \tag{79}$$

where v_{sn} and v_{sp} are the saturated drift velocities for electrons and holes, respectively. The net positive charge in the drift region is then given by

$$N^+ = N_{\text{D}} + \frac{J_p}{q v_{sp}} - \frac{J_n}{q v_{sn}} \tag{80}$$

where N_{D} is the doping concentration in the drift region.

The electric-field distribution in the drift region is determined by this charge. Unlike the steady-state forward blocking condition where the drift region charge is equal to the doping concentration (N_{D}), under FBSOA conditions the net charge is usually much larger because the hole current density is significantly larger than the electron current density. This increase in the charge in the drift region results in an increase in the electric field, which in turn leads to breakdown in the IGBT cell at voltages lower than the breakdown voltage of the edge termination. Using a simple one-dimensional analysis, Poisson's equation can be solved with a net positive charge in the drift region given by N^+. The breakdown voltage that determines the SOA for silicon one-sided abrupt junctions (BV_{SOA}) is given by[5,51]

$$BV_{SOA} = \frac{5.34 \times 10^{13}}{(N^+)^{3/4}} \tag{81}$$

We must account for the current gain of the open-base transistor to determine the maximum voltage that can be supported. Thus, the FBSOA limit is given by the criterion

$$\alpha_{pnp} M = 1 \tag{82}$$

with

$$\alpha_{pnp} = \frac{1}{\cosh(l/L_a)} \tag{83}$$

and

$$M = \left[1 - \left(\frac{V}{BV_{SOA}}\right)^n\right]^{-1} \tag{84}$$

where l is the undepleted n-base width, and n has a value of between 4 and 6. These equations indicate that the onset of avalanche breakdown occurs in the IGBT cell at a lower collector bias when the collector current is increased. Note that a higher breakdown voltage (BV_{SOA}) can be obtained by reducing the doping concentration (N_D) in the drift region. Although this cannot be done for a symmetric IGBT structure because it would lead to reach-through breakdown problems, it is possible in the asymmetric structure because the buffer layer prevents reach-through.

The reverse-biased safe operating area (RBSOA) is important during turn-off of the IGBT. Since the gate bias is zero or a negative value under these conditions, the current is transported in the drift region exclusively via holes for an n-channel IGBT. The holes add charge to the drift region, resulting in an increase in the electric field at the p-base–n-drift-region junction. Since there are no electrons in the space-charge region, the electric-field enhancement during the RBSOA conditions is worse than for the FBSOA conditions. The net charge in the space-charge region under the RBSOA conditions is given by

$$N^+ = N_D + \frac{J_C}{qv_{sp}} \tag{85}$$

where J_C is the total collector current. As in the FBSOA analysis, the avalanche breakdown limit under RBSOA conditions is given by Eq. 82 with the breakdown voltage determined by the net charge given by Eq. 85.

4.4.7 p-Channel IGBTs

In numerical and appliance controls, it is preferable to use one n-channel device and one p-channel device in parallel to form a composite AC switch to allow control of both IGBTs with a common reference terminal. For these applications, p-channel power MOSFETs have been developed to complement the n-channel devices. Because of the lower mobility of holes in silicon, p-channel power MOSFETs have a three times higher specific on-resistance than n-channel devices . However, in the IGBT, the drift region is flooded with minority carriers during forward conduction, and the concentration of the free carriers greatly exceeds the doping level. Thus, the carrier transport is determined by ambipolar diffusion and drift, which is similar for both n-channel and p-channel devices. Consequently, the on-state voltage drop of the p-channel IGBT is very close to that of the n-channel IGBT.[52]

4.4.8 High-Voltage IGBTs

The IGBT is ideally suited for scaling up the blocking voltage capability. In the power MOSFET, the on-resistance increases sharply with increasing breakdown voltage because an increase in the resistivity and thickness of the drift region is required to support the operating voltage. In contrast, for the IGBT, the drift-region resistance is drastically reduced by the high concentration of injected minority carriers during on-state current conduction. The contribution to the forward drop from the drift region then becomes dependent on its thickness but independent of its original resistivity. When the blocking-voltage capability of the IGBT is increased by increasing the drift-region width, there is a relatively small increase in the on-state voltage drop with increasing blocking-voltage capability compared with power MOSFETs. Experimental comparison of the forward conduction characteristic of symmetric-blocking 300-, 600- and 1200-V IGBTs demonstrates that the forward-conduction current density of the IGBT will decrease approximately as the square root of the breakdown voltage.[53] This moderate rate of reduction in current density has allowed the rapid development of devices with both high-current and high-voltage capability.[54] Devices are now available in power modules with blocking voltages of 2500 V and current ratings of 1000 A.

4.4.9   High-Temperature Operation of IGBTs

The on-state characteristics of the IGBT can be considered to consist of two segments: a diode-drop portion followed by a resistive portion. The diode voltage drop decreases when the temperature increases. This behavior is typical for a p–i–n diode, where the injection across the p–n junction becomes stronger with increasing temperature. At the same time, the resistance of the second segment, which is associated with the channel resistance, increases when the temperature increases. The decrease in the diode forward drop compensates for the increase in channel resistance. This results in a relatively small increase in the on-state voltage drop for the IGBT with increasing temperature.[55] In contrast, for the power MOSFET, the on-resistance increases rapidly, which requires derating the current-handling capability more severely than for the IGBT. This feature makes the IGBT well suited for applications in which a high ambient temperature is encountered. It is worth pointing out that the small positive temperature coefficient of the forward drop at higher current levels in IGBTs is beneficial because it ensures homogeneous current distribution within chips, and good current sharing when paralleling devices. The paralleling of IGBTs without matching devices or emitter ballasting has been successfully used to achieve high-current circuit performance.

One of the problems encountered when operating the IGBT at high temperatures has been the latch-up of the parasitic p–n–p–n thyristor structure inherent in the device structure, causing loss of gate-controlled current conduction. By using the asymmetric device structure with hole bypass regions, the latching-current density has been raised significantly to provide a current margin of more than a factor of 10, making the IGBTs useful in applications with ambient temperatures as high as 200°C. Devices that limit the collector current by saturation prior to latch-up have been demonstrated to operate at up to 125°C.

4.4.10   Trench-Gate IGBT structures

The UMOS-IGBT structure is illustrated in Fig. 30. As in the power MOSFET, the trench gate must extend below the junction between the p base region and the n-drift region to form a gate-bias-induced channel between the n^+ emitter and the n-drift region. The electron-current path indicated in the figure illustrates that there is no JFET or accumulation layer resistance in the UMOS structure. This reduces the resistance for the MOS current flow. In addition, the cell pitch can be made relatively small when compared with the DMOS structure, allowing a five-fold increase in channel density. The improvement in the MOS path resistance for the UMOS

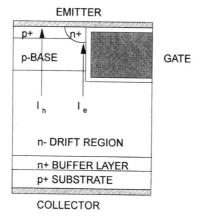

Fig. 30   Cross-section of the IGBT structure with trench gate.

structure results in a superior on-state characteristic. For devices with high minority-carrier lifetime in the drift region, the on-state voltage drop at a current density of 200 A/cm^2 was found to be 1.2 V for the UMOS structure compared with 1.8 V for the DMOS structure.[56,57] A larger difference can be expected when the minority-carrier lifetime is reduced to increase the switching speed. The latching current density for the UMOS IGBT structure is also superior to that for the DMOS structure. This is attributed to the improved hole-current flow path in the UMOS structure. As shown in Fig. 30, the hole-current flow (I_h) takes place along a vertical path in the UMOS structure, whereas in the DMOS structure, hole-current flow occurs below the n^+ emitter in the lateral direction. As a consequence, the resistance for the hole-current flow in the UMOS structure is determined only by the depth of the n^+ emitter region. A shallow p^+ layer, shown in the figure, reduces this resistance. This p^+ region is similar to the deep p^+ region required to suppress latch-up of the parasitic thyristor in the DMOS structure.

4.5   MOS-GATED THYRISTORS

The MOS-gated thyristor structure has been investigated to utilize the superior on-state characteristics of thyristors in combination with the ease of control of an MOS-gate structure. These structures are of particular interest for very-high-voltage power-switching applications because the on-state voltage drop of the IGBT increases with increasing blocking voltage capability. To control the current flow in a thyristor structure, it is necessary both to trigger the device from its off-state to its on-state and to turn off the device once it is carrying the on-state current. The turn-on of a thyristor by means of an MOS-gate structure integrated within the thyristor structure is relatively easier than the turn-off of a thyristor after it has entered into its regenerative current-conduction mode. For these devices, an important parameter is the maximum controllable current density, which is defined as the highest current density that can be switched off under MOS-gate control. A large maximum controllable current density is highly desirable to obtain a large operating current range for the device.

4.5.1 MOS-Gated Turn-On of Thyristors

To turn on a thyristor by using an MOS-gate structure, the basic concept is to supply the base drive current for one of the coupled transistors within the thyristor structure via a channel formed under the MOS-gate region.[58] Figure 31 shows a cross-section of the basic cell of an MOS-gated thyristor with the current flow paths for electrons and holes during the turn-on process. When the gate bias is zero and a positive voltage is applied to the anode, the device exhibits a high forward blocking voltage by supporting the voltage across the reverse-biased junction J_2. When a positive bias is applied to the gate electrode, electrons are supplied, as indicated by the current I_e, to the base region of the p–n–p transistor. This results in the injection of holes from the anode into the n-drift region. These holes diffuse across the n-drift region and are collected at the reverse-biased junction J_2. The current (I_h) in the p base region created by the collection of holes across junction J_2 flows into the short. However, this current flow produces a voltage drop across the resistance (R_S) of the p base region. When the voltage drop across this resistance becomes more than that of a forward-biased diode, the n^+-emitter–p-base junction J_1 becomes sufficiently forward biased to begin injection of electrons from the n^+ emitter into the p base region. This triggers the

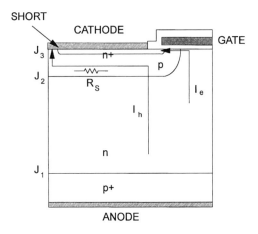

Fig. 31 MOS-gate structure for turning on a thyristor.

regenerative feedback mechanism between the two coupled transistors within the thyristor structure. The thyristor can therefore be turned on by the application of the gate voltage to the MOS electrode.

4.5.2 MOS-Controlled Thyristor

The first thyristor structure that had the ability to turn off the thyristor on-state current flow under the control of an integrated MOS-gate structure was called either the MOS-controlled thyristor (MCT) or the MOS-controlled gate turn-off thyristor (MOS-GTO).[59,60] The basic concept is to introduce an MOS-controlled short-circuit between the n^+ emitter and the p base region by forming a lateral MOSFET within the p base region. When the MOSFET gate is biased so that this short-circuit does not occur, the thyristor can be triggered into its on-state because the injection efficiency of the n^+-emitter–p-base junction is large. However, if the gate of the MOSFET is biased so that it forms a conductive path between the n^+ emitter and the p-base region, the injection efficiency of the emitter is greatly reduced. If the resulting current gain of the n–p–n transistor is sufficiently low, the thyristor can be turned off. This method for turning off the thyristor can also be regarded as raising the holding current of the thyristor to above its on-state current density.

Two basic structures for the MCT have been proposed. In the first one, shown in Fig. 32a, a p-channel MOSFET is integrated within the p base region by diffusion of an additional n region (n well) into the p base region to form the substrate for the p-channel MOSFET. This step is followed by the diffusion of a p^+ region that acts as the source of the p-channel MOSFET. This p^+ region is shorted to the n^+ cathode region of the thyristor by the cathode metalization. To obtain a low resistance for the p-channel MOSFET, the n well and the p^+ source diffusions are self-aligned using the DMOS process with the polysilicon gate as a common masking boundary. However, unlike the power MOSFET and IGBT, these diffusions must be performed into a relatively highly doped p base region. This makes it difficult to obtain a low threshold voltage for the p-channel MOSFET. When a negative bias is applied to the gate of the p-channel MOSFET, a channel is formed by the inversion of the n-well surface. This provides a path for the flow of holes from the p base region into the cathode contact that bypasses the n^+-emitter–p-base junction. The holes that flowed into the p base region when the thyristor was operating in its on-state can then be diverted via the p-channel MOSFET into the cathode electrode. If the voltage drop for the hole-current flow through the MOSFET is significantly below the on-state diode drop, the voltage across the n^+-emitter–p-base junction will become too small to inject a significant number of electrons into the p-base region. This is equivalent to a large reduction in the current gain of the n–p–n transistor or to a large increase in the holding current for the thyristor. The thyristor regenerative action is then terminated. The stored charge in the n

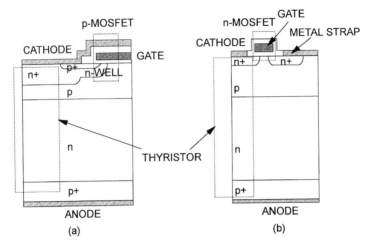

Fig. 32 Cross-sections of the MCT structures with (a) *p*-channel and (b) *n*-channel turn-off MOSFETs.

base region is now removed by the recombination of the holes and electrons, leading to a decay in the anode current.

The second MCT structure, shown in Fig. 32b, has an *n*-channel MOSFET integrated within the *p* base region. In this case, an additional n^+ region is formed adjacent to the n^+ cathode region of the thyristor to form the drain of the *n*-channel MOSFET. This region is then shorted to the *p* base region by a metal strap, which is not connected to the cathode metalization. The turn-off MOSFET in this structure has a lower resistance than in the structure of Fig. 32a because of the higher inversion-layer mobility for electrons. The *n*-channel MOSFET can be formed without the additional processing steps required in the MCT with the *p*-channel MOSFET. The channel length of the *n*-channel MOSFET is controlled by the gate length and the lateral diffusion of the n^+ regions. However, the cell size of the second structure is larger than for the first because the floating metal strap occupies a large space resulting from large thickness of the cathode metal. This results in a lower maximum controllable current density for the sescond structure.

Attempts to optimize the performance of the MCT using 1.25 μm VLSI technology have resulted in the demonstration of acceptable (250 A/cm^2) maximum controllable current density.[61–63] However, the complexity of the process for the MCT when compared with the IGBT has impeded its commercialization. The integration of the turn-off MOSFETs also results in a reduction of the injection efficiency of the n^+-emitter–*p*-base junction, which increases the on-state voltage drop. During turn-off, the current tends to localize because of the negative temperature coefficient of the on-state voltage drop. This creates current filaments during turn-off, which can

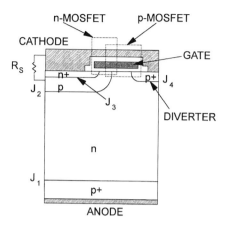

Fig. 33   Cross-section of the BRT structure.

destroy the device. Further, gate-controlled current saturation is not possible in the MCT. Thus, the MCT cannot be used to replace the IGBT in applications where short-circuit-withstand capability is required.

4.5.3   The Base-Resistance-Controlled Thyristor

In the base-resistance-controlled thyristor (BRT), a diverter region is formed adjacent to the p base region of the thyristor, as shown in Fig. 33. This device structure differs fundamentally from the MCT structure because the turn-off MOSFET is not integrated within the p base region but is formed within the n base region.[64] This has important implications not only from the point of view of device characteristics but also because it simplifies the fabrication process. The diverter is a shallow p^+ region formed adjacent to the p base region of the thyristor and connected to the cathode electrode. The p base region and the n^+ emitter region of the thyristor are formed by using the polysilicon gate as the mask. This results in the formation of a DMOS structure at the polysilicon edges. It is also important to short the p base region to the n^+ emitter at a location orthogonal to the cross-section to create a shunting resistance R_S. This ensures good forward blocking capability and a high dV/dt capability even when no gate bias is applied.

When the gate bias is zero and a positive bias is applied to the anode, the BRT operates in its forward blocking mode. In this case, junctions J_2 and J_4 are simultaneously reverse-biased. These regions act as guard rings for each other, which results in reducing the effects of junction curvature on the breakdown voltage. The forward blocking voltage is determined by the p–n–p open-base transistor breakdown as in the IGBT. The BRT can be switched from its forward blocking state to the on-state by applying a positive gate bias, which creates an inversion layer at the surface of the p base region under the gate. The electrons supplied from the n^+ emitter into the n base region then provide the base drive current to the p–n–p transistor. Holes are injected from the p^+ anode and collected at junctions J_2 and J_4. Thus, at low current levels, the device behaves like an IGBT. In this mode, the anode current can be saturated by reducing the gate bias to a value close to the threshold voltage of the n-channel MOSFET. The IGBT mode of operation can be used to control the turn-on of the BRT. When the anode current increases, sufficient hole current is collected at junction J_2 to produce a voltage drop across the shunt resistance R_S that forward biases the n^+-emitter–p-base junction to allow the n^+ emitter to inject electrons. This initiates the turn-on of the thyristor. The on-state voltage drop of the device then becomes close to that of a thyristor.

The MOS-gated turn-off of the BRT structure is achieved by applying a negative gate bias. This bias creates an inversion layer at the surface of the n-drift region under the gate electrode between the p base region and the p^+ diverter region. The hole current flowing into the p base region can now flow through an alternate path from that provided by the emitter shunt resistance R_S. If the resistance for this path for hole current flow is much smaller than the shunt resistance, the thyristor current flow is turned off because its holding current becomes larger than the on-state operating current level. The application of negative gate bias reduces the bias across the n^+-emitter–p-base junction J_1, so that the n^+ emitter stops injecting electrons. Consequently, during the turn-off process, the anode current is diverted from the n^+ emitter to the p^+ diverter region. Once the thyristor has turned off at the end of the gate ramp, the current decays because of the recombination of the holes that were injected into the n-drift region during the on-state. This produces a current tail, as in the IGBT and the MCT.

The fabrication process for the BRT is similar to the IGBT because the cell is formed using the DMOS technology. High maximum controllable current densities (1000 A/cm^2) have been achieved.[65–66] When compared with the IGBT, a fast switching speed with lower on-state voltage drop has been confirmed by using electron irradiation to control the minority-carrier lifetime.[67] However, commercialization of the BRT has not occurred because current saturation cannot be performed after latch-up of the thyristor. In the future, these devices may find applications in resonant convertors, which do not require the current saturation capability.

4.5.4 The Emitter-Switched Thyristor

The emitter-switched thyristor (EST) is a device in which the on-state current flows via a thyristor region. However, in this device, the n^+ emitter of the thyristor is not directly connected to the cathode metalization. Instead, the thyristor emitter forms the drain region of a lateral n-channel MOSFET that is integrated into the p base region of the thyristor,[68] as illustrated in Fig. 34. This MOSFET, formed by using the DMOS process with the polysilicon gate as a mask, is also used for turning on the EST. Since there is no external contact to the n^+ emitter region of the thyristor, it is referred to as a floating n^+ emitter region. The vertical thyristor formed between the n^+ emitter, the p base region, the n-drift region, and the p^+ anode region is called the main thyristor.

When a positive bias is applied to the anode with zero gate bias, the junctions J_2 and J_5 become reverse biased simultaneously. These junctions act as guard rings for each other, ameliorating the effect of junction curvature under the gate electrode. The depletion region spreads into the n^- drift region, allowing the device to support a high voltage in this forward blocking mode. When a positive gate bias is applied, an inversion layer is formed at the surface of the p base region to create the channels 1 and 2. At the same time, an accumulation layer is formed at the surface of the n-drift region under the gate electrode. When a positive bias is applied to the anode, electrons flow from the n^+ source region into the n-drift region via channel 1. The electron flow provides base drive current for the vertical p–n–p transistor formed between junctions J_1 and J_5, and junctions J_1 and J_2. Holes injected from the anode are then collected by both junction J_5 and junction J_2. The holes collected by junction J_5 flow to the cathode contact via the p^+ region on the right-hand side. The holes collected by junction J_2 flow to the cathode contact via the resistance R_p. Under these conditions, the EST operates in the IGBT mode.

When the hole current flow through the resistance R_p produces a sufficient voltage drop to forward bias junction J_3 by the potential for a forward-biased diode (V_{bi}), it begins to inject electrons and the main thyristor latches up. In the on-state, the thyristor current flows from the n^+ floating emitter through the lateral MOSFET to the cathode electrode. Thus, the thyristor current can be turned off by reducing the gate bias to zero. If the gate bias is large, the lateral MOSFET operates in its linear region and the total on-state voltage drop of the device is the sum of the voltage drop across the main thyristor and the dual-channel MOSFET.

An important feature of the EST structure is that the current saturation extends to very high voltages even after latch-up of the main thyristor.[69–71] This occurs because it is possible to pinch off the lateral MOSFET when higher anode voltages are applied. With increasing anode voltage, the potential of the n-drift region under the gate electrode increases. When this voltage becomes comparable to the gate bias voltage, the accumulation region can no longer form, and a depletion region begins to extend from junction J_2 at point A in Fig. 34. This cuts off the main thyristor current flow into the cathode electrode through the lateral MOSFET.[72] However, since the gate bias exceeds the threshold voltage, electrons can be supplied via channel 1 to the n-drift region to allow the device to operate like an IGBT with holes collected by junction J_2 and junction J_5.

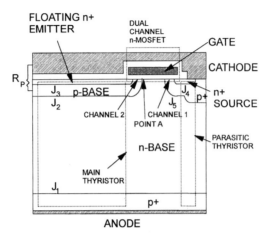

Fig. 34   Cross-section of the EST structure.

The maximum controllable current density for the EST is set by the turn-on of the parasitic thyristor within the EST structure.[70] Although the device can be operated beyond the current level at which the parasitic thyristor latches up to obtain a high-surge-current handling capability, the device cannot be turned off by reducing the gate bias to zero under these conditions. The operating range for the anode current is, therefore, determined by the latching-current density of the main thyristor and the maximum controllable current density. The parasitic thyristor latching-current density decreases with increasing length of the floated emitter region.[70] This effect is due to an increase in the total hole current that flows via the p base region of the parasitic thyristor. Thus, it is necessary to optimize the emitter length not only to minimize the on-state voltage drop[71] but to also obtain an adequate latching-current density for the parasitic thyristor.

The turn-off process for the EST is similar to that for the IGBT. When the gate bias is reduced to zero, the floating n^+ emitter immediately stops injecting electrons because it is disconnected from the cathode terminal. The anode current continues to flow via the holes within the n-drift region. This produces a current tail whose duration depends upon the lifetime of the carriers in the n-drift region. The duration of the tail can be reduced by introducing recombination centers. A reduction in the turn-off time for the EST has been demonstrated by using electron irradiation to reduce the lifetime.[67] The turn-off time can be reduced from about 3 μs for the as-fabricated devices to about 0.2 μs by using a radiation dose of 16 Mrad.

4.6 SILICON CARBIDE POWER DEVICES

As discussed earlier in this chapter, the ideal specific on-resistance for the drift region in silicon devices increases rapidly with increasing breakdown voltage. For breakdown voltages above 200 V, the ideal specific on-resistance for the n-channel silicon MOSFET becomes greater than $0.1 \, \Omega\text{-cm}^2$. This implies that the on-state voltage drop will exceed 10 V for an on-state current density of $100 \, \text{A/cm}^2$ resulting in a very high power dissipation within the devices. Although it is possible to reduce the on-state power loss by decreasing the on-state current density, this approach is undesirable because of the corresponding increase in device area, which leads to higher chip cost. From a fundamental physics viewpoint, a lower specific on-resistance for the drift layer is achievable if the semiconductor has a higher breakdown electric field strength. It can be shown[3,73] that to obtain a desired breakdown voltage (V_B), the drift region must have a doping concentration N_D of

$$N_D = \frac{\varepsilon_s \mathscr{E}_c^2}{2qV_B} \tag{86}$$

and a thickness W_D of

$$W_D = \frac{2V_B}{\mathscr{E}_c} \tag{87}$$

where $\mathscr{E}_c$ is the critical electric field at which avalanche breakdown occurs in the semiconductor. The specific on-resistance of the drift region, which was also defined as the ideal specific on-resistance, is then given by

$$R_{\text{on,sp(ideal)}} = \frac{W_D}{q\mu_n N_D} = \frac{4V_B^2}{\varepsilon_s \mathscr{E}_c^3 \mu_n} \tag{88}$$

Thus, the ideal specific on-resistance decreases inversely proportional to the mobility and as the cube of the breakdown electric field strength. The denominator ($\varepsilon_s \mathscr{E}_c^3 \mu_n$) in Eq. 88 is Baliga's figure-of-merit (BFOM) for unipolar power devices.[74,75]

By using the known material properties of semiconductors and Eq. 88, we can select those that will exhibit a lower ideal specific on-resistance when compared with silicon. The most promising semiconductor materials are GaAs, whose BFOM is 12.7 times larger than silicon, and SiC whose BFOM is 200 times larger than silicon. Vertical power MESFETs have been fabricated from GaAs[76,77] with breakdown voltages up to 200 V. SiC offers a much larger improvement in the ideal specific on-resistance and is suitable for high operating temperatures because of its large bandgap. High-performance silicon carbide power MOSFETs have not yet been demonstrated.[78,79] However, the low specific on-resistance of the drift region has been utilized successfully for the demonstration of high-voltage Schottky-barrier rectifiers with low on-state voltage drops[80–83] using a nearly ideal edge termination.[3]

4.7 SUMMARY AND FUTURE TRENDS

Power semiconductor devices are regarded as essential components in systems that require the control of power and energy. Their improved performance continues to facilitate improvements in the efficiency, size, and weight of power electronics systems. The application of MOS technology by the power device community and its assimilation into production in the 1970s has had a revolutionary impact on applications. The power bipolar transistors that were extensively used until the 1970s have been replaced by the power MOSFETs in low-voltage, high-frequency applications, and by the IGBT in medium-voltage and medium-frequency applications. In addition, many MOS-gated thyristor structures have been proposed recently. These devices are candidates to replace the GTO in high-power applications. Looking further into the future, it is likely that the silicon high-voltage bipolar devices (IGBTs and MOS-gated thyristors) will be replaced by unipolar devices based on silicon carbide.

PROBLEMS

1. Determine the doping concentration and thickness of the drift region for a silicon Schottky power rectifier to obtain a breakdown voltage of 100 V. Assume that ideal breakdown occurs at the edge termination.

2. Calculate the on-state voltage drop for the Schottky rectifier in Problem 1 at a current density of 100 A/cm^2 when a metal with barrier height of 0.8 eV is used.

3. Calculate the reverse leakage-current density at room temperature for the Schottky rectifier in Problem 1 at a bias of 80 V including the effect of the Schottky-barrier-lowering phenomenon when a metal with barrier height of 0.8 eV is used.

4. Determine the width of the n^- drift region for a p–i–n rectifier designed to support 1000 V if the doping concentration in the drift region is $5 \times 10^{13}/\text{cm}^3$.

5. Calculate the on-state voltage drop for the p–i–n rectifier in Problem 4 at a current density of 200 A/cm^2 if the lifetime in the drift region is 1 μs.

6. Consider an n-channel power MOSFET with a breakdown voltage of 500 V fabricated using the DMOS process with a gate length of 30 μm

and a polysilicon window of 16 μm. The n^+ source has a length of 5 μm and a junction depth of 1 μm. The p base region has a depth of 3 μm and a uniform doping concentration of 1×10^{17} cm^3. The gate oxide thickness is 0.1 μm. Determine the total specific on-resistance at a gate bias of 15 V based on contributions from the channel, accumulation, JFET, and drift regions.

7. Consider an n-channel UMOSFET structure with a breakdown voltage of 500 V fabricated using a mesa and trench width of 3 μm (cell pitch of 6 μm). The n^+ source has a junction depth of 1 μm. The p base region has a depth of 3 μm and a uniform doping concentration of 1×10^{17}/cm^3. The gate oxide thickness is 0.1 μm. Determine the total specific on-resistance based on contributions from the channel and drift regions.

8. Determine the doping concentration and thickness of the drift region for a symmetric blocking n-channel IGBT structure with a breakdown voltage of 500 V if the lifetime in the drift region is 1 μm.

9. Using the parameters given in Problem 8, calculate the on-state voltage drop for the IGBT at a current density of 200 A/cm^2. Assume that the IGBT has the same DMOS cell parameters given in Problem 6.

10. What is the on-state voltage drop for a p-channel IGBT with the device parameters given in Problem 9?

REFERENCES

1. B. J. Baliga, "Evolution of MOS–bipolar power semiconductor technology," *Proc. IEEE*, **76**, 409 (1988).

2. B. J. Baliga, M. S. Adler, P. V. Gray, R. Love, and N. Zommer, "The insulated gate rectifier," in *IEEE Int. Electron Devices Mtg.* 1982, p. 409.

3. B. J. Baliga, "Power semiconductor device figure of merit for high frequency applications," *IEEE Electron Dev. Lett.* **ED-10**, 455 (1989).

4. D. Alok, P. McLarty, and B. J. Baliga, "A simple edge termination for silicon carbide devices with nearly ideal breakdown voltage," *IEEE Electron Dev. Lett.* **ED-15**, 394 (1994).

5. B. J. Baliga, *Power Semiconductor Devices*, PWS, Boston, 1995.

6. S.M. Sze, *Physics of Semiconductor Devices*, John Wiley, New York, 1981.

7. L. Tu and B. J. Baliga, "On the reverse blocking characteristics of Schottky power diodes," *IEEE Trans. Electron Dev.* **ED-39**, 2813 (1992).

8. B. J. Baliga, "The pinch rectifier," *IEEE Electron Dev. Lett.* **ED-5**, 194 (1984).

9. B. J. Baliga, "Analysis of junction barrier controlled Schottky rectifier characteristics," *Solid State Electron.* **28**, 1089 (1985).

10. H. Kozaka, M. Takata, S. Murakami, and T. Yatsuo, "Low leakage current

Schottky barrier diode," in *IEEE Int. Symp. Power Semiconductor Devices and ICs*, 1992, p. 80.

11. S. Kumori, J. Ishida, M. Tanaka, M. Wakatabe, and T. Kan, "The low power dissipation Schottky barrier diode with trench structure," in *IEEE Int. Symp. Power Semiconductor Devices and ICs*, 1992, p. 66.

12. B. J. Baliga, "High voltage, junction gate field effect transistor with recessed gates," *IEEE Trans. Electron Dev.* **ED-29**, 1560 (1982).

13. M. Mehrotra and B. J. Baliga, "Very low forward drop JBS rectifiers fabricated using submicron technology," *IEEE Trans. Electron Dev.* **ED-30**, 1655 (1994).

14. M. Mehrotra and B. J. Baliga, "Trench MOS barrier Schottky (TMBS) rectifier," *Solid State Electron.* **38**, 801 (1995).

15. J. M. Fairfield and B. V. Gokhale, "Gold as a recombination center in silicon," *Solid State Electron.* **8**, 685 (1965).

16. K. P. Lisiak and A. G. Milnes, "Platinum as a lifetime control deep impurity in silicon," *J. Appl. Phys.* **46**, 5229 (1975).

17. B. J. Baliga and E. Sun, "Comparison of gold, platinum and electron irradiation for controlling lifetime in power rectifiers," *IEEE Trans. Electron Dev.* **ED-24**, 685 (1977).

18. B. J. Baliga, "Analysis of a high voltage merged *P–i–N* Schottky (MPS) rectifier," *IEEE Electron Dev. Lett.* **ED-8**, 407 (1987).

19. Y. Shimizu, M. Naito, S. Murakami, and Y. Terasawa, "High speed low-loss *p–n* diode having a channel structure," *IEEE Trans. Electron Dev.* **ED-31**, 1314 (1984).

20. M. Mehrotra and B. J. Baliga, "Comparison of high voltage power rectifier structures, in *IEEE Int. Symp. Power Semiconductor Devices and ICs*, 1993, p. 199.

21. M. N. Darwish and K. Board, "Optimization of breakdown voltage and on-resistance of VDMOS transistors," *IEEE Trans. Electron Dev.* **ED-31**, 1769 (1984).

22. V. A. K. Temple and P. V. Gray, "Theoretical comparison of DMOS and VMOS structures for voltage and on-resistance," in *IEEE Int. Electron Devices Meeting Digest*, 1979, p. 88.

23. C. G. B. Garrett and W. H. Brattain, "Physical theory of semiconductor surfaces," *Phys. Rev.* **99**, 376 (1955).

24. S. C. Sun and J. D. Plummer, "Electron mobility in inversion and accumulation layers on thermally oxidized silicon surfaces," *IEEE Trans. Electron Dev.* **ED-27**, 1497 (1980).

25. S. C. Sun and J. D. Plummer, "Modelling of the on-resistance of LDMOS, VDMOS, and VMOS power transistors," *IEEE Trans. Electron Dev.* **ED-27**, 356 (1980).

26. C. Hu, "A parametric study of power MOSFETs," in *IEEE Power Electronics Specialists Conference Record*, 1979, p. 385.

27. S. D. Kim, I. J. Kim, M. K. Han, and Y. I. Choi, "An accurate on-resistance model for low voltage VDMOS devices," *Solid State Electron.* **38**, 345 (1995).

28. K. Shenai, P. A. Piacente, R. Saia, C. S. Korman, W. Tantraporn, and B. J.

Baliga, "Ultralow resistance selectively silicided VDMOS FETs for high frequency power switching applications fabricated using sidewall spacer technology," *IEEE Trans. Electron Dev.* **ED-35**, 2459 (1988).

29. D. Ueda, H. Takagi, and G. Kano, "A new vertical power MOSFET structure with extremely reduced on-resistance," *IEEE Trans. Electron Dev.* **ED-32**, 2 (1985).

30. H. R. Chang, R. D. Black, V. Temple, W. Tantraporn, and B. J. Baliga, "Self-aligned UMOSFETs with specific on-resistance of 1 milliohm-cm^2," *IEEE Trans. Electron Dev.* **ED-34**, 2329 (1987).

31. S. Matsumoto, T. Ohno, H. Ishii, and H. Yoshino, "A high performance self-aligned UMOSFET with a vertical trench contact structure," *IEEE Trans. Electron Dev.* **ED-41**, 814 (1994).

32. B. J. Baliga, T. Syau, and P. Venkatraman, "The accumulation mode field-effect transistor," *IEEE Electron Dev. Lett.* **ED-13**, 427 (1992).

33. T. Syau, P. Venkatraman, and B. J. Baliga, "Comparison of ultralow specific on-resistance UMOSFET structures," *IEEE Trans. Electron Dev.* **ED-41**, 800 (1994).

34. Y. Baba, N. Matsuda, Y. Yanagiya, S. Hiraki, and S. Yasuda, "A study of a high voltage blocking UMOSFET with a double gate structure," in *IEEE Int. Symp. Power Semiconductor Devices and ICs*, 1992, p. 300.

35. O. Ishikawa and H. Esaki, "A high power high gain VD-MOSFET operating at 90 MHz," *IEEE Trans. Electron Dev.* **ED-34**, 1157 (1987).

36. N. Thapar and B. J. Baliga, "A comparison of high frequency cell designs for high voltage DMOSFETs," in *IEEE Int. Symp. Power Semiconductor Devices and ICs*, 1994, p. 131.

37. T. Sakai and N. Murakami, "A new VDMOSFET structure with reduced reverse transfer capacitance," *IEEE Trans. Electron Dev.* **ED-36**, 1381 (1989).

38. S. Clemente and B. R. Pelley, "Understanding the power MOSFET switching performance," in *Proc. IEEE Industrial Applications Society Meeting*, 1981, p. 763.

39. C. Hu, M. H. Chi, and V. M. Patel, "Optimum design of power MOSFETs," *IEEE Trans. Electron Dev.* **ED-31**, 1693 (1984).

40. H. R. Chang and B. J. Baliga, "Numerical and experimental analysis of 500V power DMOSFET with an atomic-lattice-layout," in *IEEE Device Research Conference*, 1989, Abstr. VB-5.

41. B. J. Baliga, M. S. Adler, R. P. Love, P. V. Gray, and N. Zommer, "The insulated gate transistor," *IEEE Trans. Electron Dev.* **ED-31**, 821 (1984).

42. J. P. Russel, A. M. Goodman, L. A. Goodman, and J. M. Nielson, "The COMFET," *IEEE Electron Dev. Lett.* **ED-4**, 63 (1983).

43. B. J. Baliga, "Switching speed enhancement in insulated gate transistors by electron irradiation," *IEEE Trans. Electron Dev.* **ED-31**, 1790 (1984).

44. B. J. Baliga, M. S. Adler, P. V. Gray, and R. P. Love, "Suppressing latch-up in insulated gate transistors," *IEEE Electron Dev. Lett.* **ED-5**, 323 (1984).

45. S. Eranen and M. Blomberg, "The vertical IGBT with an implanted buried layer," in *IEEE Int. Symp. Power Semiconductor Devices and ICs*, 1991, p. 211.

46. A. Nakagawa, H. Ohashi, M. Kurata, H. Yamaguchi, and K. Watanabe, "Non-latch-up, 1200 volt bipolar mode MOSFET with large SOA," in *IEEE Int. Electron Devices Meeting Digest*, 1984, p. 860.

47. T. P. Chow and B. J. Baliga, "Counter-doping of MOS channel (CDC)—new technique of improving suppression of latching in insulated gate bipolar transistors," *IEEE Electron Dev. Lett.* **ED-6**, 29 (1988).

48. T. P. Chow and B. J. Baliga, "The effect of MOS channel length on the performance of insulated gate transistors," *IEEE Electron Dev. Lett.* **ED-6**, 413 (1985).

49. B. J. Baliga, S. R. Chang, P. V. Gray, and T. P. Chow, "New cell designs for improving IGBT safe-operating-area," in *IEEE Int. Electron Devices Meeting Digest*, 1988, p. 809.

50. N. Thapar and B. J. Baliga, "A new IGBT structure with wider safe operating-area (SOA)," in *IEEE Symp. on Power Semiconductor Devices and ICs*, 1994, p. 177.

51. W. Fulop, "Calculation of avalanche breakdown of silicon *p–n* junctions," *Solid State Electron.* **10**, 39 (1967).

52. T. P. Chow and B. J. Baliga, "Comparison of *n* and *p* channel IGTs," in *IEEE Int. Electron Devices Meeting Digest*, 1984, p. 278.

53. T. P. Chow and B. J. Baliga, "Comparison of 300, 600, 1200 volt *n*-channel insulated gate transistors," *IEEE Electron Dev. Lett.* **ED-6**, 161 (1985).

54. A. Nakagawa and H. Ohashi, "600 and 1200V bipolar mode MOSFETs with high current capability," *IEEE Electron Dev. Lett.* **ED-6**, 378 (1985).

55. B. J. Baliga, "Temperature behavior of insulated gate transistor characteristics," *Solid State Electron.* **28**, 289 (1985).

56. H. R. Chang and B. J. Baliga, "500V *n*-channel IGBT with trench gate structure," *IEEE Trans. Electron Dev.* **ED-36**, 1824 (1989).

57. M. Harada, T. Minato, H. Takahashi, H. Nishihara, K. Inoue, and I. Takata, "600V trench IGBT in comparison with planar IGBT," in *IEEE Symp. on Power Semiconductor Devices and ICs*, 1994, p. 411.

58. B. J. Baliga, "Enhancement and depletion mode vertical channel MOS gated thyristors," *Electron. Lett.* **15**, 645 (1979).

59. V. A. K. Temple, "MOS controlled thyristors (MCTs)," in *IEEE Int. Electron Devices Meeting*, 1984, p. 282.

60. M. Stoisiek and H. Strack, "MOS GTO—a turn-off thyristor with MOS controlled emitter shorts," in *IEEE Int. Electron Devices Meeting*, 1985, p. 158.

61. F. Bauer, P. Roggwiler, A. Aemmer, W. Fichtner, R. Vuilleumier, and J. M. Moret, "Design aspects of MOS controlled thyristor elements," in *IEEE Int. Electron Devices Meeting*, 1989, p. 297.

62. M. Stoisiek, K. G. Oppermann, and R. Stengle, "A 400A/2000V MOS-GTO with improved cell design," *IEEE Trans. Electron Dev.* **ED-39**, 1521 (1992).

63. F. Bauer, T. Stockmeier, H. Lendenmann, H. Dettmer, and W. Fichtner, "Static and dynamic characteristics of high voltage (3 kV) IGBT and MCT devices," in *IEEE Int. Symp. on Power Semiconductor Devices and ICs*, 1992, p.22.

64. M. Nandakumar, B. J. Baliga, M. S. Shekar, S. Tandon, and A. Reisman, "A

new MOS-gated power thyristor structure with turn-off achieved by controlling the base resistance," *IEEE Electron Dev. Lett.* **ED-12**, 227 (1991).

65. M. Nandakumar and B. J. Baliga, "Modelling the turnoff characteristics of the base resistance controlled thyristor," *Solid State Electron.* **38**, 703 (1995).

66. M. Nandakumar, B. J. Baliga, M. S. Shekar, S. Tandon, and A. Reisman, "Theoretical and experimental characteristics of the base resistance controlled thyristor (BRT)," *IEEE Trans. Electron Dev.* **ED-39**, 1938 (1992).

67. M. Nandakumar, M. S. Shekar, and B. J. Baliga, "Fast switching power MOS-gated (EST and BRT) thyristors," in *IEEE Int. Symp. on Power Semiconductor Devices and ICs*, 1992, p. 256.

68. B. J. Baliga, "The MOS-gated emitter switched thyristor," *IEEE Electron Dev. Lett.* **ED-11**, 75 (1990).

69. M. S. Shekar, B. J. Baliga, M. Nandakumar, S. Tandon, and A. Reisman, "High voltage current saturation in emitter switched thyristors," *IEEE Electron Dev. Lett.* **ED-12**, 387 (1991).

70. M. Shekar and B. J. Baliga, "Modelling the on-state characteristics of the emitter switched thyristor," *Solid State Electron.* **37**, 1403 (1994).

71. S. Sridhar and B. J. Baliga, "Comparison of linear and circular cell dual-channel emitter switched thyristors," in *IEEE Int. Symp. on Power Semiconductor Devices and ICs*, 1995, p. 170.

72. N. Iwamuro, M. S. Shekar, and B. J. Baliga, "Forward biased safe operating area of emitter switched thyristors," *IEEE Trans. Electron Dev.* **ED-42**, 334 (1995).

73. B. J. Baliga, "Semiconductors for high voltage vertical channel field effect transistors," *J. Appl. Phys.* **53**, 1759 (1982).

74. H. Matsunami, "Semiconductor silicon carbide—expectations for power devices," in *IEEE Int. Symp. on Power Semiconductor Devices and ICs*, 1990, p.13.

75. T. P. Chow and R. Tyagi, "Wide bandgap compound semiconductors for superior high voltage power devices," in *IEEE Int. Symp. on Power Semiconductor Devices and ICs*, 1993, p. 84.

76. P. M. Campbell, R. S. Ehle, P. V. Gray, and B. J. Baliga, "150 volt vertical channel GaAs FET," in *IEEE Int. Electron Devices Meeting*, 1982, p. 258.

77. P. M. Campbell, W. Garwacki, A. R. Sears, P. Menditto, and B. J. Baliga, "Trapezoidal groove Schottky gate vertical channel GaAs FET," in *IEEE Int. Electron Devices Meeting*, 1984, p. 186.

78. B. J. Baliga, "Critical nature of oxide/interface quality for SiC power devices," *Microelectron. Eng.* **28**, 177 (1995).

79. J. W. Palmour, J. A. Edmond, H. S. Kong, and C. H. Carter, "Vertical power devices in silicon carbide," in *Inst. Phys. Conf. Ser.* No. 137, 1994, p. 499.

80. M. Bhatnagar, P. McLarty, and B. J. Baliga, "Silicon carbide high voltage (400V) Schottky barrier diodes," *IEEE Electron Dev. Lett.* **ED-13**, 501 (1992).

81. T. Kimoto, T. Urushidani, K. Kobayashi, and H. Matsunami, "High voltage (>1 kV) SiC Schottky barrier diodes with low specific on-resistances," *IEEE Electron Dev. Lett.* **ED-14**, 548 (1993).

82. R. Raghunathan, D. Alok, and B. J. Baliga, "High voltage 4H-SiC Schottky *barrier diodes," *IEEE Electron Dev. Lett.* **ED-16**, 226 (1995).

83. A. Itoh, T. Kimoto, and H. Matsunami, "High performance of high voltage 4H-SiC Schottky barrier diodes," *IEEE Electron Dev. Lett.* **ED-16**, 280 (1995).

5 Quantum-Effect and Hot-Electron Devices

S. LURYI

SUNY at Stony Brook, Stony Brook, New York

A. ZASLAVSKY

Brown University, Providence, Rhode Island

5.1 INTRODUCTION

Quantum mechanics underpins all of semiconductor physics at both the atomic level of electrons interacting with the periodic potential of the semiconductor material and at the envelope function level appropriate, for example, to metal–semiconductor contacts or metal–oxide–semiconductor interfaces. Still, the vast majority of semiconductor devices can be treated as classical systems of carriers near equilibrium. Thus, in bipolar and field-effect transistors, quantum and hot-electron effects manifest themselves either as minor corrections to the fundamentally classical operation principles or as undesirable phenomena that limit device performance and reliability. The past two decades have witnessed considerable research interest and effort in semiconductor structures that could exploit quantum and hot-electron phenomena to perform circuit functions. Even though, to date, none of these structures has evolved beyond laboratory demonstration, continuing interest in the device research community has been maintained by several mutually reinforcing factors.

First, there exists the near-universal recognition that transistor-based microelectronics, the basis of all modern computing and much of modern communications, will at some point cease to improve at the device level. The current evolution of silicon technology towards ever-denser design of

Modern Semiconductor Device Physics, Edited by S. M. Sze.
ISBN 0-471-15237-4 © 1998 John Wiley & Sons, Inc.

ever-faster devices is following a virtually one-dimensional path: the scaling of device dimensions by reducing the minimal size of lithographic features. Lithography-driven performance gains can be expected to continue for the next one or two decades. Smaller device dimensions will yield faster carrier transit times at lower operating voltages and currents, leading to higher maximum frequencies at lower power per device.[1] This evolution faces numerous technological hurdles, described elsewhere in this book—from currently unavailable deep-submicron lithographic techniques with sufficient throughput to the wiring delays and power dissipation problems anticipated in future microelectronic circuits. Another constraint is the rapidly escalating fabrication costs that may divert investment in device-level technology to the more profitable software and circuit-design arenas. However, even if both the technological and economic constraints are overcome, it appears evident that device-level performance enhancement—higher speeds, lower power consumption, or increased device functionality—will require new operational concepts. As described elsewhere in this book, the minimum critical dimension (channel length L) of a scaled Si MOSFET operated at room temperature bottoms out not too far below $L \approx 0.1\,\mu$m.

The second, related reason for the continued interest in quantum mechanical and hot-electron phenomena has been their deleterious effect on the operation of semiconductor devices even before the scaling limit is reached. Thus, as the oxide thickness d in complementary metal–oxide–semiconductor (CMOS) technology is scaled down to accommodate shorter L, electron tunneling into the oxide conduction band leads to leakage current that eventually results in gate breakdown. Further, at the smallest oxide thickness, $d \approx 30\,\text{Å}$, required near the $L \approx 0.1\,\mu$m scaling limit, direct tunneling from the MOSFET inversion layer into the gate is expected to become a limiting factor. Similarly, carrier heating near the drain junction, resulting from the high lateral electric fields, is a principal reliability issue, since hot carriers interact with the oxide and degrade the device lifetime. The anticipated importance of these constraints on the reliability of highly dense microelectronic circuitry of the next several generations has prompted much research on the phenomenology of tunneling and carrier heating in semiconductor devices, as described in Chapter 3. While much of the effort has concentrated on sidestepping these effects on the road to the scaling limit, the device know-how aimed at characterizing and controlling tunneling and carrier heating has also been applied to devices based on precisely such effects.

The final and perhaps most important factor driving active research in quantum-effect and hot-electron devices is the rapidly expanding semiconductor bandgap-engineering capability provided by modern epitaxy. Molecular-beam epitaxy (MBE) and metalorganic chemical vapor deposition (MOCVD) of III–V semiconductors, silicon, and silicon-based alloys (SiGe, SiGeC) provide exceptional control over semiconductor layer thickness, doping, and composition. This control gives the device designer unprecedented freedom in specifying regions of carrier localization and transport, tailored electric fields and potential barriers, and precise amounts of built-in strain. Thus, near-monolayer layer control has introduced size quantization, reduced dimensionality of carriers and the many attendant effects—from changes in the density of states to high carrier mobilities produced by modulation doping—into the parameter space of proposed devices. Further progress on bandgap engineering in directions other than the epitaxy axis, either via conventional processing in the deep submicron regime or via additional epitaxial regrowth on nonplanar substrates, is a subject of much current research. This research promises to extend device physics to full two- or three-dimensional quantum confinement (quantum wires and dots). Multidimensional confinement in these low-dimensional structures has long been predicted to alter significantly the transport and optical properties, compared to the bulk or planar heterostructure results.[2] More recently, the effects of charge quantization on transport in small semiconductor quantum dots[3] have stimulated much research in single-electron devices, in which the transfer of a single electron is sufficient to control the device.

Having briefly enumerated the scientific and technological factors that have been driving quantum-effect and hot-electron device research, let us turn to the performance advantages that such devices offer, at least in theory. Speed is often cited as a primary benefit. Quantum-mechanical tunneling, on which most quantum-effect devices rely, is an intrinsically fast process. Analogously, many hot-electron devices employ ballistic transport of carriers moving at velocities considerably in excess of their equilibrium thermal velocity. However, the very high speeds achieved in the active regions of the device often do not translate directly into device performance because of various delays elsewhere, for example, the RC time delays that accompany electrode biasing. Frequently, a more significant advantage is the higher functionality of quantum and hot-electron devices, that is, their capability to perform an operation with a greatly reduced device count. Higher functionality is made possible either by strong, tunable nonlinearities in their current–voltage (I–V) characteristics or by unusual electrode symmetries. As a result, these devices can perform relatively complex circuit functions, replacing large numbers of transistors or passive circuit components. Examples covered in this chapter include multistate memory and logic implementations using small numbers of tunneling devices, as well as single-device logic gates fashioned from hot-electron devices.

Finally, although quantum-effect and hot-electron devices face a long struggle with room-temperature operation and large-scale integration before they become technologically viable for general-purpose semiconductor circuitry, even today they appear poised to take over in certain niche applications. Thus, the extreme constraints on device uniformity and operating temperature inherent in single-electron devices may render them ill-suited for large-scale logic, but the robustness of charge-quantization effects in a single device at cryogenic temperatures appears ideal for extremely precise current sources in metrological applications. Similarly, the recently demonstrated quantum-cascade laser requires stringent epitaxial precision at the limit of MBE capabilities, but the absence of competing semiconductor lasers in the near-infrared makes it technologically attractive all the same.

In this chapter, the basic device structures and operating principles are discussed in Sections 5.2 and 5.3 for quantum-effect and hot-electron devices respectively. These sections also include a simple introduction to the underlying physics of quantization effects on the carrier density of states and quantum-mechanical tunneling, as well as hot-carrier production, ballistic transport, and real-space transfer. The various proposed device implementations—ranging from memories and logic circuits, to specialized applications—are presented in Section 5.4. The chapter concludes with a brief overview of the prospects of quantum-effect and hot-electron devices, incorporating both the positive impact of probable technological advances and the anticipated capabilities of the rapidly evolving silicon technology.

5.2   RESONANT-TUNNELING (RT) STRUCTURES

5.2.1   Quantum-Mechanical Tunneling

The resonant-tunneling mechanism arises from two quantum mechanical consequences of the Schrödinger equation that have no classical analog. First, if a particle is confined by some potential $V(\mathbf{r})$ on a scale comparable to its de Broglie wavelength, the particle's momentum $\hbar\mathbf{k}$ is quantized. The continuous energy spectrum $E(\mathbf{k}) = \hbar^2 k^2/2m$ corresponding to free motion (m is the particle mass) is broken up into energy subbands $E_n(\mathbf{k})$. Second, as long as the confining potential $V(\mathbf{r})$ is not infinite, the particle has a finite probability of being in the classically forbidden region, where its energy E is lower than the local value of the potential. Both of these effects are most easily illustrated in the case of one-dimensional (1D) motion in a finite potential well of width L_W and height $V(z) = V_0$ ($|z| \geq L_W/2$) shown in Fig. 1a. The 1D Schrödinger equation for the wavefunction $\chi(z)$ can be written as follows:

$$H\chi(z) = \left(\frac{-\hbar^2}{2m}\frac{d^2}{dz^2} + V(z)\right)\chi(z) = E\chi(z) \tag{1}$$

where H is the Hamiltonian and $\hbar$ is the reduced Planck's constant. Equation 1 can be solved in each of the three regions and by imposing continuity conditions on $\chi(z)$ and $d\chi/dz$ one obtains discrete energy levels E_n, as well as the explicit form of the corresponding $\chi_n(z)$. The normalized $\chi_n(z)$ are related to the probability $P(z)$ of finding the particle at some coordinate

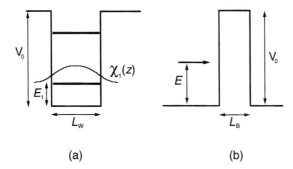

(a)   (b)

Fig. 1 (a) Finite-quantum-well potential diagram, showing the wavefunction $\chi_1(z)$ of the lowest discrete level E_1. (b) Particle of energy E incident on a single barrier.

$z = z_0$ by $P(z_0) = |\chi(z_0)|^2$. In the infinite-potential-well limit $(V_0 \to \infty)$, the eigenfunctions $\chi_n(z)$ must go to zero at $|z| = L_W/2$ and the energy levels are given by

$$E_n = \frac{\hbar^2 \pi^2 n^2}{2m \, L_W^2} \tag{2}$$

where n is an integer. In the more relevant finite-potential-well case of Fig. 1a, one finds (cf. Problem 1) that the well contains a finite number of energy levels E_n, which do not have the rapidly increasing n^2 dependence on level number (every 1D potential well contains at least one level). Furthermore, the corresponding wavefunctions $\chi_n(z)$ penetrate into the potential barriers according to

$$\chi_n(z) \approx e^{-\kappa_n|z|} \tag{3}$$

where $\kappa_n = [2m(V_0 - E_n)/\hbar^2]^{1/2}$ and the other mathematically possible solution in the barrier, $\chi_n(z) \approx e^{\kappa_n|z|}$ can be excluded on the physical grounds that it diverges as $|z| \to \infty$. Although the barrier penetration is described by an exponentially decreasing function, Eq. 3 implies that a carrier in the state characterized by $\chi_n(z)$ has a finite probability of being found in the classically forbidden barrier region $|z| > L_W/2$.

A similar treatment of a particle characterized by kinetic energy E incident from one side on a 1D potential barrier of finite height V_0 and width L_B, shown in Fig. 1b, suffices to illustrate the basic mechanism of quantum-mechanical tunneling. Classically, if $E < V_0$ the particle would be reflected regardless of barrier width, but barrier penetration analogous to Eq. 3 ensures a finite transmission probability $T(E)$ that depends on V_0 and L_B.

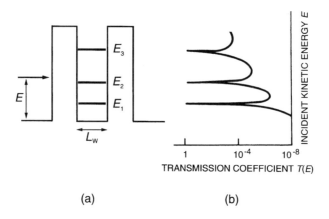

(a)   (b)

Fig. 2 (a) Particle of energy E incident on a double-barrier potential, showing quantized levels. (b) Corresponding 1D transmission coefficient T(E).

Indeed, the solutions of the Schrödinger equation to the left of the barrier, $z < 0$,

$$\chi(z) \approx A e^{ikz} + B e^{-ikz} \tag{4}$$

where A and B are constants, can be naturally associated with the incident $(\chi(z) \approx A e^{ikz})$ and reflected $(\chi(z) \approx B e^{-ikz})$ particles. The analogous solutions to the right of the barrier, $z > L_B$, can be taken as $\chi(z) \approx C e^{ikz}$, where C is a constant and we assume that the particle was originally incident from the left. Solving the Schrödinger equation in the barrier region, $0 \leqslant z \leqslant L_B$, and imposing the continuity conditions at the barrier boundaries, one associates the ratios $|B/A|^2$ and $|C/A|^2$ with reflection $R(E)$ and transmission $T(E)$ probabilities respectively,[4] with $R + T = 1$. The result is that if the incident energy E is such that $e^{-\kappa L_B} \ll 1$, where $\kappa = [2m(V_0 - E)/\hbar^2]^{1/2}$, the transmission probability is approximately (cf. Problem 2)

$$T(e) \approx e^{-2\kappa L_B} \tag{5}$$

It is apparent from Eq. 5 that in the single-barrier case the transmission probability $T(E)$ for $E < V_0$ is a monotonically increasing function of incident energy E. This rather uninspiring result changes drastically when the same particle is incident on two potential barriers separated by the well of width L_W, as shown in Fig. 2a. The explicit double-barrier transmission $T(E)$ probability can be obtained[5] by repeated application of Eq. 1, but it is instructive to consider the physics of the situation. Unless the potential barriers are very narrow or the energy of a state approaches V_0, the energy levels in the quantum well will coincide approximately with those of the finite potential well in Fig. 1a. On the other hand, semiclassically[4] a particle occupying one the energy levels E_n oscillates between the barriers with velocity $v_z = \hbar k_z/m$ and, in effect, is incident on a barrier twice in each period of oscillation $2L_W/v_z$. Every incidence involves some probability $T(E_n)$ of tunneling out of the double-barrier confining potential, making the energy levels metastable with a finite lifetime τ_n with respect to tunneling out, and hence a finite energy width $\Delta E_n = \hbar/\tau_n$ (Problem 3).

If a particle is incident on the double-barrier potential with energy E that does not coincide with one of the levels E_n, the total transmission probability is given by the product of the individual transmission probabilities of the first (emitter) and second (collector) barriers, $T(E) = T_E T_C$—an exponentially small quantity given reasonably opaque barriers with $T_E, T_C \ll 1$. On the other hand, if the incident energy matches one of the energy levels E_n, the amplitude of the wavefunction builds up in the well as the reflected waves cancel, just as in a Fabry–Perot resonator, and the resulting transmission probability[5]

$$T(E = E_n) = \frac{4T_E T_C}{(T_E + T_C)^2} \tag{6}$$

reaches unity in a symmetric structure with $T_E = T_C$. Hence, the transmission probability $T(E)$ is a sharply peaked function of incident energy, illustrated in Fig. 2b, with the energy width of the transmission peaks obtaining from finite lifetime of the discrete levels.[4]

In principle, an imaginary ideal device where monoenergetic 1D particles impinge on a double-barrier potential, whose energy levels E_n are tunable by some voltage V, would exhibit a sharply peaked current-voltage I–V characteristic, replicating the $T(E)$ shown in Fig. 2b. Several constituent parts of such a device can be implemented in semiconductors. As discussed elsewhere in this book, semiconductor heterostructures can provide the required double-barrier potential. To a good approximation, electrons and holes in direct-gap semiconductors like GaAs obey parabolic dispersions of the form $E = \hbar^2 k^2/2m^*$, differing from free electrons only by virtue of the effective mass m^* rather than the free electron mass m_0. On the other hand, monoenergetic carrier distributions and independent voltage control of $T(E)$ are more difficult to arrange. But an even more fundamental difference between the idealized 1D scenario described by Eqs. 1 to 6 and semiconductor heterostructures lies in the existence of other spatial degrees of freedom. True 1D wires, where carrier dispersion is given by Eq. 1, are difficult to fabricate and even more difficult to use, because they are limited in their current-carrying capacity. Instead, a typical semiconductor implementation of the structure in Fig. 2a has the double-barrier potential along the epitaxial direction $V(z)$, with free transverse motion in the (x, y) plane. If the in-plane motion can be separated from motion along the epitaxial (tunneling) direction, the total wavefunction $\Psi(\mathbf{r})$ of an electron in one of the metastable quantum well levels $\chi_n(z)$ depends on in-plane momentum $\mathbf{k}_\perp$ and can be written as

$$\Psi_{n,\mathbf{k}_\perp}(\mathbf{r}) = N\chi_n(z)\, e^{i\mathbf{k}_\perp \cdot \mathbf{r}_\perp} \tag{7}$$

where N is a normalization factor. Given an isotropic effective mass, the corresponding total energy is given by

$$E = E_n + \frac{\hbar^2 \mathbf{k}_\perp^2}{2m^*} \tag{8}$$

Equation 8 indicates that each of the quantized energy levels gives rise to a subband and, in contrast to the 1D situation, there are no gaps in the energy spectrum above the lowest-lying subband E_1. An electron at an energy $E > E_n$ can belong to any of the n subbands from E_1, in which case it would have a large in-plane kinetic energy, to E_n. Further, in the presence of scattering, the degenerate states belonging to different subbands become mixed and the factorization of the wavefunction in Eq. 7 generally breaks down. However, since the coupling between degenerate states belonging to different subbands involves a finite change in $\mathbf{k}_\perp$, the single-subband states $\chi_n(z)$ may be sufficiently long-lived to treat their coupling as a perturbation leading to intersubband scattering.

The in-plane motion drastically changes the effective densities of states both in the quantum well and in the regions outside the double-barrier potential, $|z| > (L_W + 2L_B)/2$. Instead of discrete levels in the well, each subband E_n contributes a constant 2D density of states, $g^{2D}(E) = m^*/\pi\hbar^2$. At the same time, in real devices the tunneling carriers arrive from carrier reservoirs outside the double-barrier potential. Typically, the states in the emitter and collector carrier reservoirs can be taken as 3D. As discussed in Appendix 5.A, the appropriate 3D density of states is

$$g^{3D}(E) = \frac{(2m^*)^{3/2}}{2\pi^2 \hbar^3}\, E^{1/2} \tag{9}$$

and one can determine the Fermi level E_F in the reservoir in terms of the 3D carrier density n_{3D}, temperature T, and the Fermi–Dirac occupational probability $f_{FD}(E)$:

$$n_{3D} = \int g^{3D}(E) f_{FD}(E - E_F)\, dE, \qquad f_{FD}(E) = (e^{-E/kT} + 1)^{-1} \tag{10}$$

Instead of the idealized monoenergetic 1D carriers, the incident particles will have a relatively broad energy distribution of at least E_F in the simplest case of degenerately doped electron tunneling reservoirs at low temperatures, where E_F separates occupied from unoccupied states in the emitter. As long as the factorization of the wavefunction into in-plane and tunneling-direction components remains valid, in-plane degrees of freedom do not complicate the situation unduly. The in-plane momentum $k_\perp$ remains a constant of motion that is conserved as the carrier tunnels from the emitter reservoir through the 2D subbands E_n with a transmission probability $T(E_z)$ that depends on the energy of motion in the tunneling direction E_z. The total tunneling current density J can be computed by integrating over the electron distribution in the emitter reservoir:

$$J = \frac{q}{2\pi\hbar} \int N(E_z) T(E_z) \, dE_z \qquad (11)$$

where $N(E_z)$ is the number of electrons with the same E_z per unit area. At a given temperature T, the quantity $N(E_z)$ is easily evaluated by integrating the product of the constant 2D density of states $g^{2D}(E)$ (see Appendix 5.A) with the Fermi–Dirac occupational probability $f_{FD}(E)$, yielding

$$N(E_z) = \frac{kTm^*}{\pi\hbar^2} \ln(1 + e^{(E_F - E_z)/kT}) \qquad (12)$$

Although these equations are a reasonable starting point for considering realistic semiconductor resonant-tunneling (RT) structures, it is important to recognize that their validity depends to a great extent on the absence of scattering. Clearly, even in the 1D picture of Fig. 2, the build-up of near-unity transmission probabilities of Eq. 6 by the Fabry–Perot mechanism requires the electron to retain phase coherence over a very large number of bounces (proportional to $(T_E + T_C)^{-1}$, the inverse of the single-barrier transmission coefficients[6]), the phases of the many multiply reflected amplitudes combining to cancel the net reflected wave. Any interaction that changes the phase of the wavefunction, whether elastic—like impurity scattering, or inelastic—like scattering by phonons or other electrons, will destroy the overall cancellation of the reflected wave. In a realistic three-dimensional structure with in-plane degrees of freedom, elastic impurity scattering relaxes $k_\perp$ conservation. More generally, scattering unavoidably mixes in-plane and tunneling direction motion, qualitatively changing the wavefunction penetration into the barrier.[7] In the absence of scattering, in-plane motion is separable from tunneling. The wavefunction penetration into the barrier then depends on the quantized energy of motion E_n in the tunneling direction, regardless of the in-plane kinetic energy $\hbar^2 k_\perp^2 / 2m^*$. Extending Eq. 3 to the case of a more general barrier potential $V(z)$, one finds

$$\Psi_n(\mathbf{r}) \approx \exp\{-\hbar^{-1} \int [2m^*(V(z) - E_n)^{1/2}] \, dz\} \qquad (13)$$

Scattering mixes in-plane motion with tunneling and barrier penetration asymptotically approaches

$$\Psi_n(\mathbf{r}) \approx \exp\{-\hbar^{-1} \int [2m^*(V(z) - E)^{1/2}] \, dz\} \qquad (14)$$

where the energy that enters into the exponential decay of the wavefunction is the total energy $E = E_n + \hbar^2 k_\perp^2 / 2m^*$. The transition from Eq. 13 to Eq. 14 is described by a pre-exponential factor that depends on the specific scattering mechanism.[7]

All of the effects associated with scattering and limited phase coherence significantly alter the idealized sharply peaked current-voltage I–V characteristic that we would obtain from the ideal transmission through the double-barrier potential illustrated in Fig. 2. The many orders of magnitude peak-to-valley ratios predicted by coherent $T(E)$ calculations have not been observed experimentally in double-barrier RT structures, even at low temperatures. In fact, in realistic semiconductor RT structures, scattering limitations and the energy width of the incident electron distributions are such that an alternative sequential tunneling model[8] predicts the I–V characteristics equally well. In this model, current transport is described by carriers tunneling into the 2D density of states in the well followed by uncorrelated tunneling out into the collector. The I–V nonlinearities arise from E and $k_\perp$ conservation without recourse to near-unity transmission coefficients of the double-barrier potential in the coherent limit. These issues, as well as other effects that become relevant in optimizing RT structures for device applications, such as maximizing peak current densities or reducing the temperature sensitivity of the I–V characteristics, are discussed in the next section.

5.2.2   Two-Terminal RT Structures

The first experimental realization of the double-barrier RT device using semiconductor heterostructures dates back to 1974, when current peaks corresponding to electrons tunneling through the lowest two subbands in a GaAs quantum well confined by $Al_xGa_{1-x}As$ barriers were observed at low temperatures.[9] Improvements in epitaxial material quality and device design since then have led to RT diodes with very sharp low-temperature I–V characteristics,[10] as illustrated in Fig. 3. The inset of Fig. 3 shows the epitaxial layer sequence of the device, with $Al_xGa_{1-x}As$ barriers, a narrow ($L_W = 56$ Å) GaAs well and heavily doped ($N_D = 2 \times 10^{17}$ cm^{-3}) n^+-GaAs electrodes. The device exhibits a strong negative differential resistance (NDR) characteristic above the peak ($V > V_P$) of the I–V curve with the peak-to-valley current ratio (PVR) reaching ~30 at low temperatures. Still, the measured PVR and overall I–V lineshape are quite different from the theoretical prediction of Fig. 2b based on the calculated coherent transmission coefficient $T(E)$. The valley current is much larger than predicted by simple theory because of nonresonant processes, such as scattering or phonon-assisted tunneling. It turns out that coherence and the Fabry–Perot model are not required to explain the I–V characteristics of realistic RT structures. It suffices to impose energy E and transverse momentum $k_\perp$ conservation on carriers tunneling into the 2D subband E_n, with the only

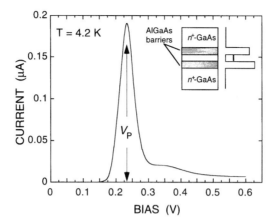

Fig. 3 Reverse-bias *I–V* characteristics of asymmetric AlGaAs/GaAs double-barrier RT structure at *T* = 4.2 K. Inset shows the layer sequence alongside the schematic conduction-band diagram (not to scale).

constraint imposed on the coherence of the oscillating wavefunction in the well being that it should be sufficient to produce a well-resolved 2D subband. The tunneling out of the well into the collector may then occur in a second step that may be completely uncorrelated with the tunneling into the well, resulting in current transport by a sequential tunneling scheme.

The sequential tunneling model[8] is illustrated in Fig. 4, using the n-$Al_x Ga_{1-x}As$/GaAs double-barrier RT structure of Fig. 3 as an example. At flatband, when no bias is applied to the device, the lowest 2D subband E_1 in the well lies above the emitter E_F, making E and $k_\perp$ conserving tunneling into E_1 states impossible. As the bias V increases, E_1 is lowered with respect to the emitter E_F, as shown in the self-consistent potential distribution of Fig. 4. Resonant tunneling becomes possible once the bias brings E_1 into alignment with the occupied states in the emitter, at which point a subset of the occupied emitter states—their measure is denoted by the supply function $N(V)$—can tunnel into the well conserving both E and $k_\perp$. At low temperature, a simple geometrical evaluation of the supply function can be constructed by noting that the occupied states in the emitter can be characterized in terms of E and $k_\perp$ as follows:

$$E = E_z + \frac{\hbar^2 k_\perp^2}{2m^*} \qquad 0 \le E_z \le E_F \qquad (15)$$

whereas the available states in the well lie on a single dispersion $E = E_1(V) + \hbar^2 k_\perp^2/2m^*$. Taking the bottom of occupied states in the emitter as the energy reference, the emitter states form a parabolic solid of revolution in E–$k_\perp$ phase space, filled up to E_F with carriers. The supply function can

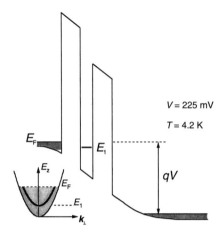

Fig. 4 Self-consistent potential distribution in the asymmetric AlGaAs/GaAs RT structure of Fig. 3 under reverse bias $V = 0.225$ V and $T = 4.2$ K. The supply function $N(V)$ is obtained from the geometric overlap between the $E(\mathbf{k}_\perp)$ dispersions in the emitter and well, as shown at lower left. (After Zaslavsky et al., Ref. 10.)

be geometrically described by the intersection of the available 2D states in the well and the occupied emitter states, see Fig. 4. It can be easily shown (cf. Problem 4) that $N(V) \approx [E_F - E_1(V)]$ as long as $E_1(V)$ does not fall below the bottom of the occupied states in the emitter, at which point the supply function drops to zero. The current into the well due to E and $\mathbf{k}_\perp$ conserving tunneling is given by

$$J = \frac{q}{2\pi\hbar^2} N(V) T_E(V) \tag{16}$$

to which one must add all other current components. Examples of additional current components are direct tunneling into collector states through both barriers $(J \approx T_E T_C)$, phonon-assisted tunneling, impurity- or interface-roughness-assisted tunneling that conserves E but not $\mathbf{k}_\perp$, and so forth. In addition, at sufficiently high V, tunneling through the second 2D subband E_2 also becomes possible. Those current components that are not cut off by $\mathbf{k}_\perp$ conservation once $E_1(V)$ is biased below the emitter states contribute to the valley current. For example, the strong electron–optical-phonon coupling in GaAs leads to a phonon-assisted replica peak when $E_1(V)$ is biased below the bottom of the occupied states in the emitter by optical-phonon energy $\hbar\omega_{opt} = 36$ meV.[11] The quantitative modeling of nonresonant current components is not well developed and typically relies on adjustable parameters.[12] This is unfortunate, since the valley current plays an important role in the minimum power dissipation of RT-based devices.

According to the sequential tunneling model, the relevant transmission coefficient that determines the current density is T_E (typically, $T_E << 1$) and the NDR is a consequence of E and $\mathbf{k}_\perp$ conservation that governs carrier tunneling into the well. In contrast, the idealized coherent model of resonant tunneling involves the total transmission coefficient $T(E_z)$ of the double-barrier potential given by Eq. 6, which goes to unity if the emitter and collector barrier transmission coefficients are equal, $T_E = T_C$, at operating bias $V \approx V_P$. Surprising though it might appear, the two models predict essentially the same I–V characteristics for realistic RT structures, in which both the bias V_P required to observe the current peak and the width of the tunneling carrier energy distribution E_F are much larger than the 2D-level widths ΔE_n.[6,13] The essential point is that whereas the total transmission coefficient $T(E_z)$ of the coherent model is exponentially large compared to the single barrier coefficients, it is also exponentially narrow, as shown in Fig. 2b. Indeed, if the incident energy E_z is close to matching a 2D subband energy E_n, the transmission coefficient of Eq. 5 can be expanded in terms of the small parameter $E_z - E_n$ as follows:[13]

$$T(E_z \approx E_n) \sim \frac{4T_E T_C}{(T_E + T_C)^2} \frac{\Delta E_n^2}{(E_z - E_n)^2 + \Delta E_n^2} \tag{17}$$

where $\Delta E_n = \hbar/\tau_n$ is the lifetime of subband E_n with respect to tunneling out. The total current through the device is obtained by averaging over Eq. 17. Since $E_F >> \Delta E_n$, the Lorentzian factor in Eq. 17 reduces to a δ-function,* $\pi \Delta E_n \, \delta(E_z - E_n)$. The δ-function, in turn, cancels one of the $(T_E + T_C)$ factors in the denominator of Eq. 17, reducing the average transmission coefficient for the carrier ensemble. To first order, the two pictures predict the same current density,[13] except in exotic limits.† Hence, the choice of coherent or sequential tunneling model might appear immaterial. However, the geometric interpretation of Fig. 4 implicit in the sequential model is useful in predicting the I–V characteristics of more complicated structures, for instance those with nonparabolic in-plane carrier dispersions $E(\mathbf{k}_\perp)$ or different dispersions in the emitter and well (cf. Problem 5). More importantly, the sequential tunneling approach provides a more natural framework for discussing three-terminal RT structures that rely on the NDR in the I–V characteristics provided by tunneling into a restricted density of states in a quantum well without an attendant second tunneling step. For this

*Here we make use of the well-known identity,

$$\lim_{\Delta \to 0} \frac{\Delta}{x^2 + \Delta^2} = \pi \delta(x)$$

†In a hypothetical device where either E_F or $qV_P < \Delta E_n$ the situation would be different, with the coherent model predicting much greater current densities.[14] Such RT structures have not been realized to date.

reason, the subsequent discussion will be based on the *sequential tunneling model*.

With the exception of direct tunneling to the collector through both barriers, all of the various current components and particularly the E and $k_\perp$ conserving term of Eq. 16 depend sensitively on the alignment of the 2D subbands with the occupied states in the emitter. Yet in a standard, vertical two-terminal implementation of RT structures, this alignment can only be controlled by changing the applied bias V, only a fraction of which lowers the 2D subbands (see Fig. 4). If one ignores the penetration of the electric field into the emitter and collector regions, one immediately obtains that $V_P^{(n)} = 2E_n$ (see Problem 4). This frequently cited result typically fails to describe realistic double-barrier RT structures, where undoped spacer regions around the double-barrier structure and relatively low electrode doping lead to significant potential drops in the emitter and collector regions. This is especially pertinent to RT devices designed for high-frequency operation, where low emitter–collector capacitance is often achieved by a large collector spacer region.[15] A self-consistent calculation of the potential distribution over the device, including the voltage drops in the accumulation and depletion regions as shown in Fig. 4, is therefore necessary to predict $V_P^{(n)}$ as a function of device parameters. An additional complication is the dynamically stored charge density σ_W in the well under bias, given by

$$\sigma_W = J\tau_n \tag{18}$$

where τ_n is the lifetime for subband E_n. The effect of σ_W is to increase the electric field in the second barrier and hence reduce the bias-induced lowering of the subbands $E_n(V)$ for a given V. In double-barrier structures with symmetric barriers, σ_W is typically small because $T_C \gg T_E$ at V_P and $\tau_n \approx T_C^{-1}$, since only tunneling out to the collector is possible under bias (see Fig. 4). If the collector barrier is made larger, σ_W can be increased and in the case of highly asymmetric barriers the result can be a bistable I–V characteristic shown in Fig. 5. The arrows in Fig. 5, which is the other bias polarity I–V of the same RT device as in Fig. 3, indicate the direction of the bias sweep. If V is increased from flatband, V_P occurs at a higher bias because of significant dynamic charge storage σ_W, whereas if the resonant alignment is approached by decreasing V from above V_P, the valley current and hence σ_W is small and the I–V characteristic switches to the high-current branch at a lower bias. In effect, the resulting intrinsically bistable I–V arises from the feedback of the dynamically stored charge in the well σ_W on the alignment of $E_1(V)$ with the occupied emitter states.[16] At least in principle, the bistable I–V offers the possibility of constructing a single-device two-state semiconductor memory.

Double-barrier structures implemented in the n-AlGaAs/GaAs material system have been very useful in clarifying the relevant physics of resonant tunneling at low temperatures, but their I–V characteristics are less suitable

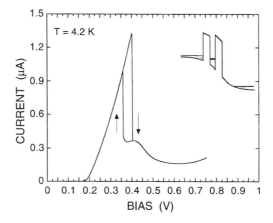

Fig. 5 Bistable forward-bias *I–V* characteristic of the same RT device as in Fig. 3. Inset shows the schematic band diagram with significant charge build-up in the well because of the larger collector barrier. Arrows indicate the direction of the bias sweep.

for real electronic devices. First, the sharp NDR characteristic offered by the RT structures needs to survive at room temperature. At $T = 300$ K, the peak current density J_P remains essentially unchanged, but the valley current is supplemented by thermionic emission over the barriers and thermally assisted tunneling through higher-lying subbands. At room temperature, both of these valley current components can significantly degrade the available PVR. Clearly, thermionic emission over the barriers can be exponentially reduced by increasing the barrier height, which in the context of AlGaAs/GaAs heterostructures implies the use of pure AlAs barriers, maximizing V_0. Yet for high-speed operation there exists the conflicting requirement of maximizing J_P, since high current densities are necessary for rapid charging of the various device and circuit capacitances—ideally $J_P \geq 10^5$ A/cm^2. The use of very narrow AlAs barriers is therefore indicated but, even so, thermally-assisted tunneling through higher-lying subbands remains a problem. For this reason, the fastest reported GaAs/AlAs double-barrier RT oscillators[17] with high $J_P(\approx 10^5$ A/cm^2) exhibit a room-temperature PVR of only 3.

A considerable improvement in the PVR and J_P figures-of-merit in two-terminal RT devices has been obtained by moving to the n-In$_{0.53}$Ga$_{0.47}$As/In$_{0.52}$Al$_{0.48}$As material system, which is lattice-matched to InP substrates. The physics of device operation is still described by Fig. 4, but the maximum barrier heights are larger and, more importantly, the lower m^* of electrons in the InGaAs well leads to higher subband separation $(E_2 - E_1)$. The room-temperature *I–V* characteristic of a double-barrier n-InGaAs/InAlAs device is shown[18] in Fig. 6, with $J_P > 10^5$ A/cm^2 and PVR ≈ 8. This RT structure included a large undoped collector spacer region

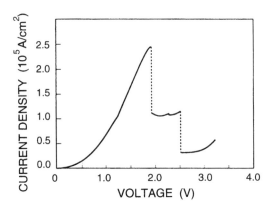

Fig. 6 *I–V* characteristics of n-$In_{0.53}Ga_{0.47}As$/AlAs RT diode, showing high peak-to-valley ratio and peak current density at room temperature. Device area is $4\ \mu m^2$. (Courtesy of E. R. Brown, 1995.)

to reduce the emitter–collector capacitance, hence the high V_P. Because of the sharp NDR, when the device is biased beyond V_P, the biasing circuit becomes unstable over a range of voltages $V_P < V \leqslant 2.5\ V$. The circuit oscillations are rectified by the RT device, leading to the characteristic discontinuous jumps in the *I–V* curve.[19] Note that the AlAs barrier layers are not lattice matched to the substrate, but since their thickness can be kept very narrow, about three monolayers for the structure of Fig. 6, they can be deposited pseudomorphically without generating large numbers of dislocations. Further improvements in the PVR of the first resonant peak can be obtained in this material system by growing a narrow InAs layer in the center of the InGaAs well, which has the effect of further separating the lowest two subbands (cf. Problem 6).

Another variant of two-terminal RT devices involves GaSb/AlSb/InAs heterostructures. These heterostructure are known as polytype because of their staggered bandgap alignment, wherein AlSb barriers separate the InAs conduction band from the GaSb valence-band edges.[20] The schematic diagram of a double-barrier polytype RT structure with GaSb electrodes, AlSb barriers, and an InAs well is shown in Fig. 7. Under bias, the current can be described in terms of holes tunneling from the GaSb emitter into the InAs well conserving E and $k_\perp$ in the usual fashion[21]—a geometrical evaluation of the supply function simply requires inverting the emitter dispersion in Fig. 4. The polytype structure represents an RT version of the Esaki tunnel diode. Its advantage lies in the bandgap blocking beyond V_P, since for $V > V_P$ the emitter states line up with the bandgap of the InAs well, as shown in Fig. 7b. As a result, impurity-assisted $k_\perp$-nonconserving tunneling into the well is completely suppressed, removing one of the major valley current contributions (the band lineup is similar to the standard tunnel

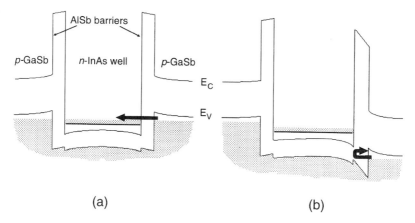

(a)   (b)

Fig. 7   Potential diagram of a polytype GaSb/AlSb/InAs RT structure near flatband (a) and under large bias $V > V_P$ (b). Arrows indicate the tunneling current of GaSb holes into the n-InAs quantum well. In (b) resonant tunneling is blocked by the InAs bandgap. (After Beresford et al., Ref. 22.)

diode). Very good PVR has been achieved in polytype structures, albeit at relatively modest peak current densities. Further, since the bandgap blocking mechanism is independent of subband quantization in the well and the electron effective mass in InAs is very small, $m^* = 0.023m_0$, RT designs with very wide quantum wells $L_W \approx 1000\,\text{Å}$ are realizable without compromising PVR.[22]

Alongside other combinations of III–V semiconductors, RT structures have been fabricated in $Si_{1-x}Ge_x/Si$ heterostructures.[23] In addition to the availability of high-quality substrates and oxides, silicon-based quantum-effect devices are interesting because of their potential integration with the dominant silicon technology. Unfortunately, the lattice mismatch hinders the epitaxy of $Si_{1-x}Ge_x$ layers with large germanium contents, and the available bandgap difference is rather small: in RT structures strained to silicon substrates, a barrier $V_0 \approx 200\,\text{meV}$ is available in the valence band, and no appreciable barrier appears in the conduction band.* Because of low V_0 and relatively small 2D subband separation in the $Si_{1-x}Ge_x$ well, where heavy and light hole branches of the dispersion give rise to separate subbands, no room-temperature NDR has been observed in p-$Si_{1-x}Ge_x/Si$ RT structures to date, although PVR ≈ 4 has been observed at cryogenic temperatures.[25] Consequently, although the strained p-$Si_{1-x}Ge_x/Si$ RT structures have been employed for spectroscopic probing of anisotropic hole dispersions[26] and

---

*A conduction band barrier can be obtained in structures strained to $Si_{1-x}Ge_x$ substrates (actually thick relaxed $Si_{1-x}Ge_x$ buffer layers grown on Si substrates). Low-temperature NDR has been observed in such n-$Si_{1-x}Ge_x/Si$ RT structures.[24]

strain relaxation in microstructures,[27] the prospects of their integration into mainstream technology appear remote.

All of the RT structures discussed thus far had been produced by epitaxial growth of a sequence of layers, with the necessary double-barrier potential arising from the bandgap offsets of the heterostructure constituents. While this approach has been the dominant one, there has been some research into fabricating lateral tunneling structures by depositing electrostatic gates on the surface of a modulation-doped 2D electron gas (2DEG) heterostructure. By applying a gate potential V_G with respect to the 2DEG, electrons can be electrostatically depleted underneath the gates (analogously to gate control of an FET). Figure 8 shows a schematic diagram of a lateral RT structure. The potential advantages include: excellent electronic properties of the 2DEG; tunability of the double-barrier potential by V_G, including control over barrier asymmetry by separate gate control of the two barriers; and planar device layout, compatible with FET technology. The main drawback is the relative weakness of the electrostatically created confining potentials in the plane of the 2DEG. The minimum geometric separation of the metal gates is set by lithographic limitations and far exceeds the monolayer control available by epitaxial techniques. Furthermore, regardless of the surface gate geometry, the double-barrier potential parameters L_B, L_W cannot be reduced below the spacer layer thickness (see Fig. 8). Finally, since the confining potential arises from the self-consistent electrostatics, the barrier height V_0 produced in the plane of the 2DEG is proportional to barrier thickness L_W—high barriers are necessarily broad. Hence, the I–V characteristics of lateral RT structures produced by electrostatic gating exhibit weak NDR and only at cryogenic temperatures.[28] If sharp confining barriers in a lateral RT structure could be produced by some means, interesting device possibilities would result. An approach using epitaxial regrowth will be discussed in the next section.

5.2.3 Three-Terminal RT Structures

All of the previously discussed double-barrier RT structures are two-terminal devices, potentially useful in oscillator and frequency multiplier circuits, but ill-suited for more general circuitry. The addition of a third terminal to control the I–V characteristics of an RT structure, either with a small current as in a bipolar transistor or a gate voltage V_G as in an FET, has been attempted in a number of schemes.

Current-controlled variants of three-terminal RT structures involve a separate contact to the quantum well that can source or sink a "base" current large enough to alter the alignment of the 2D subbands E_n and the emitter E_F. In principle, the base current can be of the same or opposite polarity as the tunneling carriers. The band diagram of a bipolar structure is illustrated in Fig. 9 for an n-type double-barrier RT device with a separate contact to a p-type quantum well. This implementation is preferable to unipolar versions

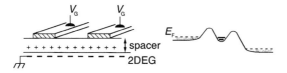

Fig. 8 Schematic diagram of a lateral RT structure, including the physical layout and the potential profile in the 2DEG plane. Lithographic resolution and the spacer thickness limit the attainable barrier and well widths.

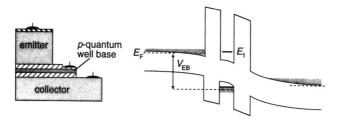

Fig. 9 Schematic device cross-section and band diagram of an RT structure with a separate contact to the *p*-type base.

both because of improved isolation between the controlling base current and the tunneling electron current and because of fabrication constraints. It is easier to contact the narrow quantum well without shorting to the nearby emitter and collector layers if the well doping is of the opposite polarity[29] (see Fig. 9). If the RT structure is biased close to V_P, a small hole current in the well can turn off the collector current I_C by biasing the 2D subband below the occupied emitter states, giving rise to negative transconductance.

A significant constraint on current-controlled three-terminal RT structures like that in Fig. 9 is the effective base resistance. To have significant 2D subband separation and hence strong NDR in the tunneling *I–V* characteristic, the quantum-well width L_W must be small. At the same time, the lateral base resistance is inversely proportional to L_W. Setting the benchmark for a truly competitive high-speed device at 1 ps, we can estimate the $R_B C$ time delays associated with charging either the emitter–well or well–collector capacitance:

$$R_B C \approx \varepsilon_s L^2 R_S / L_B \qquad (19)$$

where ε_s is the semiconductor dielectric permittivity; L is the characteristic lateral extent of the device limited by lithographic resolution to $L \geqslant 500$ Å for the foreseeable future; L_B is the emitter or collector barrier thickness; R_S is the sheet resistance of the base ($R_S = (q n_B \mu)^{-1}$ where μ is the majority carrier mobility and n_B is the charge density per unit base area). In the case of the well–collector capacitance, the barrier thickness L_B can be augmented by an undoped collector spacer, as shown in Fig. 4, but increasing the well–collector separation beyond 1000 Å introduces transit-time delays on the order of 1 ps. From Eq. 19, the resulting constraint on R_S is about 10^3 $\Omega/\square$. As in heterojunction bipolar transistors, heavy doping of the quantum well appears to resolve all difficulties, but since sharp 2D quantization requires narrow $L_W \approx 100$ Å quantum wells, doping sufficient to achieve $R_S \lesssim 10^3$ $\Omega/\square$ is problematic. First, as can be seen in Fig. 9, holes can tunnel from the quantum well into the emitter, contributing a current that will increase with emitter–well bias regardless of the 2D subband alignment with emitter states—in other words, a nonresonant current component that will reduce the PVR. The heavier hole m^* makes the corresponding tunneling transmission smaller, but if the hole density in the well exceeds the electron supply function by many orders of magnitude, the nonlinear I–V can be washed out completely. Second, the very existence of a large impurity density in the quantum well introduces substantial scattering, inhomogeneously broadening the subband energy width ΔE_n to much larger values than the lifetime broadening due to tunneling out of the well. As a result, current-controlled three-terminal devices with separately contacted quantum wells appear most promising in polytype GaSb/AlSb/InAs RT structures, where bandgap blocking of the tunneling current and low m^* in the InAs well results in good PVR even for very wide quantum wells ($L_W \approx 1000$ Å).[22] We will encounter similar structures in the discussion of hot-electron devices, where no quantization in the well is required, and operation depends on ballistic electron transport from the emitter to the collector.

Interestingly, a similar vertical structure with a separate contact to the quantum well can be employed to produce a unipolar, voltage-controlled tunneling transistor—essentially by designing the quantum well to perform the functions of a collector. Consider the schematic band diagram shown in Fig. 10, where the second barrier is designed to be so high and wide as to eliminate tunneling out of the well. The voltages are applied with respect to the quantum well contact. At some emitter bias V_E, the tunneling current into the well can be evaluated using Eq. 16: once again, it depends on the alignment of the 2D subband E_1 with the emitter E_F. This current is extracted from the well laterally. Three-terminal operation is achieved by applying a gate bias to the remaining electrode, as in Fig. 10. The gate bias V_G shifts E_1 by two mechanisms. First, the electric field in the second barrier changes the effective confining potential of the quantum well,[30] shifting E_1 down from its position at $V_G = 0$. Second, the 2DEG in the well does not screen the V_G-induced electric field completely because of the quantum capacitance effect.[31] The latter is a consequence of the Pauli exclusion principle, by which no two electrons can occupy the same quantum state. Because of this, extra kinetic energy is required to fill a given density of states with electrons. The 2D density of states in the quantum well results in a quantum capacitance

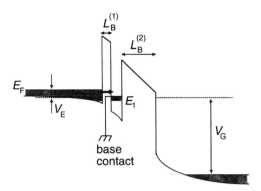

Fig. 10 Schematic band diagram of a three-terminal RT structure in which transistor action is based on a combination of quantum capacitance and the V_G-induced lowering of the 2D subband E_1 with respect to the occupied emitter states. Arrow indicates the tunneling current that is extracted laterally through the quantum well.

C_Q per unit area, given at low temperatures by

$$C_Q = m^* q^2 / \pi \hbar^2 \qquad (20)$$

As a result, it becomes energetically favorable (cf. Problem 7) for part of the V_G-induced field to penetrate into the emitter barrier, inducing additional charge in the emitter and altering the alignment of E_1 with occupied emitter states. The importance of quantum capacitance depends on the relative magnitudes of C_Q and the geometric capacitances $C^{(1,2)} = \varepsilon_s / L_B^{(1,2)}$, where $L_B^{(1,2)}$ are the barrier thicknesses in Fig. 10. In RT structures with low effective mass, gate control due to quantum capacitance can be significant. Further, negative transconductance $g_m \equiv \partial I_E / \partial V_G < 0$ is expected when, at fixed V_E, the gate bias V_G lowers E_1 below the bottom of the occupied states in the emitter and $k_\perp$ conservation cuts off the tunneling current as in a standard RT diode. The main obstacle to the fabrication of such transistors lies in implementing good lateral contacts to the quantum well and keeping the gate capacitance $C^{(2)} = \varepsilon_s / L_B^{(2)}$ large without causing significant gate leakage. To date, gate control of tunneling has been demonstrated at low temperatures[32] but with a transconductance too small to make such structures practical.

An alternative route to a three-terminal RT structure is voltage control by means of a sidewall gate electrode adjacent to the active region of a standard RT structure of sufficiently small lateral extent L that a gate bias V_G can effectively control the *I–V* curve. The vertical pillar geometry of epitaxially grown RT structures makes the fabrication rather difficult. One approach has been the self-aligned *p*-type implantation with the top metal contact of an *n*-type RT diode serving as a mask.[33] The result is a lateral

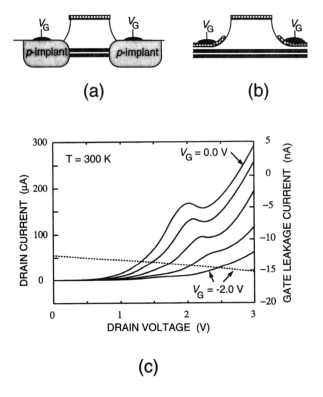

Fig. 11 Three-terminal gated RT structures with (a) implanted p–n junction gating and (b) Schottky metal gating. (c) Room temperature I–V characteristics of a three-terminal Schottky-gated GaAs/AlGaAs RT device of stripe geometry $10 \times 0.7\,\mu$m for $V_G = 0$ to 2.0 V in 0.5 V increments. Dashed line shows gate leakage for $V_G = 2.0$ V. (After Kolagunta et al., Ref. 36.)

p–n junction in the plane of the active RT region, shown in Fig. 11a. Reverse bias can then be used to deplete the RT structure from the side, controlling its effective electrical size. In addition to gate leakage currents, the difficulty with this approach is the lateral straggle of the implantation that becomes an issue for submicron lateral device diameter L. An alternative scheme involves the self-aligned deposition of an in-plane metal Schottky gate* directly adjacent to the RT diode pillar.[34] Gate bias V_G on the Schottky electrode can then be employed to deplete the effective lateral size of the RT structure.[35] By employing an undercut RT pillar profile to avoid shorting the gate to the top contact, as illustrated in Fig. 11b, structures exhibiting

---

*Low temperature operation of a p-Si$_{1-x}$Ge$_x$/Si three-terminal RT transistor with an oxide–isolated gate electrode has also been reported.[38]

room temperature control of the $I-V$ have been fabricated:[36] the $I-V$ curve of a GaAs/AlGaAs RT stripe geometry device is shown in Fig. 11c. Gate control of the resonant $I-V$ peak is achieved with reasonably small gate leakage. The reason for the observed peak position V_P shift towards higher bias as the magnitude of V_G is increased cannot be unambiguously identified. The V_G-induced lateral potential distribution will be different in the undoped active region and the doped emitter, changing the relative alignment of emitter E_F and the quantized subbands E_n, but contact series resistance could also play a role. Note that the side-gating geometry of Fig. 11a,b sacrifices the effective transconductance g_m unless the pillar diameter is extremely narrow, resulting in formidable fabrication difficulties regardless of the gate electrode fabrication technique.

A long-proposed alternative to the external gating of standard RT structure is illustrated in Fig. 12 for the GaAs/AlGaAs system.[37] The original epitaxial structure follows the double-barrier potential layer sequence, but with very large undoped spacers on both sides of the active regions. The function of these spacers is to prevent RT currents from flowing through the bulk at low source–drain voltages V. An angled interface through the double-barrier sequence is etched and an AlGaAs gate insulator is deposited, followed by a metal gate electrode. A positive gate bias V_G induces 2DEG in the undoped GaAs layers as in a standard FET, with the usual nearly triangular potential $V(x)$. In Fig. 12 we assume that only the lowest subband E_1 is occupied, which holds for moderate 2DEG densities. In the well, subband quantization arises from the AlGaAs double-barrier potential $V(z)$ in the direction of current flow combined with the FET confining potential $V(x)$ under the gate, so the lowest 1D subband E_1' lies above the Fermi level in the 2DEG. A potential difference V between the 2DEGs above and below the double-barrier potential will produce a tunneling current subject to the usual E and $\mathbf{k}_\perp$ conservation. The supply function can be determined as before, with the only difference that the conserved $\mathbf{k}_\perp = k_y$, which describes free 1D motion along the quantum wire. As a result, sharply nonlinear $I-V$ characteristics similar to standard two-terminal RT diodes is expected, but with effective gate control.

Figure 12 shows that the gate bias V_G controls the emitter 2DEG density and hence the magnitude of the RT current. More interestingly, V_G can also be used to tune V_P, because the fringing electric field penetrates into the double-barrier region, shifting E_1' with respect to E_F for the same source–drain bias V. As a result, $g_m < 0$ can be achieved. If the 2DEG depletion in the collector region is ignored, the electrostatic problem reduces to a parallel plate capacitor with a slit of width $2L_B + L_W$, and the electric-field distribution can be solved exactly by conformal mapping techniques.[37] The magnitude of the transconductance can then be explicitly calculated and, given sufficiently narrow gate insulator thickness, V_G can be nearly as effective as the source–drain voltage V in shifting the relative alignment of the emitter 2DEG and the 1D subband E_1'. In realistic devices, some

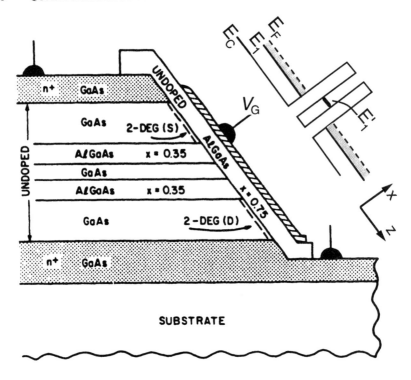

Fig. 12 Schematic cross-section of the proposed gated 2D RT transistor together with the band diagram. The current is carried by 2D electrons tunneling through 1D quantum wire subbands. Both the 2D electron-gas density at the angled surface and the relative alignment of the wire subbands E_n' with the emitter 2D electron gas can be controlled by the gate bias V_G. (After Luryi and Capasso, Ref. 37.)

depletion of the 2DEG adjacent to the collector barrier can be expected, leading to smaller fringing fields in the plane of the well for a given V_G and hence lower transconductance.[39]

The main difficulty in fabricating the three-terminal structure of Fig. 12 is the creation of a clean interface through the pregrown epitaxial structure that can support a 2DEG. The most obvious approach of etching through the double-barrier structure and subsequent epitaxial deposition of the gate layer would result in oxide formation (especially in the Al-containing barrier layers) on the interface. Proof of concept has been achieved by cleaved edge regrowth,[40] where the pregrown heterostructure is physically cleaved in the growth chamber immediately prior to deposition of the AlGaAs insulating layer on the cleaved edge. The $I-V$ curves for various V_G values of the resulting device[41] are shown in Fig. 13a. (The structure differs slightly from Fig. 12 in that modulation doping during regrowth is used to produce a 2DEG

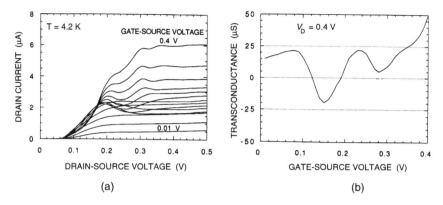

Fig. 13 (a) Three-terminal I–V_D characteristics at $T = 4.2$ K of a gated 2D RT transistor produced by cleaved edge overgrowth. Gate bias V_G is changed in 0.03 V steps. (b) Corresponding transconductance at $V_D = 0.4$ V. Device width is 300 μm. (After Kurdak et al., Ref. 41.)

under the gate even at $V_G = 0$, resulting in a "depletion mode" transistor.) Negative transconductance is indeed observed, as shown in Fig. 13b, albeit at cryogenic temperatures. Demonstration of room temperature operation and, more importantly, the fabrication of such devices by technological means, such as *in situ* etching followed by *in vacuo* transfer to the epitaxy chamber, is yet to be reported.

In addition to the severe fabrication problems faced by all of the discussed three-terminal RT devices, it is not clear that the negative transconductance they promise can be usefully applied for computation. Although it has been suggested that such devices can, in principle, perform complementary functions,[37,42] no RT transistor circuit analogous to a CMOS inverter has been demonstrated to date. The difficulty lies in the fact that in complementary CMOS transistors, current is due to carriers of opposite polarity. This makes it possible to connect the drains of two transistors, rather than the source of one to the drain of the other. Consider the CMOS inverter logic gate, shown in Fig. 14. The source of the n-channel transistor is connected to ground, the source of the p-channel transistor is connected to V_{DD}, and the same input voltage V_{IN} is applied to the gates of both transistors. As V_{IN} increases, the current in the n-channel transistor increases while the current in the p-channel transistor decreases. When the input switches between a low and a high voltage, one of the pair turns on and the other turns off. The output voltage thus switches between V_{DD}, when the n-channel device is off, and ground, when the p-channel device is off. If the input is steady, there is no current path between V_{DD} and ground, so the circuit consumes very little power. During switching, on the other hand, the transistor that is being turned on provides the necessary transient current to charge or discharge the

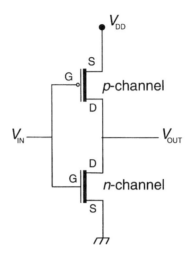

Fig. 14 Schematic diagram of a CMOS inverter. The upper and lower transistors are p-channel and n-channel, respectively. The input voltage V_{IN} is applied to both gates, and turns off one of the two transistors. As a result, V_{OUT} is either high ($V_{OUT} = V_{DD}$) or low ($V_{OUT} = 0$) depending on which transistor is on, and negligible power is dissipated (except during switching) because no current path exists from V_{DD} to ground.

output node. Note that the output node is connected to the drains of both transistors.

It might appear that by virtue of the negative transconductance exhibited by three-terminal RT structures like the one in Fig. 13, both transistors in the CMOS pair can be directly replaced by RT devices. Unfortunately, it is not always sufficient to have transconductances of complementary polarity to implement CMOS functions, at least not in a straightforward way. In RT devices, the current depends on the alignment of the emitter and quantum-well densities of states and hence on the emitter bias V_E. If the p-channel transistor in Fig. 14 were replaced by an n-type RT transistor with negative transconductance, its emitter bias would itself vary between a high and a low state, rather than remaining at a constant potential. In other words, the gate voltage V_{IN} is referenced to V_{OUT}, rather than V_{DD}. This makes it difficult to design a useful circuit.

5.2.4   Cascaded RT and Superlattice-Based Structures

The $I–V$ characteristic of a single double-barrier RT structure exhibits one or more resonant current peaks depending on the number n of quantized 2D subbands E_n. Several proposed applications of RT devices require a

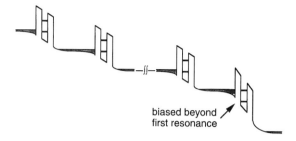

Fig. 15 Schematic band diagram of a cascaded RT structure under total bias V sufficient to bias the last RT diode beyond the first resonance.

multipeak I–V, but with the current peaks approximately equal in magnitude and regularly spaced in voltage by ΔV_P. Neither condition is fulfilled by a typical RT structure: the subband separation $(E_n - E_{n-1})$ changes with n, so the peak voltages $V_P^{(n)}$ are not evenly spaced, whereas the peak currents increase rapidly with n because the emitter transmission coefficient T_E increases exponentially as the barrier height drops. However, the desired I–V curve can be obtained from a cascaded RT structure, in which N double-barrier potentials are epitaxially grown on top of one another. If these RT potentials are separated by doped layers, the resulting band diagram is shown in Fig. 15. The distribution of the total applied bias V can be calculated self-consistently, including current continuity once the RT diodes are biased above threshold, $(V \geq N V_{th})$. As V is increased further, one of the diodes will be biased beyond V_P. In a perfect structure this would happen at the anode because of dynamic charge accumulation in the RT quantum wells. Realistically, variation in quantum-well thickness L_W or cladding layer doping can cause one of the RT diodes to have a lower V_P than the rest. Regardless of which RT diode goes off resonance first, it suddenly presents a high resistance to the biasing circuit, and the total I–V exhibits an NDR region. The crucial point is that if V is increased still further, current continuity requires that almost all of the increase drop over the off-resonance diode, until it begins to conduct through the next 2D subband E_2—this is the situation illustrated in Fig. 15. This process then is repeated with other diodes, with the result that a high-field domain consisting of diodes biased into the second resonant peak expands through the structure. As each diode is biased off resonance, another current peak appears in the I–V, for a total of N peaks that are approximately evenly spaced in V.

The maximum number of diodes that can be cascaded in this manner depends on the PVR required in the I–V characteristic. For a given RT diode that is biased beyond V_P, the other diodes act as a series resistance R_S. As discussed previously in the context of bistability, the I–V acquires hysteretic loops as R_S increases. If the R_S is sufficient to shift V_P beyond the NDR region, it also reduces the PVR. By improving RT design to increase the peak current density J_P and doping the RT regions as well as the cladding regions (which reduces the PVR of individual diodes because of increased impurity scattering, but also reduces the R_S because of other diodes in series), many RT diodes can be cascaded.[43] The room temperature I–V curve of an $N = 8$ cascaded RT structure is shown in Fig. 16. The scatter in the I_P, voltage spacing, and PVR of the eight current peaks is not great and can be attributed to monolayer variations in the barrier or well layers during epitaxial growth.

If the quantized subbands in different RT quantum wells are allowed to interact, for example, by removing the doped cladding regions in Fig. 15, the result is a superlattice (SL) of period $d = L_B + L_W$ shown, in Fig. 17. Consider the wavefunctions $\Psi(z)$ along the SL direction z. If the barriers are infinitely high, $V_0 \to \infty$, we simply have isolated quantum wells. These wells contain the usual quantized levels E_n described by wavefunctions $\chi_n^{(m)}(z)$, where m labels the quantum well. Since $\chi_n^{(m)}(z)$ do not penetrate into the barriers, each 2D subband has a degeneracy of $2N$ including spin. If the barrier height is finite, the $\chi_n^{(m)}(z)$ wavefunctions penetrate into the barriers according to Eq. 3, allowing the wavefunctions in neighboring wells to interact. The previously degenerate levels E_n will broaden into minibands of width Δ_n. In a bulk semiconductor, according to the Bloch theorem, an electronic state can be described by a product of a plane wave and a function periodic in the lattice potential. Analogously, in a superlattice, a state in the nth miniband can be described by linear combinations of wavefunctions periodic in the SL period d, $\varphi_n^{(m)}(z) \equiv \varphi_n(z - md)$ multiplied by a plane wave[44]

$$\Psi_{k_z}(z) = \sum_{m=1}^{N} e^{ik_z md} \varphi_n^{(m)}(z) \tag{21}$$

Equation 21 is a restatement of the Bloch theorem for superlattices. As long as $\Delta_n \ll (E_n - E_{n-1})$, $\varphi_n^{(m)}(z)$ are, to a good approximation, built up from combinations of $\chi_n^{(m)}(z)$. For some range of barrier parameters V_0 and L_B, only interactions with adjacent wells are significant, and the problem simplifies drastically because the periodic components of Eq. 21 can be taken as the ordinary single-well wavefunctions $\chi_n^{(m)}(z)$.* The dispersion $E(k_z)$ for motion along the SL axis becomes

$$E_n^{SL}(k_z) = E_n + S_n + 2T_n \cos(k_z d) \tag{22}$$

---

*This is known as the tight-binding approximation and is a reasonable description of semiconductor superlattices if the barriers are not too narrow. The single-well wavefunctions $\chi_n^{(m)}(z)$ from different wells are not quite orthogonal, and Eq. 22 is valid only to the extent that the overlap between $\chi_n^{(m)}(z)$ and $\chi_n^{(m+1)}(z)$ can be neglected.

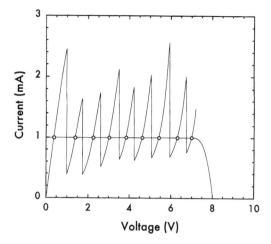

Fig. 16   Room-temperature I–V characteristic of a cascaded RT structure with $N = 8$ diodes in series. Device area is $32\ \mu m^2$. Superimposed in the I–V curve is the load line when the device is biased by an FET constant-current source, with the stable points indicated by open circles. (After Seabaugh et al., Ref. 43.)

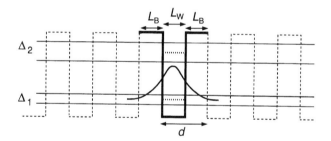

Fig. 17   Superlattice potential diagram, showing the broadening of energy levels E_n into minibands of width Δ_n. The superlattice period is $d = L_W + L_B$. Dotted lines mark the two lowest energy levels confined by the mth potential well (bold line), the corresponding wavefunction $\chi_1^{(m)}(z)$ is also shown. The potential $V_0'(z)$ used in the calculations of miniband dispersion is shown by a dashed line—it includes all wells other than the mth.

where the shift integral S_n is defined as

$$S_n \equiv \int \chi_n^{(m)}(z)\, V_0'(z)\, \chi_n^{(m)}(z)\, dz \qquad (23)$$

and the transfer integral T_n as

$$T_n \equiv \int \chi_n^{(m)}(z)\, V_0'(z)\, \chi_n^{(m+1)}(z)\, dz \qquad (24)$$

The potential $V_0'(z)$ employed in the calculation of the shift and transfer integrals, Eqs. 23 and 24, includes all potential wells other than the mth (see Fig. 17). From Eq. 22 it follows that the width of the nth miniband $\Delta_n = 4T_n$. The allowed values of k_z can be obtained by imposing periodic boundary conditions on Eq. 21: $k_z = 2\pi p/Nd$, where $p = 0, 1, 2, \ldots (N-1)$, so each miniband contains exactly $2N$ states.

The dispersion of the lowest miniband for motion along the SL direction is plotted in Fig. 18a, and the SL density of states, including the transverse degrees of freedom described by $\mathbf{k}_\perp$ is shown in Fig. 18b. It is evident that the effective mass along the SL, $m^* = \hbar^2(k_z{}^{-1}\partial E/\partial k_z)^{-1}$, is a strongly varying function of k_z: starting with a "band-edge" value $m^*{}_{SL} \equiv m^*(k_z = 0)$, the mass becomes heavier as k_z increases, diverges at $k_z = \pi/2d$ (the inflection point in $E(k_z)$, see Fig. 18a), and becomes negative thereafter.

If a constant electric field $\mathscr{E}$ is applied along the SL direction and no scattering is present, the semiclassical equation of motion $\hbar(\partial k_z/\partial t) = q\mathscr{E}$ implies that k_z changes linearly with time. Since $v(k_z)$ is periodic, carriers execute oscillatory motion. After reaching the miniband edge at $k_z = \pi/d$ they are Bragg reflected to $k_z = -\pi/d$ by the periodic SL potential (see Fig. 18a and Problem 9). These Bloch oscillations exist for any periodic potential, including that of the original semiconductor lattice a_0. They cannot be observed in bulk semiconductors because collisions typically return the carriers to the bottom of the band long before they can complete one period. That is, the scattering time τ is insufficiently long for realistic electric fields $\mathscr{E}$. A periodic potential with period $d \gg a_0$ is required to relax the constraint on τ. Early on, high-frequency ultrasonic waves propagating through the semiconductor were suggested as a possible realization of such a potential[45] and it was the pioneering suggestion of Esaki and Tsu to employ superlattices for this purpose that opened the modern era of heterostructure bandgap engineering.[46] In that celebrated paper, the effects of a finite scattering time τ on the average drift velocity v_D of electrons propagating in a 1D superlattice with dispersion given by Eq. 22 was evaluated classically:

$$v_D = \int_{t=0}^{\infty} e^{-t/\tau} a(t) \, dt \tag{25}$$

where $a(t) \equiv a[k_z(t)]$ is the acceleration of the miniband electron in the superlattice direction. Using the tight-binding approximation of Eq. 22, they obtained v_D in terms of $\mathscr{E}$, τ, SL period d, and m^*_{SL} (see Appendix 5.B),

$$v_D = \frac{\hbar}{m^*_{SL}d} \frac{\xi}{1+\xi^2} \tag{26}$$

where $\xi = q\mathscr{E}\tau d/\hbar$. The average drift velocity peaks at $\xi = 1$, that is when the electric field $\mathscr{E} = \hbar/q\tau d$. Beyond this point, increasing $\mathscr{E}$ results in a lower

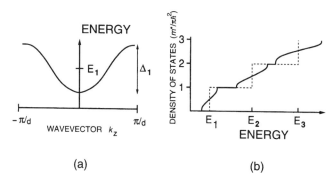

Fig. 18 (a) Model superlattice miniband dispersion and (b) the corresponding density of states. The 2D density of states is shown in (b) by a dashed line for comparison.

v_D because, on average, more and more carriers reach the negative-mass region of $E(k_z)$. As a result, the I–V characteristic should exhibit NDR.

Although it might appear that the NDR regime can be reached simply by increasing $\mathscr{E}$, this is not the case. The above analysis breaks down in high electric fields, where, because of Zener tunneling, the single-band approximation is no longer valid. It is, however, true that the constraint on τ for observing NDR is easier to achieve by a factor of 2π (Problem 9) than for Bloch oscillations. Nonetheless, these effects have proved elusive in I–V measurements, because of scattering, Zener tunneling between different minibands, and, particularly, electric-field domain formation due to space-charge instabilities associated with the nonlinear current flow through the SL.[47] For this reason, although I–V nonlinearities in SL transport have been observed and attributed to the excursion of carriers into the negative-mass regions of the dispersion,[48] these nonlinearities have not been used in devices to date.

Thus far, our discussion has considered electric fields $\mathscr{E}$ that arc weak in the sense that the carriers are essentially delocalized along the SL and obey the dispersion of Eq. 22. In higher fields, the nth miniband breaks up into a set of discrete levels, separated by energy intervals $q\mathscr{E}d$, with wavefunctions centered in different wells and extending over $\Delta_n/q\mathscr{E}d$ periods.[49,50] This so-called Wannier–Stark ladder of states, illustrated in Fig. 19a, forms for all values of $\mathscr{E}$ but becomes physically meaningful only when adjacent ladder states can be resolved: $q\mathscr{E}d > \hbar/\tau$, which is the Bloch oscillation criterion again. As soon as the extent of the Wannier–Stark wavefunctions falls below N periods, they no longer reach from one end of the superlattice to another. For any DC current to flow, some scattering process becomes necessary. The current will remain small until $\mathscr{E}$ brings into resonance Wannier–Stark states arising from different minibands, which are by then confined to individual wells:[51] $q\mathscr{E}_j d = E_j - E_1$, $(j = 2, 3, \ldots)$ At these sharply defined values of $\mathscr{E}_j$, the current can flow by sequential tunneling between different Wannier–Stark states in adjacent wells, followed by relaxation to a lower-lying state (see Fig. 19b). Ignoring possible series resistance outside the SL, the I–V curve should then exhibit peaks at $V = Nq\mathscr{E}_j d$, followed by NDR regions where current flow once again requires scattering or some inelastic mechanism.

A particularly interesting case of the latter is photon emission in the regime where $\mathscr{E} > \mathscr{E}_j$, which was proposed by Kazarinov and Suris decades ago as a system capable of voltage-tunable lasing.[51] The scheme is shown in Fig. 19c and the photon energy is $\hbar\omega = q(\mathscr{E} - \mathscr{E}_j)d$, tunable in the infrared by the applied voltage and appropriate choice of SL parameters. The problem with this exciting possibility is the same as with observing I–V peaks at $V = Nq\mathscr{E}_j d$, and even the Esaki-Tsu NDR at low $\mathscr{E}$: all of these schemes rely on a uniform electric field $\mathscr{E}$ extending through the superlattice. At the same time, current flow through the SL leads to dynamically stored space-charge densities in the various quantum wells that produce nonuniform $\mathscr{E}$. Devices that operate in the NDR regions of their I–V characteristics are particularly susceptible to the electric field breaking up into high- and low-field domains.[48] For this reason, voltage-controlled lasing illustrated in Fig. 19c has not been observed, and it is not clear whether it can be observed even in principle. Another problem with experimental measurements on current-carrying superlattices has been the impedance matching of the SL to the ohmic contacts, discussed in Appendix 5.C.

On the other hand, the alignment provided by $\mathscr{E}_j$ between different Wannier–Stark states in adjacent wells can also turn the biased SL into a lasing medium, provided that at least some fraction of the $E_2 \rightarrow E_1$ relaxation processes is radiative (cf. Fig. 19b). The voltage tunability of the emitted radiation is now lost, since $\hbar\omega = E_2 - E_1$, which is set by the SL parameters. Also, since the lower level in the radiative transition is called upon to supply the higher level in the downstream well, population inversion is difficult to achieve. On the other hand, the device need not operate in the NDR region of the I–V curve, so the problem of maintaining uniform alignment between adjacent periods of the SL becomes more tractable. Recently, infrared lasing in a conceptually similar device—the quantum cascade laser (QCL), based on intersubband transitions in a modified SL structure—has been achieved.[52] A more detailed discussion of the QCL, including its output characteristics as well as the structural design required to overcome domain formation and establish population inversion, will be discussed in Section 5.4.4.

5.2.5 RT Nanostructures and the Coulomb Blockade

If a double-barrier RT structure with well width L_W is etched into a sufficiently narrow pillar or biased to a narrow effective size by a lateral gate,

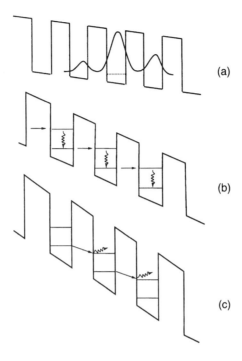

Fig. 19 (a) Schematic band diagram of a biased superlattice, in which the minibands break up into Wannier–Stark states. The lowest-lying Wannier–Stark state E_1 in the mth well is shown by the dotted line, together with the corresponding schematic wavefunction for $\Delta_1/e\mathscr{E}d \approx 3$, where Δ_1 is the miniband energy width at flatband, $\mathscr{E} = 0$. (b) Sequential tunneling through a superlattice when the excited and ground states in adjacent wells are aligned, resulting in a current peak in the *I–V* characteristic. Vertical lines represent intersubband relaxation (e.g., by phonon emission). (c) Photon-assisted sequential tunneling tunable by the electric field.

new effects come into play. The first and more obvious is the possibility of lateral size quantization in the quantum well. Current fabrication techniques can only produce structures with lateral extent $L \gg L_W$, while gate-induced electrostatic confining potentials are much weaker than the heterostructure potential $V(z)$, so lateral quantization will be much weaker. To a good approximation, each of the 2D subbands E_n in the well will give rise to a series of fully quantized, atomic-like states E_{nm}, where m labels the different states resulting from the lateral confining potential $V(x, y)$. In principle, the same lateral quantization into 1D subbands could apply to the doped emitter and collector regions. However, states in these regions are broadened by impurity scattering and their description by a 3D density of states is often a good approximation. Moreover, the confining potential in the emitter and collector regions is usually much weaker than that in the undoped well because of the screening by charged impurities.*

The RT transport resulting from alignment of the occupied emitter states with discrete quantum dot levels in the well can be treated within the usual sequential-tunneling formalism[8,53] but with a new effect. The charging energy U required to transfer even a single electron from the emitter into the well becomes significant for small L. If the charging energy is ignored, the situation is shown in Fig. 20a. Since the lateral confining potential $V(x, y)$ changes between the emitter and well, $k_\perp$ is no longer a conserved quantity, and only energy conservation holds as a tunneling selection rule. As the bias V lowers E_{11} below E_F in the emitter, tunneling through this single state becomes possible—this defines the threshold V_{th}. At higher V, additional tunneling channels open up. The resulting I–V will exhibit a rising staircase of step-like features, with bias spacing corresponding to the energy separation of the levels.[54] The strength of these features depends on the transmission coefficient of the emitter barrier $T_E(V)$ and also on the degeneracy of the E_{1m} states, which may be large if $V(x, y)$ is approximately parabolic. Finally, NDR is not expected in the I–V characteristic, since $k_\perp$ conservation no longer cuts off the tunneling through higher-lying E_{1m} states when E_{11} drops below the bottom of the occupied states in the emitter. Instead, the I–V should become nonlinear whenever the density of levels changes appreciably, for example, when E_{2m} levels arising from the second subband become accessible.

This picture of tunneling into a quantum dot would be quite unpromising from the device standpoint. However, the charging energy U associated with electronic transport through an RT nanostructure is important. If L is small, the energy $U = q^2/2C_w$ associated with the tunneling of a single electron into the well, where C_w is the effective capacitance of the quantum dot, can appreciably alter the alignment of E_{nm} with emitter E_F. This effect is illustrated in Fig. 20b. A simple, geometric estimate of the capacitance is $C_w \approx \varepsilon_s L^2/d$, where d is the effective collector barrier thickness (if the depletion in the collector electrode is negligible, $d = L_B$). For current to flow, at least one electron must tunnel into the dot. So, V_{th} shifts to higher bias by the single-electron charging energy U. The shift in the bias of other step-like features depends on the average occupation of the well by electrons, which is determined by the transmission ratio T_E/T_C of the emitter and collector barriers.[55] If $T_E/T_C << 1$ (the case for a symmetric double-barrier structure once it is biased to $V > V_{th}$), the occupation of the well by more than one electron at a time is rare and all the step-like features in the I–V corresponding to additional channels coming into resonance will be shifted

---

*The lateral confining potential $V(x, y)$ is determined by a self-consistent electrostatic potential with boundary conditions set either by the pinning of the Fermi level at the semiconductor–air interface in etched pillars or by V_G in gated structures (in addition to the Schottky barrier in metal-gated structures or the built-in p–n junction voltage in implanted structures).

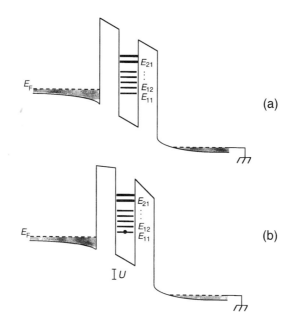

Fig. 20 (a) Schematic band diagram of a double-barrier RT nanostructure with lateral quantization. The levels E_{nm} arise from the quantization of the nth 2D subband E_n into discrete quantum dot states. (b) Coulomb blockade regime. A single electron tunneling into the well would change the emitter–well alignment by $U = q^2/2C_w$, where C_w is the dot capacitance. Tunneling into the well is cut off if U is large enough to raise E_{11} above the occupied states in the emitter, as shown.

by U. On the other hand, if $T_E/T_C \gg 1$, each available level is occupied most of the time, so the opening of every additional channel requires sufficient biasing to overcome the charging energy—this is the so-called Coulomb blockade of tunneling. For example, the second current step requires V to lower E_{12} below emitter E_F by at least $2U$ to surmount the energy barrier due to the simultaneous occupation of the well by two electrons. An additional complication is that the charging energy will vary with electron number because of electron–electron interactions in the dot and also because the effective dot size L changes. Since both the empty $(T_E/T_C \ll 1)$ and occupied $(T_E/T_C \gg 1)$ regimes are accessible in the same asymmetric double-barrier RT nanostructure by changing the bias polarity, such devices have been studied to probe the energy spectrum of quantum dots with and without electron–electron interactions.[54,55]

It is the charging energy required to change the electron occupation of the dot together with the possibility of tuning the energy alignment of the dot levels with respect to the emitter electron reservoir via a third terminal that makes RT nanostructures promising for devices. Schematically, a three-terminal nanostructure involves nothing but the addition of a gate electrode that can change the potential between the quantum dot and the emitter (see Fig. 20), but is sufficiently isolated from the dot to prevent any possibility of electron transfer from the gate. Then, if the device is biased by V_E near a voltage step corresponding to the addition of another electron to the dot, a small change in V_G can tune the occupation of the dot. This controls the current through the dot, resulting in a single-electron transistor. Because of the fabrication difficulties associated with vertical RT nanostructures, gate control of single-electron tunneling has proved easier in the planar geometry. The dot and the controlling electrode are defined by electrostatic metal gates deposited on top of a high-mobility 2D electron gas heterostructure. A top view of the gated structure is shown in the inset of Fig. 21. The outside gates are biased into deep depletion, forming a small island of 2DEG connected to the reservoirs by tunneling barriers. As we had seen in the context of planar double-barrier RT structures (see Fig. 8), these islands are necessarily large and the electrostatic barriers are wide and low, so the energy quantization in the island is weak. But, this is an advantage in the context of Coulomb-blockade devices, because the energy spectrum of the dot is now entirely defined by the charging energy U. The gate electrode can alter the effective size and capacitance of the island. As long as the island capacitance C_w is small and the island size is relatively large, $(L \approx 0.1-1 \ \mu m)$, the $I-V$ characteristic as a function of V_G should show regularly spaced steps corresponding to the adding of electrons to the island. At low temperatures very regular conductance $(G \equiv \partial I/\partial V)$ peaks have been observed in such structures;[3] an example is shown in Fig. 21.

In principle, precise single-electron control over electron occupation or the tunneling transport in small quantum dots or islands has led to many proposals of logic and memory circuits based on single-electron transistors (SETs) and other devices.[56] To some extent, single-electron devices can be considered the logical endpoint of miniaturization-driven semiconductor technology. The main difficulty from the practical standpoint is posed by the extremely stringent fabrication requirements on large-scale SET circuitry, especially at noncryogenic temperatures. Currently, SET characteristics, like the data in Fig. 21, are measured at $T < 1 \ K$, to ensure the condition $U = q^2/2C_w >> kT$. Clearly, device sizes will need to be reduced by orders of magnitude before higher temperature operation can be contemplated. For $T = 4.2 \ K$ the charging energy must be certainly larger than 1 meV. This requires a capacitance $C_w < 10^{-16} \ F$, a very stringent condition. It is imperative that there be no parallel capacitance due to leads or other electrodes. Note that a simple thin wire has an intrinsic capacitance of about $10^{-16} \ F/\mu m$. It is also not clear that semiconductor SET realizations have any advantages over metal-tunnel junctions for most proposed devices: the first observation of Coulomb-blockade phenomena[57] and the first SET with voltage gain[58] both employed small aluminum-tunnel junction capacitors.

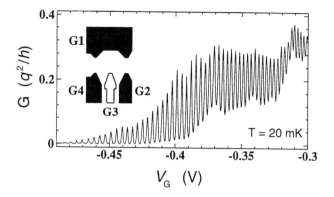

Fig. 21 Conductance of a small 2D electron gas island as a function of gate bias V_G, at $T = 20\,mK$. Inset shows a top view of the island geometry. Gates shown in black are biased into deep depletion, forming an island of submicron diameter weakly coupled to 2DEG electrodes. V_G is applied to the gate G3 shown in white. Changes in the electron occupation of the island produce regular spikes in the conductance. (Courtesy of C. J. B. Ford, 1996.)

One specific application for which the SET appears promising is the construction of precision current standards. In a gated 2DEG island, by sequentially lowering and raising the emitter and collector barriers at small V_E in the Coulomb-blockade regime, where only one excess electron can occupy the island, the transfer of one electron per cycle of barrier biasing can be achieved.[59] If the barriers are cycled at frequency f, the emitter–collector current is given by $I = qf$, making for a very precise current source. It is anticipated that such a device may provide a new metrological current standard, although single-electron transfer along a chain of small metallic islands may prove a more successful implementation.[60]

5.3 HOT-ELECTRON STRUCTURES

5.3.1 Hot Electrons in Semiconductors

In many of the resonant-tunneling structures discussed in the preceding section, electrons (or holes) are injected into the collector region with energies that are several kT above the collector Fermi energy E_F, where T is the lattice temperature. These electrons are clearly not in thermal equilibrium with the lattice and their occupation of available states is not described by the standard Fermi–Dirac function $f_{FD}(E)$. Further, their velocity distribution in the direction of current flow is strongly peaked, at least in the immediate vicinity of the collector barrier, making up a "ballistic" electron packet. As the electrons propagate into the collector, the velocity distribution broadens due to scattering, resulting in a distribution that can be taken as Maxwellian and parametrized by an effective temperature $T_e > T$. In either case, the electrons are "hot" with respect to the lattice.

Another possible reason for carrier heating is a strong electric field $\mathscr{E}$ in some region of the device. Depending on the energy relaxation time, a large fraction of the carriers can be accelerated into states of high kinetic energy. As, in the course of ongoing miniaturization, device dimensions shrink at a faster rate than various electrode voltages, the internal fields rise, and carrier heating becomes more significant. Thus, oxide damage by hot electrons accelerated by the large lateral $\mathscr{E}$ at the channel drain has become a major reliability issue as silicon MOSFETs are scaled down. Undesirable though carrier heating may be in standard silicon technology, a number of devices based on hot electrons have been proposed. In this chapter we will focus on injection devices, where the hot carriers are physically transferred between adjacent semiconductor layers.* As we shall see, although the first hot-electron injection devices[61] were proposed as far back as 1960, the abrupt heterojunction interfaces and doping profiles made possible by modern epitaxy have greatly widened hot-electron device possibilities.

Let us consider the two principal techniques for producing hot carriers by electric current—ballistic injection and electric field heating—and the resulting carrier distributions in more detail. Ballistic injection by thermionic emission from a wider-bandgap semiconductor and by tunneling into states of high kinetic energy is illustrated in Figs. 22a and 22b respectively. Immediately upon injection, the corresponding velocity distribution is also shown (cf. Problem 11). As the carriers propagate away from the injection point, their energy and velocity distribution will change and broaden. Given the initial velocity and momentum distribution $f(\mathbf{r}, v, t = 0)$ and ignoring the possibility of interband transitions (like electron–hole recombination), the evolution of $f(\mathbf{r}, v, t)$ with time as a function of spatial position can be determined by solving the Boltzmann transport equation. In the simplest case of parabolic dispersion, the Boltzmann equation has a physically transparent form:

$$\frac{\partial f}{\partial t} + v \cdot \nabla_{\mathbf{r}} f + a \cdot \nabla_v f = \left(\frac{\partial f}{\partial t} \right)_{\text{coll}} \tag{27}$$

with semiclassical equations of motion given by

$$v = \hbar \mathbf{k}/m^* \qquad m^* a = q(\mathscr{E} + v \times B) \tag{28}$$

---

*Transferred electron devices, like Gunn oscillators, employ hot-electron scattering into lower mobility satellite valleys of the semiconductor dispersion $E(\mathbf{k})$; they are covered in Chapter 6.

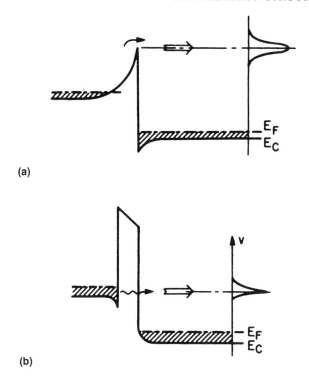

(a)

(b)

Fig. 22 Ballistic injection of electrons by (a) thermionic emission from a wider bandgap semiconductor and (b) tunneling through a barrier. The schematic velocity distribution of the injected hot electrons is shown on the right in both cases.

where $\mathscr{E}$ and $\boldsymbol{B}$ are the electric and magnetic fields, m^* is the effective mass in $E(\mathbf{k}) = \hbar^2 k^2/2m^*$, and $\boldsymbol{a}$ is the acceleration.* The collision term on the right of Eq. 27 represents all scattering processes, including phonon emission and absorption, impurity scattering, electron–electron interaction, and so on. It can be formally defined as the integral of the scattering probability $W(\mathbf{k}',\mathbf{k})$ between states characterized by wavevectors $\mathbf{k}$ and $\mathbf{k}'$ over the first Brillouin zone multiplied by the appropriate occupation probabilities. A great simplification results in the relaxation time approximation, which replaces the entire collision term by $-(f - f_0)/\tau$, where τ characterizes the time it takes for the distribution to relax to its equilibrium value f_0—the Boltzmann equation is then no longer an integral equation. This is rarely possible, however, because different characteristics of the distribution function relax at different rates. Because of this, one usually defines separately the momentum $\tau_\mathbf{k}$ and energy τ_e relaxation times in terms of $W(\mathbf{k}',\mathbf{k})$.[62] Still

---

*In the case of an arbitrary dispersion $E(\mathbf{k})$, the equations of motion become $v = \hbar^{-1}\nabla_\mathbf{k} E(\mathbf{k})$ and $\hbar \, \partial \mathbf{k}/\partial t = q(\mathscr{E} + v \times \boldsymbol{B})$, with corresponding changes in Eq. 27.

another complication is that the transition matrix element $W(\mathbf{k'},\mathbf{k})$ depends on the electron energy in different ways for different scattering mechanisms. For states of higher energy new scattering mechanisms set in: optical phonon emission, impact ionization, and so on. Finally, since the collision integral extends over the Brillouin zone, it depends on the density of states available for scattering (i.e., on the explicit form of $E(\mathbf{k})$, leading to ever greater complexity as energy increases and larger sections of the Brillouin zone become accessible), and even on $\mathbf{k}$ orientation. For this reason, Eqs. 27 and 28 are rarely tractable analytically, and numerical Monte Carlo techniques are often employed.[63]

Of course, there exists one limiting case which avoids the difficulty altogether: ballistic motion, in which collisions are negligible. This limits the critical dimension of any device to $v_z\tau_{\mathbf{k}}$, where v_z is the (high) injected electron velocity and $\tau_{\mathbf{k}}$ is the momentum relaxation time. This is the preferred operating regime of ballistic hot-electron transistors (HETs), in which electrons are injected into a narrow base layer of length L_B. Control over injection energy in heterostructures (see Fig. 22) can generate a narrow hot-electron distribution, centered on a high velocity normal to the base layer, as shown in Fig. 23. As long as $L_B < v_z\tau_{\mathbf{k}}$, a large fraction of the hot electrons will traverse the base without scattering.

The energy and velocity distribution of hot carriers created by a strong electric field $\mathscr{E}$ is necessarily rather different. Before the electric field is applied, the carriers are in equilibrium with the lattice. The field accelerates the carriers according to Eq. 28, shifting the distribution function $f(\mathbf{r},v,t)$ away from equilibrium. Since the scattering mechanisms depend on the carrier energy E, the same difficulty in solving the Boltzmann equation arises. However, at sufficiently high electron concentrations, numerous electron–electron collisions can establish a quasi-equilibrium within the electronic system that is effectively decoupled from the lattice. In this limit, one can define the effective electron temperature T_e from the average energy $\langle E \rangle$ of the electron ensemble: $\langle E \rangle = 3kT_e/2$. The resulting hot-electron distribution is shown in Fig. 23.

Evidently, the effective electron temperature T_e depends on the magnitude of the electric field in a complicated fashion given by the solution of Eqs. 27 and 28. Once again, Monte Carlo simulations are generally required, especially in the presence of heterostructure barriers that are necessary for real-space-transfer (RST) devices based on field-induced electron heating.[64] In these devices, carriers are heated by an electric field applied parallel to a heterostructure barrier V_0. If T_e becomes sufficiently high, some fraction of the electron distribution will acquire enough energy to spill over the barrier and transfer to a different region of the structure, which may have a different mobility or a separate electrical contact. Even though only electrons in the high-energy tail of the hot-carrier distribution function can surmount V_0 for a given heating field $\mathscr{E}$, they are quickly replenished (on a scale of τ_e), so the RST process can be fast and efficient.

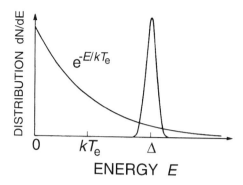

Fig. 23 Hot-electron distribution functions. In ballistic injection, the distribution function is sharply peaked at $\Delta \approx m^* v_z^2/2$, where Δ is the injection energy and v_z is the velocity normal to the barrier. In real space transfer, the distribution is approximately e^{-E/kT_e}, where T_e is the effective electron temperature.

It should be emphasized that both ballistic-injection and RST devices involve the injection of nonequilibrium carriers over (or through) heterostructure barriers into adjacent layers of the structure. Conceptually, the real distinction lies in the different hot-carrier distribution functions illustrated in Fig. 23. Operationally, the three-terminal implementations of ballistic and RST devices employ rather different controlling electrode geometries. As will become clear, there is a strong parallel between ballistic hot-electron transistors and standard bipolar transistors: the hot-electron current across the base is controlled by a smaller base current of equilibrium carriers moving in different portions of **k** space. Base transparency for ballistic carriers (i.e., the base transport factor α_T of a bipolar transistor) relies on the short time required to traverse the narrow base compared to the momentum relaxation time $\tau_{\mathbf{k}}$. Also, as in bipolar transistors, speed limitations arise from the base traversal time, $\approx L_B/v_z$ (here hot-electron transistors really shine by virtue of large injected v_z) and the $R_B C$ delay associated with charging the base–emitter and base–collector capacitances through a finite lateral base resistance.

Real-space-transfer devices, on the other hand, have no ready analog among standard transistors. Two-terminal versions, which rely on RST of hot carriers to a lower-mobility region of the structure to produce NDR in the two-terminal I–V characteristic, are essentially similar to Gunn oscillators. In three-terminal versions, RST of hot carriers to a region that can be contacted separately is employed to control the transferred current by a heating-field. As we shall see, three-terminal RST devices possess unusual terminal symmetries, arising from the insensitivity of hot-carrier distributions to the heating field polarity, that can be exploited for increased functionality.

5.3.2 Ballistic Injection Structures

Figure 24 shows a schematic band diagram of a ballistic HET based on tunneling injection and implemented in GaAs/AlGaAs.[65] Hot electrons are injected at an energy $\Delta \approx qV_{BE}$ with respect to the Fermi level in the heavily doped base (held at ground potential), traverse the base and are collected after surmounting the collector barrier, which is a function of collector bias V_{CB}. There is a clear analogy between this type of HET and the current-controlled three-terminal RT structure of Fig. 9. However, since quantization in the base region of Fig. 24a is not required, HET base width can considerably exceed the $L_B \leq 100$ Å strong-quantization condition. This leads to lower base resistance R_B and, hence, reduces the time delays associated with charging the base–emitter and base–collector capacitance. Of course, Eq. 19 for the $R_B C$ time delay still applies, so ballistic HET design involves a tradeoff between low base resistance, which requires large L_B and base doping, and high base transport factor α_T, which requires short L_B and minimal base scattering. For amplification, the highest α_T is achievable for $V_{CB} > 0$. In that configuration, electrons that have experienced scattering in the base can still be collected. If V_{BE} is much larger than the collector barrier height, the emission of one or more optical phonons still leaves the electron with enough kinetic energy to reach the collector, provided the direction of its velocity has not changed. Fortunately, optical-phonon emission by high-energy electrons is a predominantly forward-scattering process and the effective τ_k is not as short as it is for electrons just above the optical-phonon emission threshold. Still, the very high α_T required to produce competitive differential current gain $\beta \equiv \alpha_T/(1 - \alpha_T)$ is difficult to realize in ballistic HETs even with the shortest L_B allowed by Eq. 19.

First, there is the matter of quantum-mechanical reflection at the collector barrier. In our discussion of tunneling we found that given a collector-barrier height Φ_C, there is a nonzero reflection probability $R(E_z)$ for incident electrons with kinetic energy $E_z > \Phi_C$. In the case of rectangular heterostructure barriers, $R(E_z)$ remains significant unless $E_z \gg \Phi_C$, except at some special values of E_z that depend on the barrier parameters (cf. Problem 2). This difficulty can be circumvented to some extent by grading the collector barrier (Problem 12, see also Fig. 24a for an example), but reducing $R(E_z)$ to nearly zero at moderate hot-electron injection energies is problematic.*

It might appear that α_T can be improved by increasing V_{BE} and hence

---

*It should be emphasized that this envelope function approach to heterostructure barrier transmission is only valid if the heterojunction semiconductors are lattice matched and have similar dispersions in **k** space. An example is the GaAs/Al$_x$Ga$_{1-x}$As heterostructure with $x \leq 0.45$, where electrons in both materials move in the same Γ valley of the conduction band. Although frequently used, the applicability of envelope function transmission calculations to other types of heterostructures is questionable.[68]

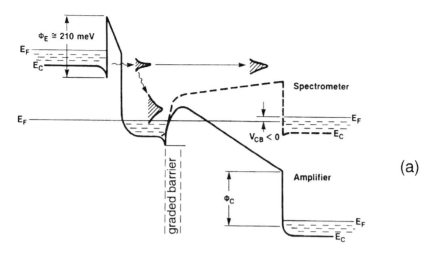

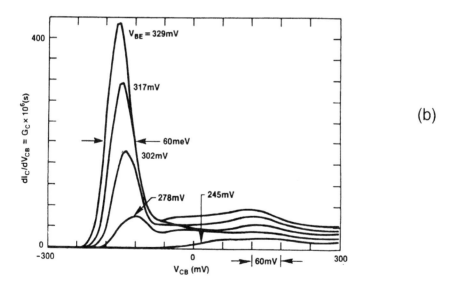

Fig. 24 (a) Schematic band diagram of a GaAs/AlGaAs ballistic-injection hot-electron structure biased as an amplifier ($V_{CB} > 0$) and a hot-electron spectrometer ($V_{CB} < 0$). Note the grading of the base–collector barrier designed to reduce quantum-mechanical reflection. (b) Measured hot-electron energy spectra for different injection energies qV_{BE}. The main peak is due to electrons arriving at the collector without any inelastic scattering. (After Heiblum et al., Ref. 65.)

the injection energy, since this should reduce the collector reflection coefficient and the base transit time simultaneously. Unfortunately, this approach runs into the second important limitation of HET performance. If the electron kinetic energy exceeds the energy of the satellite valleys in the dispersion (e.g., the L valley in GaAs, which lies 0.3 eV above the Γ conduction-band minimum), the high density of final states leads to very efficient intervalley scattering by phonons and impurities. At electron energies below the intervalley-scattering threshold, but above the optical-phonon emission threshold (36 meV in GaAs) the dominant form of phonon interaction is emission of polar optical phonons, a process that has only a weak dependence on the carrier energy. At the same time, ionized-impurity scattering in the heavily doped base is minimized with increased injection energy. Consequently, the optimum V_{BE} for high gain is just below the threshold for intervalley scattering. In GaAs/AlGaAs HETs analogous to Fig. 24a, this has limited the highest observed gain to $\beta \approx 10$ at low temperatures,[65] corresponding to $\alpha_T \approx 0.9$. Higher gains of $\beta \approx 30$ ($T = 77$ K) have been observed in similar HET structures with a narrow $L_B = 200$ Å pseudomorphic InGaAs base, because of larger the Γ–L energy separation.[66] More recently, similar structures have yielded $\beta \approx 10$ at $T = 300$ K,[67] which may be approaching the limit for HET structures grown on GaAs or InP substrates.

Compared to heterojunction bipolar transistors (HBTs), ballistic-injection HETs suffer from relatively low base-transport factors. Furthermore, the upper limit Eq. 19 places on the lateral base resistance R_B is more stringent in the case of HETs with tunneling injection, since the emitter barrier must be fairly thin to keep the tunneling current high. This renders HETs noncompetitive for most device applications. They have proved valuable for research into nonequilibrium carrier transport, however. The use of an injection HET as a hot-electron spectrometer is illustrated in Fig. 24a, and a representative set of hot-electron energy distributions is shown in Fig. 24b. The idea is to measure the collector current I_C as a function of $V_{CB} < 0$ at a fixed V_{BE} (which sets the injection energy). In a certain range of V_{CB} the collector barrier height varies linearly, $\delta\Phi_C \approx \delta V_{CB}$, and $\partial I_C/\partial V_{CB}$ is proportional to the number of carriers arriving at the collector barrier with $E_z = \Phi_C$. To the extent that V_{CB} does not affect the hot-electron energy distribution in the base, the injected distribution, and the above-barrier quantum-mechanical reflection, one can deduce the mean free path l as a function of injection energy and correlate the dynamics of energy loss with various scattering mechanisms.[65,69] For example, the main peak in Fig. 24b corresponds to hot electrons arriving at the collector barrier without a single phonon-emission event, with $l \approx 1000$ Å. This is quite remarkable considering that these electrons had to traverse not only the doped GaAs base but also the AlGaAs collector barrier. Lateral hot-electron spectrometers have been constructed in 2DEG using electrostatic barriers,[70] as in Fig. 8. The minimal scattering in 2DEG at low temperatures leads to considerably longer

$l \approx 0.5 \, \mu$m. As a result, very high $\beta \, (>100)$ was measured at $T = 4.2$ K in devices with $L_B \approx 1700$ Å. More importantly, similar structures have provided the laboratory for studying the physics of ballistic transport in small systems.[71]

Finally, it should be noted that hot-carrier injection has been employed to good effect in HBTs by designing structures with a wider-bandgap emitter, schematically illustrated in Fig. 25 for an AlInAs/InGaAs HBT lattice matched to InP. The advantages conferred by hot-electron effects are several. Electrons are injected by thermionic emission over the emitter–base barrier at an energy $\Delta \approx 0.5$ eV above the conduction band-edge in the p-InGaAs base. Here the purpose of ballistic injection is to shorten the base traversal time by replacing the relatively slow diffusive motion by faster ballistic propagation. Since there is no collector barrier to surmount, scattering does not degrade the transport factor α_T, as it does in unipolar hot-electron transistors. Further, the fact that the injected-velocity distribution is sharply peaked in the direction perpendicular to the base aids device scaling by minimizing lateral excursion into the extrinsic base region. As discussed in more detail in Chapter 1, hot-electron HBTs exhibit high gain and high-speed operation, with f_T exceeding 100 GHz at room temperature.[72] Note that in optimized transistors the base is so thin ($L_B << 1000$ Å) that the diffusive transport time across the base is also shorter than 1 ps, and hence high-speed HBT operation is not in itself evidence of ballistic transport.*

If the transport across the HBT base is truly ballistic and the velocity distribution of injected hot electrons is sufficiently narrow, $\Delta v_z / v_z << 1$; the result is a coherent transistor that is predicted to have gain above f_T.[73] When the injection from the emitter is varied at some frequency f, an electron density wave of wavelength $\lambda = v_z / f$ is set up in the base. The minority-carrier density wave is screened by majority carriers and the base remains neutral everywhere. Such is the situation with all bipolar transistors. Neglecting recombination, the base current in the HBT flows only to neutralize variations in the overall number of minority carriers. At low frequencies, $\lambda >> L_B$, the minority charge in the base increases and decreases in phase with injection. This leads to a characteristic frequency roll-off of the current gain $\beta \approx 1/f$ and the characteristic value for the $\beta = 1$ frequency cutoff $2\pi f_T = v_z / L_B$. The functional form of this roll-off begins to change when the wavelength of minority-carrier density wave becomes comparable to the base width. If L_B is an integer multiple of λ, which corresponds to f being an integer multiple of v_z / L_B, there is no change in the total minority charge as the electron density wave goes through the base. Hence, the collector-current

---

*It is an exceedingly difficult matter to demonstrate the ballistic nature of transport in a given transistor, and claims of ballistic HBT operation have always been controversial. An easier, but indirect approach involves measurements of gain in a series of transistors with variable base width L_B, which allows one to discriminate between ballistic and diffusive transport mechanisms by appealing to a theoretical model.

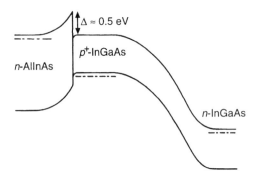

Fig. 25 Schematic band diagram for a *n–p–n* heterojunction bipolar transistor in the AlInAs/InGaAs material system lattice matched to InP. The electron injection energy $\Delta \approx 0.5$ eV corresponds to the emitter–base barrier.

modulation is accomplished, in this case, with no high-frequency base-current input, leading to $\beta \to \infty$. Obviously, in a real device β will be limited by the recombination current, scattering-induced damping of the density wave, and the finite velocity distribution width Δv_z.

Still, as long as a high degree of coherent transport is maintained, $\beta(f)$ will peak at integer multiples of $2\pi f_T$, leading to current gain above the usual cutoff frequency (we present a more detailed discussion of this effect in Appendix 5.D). Moreover, the transistor power gain has been predicted[73] to peak at multiples of πf_T, exhibiting two peaks for each peak in β. The difficulty with implementing such a transistor lies in maintaining coherence across the base. Since optical-phonon emission is a very effective scattering mechanism, the injection energy $\Delta = m^* v_z^2 / 2$ should remain below the optical-phonon energy but still much larger than kT, implying cryogenic operation. Further, device parasitics can wash out the gain peaks above f_T. To date, the proposed coherent transistor has not been experimentally characterized.

5.3.3 Real-Space Transfer Structures

Proposals of generating NDR in the two-terminal *I–V* characteristics of a device by real-space transfer (RST) between semiconductor layers of high and low mobility date back several decades.[74] These ideas received further development in the context of modulation-doped multiquantum-well GaAs/AlGaAs heterostructures,[75] and first experiments using such structures were carried out in the early 1980s.[76] A schematic illustration of the two-terminal GaAs/AlGaAs RST structure is shown in Fig. 26. If the longitudinal electric field $\mathscr{E}_x$ is small, electrons reside in the undoped GaAs quantum wells and the source–drain $I–V_D$ depends linearly on $\mathscr{E}_x$, with the slope determined by high GaAs electron mobility. However, as $\mathscr{E}_x$ is increased, the power input into the electron distribution exceeds the rate of energy loss into the lattice by phonon emission, and the electrons heat up to some field-dependent temperature T_e. At sufficiently high T_e, there will be partial transfer to the doped AlGaAs layers over the heterostructure barrier Φ, where the mobility is much lower due to heavy doping and higher m^*. The two-terminal I–V_D then exhibits NDR with the peak-to-valley ratio determined by the magnitude of the transferred electron density and the ratio of the mobilities in the GaAs and AlGaAs layers (cf. Problem 13 for an analytically tractable model). The analogy to the Gunn effect, where high-mobility carriers are scattered to low-mobility valleys in momentum space, is obvious. In fact, Gunn-effect and RST mechanisms are competing processes that depend on the relative magnitude of the barrier height Φ and the valley separation.

In addition to the interplay between transfer mechanisms, a realistic treatment of electron heating in an RST structure involves the formation of longitudinal electric-field domains, redistribution of electrons both vertically and laterally, the self-consistent electric fields $\mathscr{E}_z$ in the transfer direction, and quantum-mechanical reflections at heterostructure interfaces. As long as the electron ensemble in a given GaAs channel of Fig. 26 can be described by a local temperature $T_e(x)$, which is a function of position between source and drain, the density of the RST electron current $J(x)$ can be estimated by the thermionic-emission formula

$$J(x) \approx \frac{qn(x)v(T_e)}{L_W} e^{-\Phi/kT_e} \qquad (29)$$

where $n(x)$ is the sheet density in the quantum well, L_W is the well thickness and $v(T_e) = (kT_e/2\pi m^*)^{1/2}$. Obviously, this current depends exponentially on T_e. A semiclassical treatment of RST between two layers with a conduction-band discontinuity Φ (e.g., one of the GaAs/AlGaAs heterointerfaces in Fig. 26a) involves solving the appropriate Boltzmann equation, on either side of the junction, with the appropriate boundary conditions that include quantum reflection by the barrier. The transverse electric field $\mathscr{E}_z$ must be calculated self-consistently from the Poisson equation that includes the electron density $n(z)$ and the fixed charges that are present (e.g., ionized impurities in the AlGaAs). If the band bending due to $\mathscr{E}_z$ is much smaller than the heterostructure barrier Φ, the $\mathscr{E}_z$-dependent terms in the Boltzmann equation may be dropped, leaving $\mathscr{E}_x$ as the only electric field in the problem. Still, the collision integral on the right of Eq. 27 compels the use of Monte Carlo techniques. The results of such calculations as applied to RST structures are collected in Ref. 64.

Like two-terminal resonant-tunneling diodes, two-terminal RST structures illustrated in Fig. 26 are potentially useful as high-frequency oscillators. The figures-of-merit in this case are the speed with which electrons cycle between

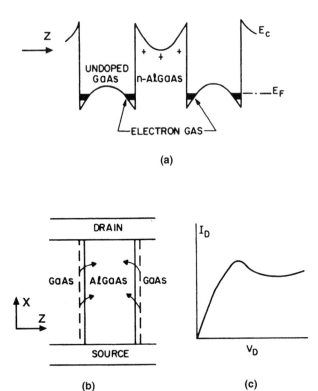

Fig. 26 Two-terminal real-space-transfer structure. (a) Schematic band diagram of a section of a multiquantum-well GaAs/AlGaAs RST structure. (b) Direction of the current flow in a field $\mathscr{E}_x$ applied between source and drain, including the RST current due to electron heating. (c) Resulting $I-V_D$ characteristic—at sufficiently high V_D a large fraction of electrons transfers into the lower-mobility AlGaAs layers, leading to negative differential resistance.

high (GaAs) and low (AlGaAs) mobility layers and the magnitude of the resulting NDR in the $I-V$ characteristic. Unfortunately, while the hot-electron transfer time to the AlGaAs can be quite short, the return process of relatively cold electrons by thermionic emission over the space-charge potential barrier of the ionized donors is much longer.[77] An additional difficulty is the formation of macroscopic traps in the AlGaAs resulting from fixed charge inhomogeneities: these potential pockets effectively collect transferred hot electrons and present a higher barrier to their return. In fact, the real-space transfer times are typically longer than the momentum-space transfer times. Further, the maximum high-frequency power that can be extracted from an RST oscillator is limited by the peak-to-valley ratio of the NDR, but a large PVR is obtained only if the mobilities of the two layers differ by orders of magnitude (Problem 13). Such a large mobility ratio is no easier to engineer in RST structures than in homogeneous multiple-valley semiconductors. For these reasons, two-terminal RST oscillators do not appear to offer any significant advantages over Gunn oscillators and have not benefited from much experimental development. What makes RST structures considerably more interesting from the device standpoint is the possibility of extracting the transferred hot-carrier current via a third terminal, resulting in an RST transistor (RSTT).[78]

Figure 27 shows a schematic cross-section and the corresponding band diagram of an RSTT device implemented in GaAs/AlGaAs. The source and drain contacts are to a high-mobility GaAs channel, whereas the collector contact is to a doped GaAs conducting layer, which is separated from the channel by a large heterostructure barrier. An electron density is induced in the source–drain channel by a sufficient positive collector bias V_C with respect to the grounded source, but no collector current I_C flows because of the AlGaAs barrier at $V_D = 0$. As V_D is increased, however, a drain current I_D begins to flow, and the channel electrons accordingly heat up to some effective temperature $T_e(V_D)$. This electron temperature determines the RST current injected over the collector barrier, and the injected electrons are swept into the collector by the V_C-induced electric field, giving rise to I_C. Thus, transistor action results from control of the electron temperature T_e in the source–drain channel, which modulates I_C flowing into the collector electrode. In contrast to the two-terminal device of Fig. 26, the RST current is removed from the drain current loop, leading to very strong NDR in the I_D–V_D curve, with room-temperature PVR reaching 160 in GaAs/AlGaAs devices[79] similar to Fig. 27. Subsequent improvements in RSTT design included structures with InGaAs channels, either lattice-matched to InP substrates or pseudomorphically strained to GaAs, taking advantage of the lower electron effective mass and higher Γ–L valley separation in InGaAs. More importantly, these devices used epitaxial rather than alloyed contacts to the source–drain channel.[80] The drain I_D and collector I_C characteristics of such an RSTT are shown in Fig. 28: the PVR in the I_D–V_D characteristics reaches 7000 at $T = 300$ K. Another version had a top-collector design with self-aligned collector regions[81] that avoid the vertical overlap between the source and drain contacts evident in Fig. 28. The result is reduced parasitic capacitance between the source, drain, and the collector: a current gain cutoff frequency $f_T > 50$ GHz was reported. Finally, there have been recent reports of a δ-doped pseudomorphic InGaAs/GaAs RSTT[82] with high channel mobility, a PVR $> 10^5$, and a transconductance of 23.5 S/mm at $T = 300$ K, as well as a silicon-compatible hole-based RSTT with SiGe channel and collector layers.[83]

As might be expected from our discussion of the RST process between two layers, theoretical modeling of RSTT devices is extremely involved, and Monte Carlo simulations are required for quantitative comparison with experiment.[64,84] Some qualitative insight can be gained by assuming that the

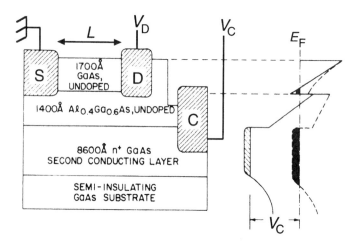

Fig. 27 Schematic cross-section and energy-band diagram of a three-terminal GaAs/AlGaAs RST transistor. In this version, there are no electrons in the channel at flatband—the channel density $n(x)$ is induced by collector bias V_C. If instead the channel consists of a modulation-doped quantum well, $n(x)$ can be nonzero at $V_C = 0$.

RST current $J(x)$ exists only within a certain domain of the concentrated longitudinal electric field along the channel. In this high-field domain we will take the electron temperature T_e as uniform and assume that the channel carriers move at their saturation velocity v_{sat}, so $I_D = qn(x)Wv_{sat}$, where W is the device width and the diffusion component of I_D is ignored. Current continuity in the source–drain loop gives $dI_D/dx = -J(x)W$. Substituting Eq. 29 for $J(x)$ one obtains that both $n(x)$ and $I_D(x)$ decrease exponentially with a characteristic length

$$\lambda = \frac{v_{sat}L_W}{v(T_e)} e^{\Phi/kT_e} \qquad (30)$$

For high T_e, λ becomes quite short, making the diffusion component non-negligible, but the exponential decay of $I_D(x)$ remains qualitatively unchanged (Problem 14). The remaining difficulty is the estimation of T_e for a given drain bias V_D. This has been done semi-analytically by assuming a uniform electric field $\mathscr{E}_x$ in the channel and taking into account only two energy-loss mechanisms: optical-phonon emission and electron RST to the collector.[85] In this model, at sufficiently high V_D the region of high T_e becomes much larger than λ and, by virtue of Eq. 30, I_D becomes vanishingly small. However, since Monte Carlo studies show that k-space transfer into lower mobility satellite valleys is a dominant scattering mechanism at high fields, simple energy-loss models are of limited applicability.

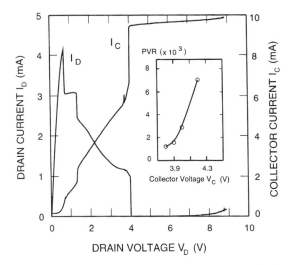

Fig. 28 Experimental real-space transfer transistor characteristics at $T = 300$ K: drain I_D and collector I_C currents vs drain voltage V_D at fixed collector $V_C = 3.9$ V. Inset shows the drain circuit PVR as a function of collector voltage. (After Mensz et al., Ref. 80.)

An interesting aspect of RSTT devices is the nature of their intrinsic speed limitations. The two contributing factors are the time-of-flight delays associated with space-charge-limited current and the finite time required to establish the hot-carrier ensemble of temperature T_e. Significantly, the relevant length entering the time-of-flight delay is not the gate length of a standard FET, because once the high-field domain is established the speed with which T_e can be modulated is not limited by the source–drain transit time. Instead, the relevant length is the extent of the high-field domain in the channel plus the thickness of the potential barrier separating the channel from the collector. Barrier thicknesses in RSTTs are $\sim 10^3$ Å, as are the high-field domains when T_e is large (Problem 14). As a result, the time-of-flight delay should be in the 1 ps range, competitive with state-of-the-art conventional transistors. As for the establishment of an effective T_e in the hot-carrier distribution, the relevant mechanisms are optical-phonon emission, **k**-space scattering, and electron–electron interaction, which might be the dominant mechanism for reaching quasi-equilibrium in the high-energy tail of the distribution. Once again, Monte Carlo simulations[84] of RSTT structures indicate that equilibration of hot-electron ensembles takes less than 1 ps, at least at high electron concentrations and operating voltages.

Let us briefly consider possible applications of RSTTs. Obviously they can be used as conventional high-speed transistors, in which case the figures-of-merit are the transconductance $g_m \equiv \partial I_C / \partial V_D$ (at fixed V_C) and current-gain cutoff frequency f_T. Like resonant-tunneling devices, RSTT combinations can be used for memory and logic elements by virtue of the strong NDR in the source–drain circuit. Further, since the source and drain contacts of an RSTT are fully symmetric, these devices have additional logic functionality. A single RSTT like that shown in Fig. 27 can perform an exclusive-OR (XOR) function, because the collector current I_C flows if source and drain are at different logic values, regardless of which is "high". This and related logic implementations will be discussed later in the chapter. Finally, by changing the doping and design of the collector region, light-emitting operation of RSTT structures has been demonstrated in the InGaAs/InAlAs material system.[86] The only change from Fig. 27 is the opposite doping in the n-InGaAs channel and the p-InGaAs active region grown on top of the p⁺-InGaAs collector. As in a standard RSTT, hot electrons are injected by RST over the InAlAs barrier, but then they recombine radiatively with holes in the active region. As long as radiative recombination in the channel is negligible, the optical output is insensitive to the parasitic leakage of collector holes into the channel. This means that the optical on–off ratio is directly determined by I_C, and hence the device works like a light-emitting diode with built-in logic functionality.

5.3.4   Resonant Hot-Electron and Bipolar Transistors

As we have seen, three-terminal RT structures in which the control electrode directly modulates the alignment of the resonant subband and the emitter are difficult to fabricate. An alternative approach is the incorporation of a double-barrier RT potential into the emitter of a hot-electron transistor.[87] A schematic band diagram of the resonant hot-electron transistor (RHET) is shown in Fig. 29. Its operation essentially combines the resonant-emitter I_E–V_{BE} characteristic with the current gain β available in the HET. Consider the collector current I_C as a function of base–emitter voltage V_{BE} at some fixed base–collector voltage V_{BC}. At small V_{BE} the emitter RT structure is below threshold, the emitter current is negligible and the collector current I_C consists of the small thermionic-emission current over the collector barrier Φ_C. At larger V_{BE} a resonant current flows through the emitter, injecting hot electrons at $\Delta \approx qV_E$ above E_F in the base. Given proper design, with $\Phi_C < \Delta \lesssim \Gamma$–$L$ energy separation in the base, a large fraction α_T of the injected electrons traverses the base and contributes to I_C. The large Γ–L separation makes InGaAs heterostructures on InP substrates advantageous for the implementation of RHETs, as discussed in Section 5.3.1. As before, the current gain $\beta \equiv \alpha_T/(1 - \alpha_T)$ is limited by the hot-electron mean free path in the heavily doped base, but room-temperature $\beta \approx 10$ has been reported in InGaAs/AlAs/InGaP RHET structures on InP substrates.[67] Finally, as V_{BE} biases the emitter RT diode beyond V_P, the emitter current drops. The corresponding PVR in I_C will approximately reproduce the PVR of the

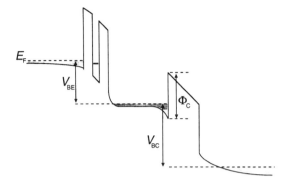

Fig. 29 Schematic band diagram of a resonant hot-electron transistor (RHET). The resonant I–V characteristic of the RT diode in the emitter–base junction is replicated in the collector current I_C, leading to a peaked I_C–V_{BE} characteristic.

emitter diode, although changes in α_T as a function of injection energy might alter this result somewhat. Peak-to-valley ratios of approximately 10 have been reported in the I_C–V_{BE} characteristics of RHETs at both $T = 300$ K and $T = 77$ K.[67,87]

Very similar characteristics can be obtained by inserting a double-barrier RT diode or several cascaded RT diodes on the emitter side of the emitter–base junction in an n–p–n bipolar transistor.[88] Such structures were fabricated in InGaAs/AlInAs: the operation is analogous to RHET, except that the emitter bias V_{BE} divides between the RT diodes in the emitter and the emitter–base n–p junction to maintain current continuity. As long as $V_{BE} < V_{bi}$ of the n–p junction, the emitter current increases as in a conventional bipolar transistor, with somewhat higher emitter series resistance resulting from the RT diodes, and the current gain β is large. Beyond flatband, $V_{BE} \geq V_{bi}$, most of the additional V_{BE} drops in the RT diodes and I_E exhibits one or more NDR regions when the diodes are biased beyond V_P. Consequently, I_C also exhibits peaks as a function of V_{BE}. The gain in the NDR regions is typically lower, because the hole current into the emitter keeps increasing with V_{BE}. Since the electron current into the base drops at $V_{BE} > V_P$, the result is lower emitter efficiency. The multipeaked I_C characteristic of a bipolar transistor with two RT diodes in the emitter has been used as a frequency multiplier. By driving the base with an AC signal of frequency f and sufficient amplitude to bias both RT diodes through their resonances, signals at $3f$ (for sawtooth input) and $5f$ (for sinusoidal input) were generated with reasonable conversion efficiency.[89]

Like RSTTs, resonant hot-electron and bipolar transistors exhibit higher logic functionality in a single device, illustrated schematically in Fig. 30. Given a common-emitter I_C-V_{BE} characteristic with reasonable PVR, shown in Fig. 30a, the output I_C can be high when $V_{BE} = V_{high} < V_P$, but low when $V_B = 0$ or $2V_{high}$ (where the RHET is in the negative transconductance regime). As a result, an exclusive-NOR (XNOR) function can be easily implemented in a single device, as shown in Fig. 30b. With the emitter grounded and two inputs to the RHET base, V_{OUT} will be high when one of the base inputs is high and low otherwise. Room-temperature XNOR gate operation with a reasonable V_{OUT} voltage swing has been demonstrated,[72] using a device layout similar to Fig. 30b. In addition to the necessary resistor network, a drawback of these designs is that unless the PVR in the I_C characteristic is very large, there is still power dissipated in the collector resistor when both base inputs are high. In a single device this added power dissipation can be minimized simply by downscaling the area and reducing I_C, but this is not possible in large circuits where I_C must charge interconnect capacitances.

5.4 DEVICE APPLICATIONS

5.4.1 RT Oscillators

Despite the high speed and functionality offered, in principle, by quantum-effect and hot-electron devices, their technological applications thus far have been few. In some cases, the main obstacle has been room-temperature operation, in others the difficulty of large-scale fabrication or the integration of nonconventional devices into standard technology. For these reasons, many of the device applications discussed below exist only as laboratory demonstrations. Although possibly relevant in the relatively distant future, when their edge over conventional devices might become compelling—at very small device dimensions L, cryogenic temperatures T, or whatever other design criteria future technology may require—few quantum and hot-electron devices offer a sufficient advantage today. One happy exception is the use of RT diodes as solid-state high-frequency oscillators. The advantages of two-terminal RT oscillators include relative ease of fabrication, reasonable output power, and high maximum oscillation frequencies f_{max} compared to competing microwave tunnel and transit-time diodes.

Figure 31 shows the simplest equivalent circuit of a two-terminal diode oscillator with a static I–V characteristic that includes an NDR region described by peak (V_P, I_P) and valley (V_V, I_V) points in both voltage and current. This equivalent circuit has been successful in the analysis of tunnel diodes with I–V characteristics similar to the RT diode of Fig. 6. The real part of the equivalent circuit impedance R_{eq} is given by

$$R_{eq} = R_S + \frac{-R_D}{1 + (\omega R_D C_D)^2} \tag{31}$$

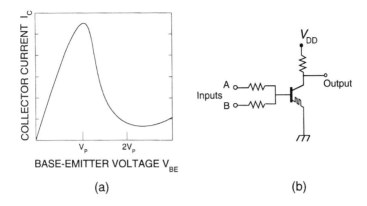

Fig. 30 (a) Schematic I_C-V_{BE} of a resonant hot-electron transistor in the common-emitter configuration. The RHET exhibits negative transconductance for $V_{BE} > V_P$ with high peak-to-valley ratio. (b) Exclusive-NOR circuit using a single RHET device, showing the double-barrier RT structure in the emitter–base junction of the RHET.

where $-R_D = (V_V - V_P)/(I_V - I_P)$ is the negative diode resistance; C_D is the diode capacitance; and R_S is the series lead resistance. For steady-state oscillation, R_{eq} must be negative, so from Eq. 31 the cutoff frequency f_{max} is found to be

$$f_{max} = \frac{1}{2\pi R_D C_D} \sqrt{\frac{R_D}{R_S} - 1} \qquad (32)$$

To increase f_{max}, the quantities to minimize are then the parasitic series resistance R_S and diode capacitance C_D. A sharp current drop after V_P and a high PVR are also helpful in minimizing R_D and hence increasing f_{max}, but there is the competing requirement of maximizing high-frequency output power P_{max}. Although the exact value of P_{max} depends on the actual I–V in the $V_P < V < V_V$ region, generally $P_{max} \approx (V_P - V_V)(I_P - I_V)$, making both PVR and a high current density essential for good oscillator performance.

By analogy with tunnel diodes, Eqs. 31 and 32 have been employed in the design of RT diode oscillators with empirical parameters (e.g., taking for C_D the measured two-terminal emitter–collector capacitance) and extended to include collector transit- and tunneling-time effects.[19] However, the equivalent circuit of Fig. 31 is physically unsatisfactory. The current flowing in an RT diode depends on the alignment of the emitter and the 2D subband in the well, with the tunneling current densities into and out of the well balancing in steady state, $J_{in} = J_{out}$. It is difficult to construct a useful

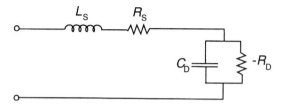

Fig. 31   Simple equivalent circuit for a two-terminal tunnel diode oscillator, including the parasitic lead resistance and inductance.

equivalent circuit for RT diodes either in the coherent or in the sequential picture. The main difficulty lies in the unknown energy distribution of the dynamically stored charge density $\sigma_W = qn_W$ in the well, which makes it impossible to describe J_{out} as a unique function of the electrostatic potential difference V_C between the well and the collector.*

A reasonable and tractable model arises if one assumes that carriers equilibrate in the well. Then, the collector current can be described by a function of σ_W and V_C, and its small variations about a steady state can be written in the form:†

$$\delta J_{out} = \frac{\delta \sigma_W}{\tau} + \frac{\delta V_C}{R_C} \tag{33}$$

where τ is the lifetime of the carriers in the well, while the collector resistance R_C reflects the dependence of tunneling rate on the well–collector potential difference due to changes in the collector barrier shape. Variation in the stored charge density σ_W and its time dependence obey Gauss' and Kirchhoff's laws:

$$\delta \sigma_W = C_E \delta V_E - C_C \delta V_C, \qquad \frac{\partial(\delta \sigma_W)}{\partial t} = \delta J_{in} - \delta J_{out} \tag{34}$$

where C_E and C_C are the emitter–well and well–collector capacitances, respectively. By definition, $\delta V_E + \delta V_C = \delta V$, the variation in total emitter–collector bias V.

---

*The Fermi level difference—which is the true meaning of "voltage"—between the well and the collector cannot even be meaningfully defined without some assumption about the electron distribution in the well.

† This model has been suggested by P. Solomon in a private communication (1995) to one of us (SL). Rigorously, parameters τ and R_C can be defined as follows: $\tau^{-1} \equiv (\partial J/\partial \sigma_W)$ at fixed V_C and $R_C^{-1} \equiv (\partial J/\partial V_C)$ at fixed σ_W. Their interpretation as lifetime and resistance is only approximate.

Not far from the tunneling resonance, J_{in} is a unique function of V_E, which determines the emitter–well alignment, $\delta J_{in} = \delta V_E / R_E$. Combining this with Eqs. 33 and 34 for J_{out}, one obtains

$$\frac{\partial(\delta \sigma_W)}{\partial t} = \frac{-\delta \sigma_W}{\tau_{eff}} + \frac{C_G \delta V}{\tau_G} \tag{35}$$

where the geometric quantities C_G and τ_G are defined as

$$\tau_G = \frac{R_E R_C (C_E + C_C)}{R_E + R_C} \qquad C_G = \frac{R_E C_E - R_C C_C}{R_E + R_C} \tag{36}$$

and τ_{eff} is given by

$$\tau_{eff} = \frac{\tau_G \tau}{\tau_G + \tau} \tag{37}$$

It is the effective time constant of Eq. 37 that determines the diode dynamics. If the applied voltage V changes abruptly by δV, the charge density in the well will evolve exponentially towards the new steady-state value $(\sigma_W + \delta \sigma_W)$ with a time constant given by τ_{eff}. The magnitude of $\delta \sigma_W$ is described by an effective capacitance C_{eff}:

$$\delta \sigma_W = C_{eff} \delta V = \left(\frac{C_G \tau_{eff}}{\tau_G} \right) \delta V \tag{38}$$

Evidently, C_{eff} can be either positive or negative, depending on the sign of C_G in Eq. 36. The value of this capacitance is irrelevant to the dynamics of the variation.

If the $\delta \sigma_W / \tau$ component of J_{out} were absent from Eq. 33, the RT diode would be in a true linear response regime and could be rigorously described by an equivalent circuit consisting of R_E paralleled with C_E in series with R_C paralleled with C_C. Solving such an equivalent circuit would give Eq. 35, but with τ_{eff} replaced by τ_G. This, however, is not a good approximation for real RT diode oscillators.* In the operating regime V_C is large, the value of τ is shorter than τ_G, and hence $\tau_{eff} \approx \tau$. This is particularly true in structures with large undoped spacer regions on the collector side of the double barrier. In such structures, the collector barrier transparency becomes only weakly

---

*Once again, the point is that these diodes are operated at high V_C and hence far from equilibrium. By contrast, near equilibrium J_{out} can be described by the Landauer formula, $\delta J_{out} = \delta V_C / R_C$, where R_C is a function of the collector barrier transmission coefficient T_C. But at high V_C it is the lifetime τ and not R_C that describes the tunneling rate of the carriers in the quantum well.

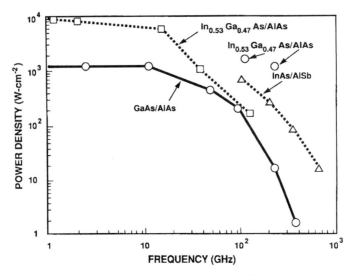

Fig. 32 Comparison of RT oscillators fabricated in different material systems. The InAs/AlSb oscillator should reach $f_{max} \approx 1$ THz. (Figure courtesy of E. R. Brown, 1995.)

dependent on V_C, effectively making R_C in Eq. 33 very large. The key parameter for high speed is τ, which should be minimized by making the collector barrier as transparent as possible while keeping the sharpness of the 2D quantization sufficient for NDR in the $I\!-\!V$ characteristic.

Figure 32 summarizes experimentally measured, room-temperature oscillator performance of high-speed RT oscillators fabricated in different material systems: GaAs/AlAs, InGaAs/InAlAs, and InAs/AlSb.[19] Although the power density P_{max} available in GaAs/AlAs is limited by the relatively low PVR at $T = 300$ K, InGaAs/InAlAs RT oscillators exhibit good output power, whereas InAs/AlSb devices show promise for submillimeter wave ($f > 300$ GHz) performance and hold the record for solid-state oscillator frequency at 712 GHz.[90] No other solid-state sources generate coherent power at submillimeter fundamental frequencies. One possible application of such devices is for low-noise local oscillators in high-sensitivity radiometers. A more detailed discussion of microwave diode performance and applications is available in Chapter 6.

5.4.2   Memories

Several approaches have been pursued in constructing memory circuits from quantum-effect and hot-electron devices. Single-device memories can be constructed from asymmetric two-terminal RT diodes with a bistable $I\!-\!V$ characteristic shown in Fig. 5 by biasing the device below V_P in the bistable

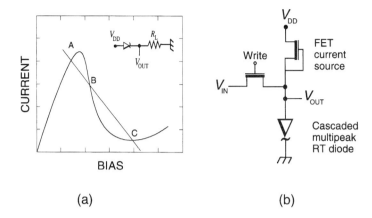

(a)   (b)

Fig. 33 (a) Schematic bistable memory made from an RT diode in series with a load resistor R_L. Output voltage $V_{OUT} = IR_L$ depends on whether the circuit is at point A or C, point B is unstable. (b) Schematic diagram of a multistate memory constructed by biasing a cascaded RT device with a multipeaked $I–V$, like that in Fig. 16, with a constant current from an FET. (After Seabaugh et al., Ref. 43.)

region and changing the memory state using voltage pulses. Alternatively, an ordinary RT diode with an NDR $I–V$ characteristic in series with a load resistor R_L can be DC biased into a regime with two stable bias points, as shown in Fig. 33a. Once again, voltage pulses can be used to change the memory state. The drawback of such memories is that at least one of the memory states corresponds to a high current through the RT diode. The resulting power dissipation is prohibitively large compared to the larger and more complex conventional memory designs. One approach for overcoming the power dissipation problem is to increase the functionality of the RT memory by employing a multistate design. As discussed in Section 5.2.3, proper design of a cascaded RT structure with N diodes results in a multipeaked $I–V$ characteristic with N current peaks of approximately equal magnitude evenly spaced by ΔV_P (see Fig. 16 for a cascaded RT structure with $N = 8$ and $\Delta V_P \approx 0.95$ V). By biasing such a structure with a constant operating current I_{OP} supplied by an FET, as shown in Fig. 33b, the output node V_{OUT} can be at any of the $N + 1$ stable voltage points. Switching between V_{OUT} states is performed by setting an input voltage via a momentarily enabled write line. As soon as the write line is disabled, the cascaded RT will adjust to the nearest stable V_{OUT} value and maintain it indefinitely, leading to an $(N + 1)$-state memory.[44] However, this type of multistate memory still dissipates power $P_{OUT} = I_{OP}V_{OUT}$, with average $\langle P_{OUT} \rangle \approx I_{OP}\Delta V_P N/2$. Minimum power dissipation requires high PVR, since $I_{OP} > I_V$, where I_V is the worst-case valley current among the N peaks, and small ΔV_P. Also, $\langle P_{OUT} \rangle$ increases with N and the accumulated series

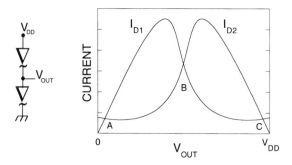

Fig. 34 Graphical construction for determining the operating points of a circuit consisting of two identical NDR devices in series. Points A and C are stable, point B is unstable.

resistance from the N RT diodes enters into the RC switching time delay. Furthermore, practical implementation of circuits based on multistate RT memories requires stringent reproducibility of characteristics between different devices. Ultimately, the quantifiable advantage of a multistate memory is the reduction of the number of elements necessary to store the same amount of information by a factor of $\log_2(N + 1)$ for an $(N + 1)$-state device replacing a binary flipflop.

A different approach is the series connection of two negative-resistance devices, which can be RT diodes, RSTTs, RHETs, or any other device with an NDR I–V characteristic. In fact, many of these devices were proposed in the 1960s with tunnel diodes in mind.[91] If the total applied bias V_{DD} exceeds roughly twice the critical voltage V_P for the onset of NDR in one device, the voltage division between the two devices becomes unstable because of current continuity. One of the two devices takes on most of the applied bias, thereby determining the voltage of the middle node V_{OUT}. This is illustrated by the load-line construction in Fig. 34: operating points A and C are stable, whereas B is unstable. As V_{DD} is ramped up beyond $2V_P$, the system will choose one of the two stable points depending on which of the devices goes into NDR first—either because of a fluctuation or, realistically, because of a slight difference in the I–V characteristics. Switching between the two states can be accomplished either by controlling the parameters of the two devices (in the case of three-terminal RT structures or RSTTs) or by changing the middle node bias via an additional electrode. Significantly, the current flowing through the two NDR devices connected in series when $V_{DD} > 2V_P$ depends on the valley current (see Fig. 34). If the PVR of the devices is large, the current will be small regardless of whether the circuit is in state A or C.

A schematic memory constructed from two RT diodes in series with an additional control electrode separated from the middle node by a tunnel

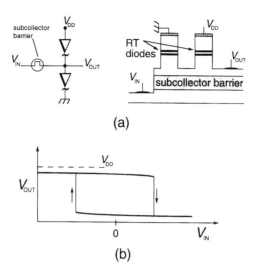

(a)

(b)

Fig. 35 (a) Circuit and cross-sectional diagrams of a memory based on two RT diodes in series with an additional subcollector terminal to control the voltage V_{OUT} of the middle node. (After Shen et al., Ref. 92.) (b) Schematic input–output characteristics: V_{OUT} voltage swing (vertical extent of the loop) depends on the diode characteristics, increasing with PVR. The V_{IN} switch points (horizontal extent of the loop) depend on the subcollector $I–(V_{\text{OUT}} - V_{\text{IN}})$ two-terminal characteristic.

barrier is shown in Fig. 35a. Such devices have been fabricated in the InAs/AlSb/GaSb material system,[92] which provides good room-temperature PVR in the $I–V$ characteristics, after their original demonstration at $T = 77$ K using InGaAs/InAlAs/InP RHETs.[93] As in Fig. 34, biasing the RT diodes in series with $V_{\text{DD}} > 2V_P$ switches one of them to the valley region, so a small current $I << I_P$ flows and V_{OUT} is close to either V_{DD} or ground. To change V_{OUT}, a voltage V_{IN} is applied to the subcollector control electrode, causing current I_C to flow between the middle mode and the subcollector. Since this current flows by tunneling through the subcollector barrier, it increases rapidly with the potential difference $|V_{\text{OUT}} - V_{\text{IN}}|$. When the subcollector current reaches I_P, V_{OUT} switches, resulting in hysteresis in the V_{OUT} vs. V_{IN} characteristic, shown schematically in Fig. 35b. The magnitude of the V_{OUT} voltage swing depends in the RT diode $I–V$ characteristic and can nearly reach V_{DD} if the PVR of the diodes is high enough. Conversely, the required switching bias V_{IN} depends on the single barrier I_C as a function of $(V_{\text{OUT}} - V_{\text{IN}})$ curve. A smaller subcollector barrier requires a smaller $|V_{\text{OUT}} - V_{\text{IN}}|$ difference to reach I_P and switch the middle node, in effect squeezing the hysteretic loop in Fig. 35b along the horizontal axis. Crucially, until a switching V_{IN} pulse is applied to the control electrode, the total current flowing through the memory is limited from below by the valley current in the RT I–V characteristic, since the additional subcollector leakage current can be made very small by designing an appropriate subcollector tunnel barrier.

Memory cells based on two RT diodes or RHETs in series, along the lines of Fig. 35, are smaller than standard CMOS designs. Thus, the RHET version[93] operating at $T = 77$ K claims an order of magnitude in area savings, while the room-temperature RT diode implementation[92] offers area savings of two to four, depending on whether the diodes are laid out horizontally or stacked vertically.

The remaining issue for large-scale memory arrays is power dissipation. Since a reasonable I_P is needed to charge up the interconnect capacitance and the standby power dissipation depends on the valley current, the relevant figure-of-merit is the available PVR. By using polytype InAs/GaSb/AlSb RT diodes, a room temperature PVR of nearly 20 has been achieved,[92] but much higher PVR appears necessary to achieve acceptable power dissipation. As a result, NDR-based memories with their exotic materials appear unlikely to challenge CMOS in high-density memory applications. On the other hand, they may be suitable for applications that require small amounts of memory and can afford higher static power consumption.

5.4.3 Logic Elements

In addition to memory devices, the use of RHETs and RSTTs for logic elements has been proposed and, in some cases, demonstrated by a number of groups. In particular, the compact XNOR functionality of RHETs, illustrated in Fig. 30b, has been employed in the design of elementary logic components, such as latches and full adders.[94] A typical building block in such designs is the three-input majority logic gate, shown in Fig. 36, which uses three RHETs. By using a four-resistor summing network connected to the emitter–base diode of the first RHET, the operating point lies below V_P in the I_C-V_{BE} characteristic if fewer than two of the three inputs is high and above V_P if two or three inputs are high (cf. Fig. 30a). The second RHET senses whether the output of the first RHET is above or below V_P. The third RHET, which is larger and delivers higher I_C, increases the output current drive of the logic gate. By combining this majority logic gate with two XNOR gates made of two RHETs each, a full adder operating at $T = 77$ K was demonstrated.[94] Room-temperature operation of a hybrid full adder incorporating bipolar transistors with and without RT diodes in the emitter–base junction has also been reported.[95] Such designs accomplish the required logic function with a reduced number of transistors. Note, however, that the reduced transistor counts available in RHET and resonant bipolar logic designs come at the expense of fabricating additional resistors. But since thin-film resistor fabrication in microelectronic technology requires additional processing steps and real estate, it is not clear that such circuits provide great area-saving advantages. Further, the impact of all these resistors on the

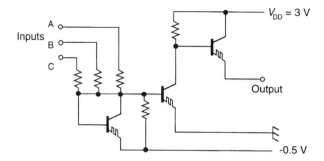

Fig. 36 A three-input majority logic gate implemented with three resonant hot-electron transistors and a resistor summing network. The last RHET before the output node is larger in area to increase I_C and, hence, the current drive of the logic gate. (After Takatsu et al., Ref. 94.)

switching speed and propagation delay in such circuits has not been characterized to date. Finally, the integration of these circuits with conventional silicon technology is problematic, whereas the possibility of a stand-alone quantum device logic circuitry built in III–V semiconductors competing with the ever-advancing silicon CMOS logic is extremely remote.

Integration of high functionality devices with conventional logic circuitry is considerably easier when they are built in Si/SiGe heterostructures. As discussed in Section 5.2.2, silicon-based RT diodes and transistors[23–25,37] perform acceptably at low temperatures only, because of the low Si/SiGe barriers. On the other hand, there has been recent progress in Si/SiGe real-space transfer devices. The drain I_D–V_D and collector I_C–V_D characteristics of a p-Si/SiGe RSTT at room temperature[83] are shown in Fig. 37. The structure of the device is identical to the GaAs/AlGaAs RSTT of Fig. 27, but with $Si_{0.7}Ge_{0.3}$ layers forming the channel and collector regions, separated by an undoped 3000 Å silicon barrier. Negative collector bias induces a hole density in the channel, whereas V_D drives a source–drain current and heats the holes. As V_C increases, the drain characteristic exhibits RST-induced NDR, with PVR slightly exceeding two at $V_C = -5.5$ V. A further increase in V_C results in increasing leakage current due to cold hole tunneling. Although the PVR is greatly inferior to that available in III–V RSTTs, it is sufficient to implement a single-device XOR gate: with $V_C = -4.0$ V and V_S, $V_D = 0$ or -4 V for low and high inputs, the gate has a 10 dB on/off ratio at $T = 300$ K and a 65 dB on/off ratio at $T = 77$ K. For a source–drain separation $L = 0.5\ \mu$m, this device had a current-gain cutoff frequency $f_T = 6$ GHz. Finally, simulations indicate that there is considerable room for improvement of the drain-circuit PVR by reducing the barrier thickness and fine-tuning some of the structural parameters.[83]

Since the input source and drain terminals in an RSTT are completely

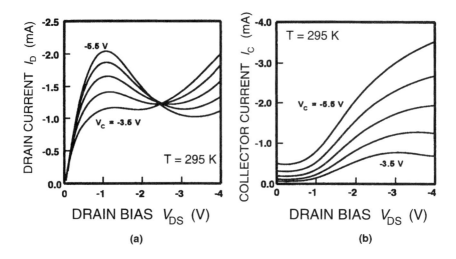

Fig. 37 Room-temperature Si/SiGe RSTT drain current (a) and collector current (b) as a function of source–drain bias V_{DS} for constant $V_C = -3.5$ to -5.5 V in -0.5 V increments. Source–drain channel length $L = 0.5 \mu m$, device width is $40 \mu m$. (After Mastrapasqua et al., Ref. 83.)

symmetric, even higher logic functionality can be obtained by increasing the number of input terminals. For example, three input terminals permit a single-device implementation of an ORNAND gate. Depending on whether the control input is high or low, the output current behaves as either a NAND or an OR function of the other two inputs.[96] The device structure is illustrated in Fig. 38a, where the control input V_3 is subject to periodic boundary conditions for ORNAND functionality. The logic operation of this device at $T = 77$ K, $V_C = -5$ V and V_{low}, $V_{high} = 0$, -3 V respectively, is shown[87] in Fig. 38b. The same device is expected to perform at room temperature if either the cold hole-leakage current or the channel length L is reduced.

In principle, Si/SiGe RSTTs are compatible with silicon microelectronics, although the epitaxial deposition of pseudomorphic SiGe layers for the active regions obviously requires additional fabrication steps and reduces the thermal budget available for subsequent processing. The tradeoff between the added fabrication complexity and the area savings due to the higher functionality will decide the future of silicon-based RSTTs. If technological evolution eventually brings silicon technology to cryogenic ($T = 77$ K) operating temperatures, the chances of silicon-based quantum-effect and hot-carrier devices will improve dramatically.*

---

*Any silicon-based heterostructure with higher barriers than the 0.2–0.3 eV available in Si/SiGe, would greatly brighten the prospects of such devices.

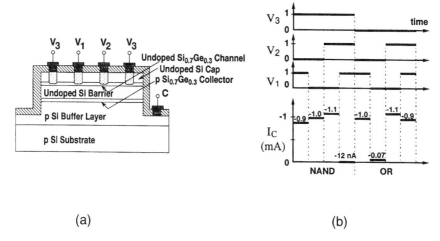

(a)   (b)

Fig. 38 (a) A Si/SiGe RSTT ORNAND gate with three inputs. The ORNAND logic requires a periodic boundary condition on V_3. Channel length between different inputs $L = 1\,\mu m$, device width is $50\,\mu m$. (b) ORNAND logic operation at $T = 77$ K: $V_C = -5$ V, inputs $V_{high} = -3$ V and $V_{low} = 0$. (After Mastrapasqua et al., Ref. 83.)

5.4.4 Quantum-Cascade Laser

As we have seen, prospects of quantum-effect or hot-electron devices replacing conventional semiconductor technologies—whether digital logic and memory chips or analog amplifiers and switches—are hampered by difficulties with room temperature operation, device reproducibility, and fabrication complexities. The advantages of novel devices, typically higher functionality and speed, have not and for the foreseeable future will not displace standard FET and bipolar technologies. On the other hand, such devices as submillimeter RT diode oscillators and Coulomb-blockade current sources are poised for success in niche applications, precisely because conventional solid-state alternatives do not exist. Yet another, potentially more significant area where quantum-effect devices are about to make their mark is solid-state laser sources in the mid-infrared, operating in the $\lambda = 4$–$12\,\mu m$ wavelength range, where current technology relies on low-power and low-yield lead-salt devices. The recently developed quantum-cascade laser (QCL),[52,97] combines resonant-tunneling and hot-electron aspects in a device structure that makes full use of heterostructure bandgap engineering. The lasing occurs in an intersubband transition and λ is tunable in the infrared region via the quantum-well design.

Figure 39 shows a partial band diagram of the QCL gain region together with its output characteristics. The entire QCL structure is grown by MBE,

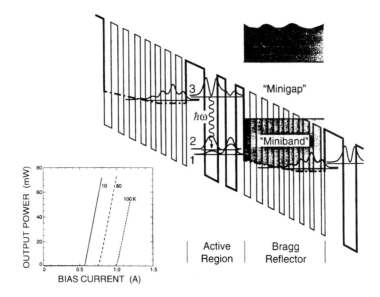

Fig. 39 Schematic conduction-band diagram of the quantum-cascade laser at a field $\mathscr{E}_z = 8.5 \times 10^4$ V/cm. The radiative $E_3 \to E_2$ transition in the coupled-quantum-well active region is shown by the wavy line. Bold lines indicate the squared moduli $|\chi(z)|^2$ of the subband wavefunctions in the active region. Also shown is the miniband structure of the superlattice Bragg reflector region. Note that the lower two states in the coupled quantum well, E_2 and E_1, line up with the SL miniband, while the E_3 state lines up with the SL minigap. The peak optical-power output from a single facet vs injection current for this laser at various heatsink temperatures T is shown at lower left (pulsed mode operation). At $T = 100$ K, the threshold current density $J_{th} = 3 \times 10^3$ A/cm^2. (After Faist et al., Ref. 97).

lattice matched to an n-InP substrate. The total gain region comprises 25 stages of an InGaAs/AlInAs coupled-quantum-well active region followed by a superlattice Bragg reflector. The undoped, coupled-quantum-well active region is designed for the following 2D subband structure under the operating applied bias of $\sim 10^5$ V/cm: an upper-lying E_3 subband with a wavefunction $|\chi^3(z)|^2$ concentrated in the first well, and two lower-lying subbands E_2 and E_1 concentrated in the first and second well respectively (see Fig. 39). The radiative transition is $E_3 \to E_2$, so the laser output energy is $\hbar\omega = E_3 - E_2$. As in all lasers, the radiative transition has to compete with other $E_3 \to E_2$ relaxation mechanisms. In the QCL, the main nonradiative relaxation mechanism involves optical-phonon emission. This process is relatively slow, however, because $\hbar\omega \gg \hbar\omega_{opt}$ and hence $E_3 \to E_2$ relaxation requires large in-plane momentum transfer. On the other hand, since $E_2 - E_1 \approx 30$ meV $\approx \hbar\omega_{opt}$, $E_2 \to E_1$ relaxation by optical-phonon emission is very fast. The superlattice (SL) downstream of the coupled quantum well completes the set of conditions necessary for population inversion between E_3 and E_2. In the InGaAs/AlInAs SL Bragg reflector region, the well and barrier widths are adjusted in pairs[98] to compensate for the applied bias and give rise to an approximately flat miniband structure, shown in Fig. 39. The sequence that contributes to photon emission is as follows: an electron in the E_3 state of a given active region relaxes radiatively via the $E_3 \rightarrow E_2$ transition, then relaxes via the $E_2 \rightarrow E_1$ transition by phonon emission, tunnels from the E_1 state into the lowest SL miniband of the Bragg reflector, and finally tunnels into the E_3 state of the next active region downstream. There the process is repeated, until the electron cascades down all of the 25 stages and is collected in the doped optical cladding layers that sandwich the active region.

In order to achieve optical gain, population inversion between the E_3 and E_2 states in the QCL active regions is needed. This requires rapid removal of electrons from the lower E_2 state and long nonradiative lifetime in the upper E_3 state. As described above, optical-phonon emission vacates the E_2 state very quickly, but is much slower to vacate the E_3 state. The other factor required for a long nonradiative lifetime in the E_3 state is the prevention of direct tunneling out of the active region. However, as shown in Fig. 39, direct tunneling out of the E_3 state into the SL region is blocked because E_3 lines up with the SL minigap. That is, the SL acts as a Bragg reflector (see discussion of superlattice miniband structure in Section 5.2.4).

Finally and crucially, the layers near the middle of the SL region are doped in the $10^{17}/cm^3$ range to provide carriers for injection into the coupled-quantum-well regions and ensure the overall charge neutrality under operating conditions, when a current density J flows through the structure. To avoid space-charge accumulation associated with J, a reservoir of fixed positive charge is needed to compensate the current-carrying electrons in each QCL stage. The role of the SL regions is best appreciated by comparing the QCL structure of Fig. 39 with the conceptually similar SL structure of Fig. 19. Even if SL were biased into resonance between adjacent wells, rather than into the NDR regime originally proposed by Kazarinov and Suris, tunneling from the higher-lying E_2 states into the continuum would work against population inversion. Also, a constant electric field in an undoped SL would be impossible to maintain in the presence of significant current. Introduction of a doped SL region was the key design innovation that led to the successful implementation of the first QCL.[53] Subsequent QCL designs relied on the SL regions to suppress tunneling from the upper level of the radiative transition into the continuum. To this end, the SL regions were designed to serve as electronic Bragg reflectors with minigap in the range of energy near the E_3 state (cf. Fig. 39). Note that since effective Bragg reflection requires very accurate grading of layer widths in the SL regions, this elegant approach places stringent demands on band-structure modeling and layer control by molecular-beam epitaxy.

The lasing characteristics shown in Fig. 39 (lower left) correspond to a $\lambda \approx 4.5\,\mu m$ laser with cleaved facets operated in pulsed mode, but continuous-mode operation at $T = 140$ K and pulsed operation at room temperature has recently been reported in an optimized QCL structure.[99] The power output is quite high, but the threshold current density J_{th} increases rapidly with temperature, reaching 3×10^3 A/cm² at $T = 100$ K. If the $E_3 \rightarrow E_2$ radiative transition is treated as an atomic two-level system, the degradation of performance at higher T can be attributed to reduced population inversion because of temperature-induced backfilling of the E_2 level from the electrons in the doped SL regions.[99] The realistic situation is certainly more complicated. A recent theoretical analysis of gain in QCL pointed to the importance of hot-electron effects in the presence of in-plane subband nonparabolicity.[100] Indeed, not only do electrons tunnel into E_3 with a considerable spread in energy of in-plane motion, but those that relax nonradiatively to the E_2 subband are initially very hot—on average, $\hbar\omega - \hbar\omega_{opt} \approx 250$ meV $= 3000$ K above the bottom of the subband for the $\lambda = 4.5\,\mu m$ transition. If the in-plane subband dispersion is nonparabolic (certainly true that far above the subband minima), $\hbar\omega = E_3 - E_2$ changes as a function of in-plane energy and therefore the gain depends on the difference between hot-electron distributions in these subbands. The shapes of these distributions are radically different in the limits of low and high sheet-carrier concentrations n_D per QCL period, provided by the doped SL regions. For $n_D \ll 10^{11}$/cm², the rate of electron-electron collisions is low and the distribution functions are not Maxwellian. The dominant scattering process is then due to optical–phonon emission within the same subband. It is reasonable to assume that the distribution of electrons tunneling into the upper E_3 subband from the SL miniband is in quasi-equilibrium with the lattice temperature T.* After a nonradiative $E_3 \rightarrow E_2$ intersubband transition, the lower subband electrons are in states of high kinetic energy. Subsequently, they cascade down emitting optical phonons and partially escaping into the SL miniband reservoir. The resulting distribution is given by a quasi-discrete ladder with the occupation probabilities decreasing towards the bottom of the E_2 subband, as if the effective temperature were negative.

The calculated gain spectra for low n_D are shown in Fig. 40a for several lattice temperatures T. The peak gain is substantial even at $T = 300$ K. Note that no overall population inversion between the E_3 and E_2 subbands is assumed, $\xi \equiv n_3/n_2 = 1$. In the absence of lasing, ξ is determined by nonradiative kinetics as the ratio of the $E_3 \rightarrow E_2$ nonradiative transition rate and the rate of carrier removal from the E_2 subband. In the low-concentration regime, the peak wavelength in the gain spectra does not depend on temperature. To our knowledge, this regime has not yet been realized experimentally.

---

*This assumption implies a sufficiently rapid energy relaxation in transport between the QCL stages.

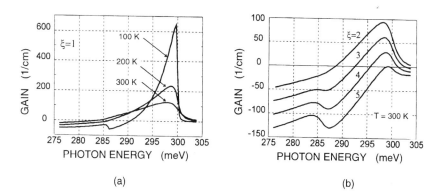

Fig. 40 (a) Quantum-cascade-laser gain spectra in the low concentration limit ($n_D \ll 10^{11}/\text{cm}^2$) vs lattice temperature. No overall inversion between the E_3 and E_2 subbands is assumed, $\xi = 1$. The calculation uses the following parameters: $E_2 = 138$ meV, $E_3 = 438$ meV, in-plane masses $m_2 = 0.051 m_0$ and $m_3 = 0.075 m_0$, $n_3 = 5 \times 10^9/\text{cm}^2$, and $J = 800$ A/cm^2. (b) Room-temperature gain spectra for different values of population inversion ξ. (After Gorfinkel et al., Ref. 100.)

On the other hand, the gain spectra calculated for the high n_D limit,[101] where it is safe to assume Maxwellian hot-electron distributions, also show a range of positive gain, but the peak gain is much lower. Moreover, the peak shifts to longer wavelengths at higher T. These effects have been observed experimentally in existing QCL structures. In the high-concentration regime, the range of positive gain for $n_2 > n_3$ arises entirely from the nonparabolicity. In fact, if quasi-Fermi levels E_{F3} and E_{F2} are introduced to characterize the hot-electron distributions in the two relevant subbands, positive gain occurs when $\hbar\omega < E_{F3} - E_{F2}$, a condition that is familiar from the theory of conventional semiconductor lasers. By contrast, the existence of positive gain in the low-concentration limit does not rely on nonparabolicity and persists to concentrations far from overall population inversion.[100] Room-temperature gain spectra calculated for several values of ξ at low n_D are shown in Fig. 40b. Implementation of the low-concentration regime appears to be a promising strategy for maximizing QCL performance.

5.5 SUMMARY AND FUTURE TRENDS

In this chapter, we have reviewed some of the recent research in the area of quantum-effect and hot-electron devices. It has not escaped the reader that although many of these devices are quite successful according to some (but not all) benchmarks, none has found large-scale commercial application to date. A decade or two ago this situation could perhaps be attributed to the immaturity of the field and the need for further development. But today, a quarter century after the first experimental demonstrations of both resonant-tunneling diodes and hot-electron transistors, this excuse is no longer available. It is imperative to confront the basic issue: what are these devices good for?

To be sure, exotic device research can be proud of its scientific accomplishments. Fascinating new physics has been discovered, with the fractional[102] quantum Hall effect serving as a prime example, and many previously obscure issues have been elucidated. The basic effects relevant to electronic devices, such as tunneling in heterostructures, ballistic transport, carrier heating, and charge injection across potential barriers, are no longer manifested by hardly discernible blips in low-temperature characteristics. They are now available as robust and reproducible phenomena, with on/off ratios quite adequate for the implementation of useful devices. Despite these successes, or perhaps precisely because of them, the general attitude toward the exotic device research has hardened into a widespread skepticism. If these devices have not made it, despite the considerable world-wide effort, why should we throw good research funding after the bad?

In our opinion, there is, indeed, little chance that either resonant-tunneling or hot-electron devices will form the basis of a successful stand-alone technology. On the other hand, they have significant potential in connection with other technologies, such as optoelectronic integrated circuits that are likely to benefit from the introduction of ultrafast functional elements, based on resonant-tunneling or hot-electron effects. The recently developed quantum-cascade laser appears particularly hopeful in this regard, since it promises good performance in the mid-infrared wavelength range ($\lambda > 4\ \mu m$). Moreover, superb frequency characteristics can be expected from this class of lasers, with modulation frequencies exceeding 100 GHz. It is generally believed that light will eventually replace electrical current as the carrier of information signals, both in computer and communications applications, currently the main drivers of innovation in semiconductor devices. Still, the "dark" age of electronics is far from over, and it is interesting to contemplate possible application of exotic devices within the context of the future evolution of circuits which operate without emitting, absorbing, or transforming light.

The evolution of semiconductor electronics has always been intimately connected with advances in materials science and technology. The first revolution in electronics, which replaced vacuum tubes with transistors, was based upon doped semiconductors and relied on newly discovered methods of growing pure crystals. Prior to the 1950s, semiconductors could not be properly termed "doped"—they were impure. Today, semiconductors routinely used in devices are cleaner, in terms of the concentration of undesired foreign particles, than the vacuum of vacuum tubes.

Subsequent evolution of transistor electronics has been associated with the progress in two areas: miniaturization of device design rules, brought about by advances in the lithographic resolution and doping by ion implantation;

and development of techniques for layered-crystal growth and selective doping, culminating in such technologies as MBE and MOCVD, that are capable of monolayer precision in doping and chemical composition.

Of these two areas, the first has definitely had a greater impact in the commercial arena, whereas the second has been mainly supplying the device physics field with new systems to explore. These roles may well be reversed in the future. Development of new and exotic lithographic techniques with nanometer resolution will set the stage for the exploration of various physical effects in mesoscopic devices, while epitaxially grown devices, particularly heterojunction transistors integrated with optoelectronic elements, will be gaining commercial ground. When and whether this role reversal will take place, will be determined perhaps as much by economic as by technical factors. We believe that most significant applications of heterostructure electronics will be associated with its use in silicon electronics.

The logic of industrial evolution will motivate new paths for a qualitative improvement of system components, other than the traditional path of a steady reduction in fine-line feature size. Miniaturization progress faces diminishing returns in the future, when the speeds of integrated circuits and the device packing densities will be limited primarily by the delays and power dissipation in the interconnects rather than individual transistors. Further progress may then require circuit operation at cryogenic temperatures and/or heavy reliance on high-bandwidth optical and electronic interconnects. Implementation of optical interconnects within the context of silicon microelectronics requires hybrid-material systems with islands of foreign heterostructures grown or grafted on silicon substrates. In terms of the old debate on silicon vs GaAs, our view is that silicon is the ultimate customer for GaAs. In this scenario, the current noncompetitiveness of quantum and hot-electron devices for general-purpose digital and analog electronics could give way to novel devices serving as small, highly functional application-specific components that add significant value to main blocks of microelectronic circuitry.

APPENDIX 5.A DENSITIES OF STATES AND FERMI INTEGRALS

Consider the phase space of a single particle in d dimensions ($d = 1, 2,$ or 3). It contains $2d$ axes, corresponding to the d coordinates and p_d momenta of the particle. It is a basic tenet of quantum statistics that a hypervolume V^d in the phase space contains $2V^d(2\pi\hbar)^d$ distinct states, where the factor of two in the numerator arises from the spin degeneracy. Thus, the density of states in the phase space is given by

$$\text{1D: } 2\mathrm{d}L \; \mathrm{d}p/(2\pi\hbar) \tag{A1a}$$

$$\text{2D: } 2\mathrm{d}A \; \mathrm{d}^2p/(2\pi\hbar)^2 \tag{A1b}$$

$$\text{3D: } 2\mathrm{d}V \; \mathrm{d}^3p/(2\pi\hbar)^3 \tag{A1c}$$

where dL, dA, and dV are elements of length, area, and volume, respectively.

Band theory of solids retains the same expressions (A1a–c). They now describe the density of states in each band. Of course, **p** is no longer the electron momentum, but the crystal momentum (in terms of the wavevector **k** used in the chapter, $\mathbf{p} = \hbar\mathbf{k}$). Inasmuch as the occupation of different states in equilibrium depends only on their energy, it is convenient to express the density of states as a function of energy. If we define $N(E)$ as the number of states in a given band with energy less than E, then the density of states for various dimensionalities d is given by

$$\text{1D: } g^{1D}(E) \equiv L^{-1} \, dN/dE \tag{A2a}$$

$$\text{2D: } g^{2D}(E) \equiv A^{-1} \, dN/dE \tag{A2b}$$

$$\text{3D: } g^{3D}(E) \equiv V^{-1} \, dN/dE \tag{A2c}$$

Note that the density of states has different units for different dimensionalities ($\text{cm}^{-d}\,\text{eV}^{-1}$ in d dimensions). Closed-form expressions for $g(E)$ can be obtained only for simplest band structures, e.g., for isotropic bands, $E(\mathbf{p}) = E(p)$. For isotropic and parabolic bands, $E = p^2/2m^*$, where m^* is some effective mass, $N(E)$ can be found explicitly by counting the states from the bottom of the band up to some crystal momentum p:

$$\text{1D: } N(E) = 2pL/(2\pi\hbar) = L(2m^*E)^{1/2}/\pi\hbar \tag{A3a}$$

$$\text{2D: } N(E) = 2\pi p^2 A/(2\pi\hbar)^2 = A(m^*E)/\pi\hbar^2 \tag{A3b}$$

$$\text{3D: } N(E) = 2(4\pi p^3/3)V/(2\pi\hbar)^3 = V(2m^*E)^{3/2}(3\pi^2\hbar^3) \tag{A3c}$$

Substituting Eqs. A3a–c into the appropriate expressions for $g(E)$ one obtains the following densities of states:

$$\text{1D: } g^{1D}(E) = (m^*/2E)^{1/2}(\pi\hbar) \tag{A4a}$$

$$\text{2D: } g^{2D}(E) = m^*/(\pi\hbar^2) \tag{A4b}$$

$$\text{3D: } g^{3D}(E) = m^{*3/2}(2E)^{1/2}/(\pi^2\hbar^3) \tag{A4c}$$

The actual density n of electrons in the system (per unit length, area, or volume, as appropriate) is found by integrating the density of states multiplied by the Fermi–Dirac occupation probability $f_{FD}(E - E_F)$, as in Eq. 10. The resulting general equation,

$$n = \int_0^\infty g(E)\, f_{FD}(E - E_F)\, dE, \qquad f_{FD}(E) = (e^{-E/kT} + 1)^{-1} \tag{A5}$$

provides a relation between n and the Fermi level E_F. In general, this relation contains a Fermi integral of order s,

$$F_s\left(\frac{E_F}{kT}\right) \equiv \frac{1}{\Gamma(s+1)} \int_0^\infty \frac{E^s\, dE}{1 + e^{(E-E_F)/kT}} \tag{A6}$$

where Γ is the gamma function: $\Gamma(1/2) = \pi^{1/2}$, $\Gamma(1) = 1$, $\Gamma(s+1) = s\Gamma(s)$. It follows from Eqs. A4a–c that the Fermi integrals for 3D, 2D, and 1D systems are of order $s = \frac{1}{2}$, 0, and $-\frac{1}{2}$ respectively. Analytic solutions of the Fermi integral are available only for integer s, so for 3D and 1D systems Eq. A5 must be evaluated numerically. In 2D, on the other hand, the density of states is constant and the appropriate Fermi integral is $F_0(\eta) = \ln(1 + e^\eta)$, yielding

$$n = \left(\frac{m^*kT}{\pi\hbar^2}\right) F_0\left(\frac{E_F}{kT}\right) = \left(\frac{m^*kT}{\pi\hbar^2}\right) \ln\left(1 + e^{E_F/kT}\right) \tag{A7}$$

where the prelogarithmic factor $(m^*kT/\pi\hbar^2)$ is the effective density of states in the 2D subband. A similar calculation gives Eq. 12. Note that the lower limit of integration in Eq. A5 refers to an appropriate zero energy point. In a bulk semiconductor this would be the bottom of the conduction band, in a quantum well this would be the bottom of a given 2D subband (cf. Eq. 12).

APPENDIX 5.B DRIFT VELOCITY IN A SUPERLATTICE WITH SCATTERING

Take a superlattice (SL) in the tight-binding approximation, such that the dispersion along the SL direction z is given by Eq. 22. Consider the motion of an electron initially at rest at $k_z = 0$ in a constant electric field $\mathcal{E}$. As a function of k_z, the band velocity is

$$v(k_z) = -\frac{2Td}{\hbar} \sin(k_z d) \tag{B1}$$

where d is the SL period and T is the transfer integral defined by Eq. 24 (one can show by direct calculation that T is a negative quantity). Since $\hbar(dk_z/dt) = q\mathcal{E}$, the acceleration $a(t)$ is given by

$$a(t) = \frac{dv}{dt} = \frac{dv}{dk_z} = -\frac{2Td^2}{\hbar^2}\cos(k_z d)\, q\mathcal{E} \tag{B2}$$

On the other hand, the effective mass $m^*(k_z)$, defined by $\hbar k_z = m^*(k_z)v(k_z)$, is

$$m^*(k_z) = -\frac{\hbar^2 k_z}{2Td \sin (k_z d)} \tag{B3}$$

Taking the $k_z \to 0$ limit of Eq. B3, we obtain the effective mass m^*_{SL} at the bottom of the miniband, $m^*_{SL} = -(\hbar^2/2Td^2)$. Note that the miniband width Δ in terms of m^*_{SL} is given by

$$\Delta = \frac{2\hbar^2}{m^*_{SL} d^2} \tag{B4}$$

Substituting Eq. B3 into Eq. B2, we have

$$a(k_z) = \frac{1}{m^*_{SL}} \cos (k_z d) q\mathscr{E} \tag{B5}$$

Finally, inserting Eq. B5 into the Esaki-Tsu expression for the average drift velocity v_D (Eq. 25), one finds[46]

$$v_D = \int_{t=0}^{\infty} e^{-t/\tau} a[k_z(t)] \, dt = \frac{q\mathscr{E}}{m^*_{SL}} \int_{t=0}^{\infty} e^{-t/\tau} \cos \left(\frac{q\mathscr{E}d}{\hbar} t \right) dt$$

$$= \frac{q\mathscr{E}\tau}{m^*_{SL}} \frac{1}{1 + (q\mathscr{E}\tau d/\hbar)^2} \tag{B6}$$

This result is equivalent to Eq. 26. Note that, in the absence of scattering ($\tau \to 0$), the average drift velocity goes to zero. The particle is localized and performs purely oscillatory motion. These are the famous Bloch oscillations discussed in Section 5.2.4.

APPENDIX 5.C   CONTACTS AND SUPERLATTICES

Consider a superlattice (SL) of N identical periods sandwiched between doped contact electrodes. Suppose a voltage is applied between the electrodes and a current flows through the SL. Given a sufficiently small current, the space-charge effects associated with the current can be neglected. One might imagine that a uniform electric field exists across the SL, as shown in Fig. 41a. The real situation is shown in Fig. 41b: most of the applied voltage drops over the first and last barriers, while the superlattice in-between exerts little if any influence on the I–V characteristic of the device. This is a manifestation of the dramatic difference between coherent transmission and incoherent decay in quantum mechanics.

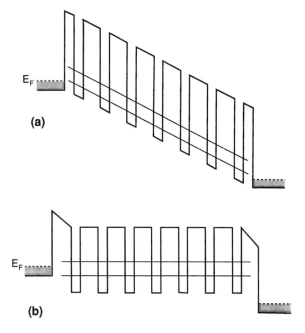

(a)

(b)

Fig. 41 Schematic illustration of a superlattice sandwiched between doped electrode layers under applied bias. (a) Experimentally desirable situation of a uniform electric field extending through the superlattice, which requires the first and last barriers to be narrower. (b) Electric-field distribution that arises if the superlattice barriers are identical, because the first and last barriers present a larger impedance to the current.

In order to appreciate this difference, consider two apparently related processes illustrated in Fig. 42. Figure 42a shows a symmetric coupled-quantum-well system, with the wells separated by a tunneling barrier of height V_0 and width L_B. In the absence of interwell tunneling (e.g., in the $V_0 \to \infty$ limit), each well would contain a quantized level E_0 as shown. Tunneling between the wells splits E_0 into symmetric and antisymmetric states. The doublet splitting $\hbar\omega$ between the symmetric (lower) and antisymmetric (upper) states is, approximately, $\hbar\omega \approx E_0 e^{-\kappa L_B}$, where $\kappa = [2m(V_0 - E_0)/\hbar^2]^{1/2}$ and m is the electron mass. If, at time $t = 0$, an electron is placed in the left well, it will oscillate between the wells with a characteristic frequency ω. After a "short" time $\tau_1 = \pi/\omega$, the electron will be in the right well with unity probability.

Next, consider the escape process of an electron initially placed in the metastable state E_0 of a single quantum well separated from the continuum by the same barrier of height V_0 and width L_B, as shown in Fig. 42b. The possibility of escape means that the state has a finite lifetime τ_2 and hence a finite energy width $\Gamma = \hbar/\tau_2 \approx E_0 e^{-2\kappa L_B}$ (cf. Eq. 5). Since $e^{-\kappa L_B}$ is the small parameter in the tunneling problem, typically $\Gamma \ll \hbar\omega$. Therefore, the lifetime τ_2 can be "long", perhaps orders of magnitude longer than τ_1.

The energy splitting $\hbar\omega$ in the coupled-quantum-well system is similar to the miniband width Δ in the superlattice problem. It describes the resonant transmission rate between discrete states, which is much faster than the seemingly analogous incoherent decay process into the continuum. In order to achieve a constant electric field in the superlattice, shown in Fig. 41a, the first and last barriers of the superlattice must be impedance-matched by making the $2\kappa L_B$ of the first and last barriers equal to the κL_B of the internal superlattice barriers. To first order, this can be achieved by making the first and last barriers narrower by a factor of approximately two. Failure to do so has been a common problem in many experimental studies of superlattice transport.[103]

APPENDIX 5.D   COHERENT TRANSISTOR BASE TRANSPORT

In general, every bipolar or hot-electron ballistic transistor is characterized by a base transport factor α which is a complex function of frequency ω:

$$\alpha \equiv \left(\frac{\partial I_C}{\partial I_E}\right)_{V_{BC}} = e^{-i\omega\tau}|\alpha| \qquad (D1)$$

The time τ, which enters into the phase of α, is the base transit time. In practice, all transistors operate at frequencies sufficiently low that $\omega\tau \ll 1$. In a well-designed bipolar transistor the deviation of $|\alpha|$ from unity is negligible at low frequencies, since the base is much narrower than the diffusion length. Consider the case $|\alpha| = 1$ more closely. The complex current gain $\beta(\omega)$ becomes

$$\beta = \frac{\alpha}{1-\alpha} = \frac{e^{-i\omega\tau}}{1-e^{-i\omega\tau}} = \frac{e^{-i\omega\tau/2}}{2i}\frac{1}{\sin(\omega\tau/2)} \qquad (D2)$$

At low ω, the frequency dependence of the current gain is, therefore,

$$|\beta| = \frac{1}{2\sin(\omega\tau/2)} \approx \frac{1}{\omega\tau} \qquad (D3)$$

The magnitude of the current gain rolls off as ω^{-1}. This type of roll-off (typically referred to as 10 dB per decade or 3 dB per octave) is normally observed in microwave characterization of transistors. Extrapolating Eq. D3 to unity gain, one obtains the cutoff frequency $f_T = (2\pi\tau)^{-1}$.

Note, however, that Eqs. D2 and D3 predict regions of high gain above f_T. These are the "coherent" gain peaks, corresponding to integer numbers

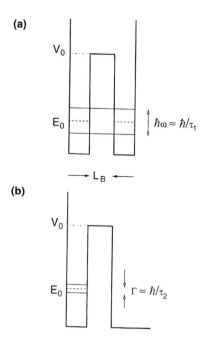

Fig. 42 Coherent oscillation vs incoherent decay. The coupled quantum wells (a) and the single quantum well (b) are confined by the same tunneling barrier, but the coherent oscillation period in the coupled-quantum-well system is much shorter than the lifetime of the metastable state in the single quantum well.

of minority-carrier density wave periods in the base. The necessary condition for observing these peaks is the persistence of near-unity $|\alpha|$ at high frequencies. In fact, all that is required for $|\beta| > 1$ at $f = 2\pi f_T p$, where p is an integer, is $|\alpha| > 0.5$ at that frequency.[73] This condition, however, is extremely difficult to realize. If the base transport is diffusive, $|\alpha| \ll 1$ above f_T. As discussed in Section 5.3.2, ballistic transport offers one possibility of circumventing this problem. Another possibility is to replace random diffusive transport in the base by directed drift in specially graded base structures.[104]

PROBLEMS

1. Solve the finite-potential-well problem to find the bound energy levels $E_n < V_0$ and wavefunctions $\chi_n(z)$ as a function of well width L, barrier height V_0 and particle mass m^* for the symmetric quantum well illustrated in Fig. 1a. Find the number of levels contained in a potential well with $L = 100$ Å and $V_0 = 300$ meV if the particle mass

$m^* = 0.067m_0$, where m_0 is the free-electron mass. These parameters correspond approximately to electrons confined in a GaAs quantum well by $Al_{0.35}Ga_{0.65}As$ heterostructure barriers.

2. a) Derive the transmission probability $T(E)$ for a particle of mass m^* incident on a potential barrier of height V_0 and width L_B, as in Fig. 1b. Make the approximation leading to Eq. 6 and find explicitly the pre-exponential factor. Evaluate $T(E)$ for a GaAs electron of mass $m^* = 0.067m_0$, moving with kinetic energy $E = 50$ meV, which is incident on a 100-Å-thick $Al_{0.35}Ga_{0.65}As$ barrier ($V_0 \approx 300$ meV).

b) What if the incident particle's kinetic energy $E > V_0$? You should find that as long as E is not too much larger than V_0, complete transmission, $T(E) \approx 1$, occurs only for certain values of E that depend on L_B and V_0. Note that, classically, such a particle is always transmitted.

3. Estimate the energy broadening ΔE_1 and ΔE_2 for the lowest two levels in a model symmetric double-barrier potential resulting from tunneling out of the well. The parameters are well width $L = 100$ Å, barrier thickness $L_B = 70$ Å, barrier $V_0 = 300$ meV, and $m^* = 0.067m_0$. To find the lifetimes, consider the electron to be a semiclassical particle bouncing back and forth inside the confining double-barrier potential with a tunneling probability of escaping from the well given by Eq. 6.

4. a) Given the model band diagram shown in Fig. P1 assume the applied bias V drops linearly in the barrier and well, with no electric-field penetration into the doped electrodes. Calculate the supply function $N(V)$ as a function of V. You can assume $T = 0$ and $m^* = 0.067$ in the emitter and well layers.

b) Calculate the corresponding $N(V)$ in the case of 2D carriers tunneling into 1D quantum wire subbands (see Fig. 11 for one experimental version of such a system).

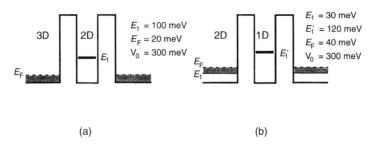

(a)   (b)

Fig. P1.

5. Real semiconductor dispersion $E(\mathbf{k})$ are generally nonparabolic and often anisotropic, so the effective mass m^* varies with energy and with

the direction of **k** with respect to the crystallographic axes. As a simple analog of the real situation, use the 3D to 2D double-barrier RT band diagram of the preceding problem, but assume the effective mass (and hence the in-plane $E(\mathbf{k}_\perp)$ curvature) to be different in the emitter and well, $m_E^* \neq m_W^*$. Which way does the resonant current peak shift as a function of effective mass mismatch (i.e., if $2m_E^* = m_W^*$)? To estimate the possible magnitude of the effect, suppose $m_E^* = 0.4$ but the 2D subband in the well is dispersionless ($m_W^* \to \infty$, which is not far from the truth for the in-plane dispersion of the light-hole subband in some p-type RT structures): how far will V_P shift from the result expected if $m_E^* = m_W^*$?

6. Suppose that the valley current of the first resonant peak in a two-terminal RT device is completely due to the thermally assisted tunneling through the second subband, making the PVR $\approx e^{(E_2 - E_1)/kT}$. If you have the option of introducing an additional layer that has a conduction band discontinuity of Δ into the middle of the well, $L_W/4 \leqslant z \leqslant 3L_W/4$, as shown in Fig. P2, how much will the PVR improve? Treat the additional layer as a first-order perturbation to the infinite well wavefunctions $\chi_n(z) \approx \sin(n\pi z/L_W)$. This trick has been used in n-In$_{0.53}$Ga$_{0.47}$As/AlAs double-barrier RT structures, with a narrow InAs layer in the In$_{0.53}$Ga$_{0.47}$As well.[105]

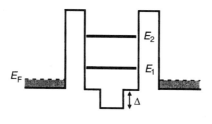

Fig. P2.

7. Consider the electrostatic problem of the three-plate capacitor in which the middle plate represents a 2D electron gas (2DEG). The top (emitter) plate is at a voltage V_E, which corresponds to some charge density σ_E per unit area, whereas the bottom (collector) plate and the 2DEG are grounded (shown on the left in Fig. P3). Let the charge densities induced at the bottom and middle plates be σ_C and σ_{2D} respectively, with $(\sigma_C + \sigma_{2D}) = -\sigma_E$.
a) If one ignores the kinetic energy of the 2DEG, show by minimizing the energy stored in the electric fields between the plates ($E \equiv \varepsilon_s \mathscr{E}^2/2$,

where ε_s is the dielectric permittivity and $\mathscr{E}$ is the field) that all of the induced charge goes onto the middle plate: $\sigma_C = 0$, $\sigma_{2D} = -\sigma_E$.

b) Now include the 2DEG kinetic energy E_{2D} as a function of σ_{2D} (take $T = 0$). By minimizing the total energy, show that the ratio $\sigma_C/\sigma_{2D} = C_C/C_Q$, where $C_C = \varepsilon_s/d_2$ is the geometric collector capacitance per unit area and C_Q is the quantum capacitance of Eq. 20. As a result, show that the equivalent circuit for the capacitance seen at the emitter node is as shown in Fig. P3 (right).

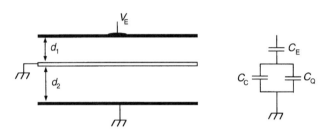

Fig. P3.

8. Predict the approximate $I\text{--}V$ peak positions of a vertically cascaded RT structure with $N = 5$ RT diodes in series, if each diode has barriers $L_B = 70\,\text{Å}$ and well width $L = 100\,\text{Å}$. We have already learned in Problem 1 that the subband energies E_1 and E_2 in each RT potential coincide with those of an $L = 100\,\text{Å}$ quantum well. Take the Fermi level in the cladding regions $E_F = 20\,\text{meV}$, $T = 0$, and assume there is no potential drop in the electrodes outside the barriers.

9. Suppose you have a superlattice with period $d = 100\,\text{Å}$ and lowest miniband width $\Delta_1 = 10\,\text{meV}$, with dispersion described by Eq. 22. What is the minimum scattering time τ for a carrier to complete one period and what is the corresponding Bloch oscillation frequency f for an electric field $\mathscr{E} = 10^3\,\text{V/cm}$? Is the required scattering time τ realistic? (A crude yardstick might be the low-field mobility $\mu = e\tau/m^*$, which in high-purity bulk GaAs at low temperature might reach $10^5\,\text{cm}^2/\text{V-s}$.)

10. Consider a cylindrical quantum dot like that in Fig. 20 in which the in-plane confining potential is parabolic, $V(\mathbf{r}_\perp) = m^*\omega^2\mathbf{r}_\perp^2/2$. We define ΔE as the energy separation between adjacent low-lying quantum dot levels, $\delta E = \hbar\omega = E_{1(m+1)} - E_{1m}$. We have seen that ΔE together with the charging energy $U = q^2/2C_w$ determine the spacing of the current steps near threshold. Show that $U/\delta E$ is independent of the effective radius R of the dot if C_w is the geometric capacitance and derive the condition on d (collector barrier thickness) for observing single-electron charging.

11. Calculate the effective energy width ΔE_z of the injected hot-electron distribution in Fig. 22b at zero temperature, in terms of the emitter E_F, barrier height Φ and barrier width L_B. Use the approximate transmission coefficient $T(E_z)$ of Eq. 5. Compare ΔE_z at $T = 0$ and $T = 300$ K for typical GaAs/AlGaAs HET parameters: $E_F \approx 50$ meV, $\Phi = 250$ meV, $L_B = 100$ Å.

12. Reflection of a 1D particle of energy E by a rectangular barrier of height Φ was considered in Problem 2b for the case of $E > \Phi$. One finds that unless $E >> \Phi$, $T(E) \neq 1$ (and hence $R(E) \neq 0$) except at discrete values of E that depend on the barrier width. In order to minimize $R(E)$, consider a smooth wall barrier of height Φ graded over some characteristic distance a: $V(z) = \Phi(1 + e^{-z/a})^{-1}$.

a) Show that the reflection coefficient $R(E)$ is given by

$$R(E) = \frac{\sinh^2[\pi a(k - q)]}{\sinh^2[\pi a(k + q)]}$$

where $E = \hbar^2 k^2/2m$ and $E - \Phi = \hbar^2 q^2/2m$. Of course, the rectangular barrier result should reappear in the $a \to 0$ limit.

b) Consider the case $ka >> 1$, which corresponds to a being a number of lattice constants. Show that $R(E) \to e^{-4\pi qa}$ and, hence, reflection can be almost completely eliminated even at relatively low E of incident electrons as long as a is sufficiently large.

13. Consider a simple model of RST that neglects the electric field associated with transferred electrons. Assume a periodic multilayer structure like that in Fig. 26 with narrow-gap layers of thickness d_1 and wide-gap layers of thickness d_2. The effective masses and mobilities in the layers are m_1 and m_2, and μ_1 and μ_2 respectively, with $\mu_1 > \mu_2$. Take the rate of energy loss to the lattice proportional to $(T_e - T)/\tau$ per electron, where τ is the same for both layers. Further, assume that the effective electron temperature T_e is the same for both layers, so that no energy is transferred on average as electrons jump between layers. The total density of electrons is fixed, $n = n_1 + n_2$ is constant.

a) Derive the energy balance equation.

b) Express the ratio n_1/n_2 in terms of T_e and the barrier height Φ.

c) Derive the current field characteristics in a parametric form $\mathscr{E} = \mathscr{E}(T_e)$, $J = J(T_e)$ and plot $J(\mathscr{E})$ for a few values of the parameters. What is necessary to achieve high NDR in the source–drain current-field characteristic?

14. In the derivation of Eq. 30 for the transformation of channel current I_D into RST collector current, the diffusion component of I_D was ignored. Keeping this term would make $I_D = qn(x)Wv_{sat} + qWDdn(x)/dx$, where D is the diffusivity and W is the width of the device. Find the new

characteristic decay length λ' in terms of λ given by Eq. 30 and the ratio D/v_{sat}. Find the magnitude of the correction by estimating the decay lengths λ and λ' for a GaAs/AlGaAs RSTT structure with GaAs quantum well $L_W = 200$ Å, heterostructure barrier $\Phi = 0.2$ eV, $T_{el} \approx 1500$ K, and $D = 100$ cm²/s (appropriate for hot electrons).

REFERENCES

1. At the time of writing, the Semiconductor Research Council roadmap predicts continuous improvement in device performance to the year 2015, at which point the minimum lithographic size would reach below 1000 Å and the DRAM memory size would reach 16 Gb. See J. F. Freedman, "Comments on the National Technology Roadmap for semiconductors," in *Future Trends in Microelectronics: Reflections on the Road to Nanotechnology*, S. Luryi, J. Xu, and A. Zaslavsky, Eds., Kluwer, Dordrecht, 1996.

2. H. Sakaki, "Scattering suppression and high-mobility effect of size-quantized electrons in ultrafine semiconductor wire structures," *Jap. J. Appl. Phys.* **19**, L735 (1980);
   Y. Arakawa and H. Sakaki, "Multidimensional quantum well laser and temperature dependence of its threshold current," *Appl. Phys. Lett.* **40**, 439 (1982).

3. U. Meirav, M. A. Kastner, and S. J. Wind, "Single-electron charging and periodic conductance resonances in GaAs nanostructures," *Phys. Rev. Lett.* **65**, 771 (1990);
   L. P. Kouwenhoven, N. C. van der Vaart, A. T. Johnson, W. Kool, C. J. P. M. Harmans, J. G. Williamson, A. A. M. Staring, and C. T. Foxon, "Single electron charging effects in semiconductor quantum dots," *Z. Phys. B* **85**, 367 (1991).

4. This problem is treated in all textbooks on quantum mechanics. For a particularly thorough discussion, see C. Cohen-Tannoudji, B. Diu, and F. Laloë, *Quantum Mechanics*, Vol. 1, Ch. I, Wiley–Interscience, New York, 1977.

5. B. Ricco and M. Ya. Azbel, "Physics of resonant tunneling: the one-dimensional double-barrier case," *Phys. Rev. B* **29**, 1970 (1984).

6. M. Jonson and A. Grincwajg, "Effect of inelastic scattering on resonant and sequential tunneling in double-barrier heterostructures," *Appl. Phys. Lett.* **51**, 1729 (1987).

7. S. V. Meshkov, "Tunneling of electrons from a two-dimensional channel into the bulk," *Zh. Eksp. Teor. Fiz.* **91**, 2252 (1986) [*Sov. Phys. JETP* **64**, 1337 (1986).]

8. S. Luryi, "Frequency limit of double-barrier resonant-tunneling oscillators," *Appl. Phys. Lett.* **47**, 490 (1985).

9. L. L. Chang, L. Esaki, and R. Tsu, "Resonant tunneling in semiconductor double barriers," *Appl. Phys. Lett.* **24**, 593 (1974).

10. A. Zaslavsky, D. C. Tsui, M. Santos, and M. Shayegan, "Magnetotunneling in double-barrier heterostructures," *Phys. Rev. B* **40**, 9829 (1989).

11. V. J. Goldman, D. C. Tsui, and J. E. Cunningham, "Evidence for LO-phonon-emission-assisted tunneling in double-barrier heterostructures," *Phys. Rev. B* **36**, 7635 (1987).

12. N. S. Wingreen, K. W. Jacobsen, and J. W. Wilkins, "Resonant tunneling with electron-phonon interaction: an exactly solvable model," *Phys. Rev. Lett.* **61**, 1396 (1988);
    F. Chevoir and B. Vinter, "Calculation of phonon-assisted tunneling and valley current in a double-barrier diode," *Appl. Phys. Lett.* **55**, 1859 (1989).

13. M. C. Payne, "Transfer Hamiltonian description of resonant tunneling," *J. Phys. C* **19**, 1145 (1986);
    T. Weil and B. Vinter, "Equivalence between resonant tunneling and sequential tunneling in double-barrier diodes," *Appl. Phys. Lett.* **50**, 1281 (1987).

14. An extensive discussion is available in M. Büttiker, "Coherent and sequential tunneling in series barriers," *IBM J. Res. Develop.* **32**, 63 (1988).

15. See, for example, S. K. Diamond, E. Ozbay, M. Rodwell, D. M. Bloom, Y. C. Pao, E. Wolak, and J. S. Harris, "Fabrication of 200-GHz f_{max} resonant-tunneling diodes for integrated circuit and microwave applications," *IEEE Electron Dev. Lett.* **10**, 104 (1989).

16. V. J. Goldman, D. C. Tsui, and J. E. Cunnigham, "Observation of intrinsic bistability in resonant-tunneling structures," *Phys. Rev. Lett.* **58**, 1256 (1987).

17. E. Ozbay and D. M. Bloom, "110-GHz monlithic resonant-tunneling-diode trigger circuit," *IEEE Electron Dev. Lett.* **12**, 480 (1991).

18. E. R. Brown, C. D. Parker, A. R. Kalawa, M. J. Manfra, and K. M. Molvar, "A quasioptical resonant-tunneling-diode oscillator operating above 200 GHz," *IEEE Trans. Microwave Theory Tech.* **41**, 720 (1993).

19. E. R. Brown, "High-speed resonant-tunneling diodes," in *Heterostructures and Quantum Devices*, N. G. Einspruch and W. R. Frensley, Eds., Academic Press, New York, 1994.

20. M. Sweeny and J. Xu, "Resonant interband tunneling diodes," *Appl. Phys. Lett.* **54**, 546 (1989);
    J. R. Söderstrom, D. H. Chow, and T. C. McGill, "New negative differential resistance device based on resonant interband tunneling," *Appl. Phys. Lett.* **55**, 1094 (1989).

21. E. E. Mendez, J. Nocera, and W. I. Wang, "Conservation of momentum, and its consequences, in interband resonant tunneling," *Phys. Rev. B* **45**, 3910 (1992).

22. R. Beresford, L. Luo, K. Longenbach, and W. I. Wang, "Resonant interband tunneling through a 110 nm InAs quantum well," *Appl. Phys. Lett.* **56**, 551 (1990).

23. H. C. Liu, D. Landheer, M. Buchanan, and D. C. Houghton, "Resonant tunneling in $Si/Si_{1-x}Ge_x$ double-barrier structures," *Appl. Phys. Lett.* **52**, 1809 (1988).

24. Z. Matutinovic-Krstelj, C. W. Liu, X. Xiao, and J. C. Sturm, "Evidence for phonon-absorption-assisted electron resonant tunneling in $Si/Si_{1-x}Ge_x$ diodes," *Appl. Phys. Lett.* **62**, 603 (1993).

25. G. Schuberth, G. Abstreiter, E. Gornik, F. Schäffler, and J. F. Luy, "Resonant tunneling of holes in $Si/Si_{1-x}Ge_x$ quantum-well structures," *Phys. Rev. B* **43**, 2280 (1991).

26. U. Gennser, V. P. Kesan, D. A. Syphers, T. P. Smith III, S. S. Iyer, and E. S. Yang, "Probing band structure anisotropy in quantum wells via magnetotunneling," *Phys. Rev. Lett.* **67**, 3828 (1991).

27. A. Zaslavsky, K. R. Milkove, Y. H. Lee, B. Ferland, and T. O. Sedgwick, "Strain relaxation in silicon-germanium microstructures observed by resonant tunneling spectroscopy," *Appl. Phys. Lett.* **67**, 3921 (1995).

28. S.-Y. Chou, D. R. Allee, R. F. W. Pease, and J. Harris, Jr., "Observation of electron resonant tunneling in a lateral dual-gate resonant tunneling field-effect transistor," *Appl. Phys. Lett.* **55**, 176 (1989);
    K. Ismail, D. A. Antoniadis, and H. I. Smith, "Lateral resonant tunneling in a double-barrier field effect transistor," *Appl. Phys. Lett.* **55**, 589 (1989).

29. M. A. Reed, W. R. Frensley, R. J. Matyi, J. N. Randall, and A. C. Seabaugh, "Realization of a three-terminal resonant tunneling device: the bipolar quantum resonant tunneling transistor," *Appl. Phys. Lett.* **54**, 1034 (1989).

30. A. R. Bonnefoi, D. H. Chow, and T. C. McGill, "Inverted base-collector tunnel transistors," *Appl. Phys. Lett.* **47**, 888 (1985).

31. S. Luryi, "Quantum capacitance devices," *Appl. Phys. Lett.* **52**, 501 (1988).

32. F. Beltram, F. Capasso, S. Luryi, S. N. G. Chu, and A. Y. Cho, "Negative transconductance via gating of the quantum well subbands in a resonant tunneling transistor," *Appl. Phys. Lett.* **53**, 219 (1988).

33. C. J. Goodings, H. Mizuta, J. Cleaver, and H. Ahmed, "Variable-area resonant tunneling diodes using implanted in-plane gates," *J. Appl. Phys.* **76**, 1276 (1994).

34. T. K. Woodward, T. C. McGill, and R. D. Burnham, "Experimental realization of a resonant tunneling transistor," *Appl. Phys. Lett.* **50**, 451 (1987).

35. M. Dellow, P. H. Beton, M. Henini, P. C. Main, L. Eaves, S. P. Beaumont, and C. D. W. Wilkinson, "Gated resonant tunneling devices," *Electron. Lett.* **27**, 134 (1991);
    P. Guéret, N. Blanc, R. Germann, and H. Rothuizen, "Confinement and single-electron tunneling in Schottky-gated, laterally squeezed double-barrier quantum-well heterostructures," *Phys. Rev. Lett.* **68**, 1896 (1992).

36. V. R. Kolagunta, D. B. Janes, G. L. Chen, K. Webb, M. R. Melloch, and C. Youtsey, "Self-aligned sidewall-gated resonant tunneling transistors," *Appl. Phys. Lett.* **69**, 374 (1996).

37. S. Luryi and F. Capasso, "Resonant tunneling of two dimensional electrons through a quantum wire," *Appl. Phys. Lett.* **47**, 1347 (1985); *erratum, ibid.* **48**, 1693 (1986).

38. A. Zaslavsky, K. R. Milkove, Y. H. Lee, K. K. Chan, F. Stern, D. A. Grützmacher, S. A. Rishton, C. Stanis, and T. O. Sedgwick, "Fabrication of three-terminal resonant tunneling devices in silicon-based material," *Appl. Phys. Lett.* **64**, 1699 (1994).

39. A. Zaslavsky, D. C. Tsui, M. Santos, and M. Shayegan, "Resonant tunneling of two-dimensional electrons into one-dimensional subbands of a quantum

wire," *Appl. Phys. Lett.* **58**, 1440 (1991).

40. L. N. Pfeiffer, K. W. West, H. L. Stormer, J. P. Eisenstein, K. W. Baldwin, D. Gershoni, and J. Spector, "Formation of a high-quality two-dimensional electron gas on cleaved GaAs," *Appl. Phys. Lett.* **56**, 1697 (1990).

41. Ç. Kurdak, D. C. Tsui, S. Parihar, M. B. Santos, H. Manoharan, S. A. Lyon, and M. Shayegan, "Surface resonant tunneling transistor: a new negative transconductance device," *Appl. Phys. Lett.* **64**, 610 (1994).

42. F. Capasso, K. Mohammed, and A. Y. Cho, "Resonant tunneling through double barriers, perpendicular quantum transport phenomena in superlattices, and their device applications," *IEEE J. Quantum Electron.* **QE-22**, 1853 (1986).

43. A. C. Seabaugh, Y.-C. Kao, and H.-T. Yuan, "Nine-state resonant tunneling diode memory," *IEEE Electron Dev. Lett.* **13**, 479 (1992).

44. For a complete discussion see G. Bastard, *Wave Mechanics Applied to Semiconductor Heterostructures*, Ch. I, Wiley, New York, 1988.

45. L. V. Keldysh, "Effect of ultrasound on the electron spectrum of a crystal," *Fiz. Tverd. Tela* **4**, 2265 (1962) [*Sov. Phys. Solid. State* **4**, 1658 (1963)].

46. L. Esaki and R. Tsu, "Superlattice and negative differential conductivity in semiconductors," *IBM J. Res. Develop.* **14**, 61 (1970).

47. L. Esaki and L. L. Chang, "New transport phenomenon in a semiconductor "superlattice'," *Phys. Rev. Lett.* **33**, 495 (1974);
K. K. Choi, B. F. Levine, R. J. Malik, J. Walker, and C. G. Bethea, "Periodic negative conductance by sequential resonant tunneling through an expanding high-field superlattice domain," *Phys. Rev. B* **35**, 4172 (1987).

48. A. Sibille, J. F. Palmier, H. Wang, and F. Mollot, "Observation of Esaki-Tsu negative differential velocity in GaAs/AlAs superlattices," *Phys. Rev. Lett.* **64**, 52 (1990);
H. T. Grahn, K. von Klitzing, K. Ploog, and G. Döhler, "Electrical transport in narrow-miniband semiconductor superlattices," *Phys. Rev. B* **43**, 12094 (1991).

49. H. M. James, "Electronic states in perturbed periodic systems," *Phys. Rev.* **76**, 1611 (1949).

50. The extent of Wannier–Stark wavefunctions over a finite number of SL periods allows for their observation by photocurrent measurements, see E. E. Mendez, F. Agulló-Rueda, and J. M. Hong, "Stark localization in GaAs-GaAlAs superlattices under an electric field," *Phys. Rev. Lett.* **60**, 2426 (1988).

51. R. Kazarinov and R. Suris, "Possibility of the amplification of electromagnetic waves in a semiconductor with a superlattice," *Fiz. Tekh. Poluprovodn.*, **5**, 797 (1971) [*Sov. Phys. Semicond.* **5**, 707 (1971).

52. J. Faist, F. Capasso, D. L. Sivco, C. Sirtori, A. L. Hutchinson, and A. Y. Cho, "Quantum cascade laser," *Science* **264**, 533 (1994).

53. H. C. Liu and G. C. Aers, "Resonant tunneling through one-, two-, and three-dimensionally confined quantum wells," *J. Appl. Phys.* **65**, 4908 (1989).

54. B. Su, V. J. Goldman, and J. E. Cunningham, "Observation of single-electron charging in double-barrier heterostructures," *Science* **255**, 313 (1992); "Single-electron tunneling in nanometer-scale double-barrier heterostructure devices," *Phys. Rev. B* **46**, 7644 (1992).

55. T. Schmidt, M. Tewordt, R. H. Blick, R. J. Haug, D. Pfannkuche, K. von Klitzing, A. Förster, and H. Luth, "Quantum-dot ground states in a magnetic field studied by single-electron tunneling spectroscopy on double-barrier heterostructures," *Phys. Rev. B* **51**, 5570 (1995).

56. An extensive discussion is available in H. Grabert and M. H. Devoret, Eds., *Single Charge Tunneling: Coulomb Blockade Phenomena in Nanostructures*, Plenum Press, New York, 1992. In particular, the chapter by D. V. Averin and K. K. Likharev is devoted to device applications.

57. T. A. Fulton and G. J. Dolan, "Observation of single-electron charging effects in small junctions," *Phys. Rev. Lett.* **59**, 109 (1987).

58. G. Zimmerli, R. L. Kautz, and J. M. Martinis, "Voltage gain in the single-electron transistor," *Appl. Phys. Lett.* **61**, 2616 (1992).

59. L. P. Kouwenhoven, A. T. Johnson, N. C. van der Vaart, C. J. Harmans, and C. T. Foxon, "Quantized current in a quantum-dot turnstile using oscillating tunnel barriers," *Phys. Rev. Lett.* **67**, 1626 (1991).

60. H. Pothier, P. Lafarge, C. Urbina, D. Esteve, and M. H. Devoret, "Single-electron pump based on charging effects," *Europhys. Lett.* **17**, 249 (1992); J. M. Martinis, M. Nahum, and H. D. Jensen, "Metrological accuracy of the electron pump," *Phys. Rev. Lett.* **72**, 904 (1994).

61. C. A. Mead, "Tunnel-emission amplifiers," *Proc. IRE* **48**, 359 (1960).

62. K. Seeger, *Semiconductor Physics*, 2nd ed., Springer-Verlag, Berlin, 1982; E. Schöll, "Theory of oscillatory instabilities in parallel and perpendicular transport in heterostructures," in *Negative Differential Resistance and Instabilities in 2D Semiconductors*, N. Balkan, B. K. Ridley, and A. J. Vickers, Eds., Plenum, New York, 1993, pp. 37–51.

63. P. J. Price, "Monte Carlo calculation of electron transport in solids," in *Semiconductors and Semimetals*, Vol. 14, Academic Press, New York, 1979, pp. 249-308; C. Moglestue, *Monte Carlo Simulation of Semiconductor Devices*, Chapman & Hall, London, 1993.

64. A thorough discussion of real-space transfer effects is available in a review article by Z. S. Gribnikov, K. Hess, and G. A. Kosinovsky, "Nonlocal and nonlinear transport in semiconductors: real-space transfer effects," *J. Appl. Phys.* **77**, 1337 (1995).

65. M. Heiblum, M. I. Nathan, D. Thomas, and C. M. Knoedler, "Direct observation of ballistic transport in gallium arsenide," *Phys. Rev. Lett.* **55**, 2200 (1985); M. Heiblum, I. Anderson, and C. M. Knoedler, "DC performance of ballistic tunneling hot-electron-transfer amplifiers," *Appl. Phys. Lett.* **49**, 207 (1986).

66. K. Seo, M. Heiblum, C. M. Knoedler, J. Oh, J. Pamulapati, and P. Bhattacharya, "High-gain pseudomorphic InGaAs base ballistic hot-electron devices," *IEEE Electron Dev. Lett.* **10**, 73 (1989).

67. T. S. Moise, A. C. Seabaugh, E. A. Beam III, J. N. Randall, "Room-temperature operation of a resonant-tunneling hot-electron transistor based integrated circuit," *IEEE Electron. Dev. Lett.* **14**, 441 (1993).

68. A. A. Grinberg and S. Luryi, "Electron transmission across interface of different

one-dimensional crystals," *Phys. Rev. B* **39**, 7466 (1989).

69. J. R. Hayes, A. F. J. Levi, and W. Wiegmann, "Hot electron spectroscopy," *Electron. Lett.* **20**, 851 (1984);
A. F. J. Levi, J. R. Hayes, P. M. Platzmann, and W. Wiegmann, "Injected hot-electron transport in GaAs," *Phys. Rev. Lett.* **55**, 2071 (1985).

70. A. Palevski, M. Heiblum, C. P. Umbach, C. M. Knoedler, A. N. Broers, and R. H. Koch, "Lateral tunneling, ballistic transport, and spectroscopy of a two-dimensional electron gas," *Phys. Rev. Lett.* **62**, 1776 (1989);
A. Palevski, C. P. Umbach, and M. Heiblum, "A high gain lateral hot-electron device," *Appl. Phys. Lett.* **55**, 1421 (1989).

71. J. Spector, H. L. Stormer, K. W. Baldwin, L. N. Pfeiffer, and K. W. West, "Ballistic electron transport beyond 100 μm in 2D electron systems," *Surf. Sci.* **228**, 283 (1990);
A. Yacoby, U. Sivan, C. P. Umbach, and J. M. Hong, "Hot ballistic transport and phonon emission in a two-dimensional electron gas," *Phys. Rev. Lett.* **66**, 1938 (1991).

72. For example, see J. Song, B. W.-P. Hong, C. J. Palmstrom, B. P. van der Gaag, and K. B. Chough, "Ultra-high-speed InP/InGaAs heterojunction bipolar transistors," *IEEE Electron Dev. Lett.* **15**, 94 (1994).

73. A. A. Grinberg and S. Luryi, "Coherent transistor," *IEEE Trans. Electron. Dev.* **40**, 1512 (1993).

74. Z. S. Gribnikov, "Negative differential conductivity in a multilayer heterostructure," *Fiz. Tekh. Poluprovodn.* **6**, 1380 (1972) [*Sov. Phys. Semicond.* **6**, 1204 (1973)].

75. K. Hess, H. Morkoç, H. Shichijo, and B. G. Streetman, "Negative differential resistance through real-space electron transfer," *Appl. Phys. Lett.* **35**, 469 (1979).

76. M. Keever, H. Shichijo, K. Hess, S. Banerjee, L. Witkowski, H. Morkoç, and B. G. Streetman, "Measurements of hot-electron conduction and real-space transfer in GaAs/Al$_x$Ga$_{1-x}$As heterojunction layers," *Appl. Phys. Lett.* **38**, 36 (1981).

77. N. Z. Vagidov, Z. S. Gribnikov, and V. M. Ivastchenko, "Modeling of electron transport in real space in GaAs/Al$_x$Ga$_{1-x}$As heterostructures (with low and high values of x)," *Fiz. Tekh. Poluprovodn.* **24**, 1087 (1990) [*Sov. Phys. Semicond.* **24**, 684 (1990)].

78. A. Kastalsky and S. Luryi, "Novel real-space hot-electron transfer devices," *IEEE Electron. Dev. Lett.* **4**, 334 (1983);
S. Luryi, A. Kastalsky, A. C. Gossard, and R. H. Hendel, "Charge injection transistor based on real-space hot-electron transfer," *IEEE Trans. Electron. Dev.* **31**, 832 (1984).

79. A. Kastalsky, R. Bhat, W. K. Chan, and M. Koza, "Negative-resistance field-effect transistor grown by organometallic chemical vapor deposition," *Solid State Electron.* **29**, 1073 (1986).

80. P. M. Mensz, P. A. Garbinski, A. Y. Cho, D. L. Sivco, and S. Luryi, "High transconductance and large peak-to-valley ratio of negative differential conductance in three-terminal InGaAs/InAlAs real-space transfer devices," *Appl. Phys.*

Lett. **57**, 2558 (1990).

81. M. R. Hueschen, N. Moll, and A. Fischer-Colbrie, "Improved microwave performance in transistors based on real-space electron transfer," *Appl. Phys. Lett.* **57**, 386 (1990);
G. L. Belenky, P. A. Garbinski, S. Luryi, P. R. Smith, A. Y. Cho, R. A. Hamm, and D. L. Sivco, "Microwave performance of top-collector charge injection transistors on InP substrates," *Semicond. Sci. Technol.* **9**, 1215 (1994).

82. C. L. Wu, W. C. Hsu, M. S. Tsai, and H. M. Shieh, "Very strong negative differential resistance real-space transfer transistor using a mulitple Δ-doping GaAs/InGaAs pseudomorphic heterostructure," *Appl. Phys. Lett.* **66**, 739 (1995).

83. M. Mastrapasqua, C. A. King, P. R. Smith, and M. R. Pinto, "Charge injection transistor and logic elements in $Si/Si_{1-x}Ge_x$ heterostructures," in *Future Trends in Microelectronics: Reflections on the Road to Nanotechnology*, S. Luryi, J. Xu, and A. Zaslavsky, Eds., Kluwer, Dordrecht, 1996.

84. I. G. Kizilyalli and K. Hess, "Physics of real-space transfer transistors," *J. Appl. Phys.* **65**, 2005 (1989).

85. A. A. Grinberg, A. Kastalsky, and S. Luryi, "Theory of hot-electron injection in CHINT/NERFET devices," *IEEE Trans. Electron. Dev.* **34**, 409 (1987).

86. M. Mastrapasqua, S. Luryi, G. L. Belenky, P. A. Garbinski, A. Y. Cho, and D. L. Sivco, "Multi-terminal light emitting logic device electrically reprogrammable between OR and NAND functions," *IEEE Trans. Electron. Dev.* **40**, 1371 (1993);
G. L. Belenky, P. A. Garbinski, S. Luryi, M. Mastrapasqua, A. Y. Cho, R. A. Hamm, T. R. Hayes, E. J. Laskowski, D. L. Sivco, and P. Smith, *J. Appl. Phys.* **73**, 8618 (1993).

87. N. Yokoyama, K. Imamura, S. Muto, S. Hiyamizu, and H. Nishi, "A new functional resonant-tunneling hot electron transistor (RHET)," *Jap. J. Appl. Phys.* **24**, L-853 (1985);
N. Yokoyama, K. Imamura, H. Ohnishi, T. Mori, S. Muto, and A. Shibatomi, "Resonant-tunneling hot electron transistor (RHET)," *Solid State Electron.* **31**, 577 (1988).

88. For a review of resonant tunneling bipolar transistor research see F. Capasso, S. Sen, and F. Beltram, "Quantum-effect devices," in *High-Speed Semiconductor Devices*, S. M. Sze, Ed., Wiley, New York, 1990, pp. 465–520.

89. S. Sen, F. Capasso, A. Y. Cho, and D. L. Sivco, "Multiple state resonant tunneling bipolar transistor operating at room temperature and its application as a frequency multiplier," *IEEE Electron. Dev. Lett.* **9**, 533 (1988).

90. E. R. Brown, J. R. Söderstrom, C. D. Parker, L. J. Mahoney, K. M. Molvar, and T. C. McGill, "Oscillations up to 712 GHz in InAs/AlSb resonant-tunneling diodes," *Appl. Phys. Lett.* **58**, 2291 (1991).

91. W. F. Chow, *Principles of Tunnel Diode Circuits*, Wiley, New York, 1964.

92. J. Shen, G. Kramer, S. Tehrani, H. Goronkin, and R. Tsui, "Static random access memories based on resonant interband tunneling diodes in the InAs/GaSb/AlSb material system," *IEEE Electron. Dev. Lett.* **16**, 178 (1995).

93. T. Mori, S. Muto, H. Tamura, and N. Yokoyama, "A static random access memory cell using a double-emitter resonant-tunneling hot electron transistor for gigabit-plus memory applications," *Jap. J. Appl. Phys.* **33**, 790 (1994).

94. M. Takatsu, K. Imamura, H. Ohnishi, T. Mori, T. Adachibara, S. Muto, and N. Yokoyama, "Logic circuits using resonant-tunneling hot-electron transistors (RHET's)," *IEEE J. Solid-State Circ.* **27**, 1428 (1992);
N. Yokoyama, H. Ohnishi, T. Mori, M. Takatsu, S. Muto, K. Imamura, and A. Shibatomi, "Resonant hot electron transistors," in *Hot Carriers in Semiconductor Nanostructures: Physics and Applications*, J. Shah, Ed., Academic Press, New York, 1992, pp. 443–467.

95. A. C. Seabaugh and M. A. Reed, "Resonant-tunneling transistors," in *Heterostructures and Quantum Devices*, N. G. Einspruch and W. R. Frensley, Eds., Academic Press, New York, 1994.

96. S. Luryi, P. Mensz, M. R. Pinto, P. A. Garbinski, A. Y. Cho, and D. L. Sivco, "Charge injection logic," *Appl. Phys. Lett.* **57**, 1787 (1990);
K. Imamura, M. Takatsu, T. Mori, Y. Bamba, S. Muto, and N. Yokoyama, "Proposal and demonstration of multi-emitter HBTs," *Electron. Lett.* **30**, 459 (1994).

97. J. Faist, F. Capasso, C. Sirtori, D. L. Sivco, A. L. Hutchinson, and A. Y. Cho, "Vertical transition quantum cascade laser with Bragg confined excited state," *Appl. Phys. Lett.* **66**, 538 (1995).

98. C. Sirtori, J. Faist, F. Capasso, D. L. Sivco, and A. Y. Cho, "Narrowing of the intersubband absorption spectrum by localization of continuum resonances in a strong electric field," *Appl. Phys. Lett.* **62**, 1931 (1993).

99. J. Faist, F. Capasso, C. Sirtori, D. L. Sivco, J. N. Baillargeon, A. L. Hutchinson, and A. Y. Cho, "High power mid-infrared ($\lambda \sim 5\ \mu$m) quantum cascade lasers operating above room temperature," *Appl. Phys. Lett.* **68**, 3680 (1996).

100. V. B. Gorfinkel, S. Luryi, and B. Gelmont, "Theory of gain spectra for quantum cascade lasers and temperature dependence of their characteristics at low and moderate carrier concentrations," *IEEE J. Quantum Electron.* **32**, 1995 (1996).

101. B. Gelmont, V. B. Gorfinkel, and S. Luryi, "Theory of the spectral line shape and gain in quantum wells with intersubband transitions," *Appl. Phys. Lett.* **68**, 2171 (1996).

102. D. C. Tsui, H. L. Störmer, and A. C. Gossard, "Two-dimensional magnetotransport in the extreme quantum limit," *Phys. Rev. Lett.* **48**, 1559 (1982).

103. L. V. Iogansen, "Errors in papers on resonant electron tunneling in finite superlattices," *Pis'ma Zh. Tekh. Fiz.* **13**, 1143 (1987) [*Sov. Tech. Phys. Lett.* **13**, 478 (1987)].

104. S. Luryi, A. A. Grinberg, and V. B. Gorfinkel, "Heterostructure bipolar transistor with enhanced forward diffusion of minority carriers," *Appl. Phys. Lett.* **63**, 1537 (1993).

105. T. P. E. Broekaert, W. Lee, and C. G. Fonstad, "Pseudomorphic In$_{0.53}$Ga$_{0.47}$As/AlAs/InAs resonant tunneling diodes with peak-to-valley current ratios of 30 at room temperature," *Appl. Phys. Lett.* **53**, 1545 (1988).

6 Active Microwave Diodes

HERIBERT EISELE and GEORGE I. HADDAD
The University of Michigan, Ann Arbor

6.1 INTRODUCTION

Two-terminal devices were the first to be used in solid-state microwave oscillators, and major technological challenges had to be mastered on the way from the inception of the device structures to the first experimental results in the early 1960s (see overview in Ref. 1). Device structures and circuits were developed for both amplifier and oscillator applications. However, rapid progress in high-speed and high-frequency three-terminal devices with excellent noise performance practically eliminated two-terminal devices from all low-noise preamplifier applications up to the high millimeter-wave frequencies. Additionally, oscillators with three-terminal devices continue to reach higher and higher frequencies and offer similar or even higher RF output power levels and DC-to-RF conversion efficiencies compared with the most powerful two-terminal devices. Nonetheless, in system applications, two-terminal devices still generate the highest power levels per device, per device area, or per volume of the circuit, especially at millimeter-wave and higher frequencies. Besides two-terminal devices, tubes are still employed in many high-power amplifiers for systems applications because three-terminal devices have met with only limited success for this type of application. Two-terminal devices are still considered the prime candidate for high RF power generation or low-noise local oscillators. Additionally, pulsed operation of these devices overcomes thermal limits and increases peak RF power levels by more than an order of magnitude. In this chapter we will focus on the fundamental power capabilities of these devices and discuss the properties and performance of those devices that can be found in or are potential candidates for widespread system applications above

Modern Semiconductor Device Physics, Edited by S. M. Sze.
ISBN 0-471-15237-4 © 1998 John Wiley & Sons, Inc.

TABLE 1   Band Designations, Frequency Ranges, and Inside Dimensions of Commonly Used Rectangular Waveguides for Frequencies above 10 GHz

Waveguide Band Designation	Frequency Range (GHz)	Rectangular Waveguide (EIA)	Inside Dimensions (mm)
X	8.2–12.4	90	22.86 × 11.43
Ku	12.4–18	62	15.799 × 7.899
K	18–26.5	42	10.668 × 5.334
Ka	26.5–40	28	7.112 × 3.556
Q	33–50	22	5.690 × 2.845
U	40–60	19	4.775 × 2.388
V	50–75	15	3.759 × 1.880
E	60–90	12	3.099 × 1.549
W	75–110	10	2.540 × 1.270
F	90–140	08	2.032 × 1.016
D	110–170	06	1.651 × 0.826
G	140–220	05	1.295 × 0.648
Y	170–260	04	1.092 × 0.546
J	220–235	03	0.864 × 0.432

30 GHz, that is, millimeter-wave and higher frequencies. Table 1 lists the band designations, frequency ranges and inside dimensions for the most common waveguides. These band designations are widely used, and their use is often independent of whether waveguide circuits are employed or not.

6.1.1   An Oscillator with a Two-Terminal Device

Figure 1a shows a simplified equivalent circuit of a two-terminal negative-resistance device connected to a resonant circuit as a load. The admittance per unit area Y_D of the two-terminal device is represented by the negative conductance G_D and the susceptance jB_D, which results mainly from the "cold capacitance" C_C of the device. The series resistance $R_s = \rho_s/A$ for a device with area A represents the combined losses of the device (contacts, package, etc.) and of the load circuit. ρ_s denotes an equivalent specific contact resistance. The oscillation frequency f_o ($\omega_o = 2\pi f_o$) is just slightly above the resonant frequency of the load, where

$$\frac{1}{\omega_o C_R} < \omega_o L_R \tag{1}$$

At this frequency, we can further simplify the equivalent circuit of the device (see Fig. 1b) using

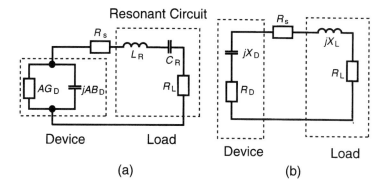

Fig. 1 Simplified equivalent circuit of an oscillator with a two-terminal device connected to a load.

$$R_D = \frac{G_D}{A(G_D^2 + B_D^2)} \tag{2}$$

and

$$X_D = \frac{-1}{\omega_o C_d} = \frac{-B_D}{A(G_D^2 + B_D^2)} \tag{3}$$

as well as

$$X_L = \omega_o L_L = \omega_o \left(L_R - \frac{1}{\omega_o^2 C_R} \right) \tag{4}$$

The admittance Y_D of the device is a nonlinear function of several parameters, such as oscillation frequency f_o, bias at the operating point, amplitude of the RF signal at the device, and device temperature. The oscillation condition ($Z_D = R_D + jX_D \equiv Z_L = R_L + jX_L$), where Z_L is the load impedance as seen by the device, is fulfilled at f_o for

$$R_D + R_s + R_L = 0 \quad \text{(real part)} \tag{5a}$$

and

$$\frac{-1}{\omega_o C_d} + \omega_o L_L = 0 \quad \text{(imaginary part)} \tag{5b}$$

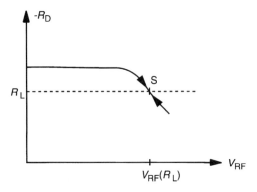

Fig. 2 Qualitative dependence of the negative resistance in a two-terminal device as a function of the RF voltage.

The values for the loaded Q factor of the resonant circuit

$$Q = \frac{\omega_o L_R}{R_L} \qquad (6)$$

depend on the circuit structure and, typically, range from 20 to 200. In most applications at millimeter-wave frequencies, the resonant circuit of the load consists of a waveguide resonant cavity with tuning elements, but oscillators using microstrip-line circuits have also been demonstrated.[2] At lower frequencies, circuits consisting of lumped elements or coaxial line are more common and examples are shown in Ref. 1.

When the DC bias to the device is turned on, no oscillations are present, and the absolute value of the device resistance $|R_D|$ as shown in Fig. 2 is higher than the load resistance R_L. Therefore, noise components, which are present in any device all the time, are amplified and filtered by the resonant circuit. As a consequence, the RF voltage V_{RF} at the resonance frequency grows and the negative resistance of the device decreases until the oscillation conditions of Eq. 2 are fulfilled, i.e., RF amplitude and oscillation frequency reach the stable operating point (S) of Fig. 2. The RF power that is generated in the device at the RF voltage V_{RF} is

$$P_{RF}(\text{Gen}) = -\tfrac{1}{2} A G_D V_{RF}^2 \qquad (7)$$

Commonly, two-terminal devices require a low load resistance R_L, and the chosen value is a compromise, particularly at millimeter wave frequencies: The lower R_L is, the larger the area A of the device and the higher the generated RF power. However, losses in the circuit that transforms the actual load with the high characteristic impedance of a coaxial line, e.g., $50\,\Omega$, into the load impedance Z_L at the device or provides the coupling from the output waveguide to the device, increase with lower R_L. These losses correspond to a much higher series resistance R_s in the equivalent circuit and thus reduce the RF output power $P_{RF}(R_L)$ that is actually delivered to the load:

$$P_{RF}(R_L) = \tfrac{1}{2}A^2(G_D^2 + B_D^2)V_{RF}^2 R_L \qquad (8)$$

The oscillation condition in Eq. 5a, together with Eqs. 2, 3, and 4, determines the device area A:

$$A = \frac{1}{R_L}\left(\frac{-G_D}{G_D^2 + B_D^2} - \rho_s\right) \qquad (9)$$

and if

$$\frac{-G_D}{G_D^2 + B_D^2} < \rho_s$$

oscillations cease and no RF power is generated.

The real part of the load impedance $(= R_L)$ is typically in the range 0.25–10 Ω. Therefore, impedance matching as well as frequency-stable operation require a high Q, low-loss resonant circuit as well as fabrication techniques that minimize series resistances and losses in undepleted device layers, the contacts, and the package. A load resistance R_L of 1 Ω is widely assumed as a reference to compare the performance of different two-terminal devices, and often gives very good agreement between predicted and measured performance of various two-terminal devices in waveguide cavities.

6.1.2 Fabrication Technologies

As we will see later in this chapter for different devices, DC-to-RF conversion efficiencies range from approximately more than 20% down to less than 1%. As a consequence, most of the DC input power needs to be dissipated as heat in the device. In most cases, the contact near the active region of the device also acts as the heatsink and, therefore, two-terminal devices for millimeter-wave frequencies generally are mesa-type devices. Additionally, operation at these frequencies requires thin devices to reduce losses in the substrate resulting from the skin effect. The integral heatsink technology is the most widespread for devices at millimeter-wave frequencies. To reduce losses in the substrate, most of it needs to be removed during fabrication.

In early fabrication technologies, vapor-phase epitaxy provided the layer structures, and doping profiles were defined by dopant diffusion and ion implantation. As a first step in processing, a few small holes across the sample were etched through the epitaxial layers down into the substrate. An appropriate depth of the holes was chosen to gage the thickness during substrate removal. The advent of more advanced growth techniques, such as molecular beam epitaxy (MBE), metalorganic chemical vapor deposition (MOCVD), and chemical beam epitaxy (CBE), allows the incorporation of a lattice-matched stop-etch layer between the substrate and the epitaxial layers of the device. This way, the substrate is completely removed, and precise control of mesa height and diameter is achieved. Fabrication technologies for substrateless devices on integral heatsinks or on diamond heatsinks for better heat removal were developed and described in the literature. These selective etching technologies have been established for silicon,[3] GaAs,[4,5] and InP[6,7] employing as etch-stop layers heavily p^+-doped silicon,[8] lattice-matched $Ga_xAl_{1-x}As$, and $In_{0.53}Ga_{0.47}As$ layers, respectively. Improved yield, reproducibility, and performance characterize these substrateless devices.

Figure 3 summarizes the basic steps of these fabrication technologies. The batch fabrication of GaAs IMPATT (impact ionization avalanche transit-time) or TUNNETT (tunnel injection transit-time) diodes on integral heatsinks serves as an example.[9] Most of the heat in such a diode is generated near the p–n junction (see Section 6.2) and, typically, the heavily p^+-doped region is chosen to be the top epitaxial layer. In the first step, the metalization for the p ohmic contact (Ti/Pt/Au) is evaporated or sputtered onto the surface. A thick gold layer is then electroplated onto this metalization to form the integral heatsink. The sample is mounted on a carrier to provide additional mechanical support and to protect the heatsink during the subsequent processing steps. The substrate is removed in a selective etchant of H_2O_2:NH_4OH 1:19,[4,5] which does not significantly attack the $Ga_xAl_{1-x}As$ etch-stop layer if $x > 0.4$. Ohmic contacts are difficult to form on $Ga_xAl_{1-x}As$. Therefore, this $Ga_xAl_{1-x}As$ layer is selectively removed in HF, which does not attack GaAs. A photolithography step defines the openings on this GaAs surface, where the metalization (Ni/Ge/Au/Ti/Au) for the n ohmic contacts on the heavily n^+-doped layer is deposited. Excess metal outside the contacts is lifted off with the photoresist and, using another photolithography step, the contacts are selectively electroplated with several microns of gold to form a good bonding pad. The contact pad acts as a mask when the mesa of the diode is etched in a nonselective etch. After the sample has been removed from the carrier, the contacts are annealed, and the sample is diced into individual diodes. Diodes are then mounted in packages for appropriate RF circuits.

Figure 4a shows a typical package for IMPATT diodes and Gunn devices. It consists of a gold-plated threaded copper puck, which can be screwed into the RF circuit, an alumina ring, and a top lid for a hermetic seal. The device is soldered or thermocompression bonded onto a pedestal inside the ring, and gold straps are thermocompression bonded to the device and the top metalization of the alumina ring. The height and diameter of the ring depend on the operating frequency as well as the device, and typical values are given

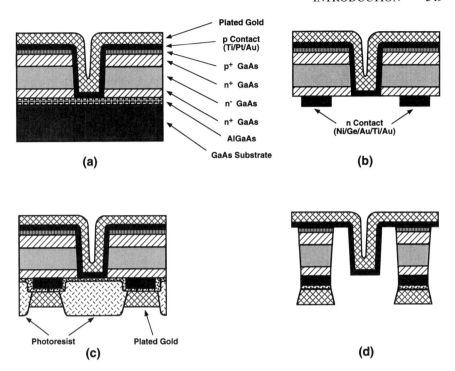

Fig. 3 Steps in the fabrication of GaAs IMPATT or TUNNETT diodes on integral heatsinks. (a) Island definition, p-ohmic evaporation and gold plating of heatsink ($\approx 20 \, \mu$m). (b) Substrate thinning, etch-stop layer removal, and n-ohmic evaporation. (c) Gold plating of ohmic contacts. (d) Final diodes after annealing and mesa etch. (After Kamoua et al., Ref. 6, and Eisele et al., Ref. 9.)

in Fig. 4a. This type of package is used up to frequencies of 94 GHz, and its parasitic elements can be approximated by lumped elements, as illustrated in Fig. 4b. Different ribbon configurations are chosen to minimize the influence of the parasitic inductance, which is the highest for just one gold strap across and the lowest for the "star" configuration. The useful frequency range of the package can be extended to 140 GHz and higher if the alumina ring is replaced by a quartz ring. However, many devices at frequencies above 100 GHz are still in their experimental stage and, for that reason, commonly, a low-parasitic open package with two or four standoffs at the highest millimeter- and up to submillimeter-wave frequencies is employed.[7,10–13]

6.1.3 Microwave Circuits

Many different circuit configurations for oscillators with two-terminal devices were investigated. At millimeter-wave frequencies, waveguide circuits are

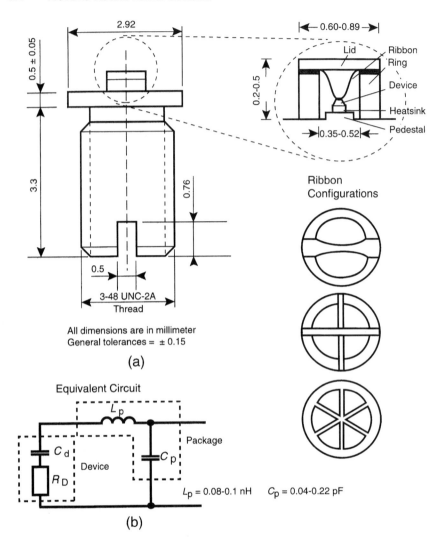

Fig. 4 (a) Hermetically sealed package for millimeter-wave two-terminal devices. (b) Equivalent circuit of the parasitic elements.

quite common. An overview of typical configurations for waveguide circuits[2,11] is shown in Fig 5. Examples of oscillator circuits using coaxial lines at microwave frequencies can be found in Ref. 1.

6.1.4 Noise

The noise in the output spectrum of an oscillator consists of fluctuations in amplitude (AM noise) and oscillation frequency (FM noise). The stable

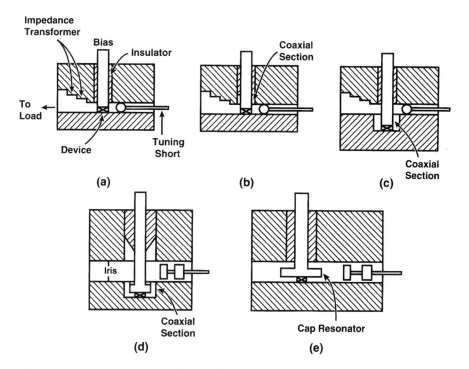

Fig. 5 Examples of waveguide circuits for two-terminal device oscillators. (After Kuno, Ref. 11.)

operating point S of Fig. 2 acts as a hard limit for the amplitude of the oscillations. Therefore, FM noise dominates in two-terminal devices. It can be described as effective frequency modulation Δf_{rms} vs frequency f_m off the oscillation frequency f_o. It corresponds to the noise-to-carrier ratio $N/C|_{FM}$ as, for example, seen on a spectrum analyzer:

$$\left.\frac{N}{C}\right|_{FM} = \frac{\Delta f_{rms}^2}{2f_m^2} \tag{10}$$

To compare the noise performance of oscillators of different two-terminal devices, the FM noise measure M is more appropriate:

$$M = \frac{\Delta f_{rms}^2 Q^2}{f_o^2 k T_0 B} P_{RF} \tag{11}$$

where T_0 is the absolute temperature and B is the bandwidth. The loaded Q factor of the resonant circuit is determined in a waveguide setup, as illustrated in Fig. 6a. The sweep oscillator injects a signal at the power level

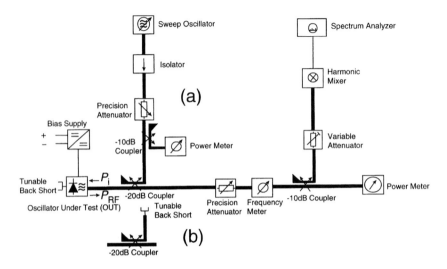

Fig. 6 Waveguide test setup to determine the injection locking range Δf_s and Q factor of an oscillator. (a) Injection locking with a sweep oscillator. (b) Self-injection locking.

P_i into the oscillator under test (OUT), and the maximum continuous frequency range Δf_{fs} over which the OUT remains injection locked with the sweep oscillator is determined:[14]

$$Q = \frac{2f_0}{\Delta f_s}\sqrt{\frac{P_i}{P_{RF}}}$$ (12)

An alternative is also shown in Fig. 6b. The signal from the OUT is reflected at a tunable low-loss back short and injected back as P_i through the coupler into the OUT. If the position of the back short is moved by more than half of the guide wavelength λ_g, the oscillation frequency continuously changes from a lower to an upper limit. This maximum tuning range, that is, the self-injection locking range, is now Δf_s in Eq. 12.

6.2 TRANSIT-TIME DIODES

This group comprises diodes with widespread system applications, for example, IMPATT diodes as the most powerful two-terminal devices, as well as BARITT (barrier injection transit-time) and TUNNETT diodes as very low-noise devices for local oscillators and very sensitive self-oscillating mixers. Some more exotic devices, for example, QWITT (quantum-well injection transit-time) devices[15] and DOVATT (double-velocity avalanche

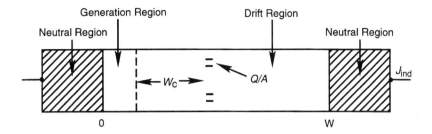

Fig. 7 Schematic longitudinal section of a negative-resistance transit-time diode.

transit-time) diodes[15] also belong to this group, but, beyond the basic principles of operation, are not covered here.

6.2.1 Principles of Operation

Transit-time diodes share a common mechanism that dynamically creates a negative resistance. As shown in Fig. 7, a packet of carriers (e.g., electrons) is generated in a confined and narrow zone (generation region) and injected into the adjacent fully depleted zone (drift region). Several mechanisms can generate and inject carriers. These include the following:

1. Thermionic emission over a barrier: This barrier can be formed by a *p–n* junction or Schottky junction in forward bias or by a heterojunction of a layer with a wider bandgap than in the neutral and drift regions. This would result in a BARITT diode.
2. Tunneling through a barrier: At very high electric fields, electrons can tunnel from the valence band to the conduction band of a reverse-biased p^+-n^+ junction if the distance is short and energy states are available. Also, tunneling through a thin heterojunction barrier or resonant tunneling through a double barrier can be employed. This would result in a TUNNETT diode or QWITT device, respectively.
3. Avalanche multiplication through impact ionization: Electrons or holes at high-enough energy levels in a strong electric field create new electron–hole pairs. This would result in an IMPATT diode.
4. At very high frequencies, both tunneling and avalanche mechanisms are present, and thus a mixed mode results. This would result in a MITATT (mixed tunneling–avalanche transit-time) diode.

In the drift region, this pulse of charge Q at the location W_c propagates under a high electric field at the drift velocity v_Q and induces a current flow in the external circuit. The corresponding current density J_{ind} in the external circuit is given by the Ramo–Shockley theorem:[1]

$$J_{ind} = \frac{Q}{W}\left(v_Q - \frac{W_c}{W}\frac{dW}{dt}\right)$$ (13)

Under ideal conditions, the diode is always punched through, and the electric field in the drift region is always high enough so that the carriers maintain their saturated velocity $v_Q = v_s$. The velocity-field characteristics for GaAs and InP can be found as examples in Section 6.4, and in these semiconductor materials, the condition for a saturated velocity is satisfied at $\mathscr{E} \gg \mathscr{E}_{th}$. Thus Eq. 13 reduces to

$$J_{ind} = \frac{Q}{W}v_s$$ (14)

This common mechanism allows us to use a simple unified large-signal analysis to predict the basic power generation capabilities of transit-time diodes. Assuming such large-signal conditions, we can approximate the current and voltage waveforms for all transit-time diodes, as illustrated in Fig. 8. The voltage V_T across the diode is given by

$$V_T = V_{DC} + V_{RF}\sin(\omega t)$$ (15)

where V_{DC}, V_{RF}, and ω denote the DC voltage, the amplitude of the RF voltage, and the circular operating frequency, respectively. The current pulse is represented by the injection phase angle Θ_M and the width angle Θ_W of the pulse. The induced current pulse is represented by the current density J_{max} and the transit angle in the drift region Θ_D $(=\omega W/v_s)$.

As we will discuss later for each type of diode, the properties of the injection mechanism determine the phase Θ_M at which carriers are generated and injected during the RF cycle. The RF power that is generated in the two-terminal diode with an area A is determined from the integral

$$P_{RF} = -\frac{A}{2\pi}\int_0^{2\pi} J_{ind}(\omega t)V_{RF}\sin(\omega t)\,d(\omega t)$$ (16)

which simplifies to

$$P_{RF} = AV_{RF}J_{DC}\frac{\sin(\Theta_W/2)}{(\Theta_W/2)}\frac{\cos(\Theta_M + \Theta_D) - \cos\Theta_M}{\Theta_D}$$ (17)

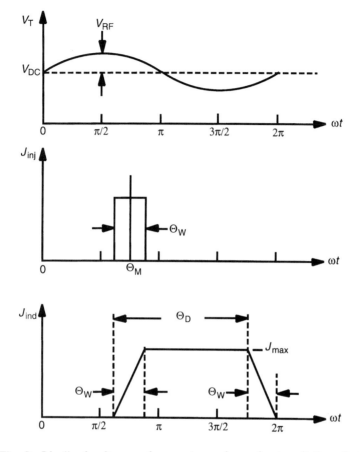

Fig. 8   Idealized voltage and current waveforms for transit-time diodes.

using

$$J_{DC} = \frac{1}{2\pi} \int_0^{2\pi} J_{ind}(\omega t)\, d(\omega t) = \frac{J_{max}}{2\pi} \Theta_D \qquad (18)$$

The intrinsic DC-to-RF conversion efficiency η of the diode is given by

$$\eta = \frac{P_{RF}}{P_{DC}} = \frac{V_{RF}}{V_{DC}} \frac{\sin(\Theta_W/2)}{(\Theta_W/2)} \frac{\cos(\Theta_M + \Theta_D) - \cos\Theta_M}{\Theta_D} \qquad (19)$$

indicating that a sharp pulse for the carrier injection ($\Theta_W \rightarrow 0$) is desired in any transit-time diode. In an IMPATT diode, the maximum of the generated carriers occurs approximately when V_{RF} goes through zero. Therefore, $\Theta_M = \pi$ and optimum η is obtained for $\Theta_D \approx 0.7420\pi$, which is close to $\frac{3}{4}\pi$.

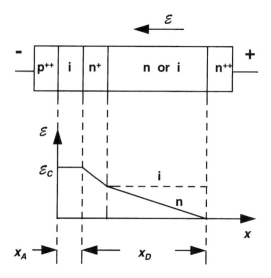

Fig. 9 Schematic layer sequence and electric-field profile for single-drift transit-time diodes.

In a TUNNETT or BARITT diode the maximum of the generated carriers occurs approximately when V_{RF} reaches its maximum. Therefore, $\Theta_M = \pi/2$ and optimum η is obtained for $\Theta_D \approx 1.4303\pi$, which is close to $\frac{3}{2}\pi$.

6.2.2 Estimated Power Generation Capabilities

To arrive at a first-order estimate for the power-generating capabilities of IMPATT and TUNNETT diodes, we assume a basic diode structure, as shown in Fig. 9, where a narrow n^+-doping spike separates two intrinsic layers. A critical electric field $\mathscr{E}_C < 1\,\mathrm{MV/cm}$ and an avalanche region $x_A > 50\,\mathrm{nm}$ are typical in an IMPATT diode, whereas $\mathscr{E}_C > 1.5\,\mathrm{MV/cm}$ and $x_A < 30\,\mathrm{nm}$ are characteristic of a TUNNETT diode. A MITATT diode is characterized by $1\,\mathrm{MV/cm} < \mathscr{E}_C < 1.5\,\mathrm{MV/cm}$ and $30\,\mathrm{nm} < x_A < 50\,\mathrm{nm}$. The corresponding approximate field profile for the unified large-signal analysis at the DC bias V_{DC} is illustrated in Fig. 10:

$$V_{DC} = (\mathscr{E}_m - \mathscr{E}_D)x_A + \mathscr{E}_D(x_A + x_D) \tag{20}$$

The optimum transit angle Θ_D determines the length x_D of the drift region:

$$x_D = \frac{\Theta_D\, v_s}{2\pi\, f_o} \tag{21}$$

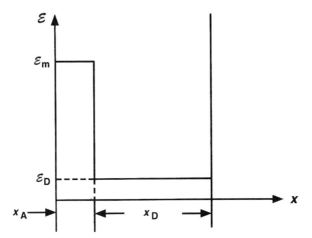

Fig. 10 Approximate field profile for a unified large-signal analysis of transit-time diodes.

The admittance $\mathbf{Y}_D$ of the diode can be derived from the fundamental component of the induced current density $\mathbf{J}_1$ and the RF voltage V_{RF} across the terminals,

$$
\begin{aligned}
J_1(\omega t) = & \frac{1}{\pi}\left(\int_0^{2\pi} J_{\text{ind}}(\omega t)\cos(\omega t)\,d(\omega t)\right)\cos(\omega t) \\
& + \frac{1}{\pi}\left(\int_0^{2\pi} J_{\text{ind}}(\omega t)\sin(\omega t)\,d(\omega t)\right)\sin(\omega t) \\
= & \frac{J_{\text{max}}}{\pi}\frac{\sin(\Theta_W/2)}{(\Theta_W/2)}\{[-\sin\Theta_M + \sin(\Theta_M + \Theta_D)]\cos(\omega t) \\
& + [\cos\Theta_M - \cos(\Theta_M + \Theta_D)]\sin(\omega t)\}
\end{aligned} \tag{22}
$$

Substituting for J_{max} in terms of J_{DC} yields

$$
\begin{aligned}
J_1(\omega t) = & \frac{2J_{DC}}{\Theta_D}\frac{\sin(\Theta_W/2)}{(\Theta_W/2)}\{[-\sin\Theta_M + \sin(\Theta_M + \Theta_D)]\cos(\omega t) \\
& + [\cos\Theta_M - \cos(\Theta_M + \Theta_D)]\sin(\omega t)\}
\end{aligned} \tag{23}
$$

For an IMPATT diode, $\Theta_M \approx \pi$, $\Theta_D \approx 0.75\pi$, and, with $\Theta_W \to 0$, the component at the fundamental frequency becomes

$$
J_1(\omega t) = \frac{8J_{DC}}{3\pi}\left[-\frac{1}{2}\sqrt{2}\cos(\omega t) - \left(1 + \frac{1}{2}\sqrt{2}\right)\sin(\omega t)\right] \tag{24}
$$

and the admittance per unit area Y_1 is

$$Y_1 = \frac{-J_1}{jV_{RF}} = \frac{8J_{DC}}{3\pi V_{RF}}\left(-1 - \frac{1}{2}\sqrt{2} - j\frac{1}{2}\sqrt{2}\right) \tag{25}$$

The total admittance of the diode Y_D also must include the depletion-layer capacitance (cold capacitance) C_C, which per unit area is

$$C_C = \frac{\varepsilon_s}{x_D + x_A} \tag{26}$$

resulting in

$$Y_D = G_D + jB_D = \frac{8J_{DC}}{3\pi V_{RF}}\left(-1 - \frac{1}{2}\sqrt{2} - j\frac{1}{2}\sqrt{2}\right) + \frac{j\omega\varepsilon_s}{x_D + x_A} \tag{27}$$

For a TUNNETT diode, $\Theta_M \approx 0.5\pi$, $\Theta_D \approx 1.5\pi$, and, with $\Theta_W \to 0$, the component at the fundamental frequency becomes

$$J_1(\omega t) = \frac{4J_{DC}}{3\pi}[-\cos(\omega t) - \sin(\omega t)] \tag{28}$$

and the conductance per unit area Y_1 is

$$Y_1 = \frac{-J_1}{jV_{RF}} = \frac{4J_{DC}}{3\pi V_{RF}}(-1 - j) \tag{29}$$

Again, the depletion-layer capacitance C_C must be included resulting in

$$Y_D = G_D + jB_D = \frac{4J_{DC}}{3\pi V_{RF}}(-1 - j) + \frac{j\omega\varepsilon_s}{x_D + x_A} \tag{30}$$

To estimate V_{RF} and J_{DC}, we approximate the electric-field profile at V_{DC} as shown in Fig. 10. For an IMPATT diode, x_A is the length of the avalanche region, and $\mathscr{E}_m$ denotes the maximum electric field in this region. For a TUNNETT diode, x_A is the length of the carrier generation region, and $\mathscr{E}_m$ denotes the maximum electric field in this region. In the IMPATT diode, the breakdown condition for field-dependent ionization rates $\alpha(\mathscr{E})$ is[1]

$$\int_0^{x_A + x_D} \alpha[\mathscr{E}(x)]\, dx = 1 \tag{31}$$

It is important to keep the carrier generation in the drift region very low since it disturbs the phase relationships of the waveforms of Fig. 8. Therefore, we assume

$$\int_{x_A}^{x_D+x_A} \alpha[\mathscr{E}(x)]\,dx = 0.1 \tag{32}$$

and

$$\int_0^{x_A} \alpha[\mathscr{E}(x)]\,dx = 0.9 \tag{33}$$

follows for the avalanche region of the IMPATT diode, with the ionization rate $\alpha(\mathscr{E}) = A_i e^{-(b/\mathscr{E})m}$. For GaAs and $T = 500$ K, $A_i = 3.85 \times 10^5$ cm^{-1}, $b = 6.85 \times 10^5$ V/cm, and $m = 2$, which gives

$$A_i e^{-(b/\mathscr{E}_D)^2} x_D = 0.1 \tag{34}$$

and, therefore,

$$A_i e^{-(b/\mathscr{E}_m)^2} x_A = 0.9 \tag{35}$$

Equation 36 follows from Eq. 34 and determines the electric field in the drift region of the IMPATT or TUNNETT diode:

$$\mathscr{E}_D = \frac{b}{\sqrt{\ln\left(\dfrac{A_i x_D}{0.1}\right)}} \tag{36}$$

whereas

$$x_A = \frac{0.9}{A_i} e^{(b/\mathscr{E}_m)^2} \tag{37}$$

determines the length of the avalanche region as $x_A = 48$ nm in the IMPATT diode for $\mathscr{E}_m = 800$ kV/cm. Experimental results[9] have shown that, in the GaAs TUNNETT diode, the electric field sufficient for tunneling is typically around $\mathscr{E}_m = 1.8$ MV/cm and the equivalent length of the generation region is $x_A = 15$ nm.

As the charge pulse Q leaves the generation region and starts traveling through the drift region, the space charge of the electrons changes the electric field in the drift region. This situation is illustrated in Fig. 11. The field step $\Delta \mathscr{E}_c$ in the active region with a carrier concentration $n(x)$ can be calculated from Poisson's equation

$$\Delta \mathscr{E}_c(x) = \mathscr{E}(x) - \mathscr{E}(0) = -\frac{q}{\varepsilon_s} \int_0^x n(x')\,dx' \tag{38}$$

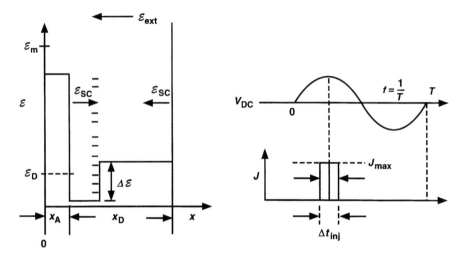

Fig. 11 Approximate electric field and current pulse for estimating the space-charge effect and J_{DC}.

and in the drift region from

$$\Delta \mathscr{E} = \Delta \mathscr{E}(x) = \mathscr{E}(x_D + x_A) - \mathscr{E}(x_A) = -\frac{q}{\varepsilon_s} \int_{x_A}^{x_D + x_A} n(x') \, dx' \qquad (39)$$

with the total charge Q in the drift region,

$$Q = A J_{max} \Delta t_{inj} = qA \int_{x_A}^{x_D + x_A} n(x') \, dx' \qquad (40)$$

$$\Delta \mathscr{E} = \frac{-q}{\varepsilon_s} \frac{J_{max} \Delta t_{inj}}{q} \qquad (41)$$

$|\Delta \mathscr{E}|$ must be smaller than $\mathscr{E}_D$ in the drift region, and this condition limits the bias current density,

$$J_{DC} = \frac{J_{max} \Delta t_{inj}}{T} = J_{max} \Delta t_{inj} f \qquad (42)$$

to

$$J_{DC} \leq \varepsilon_s \mathscr{E}_D f \qquad (43)$$

TABLE 2   Important Diode Parameters for Different Center Frequencies f in IMPATT and TUNNETT Diodes

f (GHz)	x_D (μm)	$\mathscr{E}_D$ (kV/cm)	J_{DC} (kA/cm^2)	V_{DC} (V)	V_{RF} (V)	V_{RF}/V_{DC}
			IMPATT Diodes			
33	0.57	295	10.3	20.7	10.3	0.5
44	0.43	303	15.2	16.8	8.4	0.5
60	0.31	313	21.4	13.7	6.8	0.5
			TUNNETT Diodes			
100	0.38	307	35.0	14.2	7.9	0.55
150	0.25	321	54.8	10.7	5.6	0.52
200	0.19	331	75.5	8.9	4.4	0.5

In the IMPATT diode, the total voltage V_T goes below V_{DC}, and the electric field everywhere in the active region decreases as the charge pulse propagates through the drift region. For the carriers to remain at the saturated drift velocity v_s, an electric field $\mathscr{E} \geq \mathscr{E}_s$ must be maintained throughout the drift region. This limits the RF voltage V_{RF} to approximately $0.5V_{DC}$. In the TUNNETT diode, the RF voltage reaches its minimum when the charge pulse has traveled approximately two-thirds of the drift region, therefore,

$$V_{RF} = V_{DC} - V_{min} = (\tfrac{2}{3}\mathscr{E}_D - \mathscr{E}_s)(x_A + x_D) \qquad (44)$$

Table 2 summarizes the important diode parameters for IMPATT and TUNNETT diodes at different frequencies, assuming for GaAs $\mathscr{E}_s = 3$ kV/cm and an effective drift velocity $v_s = 5 \times 10^6$ cm/s.

Now, we know all the parameters for either Eq. 27 or 30 to determine the admittance of the diode as well as for Eq. 7 to estimate the RF power that is generated in the diode. As we have seen from Eqs. 8 and 9, the total series resistance R_s reduces the RF power that is actually delivered to the load. The series resistances of the undepleted regions, the skin effect in the buffer and contact regions of the diode, losses in the package, and the impedance transformation circuit all contribute to the total series resistance R_s. However, at millimeter-wave frequencies, this series resistance is dominated by the resistance of the ohmic contacts on the heavily doped p- and n-type contact regions. For all the calculations in Tables 3 through 8, therefore, this specific series resistance was assumed to be 1.5×10^{-6} Ω-cm^2 as reported for GaAs W-band IMPATT diodes.[16] The results are tabulated for $R_L = 1.0\,\Omega$, $1.5\,\Omega$, and $2.0\,\Omega$. More than 50% of the DC input power is dissipated as heat in the diode, and the heat flow resistance R_{th} is an estimate for the increase in the operating junction temperature above ambient. The analysis of the heat-flow resistances is based on the spreading approximation[17] and assumes the metalization schemes as used in the fabrication of GaAs TUNNETT diodes.[18] Figure 12 shows a comparison of predicted values for GaAs TUNNETT diodes with measured values for GaAs IMPATT diodes of similar metalization schemes. Except for very small TUNNETT diodes at 200 GHz, where the diode diameter approaches the same order of magnitude as the metalization thicknesses, the diodes have to be mounted on diamond heatsinks (thermal conductivity of diamond at 200°C: 11 W/cm-K) to maintain an operating junction temperature below 200°C. Operation on integral heatsinks, that is, on gold-plated copper blocks or studs (thermal conductivity of copper: 3.9 W/cm-K), requires reduced DC input-power levels and optimum RF-power levels. Consequently, high DC-to-RF conversion efficiencies cannot be reached.

So far, we have neglected, in this large-signal analysis, to consider that a negative resistance must be present at V_{DC} to permit oscillations. As seen in Fig. 1, the required impedance levels are higher than under large-signal conditions. We use the small-signal theory[1] for IMPATT diodes to estimate the corresponding current densities. The small-signal negative resistance R_D

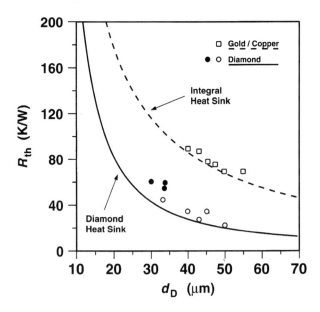

Fig. 12 Predicted heat-flow resistances R_{th} of diodes on diamond and integral heatsinks as a function of the diode diameter. Measured heat-flow resistances of GaAs W-band (●), V-band (○) IMPATT diodes on diamond heatsinks, and GaAs V-band IMPATT diodes on integral heatsinks (□). After Eisele and Haddad, Ref. 18.

TABLE 3   Predicted Capabilities of IMPATT Diodes at a Center Frequency of 33 GHz

R_L (Ω)	d_D (μm)	I_{DC} (mA)	R_s (mΩ)	P_{DC} (W)	P_{RF} (W) Gen.	P_{RF} (W) Load	η (%)	Diamond R_{th} (K/W)	Diamond ΔT (K)	Copper R_{th} (K/W)	Copper ΔT (K)
1.0	124	1359	12.3	28.1	10.2	10.1	35.8	5.85	164	15.1	422
1.5	102	905.7	18.4	18.7	6.78	6.70	35.8	7.57	142	19.5	364
2.0	88.3	679.3	24.5	14.0	5.08	5.03	35.8	9.14	128	22.9	322

is given by

$$R_D = \frac{x_d}{\omega \varepsilon_s A} \frac{1 - [(\cos \Theta_D)/\Theta_D]}{1 - (\omega^2/\omega_a^2)} \tag{45}$$

where ω_a is the avalanche resonance frequency and is given by

$$\omega_a^2 = \frac{3 J_{DC} v_s}{\varepsilon_s} \frac{d\alpha}{d\mathscr{E}} \tag{46}$$

Current densities of 6.8, 12, and 22.4 kA/cm^2 are necessary to obtain at 33, 44, and 60 GHz, respectively, an absolute value of R_D that is 50% higher than the value under large-signal conditions ($\alpha' = 0.25/V$). We now assume that the large-signal current densities as shown in Table 2 are partly superimposed on the DC bias-current densities. Therefore, P_{DC} and the rise in the operating junction temperatures are much higher than predicted in Tables 3–5, and the corresponding DC-to-RF conversion efficiencies decrease. In TUNNETT diodes we can define an equivalent effective avalanche resonance frequency[19]

$$\omega_{a,eff}^2 = \omega_a^2 + \frac{3 v_s}{\varepsilon_s} \frac{dg_T}{d\mathscr{E}} \tag{47}$$

where g_T denotes the tunneling generation rate. The second term of Eq. 47 is dominant in TUNNETT diodes, and the corresponding model for the small-signal impedance[19] predicts a negative resistance at significantly reduced current densities. Therefore, the predictions for TUNNETT diodes at 100, 150, and 200 GHz in Tables 6–8 are less affected than the ones for IMPATT diodes at 33, 44, and 60 GHz.

Still-higher RF power levels are obtained if both electrons and holes are allowed to contribute to the negative resistance of the diode. The schematic layer sequence for such a double-drift structure is illustrated in Fig. 13. The impedance level Z_D and the bias voltage V_{DC} are approximately doubled, and twice the device area can be matched into the same load resistance R_L. At twice the area, the bias current $I_{DC} = A J_{DC}$ is doubled, too, and, theoretically, four times the RF output power of a single-drift structure can be obtained ($P_{RF} = \eta V_{DC} I_{DC}$). In practical diode structures, however, thermal and electrical limits reduce the improvement in RF output power to a factor of two.

In summary, the predictions of Tables 3–8 show that a unified large-signal analysis can be employed as a simple tool to estimate the fundamental power capabilities of transit-time diodes. At higher frequencies, however, the time for the avalanche to form or the transit time through the avalanche region takes up a significant portion of the RF cycle. Additionally, carrier dynamics as well as quantum-mechanical effects in the much shorter tunneling and drift

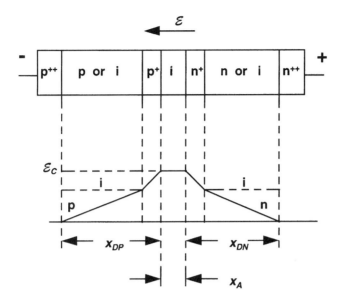

Fig. 13 Schematic layer sequence and electric-field profile for double-drift transit-time diodes.

regions become increasingly important and need to be considered. Therefore, more sophisticated simulation programs are employed as a tool to predict the performance of transit-time diodes.[20] These predictions were found to be in excellent agreement with experimental results for the same diode structures.[21]

6.2.3 Performance of Diodes under CW Operation

The highest power levels from any single-device solid-state oscillator in CW operation were reported from IMPATT diodes. RF power levels (and corresponding DC-to-RF conversion efficiencies) of 2.5 W (10%) at 44 GHz,[13] 2.5 W (13.5%) at 60 GHz,[13] and 1.06 W (10%) at 97 GHz[3] were measured with Si IMPATT diodes. Early work on Si IMPATT diodes already yielded very high RF power levels above 200 GHz, for example, 50 mW at 245 GHz,[12] 12 mW at 255 GHz,[12] 7.5 mW at 285 GHz,[10] and 0.2 mW at 361 GHz.[10] GaAs IMPATT diodes typically show better performance than Si IMPATT diodes below 60 GHz, and RF power levels (and corresponding DC-to-RF conversion efficiencies) of 3.0 W (22%) at 33 GHz,[22] 2.7 W (18.6%) at 44 GHz,[22] and 1.7 W (15%) at 60 GHz[13] were measured. For both Si and GaAs IMPATT diodes, doping profiles (shown schematically in Fig. 14) were designed, optimized, and grown by MBE or VPE to reach the highest RF power levels.

TABLE 4 Predicted Capabilities of IMPATT Diodes at a Center Frequency of 44 GHz

R_L (Ω)	d_D (μm)	I_{DC} (mA)	R_s (mΩ)	P_{DC} (W)	P_{RF} (W) Gen.	P_{RF} (W) Load	η (%)	Diamond R_{th} (K/W)	Diamond ΔT (K)	Copper R_{th} (K/W)	Copper ΔT (K)
1.0	93.2	1037	22.0	17.4	6.32	6.19	35.5	8.51	148	21.6	376
1.5	76.1	691.2	33.0	11.6	4.21	4.13	35.5	11.1	129	27.3	317
2.0	65.9	518.4	44.0	8.72	3.16	3.09	35.5	13.6	118	32.2	281

TABLE 5 Predicted Capabilities of IMPATT Diodes at a Center Frequency of 60 GHz

R_L (Ω)	d_D (μm)	I_{DC} (mA)	R_s (mΩ)	P_{DC} (W)	P_{RF} (W)		η (%)	Diamond		Copper	
					Gen.	Load		R_{th} (K/W)	ΔT (K)	R_{th} (K/W)	ΔT (K)
1.0	67.8	772.8	41.5	10.6	3.83	3.69	34.9	13.0	138	31.2	329
1.5	55.4	515.2	62.3	7.05	2.55	2.46	34.9	17.3	122	39.5	278
2.0	48.0	386	83.1	5.28	1.91	1.84	34.9	21.3	112	46.9	248

TABLE 6 Predicted Capabilities of TUNNETT Diodes at a Center Frequency of 100 GHz

| R_L (Ω) | d_D (μm) | I_{DC} (mA) | R_s (mΩ) | P_{DC} (W) | P_{RF} (mW) | | η (%) | Diamond | | Copper | |
					Gen.	Load		R_{th} (K/W)	ΔT (K)	R_{th} (K/W)	ΔT (K)
1.0	26.4	191.4	274	2.72	320	255	9.35	52.6	143	96.6	262
1.5	21.5	127.6	411	1.82	213	170	9.35	72.7	132	124	226
2.0	18.7	95.71	548	1.36	160	127	9.35	91.8	125	149	203

TABLE 7 Predicted Capabilities of TUNNETT Diodes at a Center Frequency of 150 GHz

R_L (Ω)	d_D (μm)	I_{DC} (mA)	R_s (Ω)	P_{DC} (W)	P_{RF} (mW) Gen.	P_{RF} (mW) Load	η (%)	Diamond R_{th} (K/W)	Diamond ΔT (K)	Copper R_{th} (K/W)	Copper ΔT (K)
1.0	14.8	93.6	0.878	1003	111	61.0	6.08	135	136	203	203
1.5	12.0	62.4	1.32	669	74.0	40.7	6.08	190	127	266	178
2.0	10.4	46.8	1.76	502	55.5	30.5	6.08	243	122	323	162

369

TABLE 8 Predicted Capabilities of TUNNETT Diodes at a Center Frequency of 200 GHz

R_L (Ω)	d_D (μm)	I_{DC} (mA)	R_s (Ω)	P_{DC} (mW)	P_{RF} (W) Gen.	P_{RF} (W) Load	η (%)	Diamond R_{th} (K/W)	Diamond ΔT (K)	Copper R_{th} (K/W)	Copper ΔT (K)
1.0	6.96	28.7	3.94	256	26.9	5.75	2.25	487	125	575	147
1.5	5.69	19.2	5.91	171	17.9	3.83	2.25	694	118	778	133
2.0	4.92	14.4	7.89	128	13.4	2.88	2.25	892	114	970	124

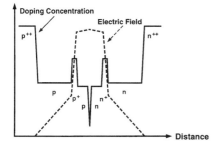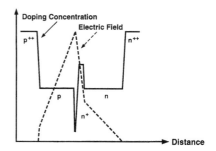

Fig. 14 Schematic doping profiles for high-power, high-efficiency, double-drift IMPATT diodes.

GaAs IMPATT diodes were originally thought to be limited to frequencies below 60 GHz because they showed much lower efficiencies above 60 GHz than at lower frequencies (e.g., 33 GHz). However, experimental results at W-band from a single-drift flat-profile structure with 320 mW (6%) at 95 GHz[16] and at D-band from a double-drift structure with 100 mW (5%) at 144 GHz[23] indicate excellent performance at frequencies above 60 GHz. Si IMPATT diodes, however, show better performance, which can be attributed to higher saturated velocities in the drift regions. These saturated velocities are in the range of 6–8×10^6 cm/s for silicon as compared to 3.5–5×10^6 cm/s for GaAs. GaAs also has ionization rates with a more pronounced tendency to saturate at high electric fields than does silicon.[16,24] As an example, an RF power of 100 mW with a corresponding DC-to-RF conversion efficiency of 5% at 144 GHz was reported from GaAs double-drift Read low–high–low IMPATT diodes[23] as compared to 300 mW with 8.1% at 138.5 GHz and 200 mW with 6.2% at 148 GHz from silicon double-drift Read low–high–low IMPATT diodes.[25] Figure 15 summarizes the state-of-the-art performance of transit-time diodes including BARITT diodes[26,27] and MITATT diodes.[28,29]

6.2.4 Noise

The statistical nature of the avalanche process leads to a strong shot-noise component in the RF current of the IMPATT diode. Therefore, oscillators with IMPATT diodes are generally thought of as noisy, which is also confirmed by experimental results at centimeter-wave frequencies. At millimeter-wave frequencies, however, the avalanche region takes up a larger fraction of the active region and the statistical fluctuations in the injected current are averaged out.[30] Additionally, the avalanche region is no longer swept free of carriers during the RF cycle,[31] and the avalanche process starts from more residual electrons and holes in the avalanche region, resulting in

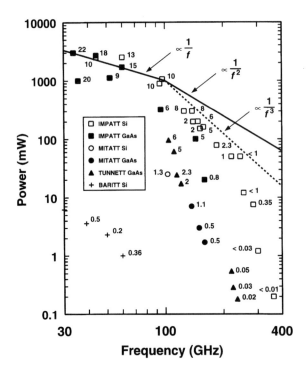

Fig. 15 State-of-the-art RF power levels from transit-time diodes under CW operation in the frequency range 30–400 GHz. Numbers next to the symbols denote DC-to-RF conversion efficiencies (%).

fewer fluctuations. Both effects are expected to lower the noise in the oscillator, and experimental results confirm these predictions. GaAs single-drift flat-profile IMPATT diodes with long effective avalanche regions show FM noise measures M comparable to InP Gunn devices at W-band frequencies.[16] The FM noise measure of Si IMPATT diodes is at least 6–10 dB higher than that of GaAs IMPATT diodes. However, Si IMPATT diodes have considerably smaller 1/f-noise components[32] and, therefore, are well suited for Doppler RADAR applications like collision avoidance in automobiles. Free-running oscillators with TUNNETT diodes have the lowest FM noise of any oscillator made with two-terminal devices, as shown in Fig. 16. The RF power levels of 100 mW at 104 GHz and 64 mW at 111 GHz from GaAs TUNNETT diodes[18] compare favorably with those from GaAs Gunn devices,[33] whereas the DC-to-RF conversion efficiencies, 5.9% at 104 GHz and 5.3% at 111 GHz, are approximately twice as high as those of GaAs Gunn devices. In Fig. 16, we distinguish between two different operating modes: in the small-signal mode, the measured RF power does not exceed 10% of the maximum available RF power, whereas in the

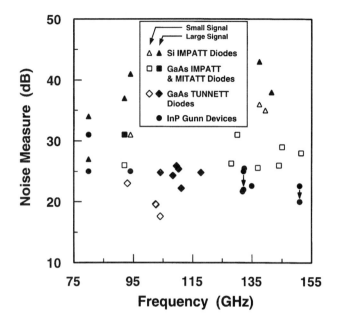

Fig. 16 Comparison of the FM noise measure M in free-running oscillators with different two-terminal devices at millimeter-wave frequencies from 75 to 155 GHz.

large-signal mode at least 80% of the maximum available RF power is measured.

6.2.5 Performance of Diodes under Pulsed Operation

Driving devices with pulses overcomes thermal limits, and significantly higher RF power levels are obtained. Examples for individual Si IMPATT diodes are 28 W at 35 GHz,[13] 42 W (8%) at 96 GHz,[3] and 5.6 W at 140 GHz,[12,13] with short pulses in the order of 100 ns (50–200 ns) duration and a duty cycle in the range 0.25–1.0%. Earlier work[12] resulted in 6.5 W at 130 GHz, 1 W at 217 GHz, and 620 mW at 240 GHz for a 100-ns pulsewidth and a 25-kHz repetition rate. Examples for individual GaAs IMPATT diodes driven with long pulses of 300–500 ns duration are 8.6 W (16%) at 35 GHz,[22] 16 W (15%) at 40 GHz (with 5% duty cycle),[34] and 1.7 W (10%) at 94 GHz.

6.3 RESONANT-TUNNELING DIODES

Anomalous I–V characteristics with a broadband negative resistance have long been known from experience with heavily doped p–n junctions in germanium, GaAs, and others. In 1958, Esaki found that interband tunneling was an explanation for this phenomenon.[35] Some of the first negative-resistance oscillators were built with tunnel diodes, and RF power generation was demonstrated up to millimeter-wave frequencies. In 1974, another tunneling mechanism was demonstrated experimentally,[36] namely, resonant tunneling through a double heterobarrier. This mechanism works as follows. A thin layer (<20 nm) of a semiconductor material with a lower bandgap (e.g., GaAs, InAs, and InGaAs), the so-called quantum well, is sandwiched and confined between two very thin (<10 nm) layers of a semiconductor material with a wider bandgap (e.g., AlGaAs or AlAs, AlSb, and AlAs). Figure 17 shows the layer structure and the current–voltage characteristics of an experimentally investigated resonant-tunneling diode (RTD).[37]

If the thickness of the quantum well approaches the order of the de Broglie wavelength of electrons, electrons in the well are confined to discrete energy levels (E_1, E_2, etc.), as shown in Fig. 18a for thermal equilibrium at a bias voltage $V = 0$. As the applied bias voltage is increased, an accumulation region forms near the barrier at the cathode side and a depletion region forms near the barrier at the anode side. Only a few electrons can tunnel through the double barrier. Once the bias reaches a value where the conduction band's occupied energy states on the cathode side line up with empty states at E_1 in the well, resonance occurs. At this point, many electrons can tunnel through the left barrier into the well and subsequently through the right barrier into unoccupied states in the conduction band of the anode side. The current peak I_p in the current–voltage characteristic as shown in Fig. 18d occurs for

$$V_p > \frac{2E_1}{q} + \Delta V_{da} \tag{48}$$

where ΔV_{da} describes the voltage drop in the accumulation and depletion regions near the barrier.

As V is increased further, the conduction-band edge on the left rises above E_1, and the number of electrons that can tunnel through the barriers is significantly reduced. The valley current I_V at V_V results from excess current components, which increase with bias voltage. Tunneling via an upper valley in the barrier material (e.g., X-valley in AlAs) contributes to this current flow as well as phonon-assisted or impurity-assisted tunneling (inelastic scattering). Typical peak-to-valley current ratios I_p/I_v range from approximately 2 to 24 for different material systems at $T = 300$ K. Resonant tunneling is a very fast quantum mechanical process, and both experimental results[38] and simulations[39] indicate response time constants of less than 0.3 ps. Therefore, we can neglect this time constant in a simplified analysis of the power generating capabilities of RTDs. We also neglect the transit time τ_D through the total depletion region of the diode, which reaches from the n-doped cathode to the n-doped anode region:

$$\tau_D = \frac{W}{v_Q} \tag{49}$$

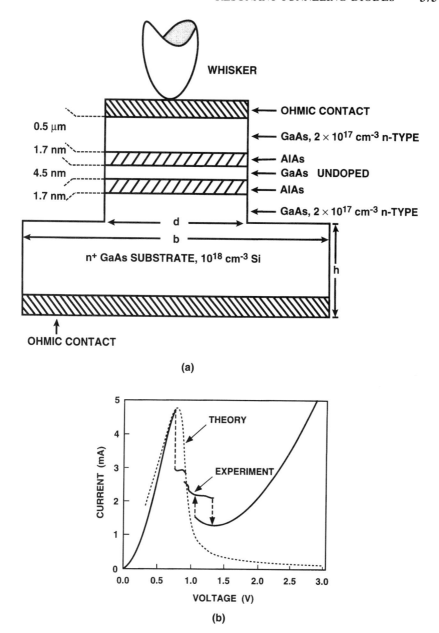

Fig. 17   (a) Cross-section of a mesa-type AlAs/GaAs/AlAs resonant-tunneling diode. (b) Measured and theoretical current–voltage characteristics. (After Brown et al., Ref. 37.)

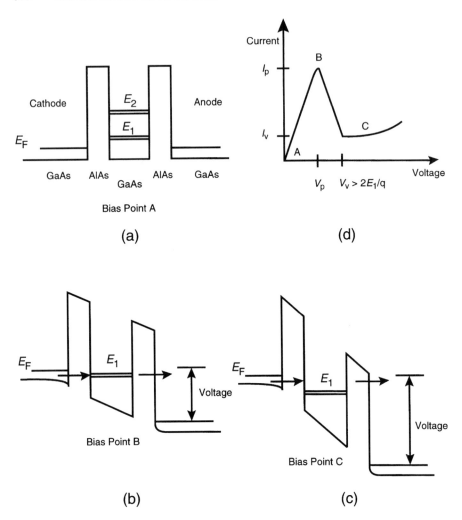

Fig. 18 (a)–(c) Bias-dependent band diagrams at three bias points A, B, and C. (d) Current–voltage characteristic for a GaAs/AlAs double heterobarrier structure.

In contrast to the other two-terminal devices of this chapter, the negative resistance of the RTD is present over a very broad frequency range starting at $f = 0\,\text{Hz}$. Therefore, the oscillator circuit as shown in Fig. 1a needs to be expanded for the bias circuitry, and is illustrated in Fig. 19.

For simplicity, we linearize the current–voltage characteristic for an analysis of the power generating capabilities as shown in Fig. 20, where V_p and V_v denote the peak and valley voltages with the peak and valley current

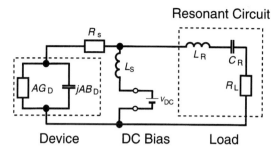

Fig. 19 Simplified equivalent circuit of an oscillator with a resonant-tunneling diode connected to a bias circuit and a load.

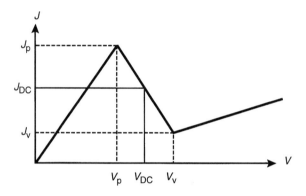

Fig. 20 Linearized current–voltage characteristic of a resonant-tunneling diode.

densities J_p and J_v, respectively. If the diode is biased at $V_{DC} = \frac{1}{2}(V_p + V_v)$, that is, $J_{DC} = \frac{1}{2}(J_p + J_v)$, the maximum RF voltage V_{RF} becomes

$$V_{RF} = \frac{1}{2}\Delta V = \frac{1}{2}(V_v - V_p) \tag{50}$$

The large-signal conductance per unit area G_D is determined by

$$G_D = \frac{J_v - J_p}{V_v - V_p} \tag{51}$$

whereas the diode susceptance per unit area B_D is represented by the capacitance per unit area C_D of the depleted region

$$B_D = \omega C_D = \omega \frac{\varepsilon_s}{W} \tag{52}$$

We now apply Eqs. 9 and 8 to determine the diode area A and RF output power P_{RF}, respectively. At low frequencies P_{RF} approaches the limit

$$P_{RF} = \tfrac{1}{8}A(J_p - J_v)(V_v - V_p) \tag{53}$$

with a corresponding DC-to-RF conversion efficiency η,

$$\eta = \frac{P_{RF}}{P_{DC}} = \frac{1}{2}\frac{(J_p - J_v)(V_v - V_p)}{(J_p + J_v)(V_v + V_p)} = \frac{1}{2}\frac{\left(\dfrac{J_p}{J_v} - 1\right)\left(\dfrac{V_v}{V_p} - 1\right)}{\left(\dfrac{J_p}{J_v} + 1\right)\left(\dfrac{V_v}{V_p} + 1\right)} \tag{54}$$

Examples for $V_v/V_p = 2$, $J_p/J_v = 4$ and $V_v/V_p = 2$, $J_p/J_v = 19$ are $\eta = 10\%$ and $\eta = 15\%$, respectively. So far, we only have considered how the RF circuit with the load resistance R_L limits the RF power. We know from Esaki tunnel diodes that the inherent broadband negative resistance gives rise to bias-circuit oscillations. As we will see from the following analysis, these instabilities further limit the RF output power of RTDs. Since current–voltage characteristics are quite similar to tunnel diodes, we can apply Hines's analysis for short-circuit stability.[40] This analysis postulates

$$\frac{L_s(A^2 G_D^2)}{(AC_D)} < -R_s(AG_D) < 1 \tag{55}$$

for the bias circuit configuration of Fig. 19. The first inequality of Eq. 55 reduces to a limit for the diode area A

$$A < -\frac{\rho_s C_D}{L_s G_D} \tag{56}$$

This area A is much smaller, and rearranging Eq. 9 shows that a much larger load resistance R_L than the previously assumed $1\,\Omega$ must be presented to the diode to fulfill the oscillation condition of Eq. 5,

$$R_L = \frac{1}{A}\left(\frac{-G_D}{G_D^2 + B_D^2} - \rho_s\right) \tag{57}$$

As a consequence, this stability condition severely limits P_{RF} in Eq. 8 to

$$P_{RF}(R_L) = -\tfrac{1}{2}A[G_D + \rho_s(G_D^2 + B_D^2)]V_{RF}^2 \tag{58}$$

The current–voltage characteristics of the experimentally investigated AlAs/GaAs/AlAs RTD as shown in Fig. 17 can be approximated by $\Delta V = 0.5\,V$, $J_p = 40\,kA/cm^2$, $J_p/J_v = 3.5$, and $W = 70\,nm$. Using these parameters, we can compare RF power levels and diode areas for the two

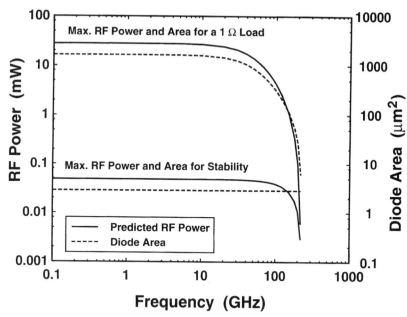

Fig. 21 Top curves: predicted RF output power and diode area for matching into $R_L = 1\ \Omega$. Bottom curves: predicted RF output power and diode area for obtaining stability with $L_s = 0.1$ nH. Linearized current–voltage characteristics for the diode of Fig. 17 were assumed.

cases illustrated in Fig. 21: (a) where only the RF circuit limits P_{RF}, and bias oscillations may occur, and (b) where stability in the bias circuit limits P_{RF}, and bias oscillations cannot occur. In both cases, $\rho_s = 1 \times 10^{-6}\ \Omega\text{cm}^2$ was assumed to account for losses mainly in the ohmic contacts.

Figure 22 summarizes the best experimental results that were obtained from RTDs in the AlAs/GaAs/AlAs, AlAs/InGaAs/AlAs, and AlSb/InAs/AlSb material systems.[37,41–43] As indicated in Fig. 17a, RTDs in oscillator circuits up to 712 GHz are generally unpackaged and contacted with whisker contacts similar to Schottky mixer and varactor diodes.[2] Therefore, these oscillator circuits differ from the ones illustrated in Fig. 5. Measured RF power levels from AlAs/GaAs/AlAs RTDs in the frequency range 30–200 GHz agree well with the predictions of Fig. 21 when stability in the bias circuit is maintained. The highest oscillation frequencies, up to 712 GHz, were reported from the AlSb/InAs/AlSb material system.[43] An RTD was successfully employed as the local oscillator in a submillimeter-wave receiver with a SIS (superconductor–insulator–superconductor) junction,[44] and promising approaches to improving the stability of the bias circuit were reported.[45] However, because of the small voltage swing ΔV, RF power levels from RTDs are low compared to the requirements in system applications,

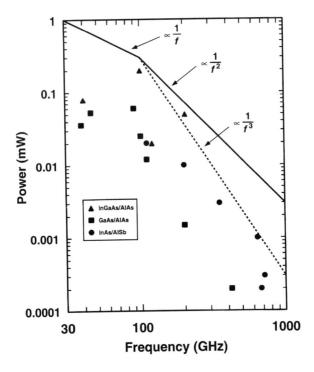

Fig. 22   State-of-the-art RF power levels from resonant-tunneling diodes in the frequency range 30–1000 GHz.

and several attempts at power combining were reported. As an example, 25 parallel-connected RTDs yielded the RF output power of 5 mW at an oscillation frequency of 1.18 GHz.[46] On the other hand, the small voltage swing ΔV makes the RTD a prime candidate for high-speed high-complexity logic circuits, where one RTD in a multiple-state circuit can replace several conventional transistors.[47]

6.4   TRANSFERRED-ELECTRON DEVICES

6.4.1   Principles of Operation

Transferred-electron devices (TEDs) utilize specific bulk-material properties of certain semiconductors. They are unipolar devices and, generally, do not exhibit the distinctive diode characteristic of p–n junctions as seen in IMPATT and TUNNETT diodes. TEDs in oscillator applications are characterized by low noise and medium RF output power. Therefore, they are well suited for local oscillators in receivers and transmitters. As bulk devices, TEDs require a particular band structure of the semiconductor material, which can be found in several materials, mainly in the groups of III–V and II–VI compound semiconductors (see Ref. 1, p. 648 for a list). These semiconductors have more than one valley in the conduction band and meet the following criteria, which were proposed independently by Ridley and Watkins[48] as well as by Hilsum[49] (RWH):

1. At least two valleys must be present in the conduction band.
2. The minimum (minima) of the upper valley(s) must be several times the thermal energy of electrons above the minimum of the lowest (= main) valley in the conduction band for electrons to initially reside in the lowest valley.
3. The energy difference (ΔE) between the minimum (minima) of the upper valley(s) and the minimum of the main valley in the conduction band must be less than the energy bandgap E_g to avoid the onset of significant impact ionization in such a device.
4. The transfer from one conduction band valley to another must require much less time than one period of the intended operation frequency.
5. The effective masses and densities of states in the upper valley(s) must be considerably higher than in the main valley. As a consequence of the higher effective masses, mobilities in the upper valley(s) must be much lower than in the main valley.

In a simplified band structure shown in Fig. 23, electrons initially reside in the main valley of the conduction band. When electrons acquire more energy, for example, by applying an electric field $\mathscr{E}$, most of them still remain in the main valley if $\mathscr{E} < \mathscr{E}_h$, where $\mathscr{E}_h$ is referred to as the threshold electric field. When electrons acquire even more energy (for $\mathscr{E} > \mathscr{E}_h$), many of them are scattered ("transferred") into the upper valley. A higher effective mass in the upper valley reduces the mobility (see Ref. 1), and the average drift velocity of electrons v decreases, as shown in Fig. 24. At large energies, that is, high electric fields ($\mathscr{E} >> \mathscr{E}_h$), most of the electrons are transferred to the upper valley, where they have a higher effective mass and lower mobility. After reaching the minimum value, the average drift velocity again increases for higher electric fields. The decrease in the average drift velocity for $\mathscr{E} > \mathscr{E}_h$ generates a region of negative differential mobility. If electrons in the upper valley reach a region where the electric field $\mathscr{E}$ drops below $\mathscr{E}_h$, they lose energy and are scattered back to the main valley.

Gunn was the first to observe experimentally current oscillations in bulk GaAs and InP, which were subsequently explained by this transferred-electron effect,[50,51] and, therefore, the name Gunn device quickly became common for this type of device. Out of more than ten semiconductor materials known for the transferred-electron effect, only GaAs and InP have

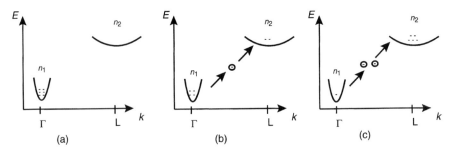

Fig. 23 Simplified energy-band diagram for a direct two-valley semiconductor showing electron transfer for (a) $\mathscr{E} < \mathscr{E}_h$, (b) $\mathscr{E} > \mathscr{E}_h$, and (c) $\mathscr{E} >> \mathscr{E}_h$.

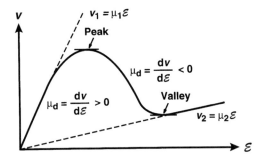

Fig. 24 Velocity-electric-field profile for the two-valley semiconductor of Fig. 23.

found widespread use in system applications. GaAs and InP have three valleys in the conduction band, and at the doping concentrations required for operation at millimeter-wave frequencies, high-field mobilities are considerably lower than the low-field mobilities. As a consequence, the electron drift velocity v monotonically decreases for electric fields above $\mathscr{E}_{th}$. Figure 25 shows simplified band-structure diagrams for GaAs and InP, and Fig. 26, their velocity-electric-field profiles. Table 9[52–54] summarizes the relevant material characteristics of GaAs and InP. The finite time it takes for electrons to gain or lose energy in an electric field accounts for a fundamental physical frequency limit in these devices.

Except for some rare conditions in devices at lower microwave frequencies, the bulk negative differential mobility alone does not cause a static negative differential resistance to be used for RF power generation as, for example, seen in resonant-tunneling diodes (see Fig. 17b). A mechanism based on the negative differential mobility causes a dynamic negative resistance as shown next. In a region of bulk semiconductor material under uniform conditions (doping concentration N_D, electric field $\mathscr{E}$, average differential mobility $\bar{\mu}_d$), any space-charge inhomogeneity $Q(x,t)$ traveling at the velocity v grows or decays following an exponential law that can be derived from Maxwell's

TABLE 9   Semiconductor Material Characteristics Relevant to GaAs and InP TEDs (at a temperature of 300 K unless noted otherwise)

Property	Semiconductor	
	GaAs	InP
Energy gap (eV)	1.42	1.34
Low-field mobility (at 500 K) (cm^2/V-s)	5000	3000
Thermal conductivity (W/cm-K)	0.46	0.68
Velocity peak-to-valley ratio	2.2	3.5
Threshold field $\mathscr{E}_{th}$ (kV/cm)	3.5	10.5
Breakdown field (at $N_D = 10^{16}$/cm^3) (kV/cm)	400	500
Effective transit velocity v_T (cm/s)	0.7×10^7	1.2×10^7
Temperature dependence of v_T (K^{-1})	0.0015	0.001
Diffusion coefficient–mobility ratio at $2\mathscr{E}_{th}$ (cm^2/s)	72	142
Energy relaxation time due to collisions (ps)	0.4–0.6	0.2–0.3
Intervalley relaxation time (ps)	—	0.25
Acceleration–deceleration time (ps) (Inertial energy time constant)	1.5	0.75

After Wandinger, Ref. 52; Fank et al., Ref. 53; Eddison, Ref. 54.

equations:

$$Q(x, t) = Q(x - vt, 0) \exp\left(-\frac{t}{\tau}\right) \qquad (59)$$

where

$$\tau = \frac{\varepsilon_s}{\sigma} = \frac{\varepsilon_s}{|q| N_D \bar{\mu}_d} \qquad (60)$$

and

$$\bar{\mu}_d = \frac{dv}{d\mathscr{E}} \qquad (61)$$

At low electric fields $\mathscr{E}$, where $\bar{\mu}_d > 0$, the charge inhomogeneity decays with $\tau = \tau_D$, the dielectric relaxation time, and at higher electric fields, where $\bar{\mu}_d < 0$, a charge inhomogeneity can grow. This charge inhomogeneity reaches a significant level only if the growth factor for the maximum traveled distance l, the device length, is very large. Therefore the condition

$$\frac{l}{v\tau} = \frac{l|q|N_D\bar{\mu}_d}{\varepsilon_s v} > 1 \qquad (62)$$

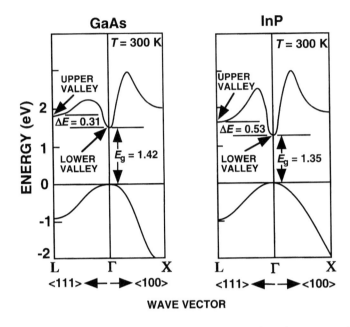

Fig. 25 Simplified band diagram for the three-valley semiconductor materials GaAs and InP.

must be satisfied, which corresponds to

$$N_D l > 1 \times 10^{12}/cm^2 \qquad (63)$$

for both GaAs and InP.

Typical Gunn devices at millimeter wave frequencies have $N_D l$ products between $1 \times 10^{12}/cm^2$ and $3 \times 10^{12}/cm^2$, and doping concentrations N_D in the active region exceed $10^{15}/cm^3$. For a doping concentration of $N_D > 10^{15}/cm^3$, space-charge inhomogeneities typically grow into so-called dipole domains, where accumulation and depletion layers are lumped together. Figure 27 shows the carrier distribution and electric-field profile for such a dipole domain under uniform conditions. Electrons in the low-field region travel at a constant v_T for a constant electric field $\mathscr{E}_1$. Electrons in region "a" are accelerated by the higher electric field until they reach region "b," where they are transferred to the upper valley and slow down to be trapped in this accumulation region. Electrons in region "c" lose energy and are transferred back to the lower valley. Their average velocity now is higher than the average velocity in region "b," thus region "c" is depleted of electrons. After a domain forms at the cathode, grows, and propagates through the active region, the voltage drop across the domain increases and, under a constant bias voltage, lowers the voltage drop outside the domain. This voltage drop

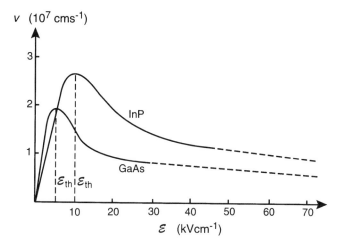

Fig. 26 Velocity–field profile for the three-valley semiconductor materials GaAs and InP.

is equivalent to a reduction in the electric field $\mathscr{E}_1$ outside the domain and generally prevents formation of new domains in the active region. It also limits the growth of the existing domain because fewer electrons are trapped in the accumulation layer or escape the depletion layer. Domains reaching the anode collapse and induce a current flow in the external circuit. The phase difference between current and voltage causes a dynamic negative resistance and generates RF power in an appropriate circuit.

Distinctive modes of operation have been investigated and described for transferred-electron devices at microwave frequencies.[55] However, as will be shown next, at millimeter-wave frequencies, finite intervalley transfer times and domain formation times reach a significant fraction of the RF cycle. In such a case domains form, grow, and suppress formation of new domains, but may never reach the stable state before they reach the anode as described above. Therefore, modes get blurred and devices generally operate in a near transit-time mode, where the operating frequency f_{op} is given by

$$f_{op} = \frac{v_T}{l} \tag{64}$$

The effective transit velocity $v_T = v(\mathscr{E}_1) = v_D(\mathscr{E}_h)$ can be determined from Butcher's equal-area rule,[56]

$$\int_{\mathscr{E}_1}^{\mathscr{E}_h} [v(\mathscr{E}) - v_D]\, d\mathscr{E} = 0 \tag{65}$$

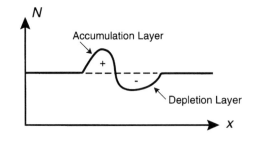

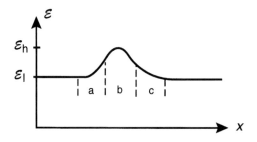

Fig. 27   Carrier concentrations and electric-field profile for a dipole domain.

for a constant diffusion coefficient throughout the active region (illustrated in Fig. 28).

If the operating frequency f_{op} differs somewhat from Eq. 64, the domain reaches the anode prematurely or is delayed. Similar to the operation of transit-time diodes, the current pulse from the collapsing domain still causes a negative resistance and generates RF power. Therefore, operation over a broad bandwidth can be achieved. Additionally, higher bias voltages increase electric fields in the device, and higher electric fields reduce the domain velocity v_D, as seen in Fig. 28. Higher electric fields also reduce the dead space, thus increasing the portion of the active region where domains travel.

Figure 29 gives an overview of typical structures and schematic doping profiles for transferred-electron devices that yielded excellent RF performance. The three-zone flat-doping and the two-zone flat-doping structures were the first to be exploited because they are easy to grow in more classical growth systems, such as liquid-phase or vapor-phase epitaxy (LPE or VPE). The three-zone flat-doping structure[57] consists of the n^--doped active region sandwiched between the highly doped n^+ regions for the ohmic contacts. Since low-ohmic alloyed contacts can be formed on n-type GaAs and InP, the highly doped region on the cathode side can be omitted and just a

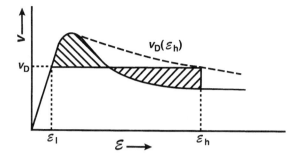

Fig. 28 Equal-area rule for transferred-electron devices.

two-zone, flat-doping structure needs to be grown by VPE or LPE. The advent of advanced growth techniques such as MOCVD, metalorganic molecular-beam epitaxy (MOMBE), and CBE allowed more complicated structures to be grown. Using these growth techniques, graded doping profiles and heterojunction barriers, as shown in Fig. 29, can be incorporated into the device structures and tailored suitably to optimize device perfor- mance at a particular frequency or to extend the frequency limit of transferred-electron devices. Computer simulations revealed that in a three-zone, flat-doping structure, "cold" electrons at low energies entering the active region from the contact zone at the cathode require some time to acquire enough energy to transfer to the upper valley. The results of such Monte Carlo simulations[6] at a frequency of 95 GHz are illustrated in Fig. 30 for a three-zone flat-doping structure in InP with a doping of $1 \times 10^{16}/\text{cm}^3$ in the active region. The finite energy relaxation times, which are shown in Fig. 31 as a function of the electron energy in GaAs and InP, create a huge "dead space" at the beginning of the 1.7-μm-long active region. The resistance of the device $R(x)$, that is, the real part of $Z(x)$, as a function of the position x,

$$Z(x) = R(x) + jX(x) = \frac{\displaystyle\int_0^x \mathscr{E}(x')\,dx'}{\dfrac{A}{l}\displaystyle\int_0^x J(x')\,dx'} \qquad (66)$$

remains positive for a large fraction of the active region and contributes to losses in this region, whereas a negative resistance contributes to the RF power generation only for a small fraction of the active region. Additionally, the peak electric field occurs near the anode, and at a high DC bias the electric field may reach values for the onset of avalanche breakdown. The energy-dependent energy relaxation times of Fig. 31 lead to effective

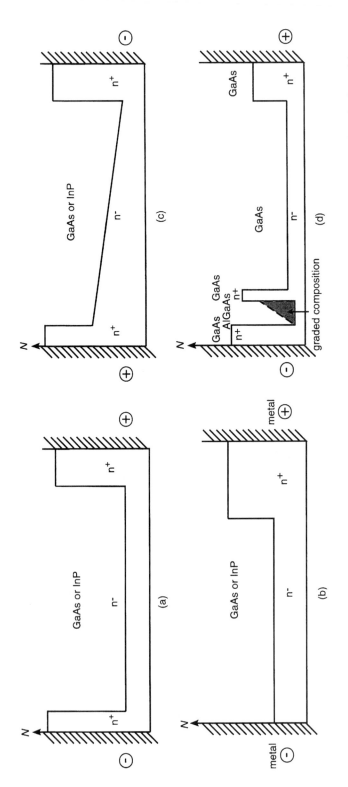

Fig. 29 Different device structures for transferred-electron devices. (a) Three-zone flat doping. (b) Two-zone flat doping. (c) Three-zone graded doping. (d) Heterojunction-barrier cathode.

388

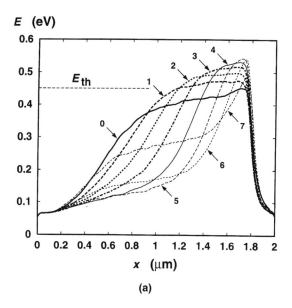

(a)

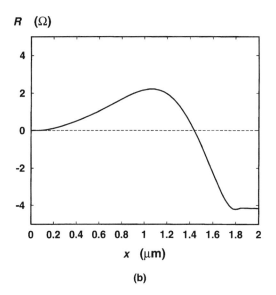

(b)

Fig. 30 Evolution of average electron energy E (a) and diode resistance R (b) as a function of position x (active region from 0.1 to 1.8 μm): $f = 95$ GHz, $V_{rf} = 1.0$ V, $V_{bias} = 5.0$ V, $I_{bias} = 474$ mA, $T = 500$ K. The graphs in part (a) show the electron energy profile at $\omega t = n\pi/8$, $n = 0 \ldots 7$, during one RF cycle.

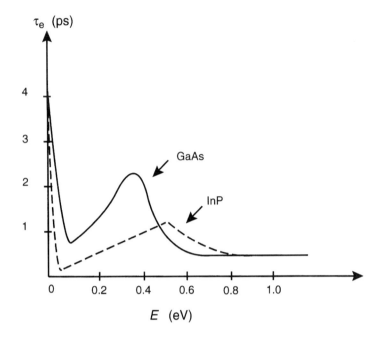

Fig. 31 Energy relaxation times in GaAs and InP as a function of the electron energy. (After Rolland et al., Ref. 24.)

transfer-time constants, as shown in Table 9 for GaAs and InP. Fundamental frequency limits of 100 GHz for GaAs and 200 GHz for InP transferred-electron devices are estimated from these effective transfer-time constants. In the following, we will discuss some solutions that help reduce the dead-space region or extend the useful frequency range close to or even beyond these fundamental frequency limits.

6.4.2 Ohmic Cathode Contacts

Transferred-electron devices are typically operated in a full-height waveguide cavity with a resonant cap on top of the device package. This configuration is shown in Fig. 5e. Modifications of this configuration include the use of a reduced-height waveguide or a mechanism for adjusting the position of the resonant cap and device package with respect to the bottom of the waveguide. Fundamental-mode operation of Gunn devices in a reduced-height post-coupled waveguide cavity was reported up to millimeter-wave frequencies, for example, for a GaAs Gunn device at 84 GHz.[58] RF power levels (and corresponding DC-to-RF conversion efficiencies) of 420 mW (6%) at 35 GHz,[13] 280 mW at 45 GHz,[13] 150 mW at 60 GHz, and 110 mW (2.8%) at 70 GHz[33] were reported from GaAs Gunn devices in fundamental mode. The frequency limit is characterized by a sharp decline in the DC-to-RF conversion efficiency of devices operating in the fundamental mode.

However, this frequency limit of transferred-electron devices can be extended by the extraction of higher harmonics from this inherently nonlinear device. Second-harmonic power extraction was proven most successful in a slightly modified version of a resonant-cap, full-height waveguide cavity. The size of the waveguide is appropriate for the second-harmonic frequency, but impedes propagation at the fundamental frequency. If in such a circuit the fringe capacitance of the cap and the device capacitance together resonate with the inductance of the bias post (see Fig. 5e) at half the output frequency, this signal cannot propagate, and the device is mainly reactively terminated at the fundamental frequency. This reactive termination causes a large voltage swing in the device and, as a result, strong nonlinear operation. A cap of appropriate size together with the coaxial post provides impedance matching into the waveguide at the second harmonic, and a back short at one side of the cavity provides power tuning . Higher frequency stability than in fundamental-mode operation is observed because the resonant circuit at the fundamental frequency is highly decoupled from the load at the second-harmonic frequency. However, more complicated circuits with precise mechanical dimensions are necessary if wide-range frequency tuning is to be implemented. As an example, for second-harmonic power extraction, RF power levels (and DC-to-RF conversion efficiencies) of 123 mW (3.1%) at 83 GHz and 96 mW (2.7%) at 94 GHz were measured with GaAs Gunn devices.[33]

The higher threshold electric field of 10.5 kV/cm in InP (see Table 9) requires higher bias voltages than those for GaAs devices of the same length. Therefore, RF power levels are thermally limited at lower frequencies, where long active regions need to be used. The advantages of InP can be seen clearly at millimeter-wave frequencies with short active regions, where lower inertial energy time constants lead to a higher fundamental frequency limit of approximately 200 GHz. In addition to the higher frequency limit, a larger valley separation of 0.53 eV in InP (see Fig. 25) reduces the temperature dependence of the transfer mechanism as well as the temperature dependence of the effective transit velocity (see Table 9). As a result, the DC-to-RF conversion efficiencies and oscillation frequencies are less temperature dependent in InP Gunn devices. RF power levels (and corresponding DC-to-RF conversion efficiencies) of 200 mW (5%) at 80 GHz and 150 mW (3.5%) at 94 GHz[59] in fundamental mode as well as 7 mW at 180 GHz and 3.2 mW at 206 GHz[60] in second-harmonic mode were reported from similar InP Gunn devices on integral heatsinks.

6.4.3 Current-Limiting Cathode Contacts

A partially annealed ohmic contact significantly reduces the typical Schottky barrier height of metals on the semiconductors GaAs and InP (0.6–0.9 eV), but still leaves a small barrier (< 200 meV). If such a contact is formed on the cathode side of the two-zone structure (see Fig. 29b) and is reverse biased, this barrier causes a high-field region at the cathode contact of the device under bias. Electrons that are injected over this barrier into the active region have a higher energy and, under this high electric field, transfer faster into the upper valleys. This faster transfer reduces the dead space. The shallow Schottky barrier also limits the current flow into the active region at the cathode. Thermionic emission and thermionic–field emission contribute to the current flow, and, in this case, the current density as a function of the voltage V_c across the barrier can be approximated by

$$J_c(V_c) = J_r\left[\exp\left(-\frac{qV_c}{nkT}\right) - \exp\left(\frac{(1-n)qV_c}{kT}\right)\right] \qquad (67)$$

where $J_r = A^*T^2\exp(-\phi_{Bn}/kT)$ (see Ref. 1). Current limiting as a boundary condition at the cathode causes the electrons in the active region of the device to approximate the current valley condition[24,61] for a saturated electron velocity v_s,

$$\overline{J(t)} = J_0 = nqv_s \qquad (68)$$

with large space-charge waves superimposed on an almost constant electric field throughout the active region.[24] This mode of operation yields very high RF power levels in the fundamental mode as well as the second-harmonic mode. Corresponding DC-to-RF conversion efficiencies are typically the highest reported to date. RF power levels (and DC-to-RF conversion efficiencies) of more than 500 mW (15%) at 35 GHz, more than 350 mW (13%) at 44 GHz, and 380 mW (10.6%) at 57 GHz[13,62] in fundamental mode as well as 175 mW (7%) at 94 GHz and 65 mW (2.6%) at 138 GHz[63] in second-harmonic mode were achieved using this technology. These devices are on integral heatsinks, and still higher RF power levels are expected from devices on diamond heatsinks. Devices on integral heatsinks with very high RF power levels are also commercially available.

As we can see from Eq. 67, a strong temperature dependence is inherent in the current flow through a shallow Schottky barrier in the reverse direction.[1] However, the high-efficiency mode reduces DC input requirements and provides higher impedance levels. As a consequence, larger device diameters can be used, which have lower thermal resistances and, therefore, the active-layer temperature is also lower. The typical temperature increase remains below 100 K at the maximum RF output power,[62,64] and the low operating active-layer temperatures ensure reliability and excellent temperature stability over wide temperature ranges of $-30°C$ to $+70°C$ for devices at 56 GHz[62] and 94 GHz[62,63] as well as of $0°C$ to $50°C$ for devices at

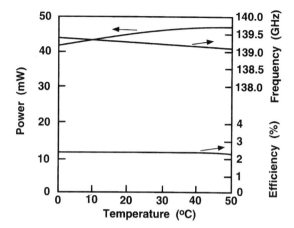

Fig. 32 Measured RF performance of a D-band InP Gunn device as a function of the ambient temperature. (After Crowley et al., Ref. 64.)

140 GHz.[64] The temperature-dependent performance of a D-band Gunn device is shown as an example in Fig. 32.

6.4.4 Graded Active Region

A doping profile with a lower doping concentration N_D at the cathode and a linear grading toward a higher doping concentration at the anode significantly decreases the peak electric field near the anode and increases the electric field near the cathode.[6] Both effects are beneficial to the device operation, and enhance the DC-to-RF conversion efficiency and RF output power of the device. A lower electric field near the anode reduces the power dissipation in this region and allows higher bias voltages without the onset of impact ionization and avalanche breakdown. A higher electric field near the cathode causes a larger fraction of the electrons to transfer to the upper valleys over a shorter distance, which is equivalent to a shorter dead-space region. A higher fraction of electrons in the upper valleys lowers the average electron velocity throughout the active region and, as a consequence, the graded-profile structure operates at lower current densities compared to a flat-profile structure of a similar doping concentration. A lower average electron velocity and shorter dead space actually decrease the optimum operating frequency for the same device length. However, more efficient operation extends the upper frequency limit and allows shorter active regions. Structures with graded doping profiles were investigated in InP and GaAs.

In the GaAs material system, RF power levels (and DC-to-RF conversion efficiencies) of 345 mW (6.8%) at 31.2 GHz and 325 mW (6.6%) at 34.9 GHz

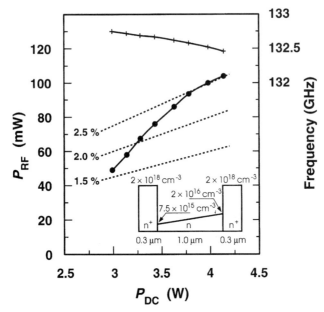

Fig. 33 Bias-dependent RF characteristics of a D-band InP Gunn device (●: output power; +: oscillation frequency; - - -: lines of constant efficiency). (After Eisele and Haddad, Ref. 67.)

from devices on integral heatsinks[65] and 116 mW (4.5%) from devices on diamond heatsinks[66] were reported. The devices on diamond heatsinks exhibited very low operating active-layer temperatures below 150°C and showed fundamental-mode operation up to 84 GHz (33 mW and 1.7%).[66] A properly designed[6] graded doping profile in InP yielded the highest RF power levels reported for any Gunn devices to date. Fundamental-mode operation was demonstrated up to 163 GHz, with RF power levels of 185 mW at 102 GHz, 134 mW at 131.5 GHz, and 62 mW at 151 GHz, and the best DC-to-RF conversion efficiencies up to 2.5% around 132 GHz were obtained from devices on diamond heatsinks.[67] In the second-harmonic mode, an RF output power of more than 0.3 mW was measured at a frequency of 283 GHz.[7] As illustrated in Fig. 33, the InP Gunn devices on diamond heatsinks allow single-mode operation over a wide range of DC input power levels. Excellent tuning behavior was also observed, which can be expected from operation in the fundamental mode. This tuning behavior over a range of more than 4.5 GHz is shown in Fig. 34 for the device of Fig. 33 near maximum bias. The oscillation frequency changes almost linearly with the position of the back short, which is the only tuning element in this full-height waveguide resonant-cap cavity (see Fig. 5e for a schematic).

The graded doping profile improves the DC-to-RF conversion efficiencies, and, similar to a device with a current-limiting cathode contact, the reduced

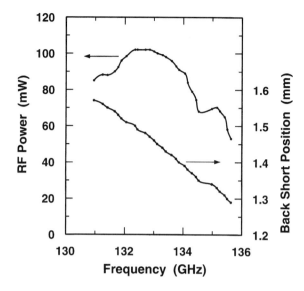

Fig. 34 Mechanical tuning characteristic for the D-band InP Gunn device of Fig. 33 close to maximum applied bias.

DC input power leads to a lower operating active-layer temperature, in particular on a diamond heatsink.[66,67]

6.4.5 Injection over a Homo- or Heterojunction Barrier

Injection of hot electrons over a barrier reduces the dead space in the active region. Several concepts, for example, planar-doped barrier, camel cathode, and heterojunction barriers, have been investigated in GaAs to improve efficiency, but also to eliminate cold-start problems in GaAs Gunn devices. Injection over a heterojunction barrier was proven to be the most successful and was experimentally investigated in the AlGaAs/GaAs system with improved efficiencies and upper frequency limit.[58,68] Figure 35 shows the band diagram of isotype heterojunctions in the lattice-matched GaAs/Al$_x$Ga$_{1-x}$As and InP/In$_x$Ga$_{1-x}$As$_y$P$_{1-y}$ material systems. In both material systems, layers can be grown lattice matched over a wide composition range and, thus, bandgap or conduction-band offset can be suitably tailored.

Linear composition grading from the GaAs to the wide bandgap AlGaAs eliminates the first barrier, and the doping spike as shown in Fig. 29d at the beginning of the active region (GaAs) reduces or eliminates the notch at the interface from the AlGaAs layer to the GaAs region. At the proper bias level, "hot" electrons are now ballistically launched into the active region from an

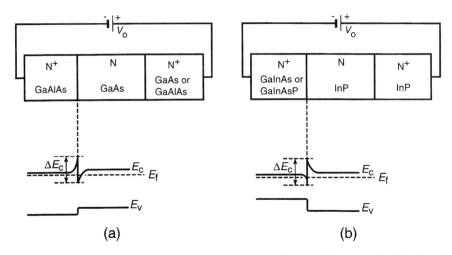

Fig. 35 Band diagram of isotype heterojunctions in GaAs/AlGaAs and InP/InGaAsP at zero bias. (After Friscourt et al., Ref. 61.)

approximately 200-meV-high step in the conduction band. The optimization of such a design requires advanced simulation schemes such as ensemble Monte Carlo techniques. Using a reduced-height post-coupled waveguide cavity (see Fig. 5b for a schematic) for the oscillator, an RF output power of 71 mW with a corresponding DC-to-RF conversion efficiency of 2.8% was measured in the fundamental mode at 77.6 GHz. Low-noise operation of such a Gunn device was achieved at least up to 84 GHz.[58] As a characteristic of fundamental-mode operation,[67] a wide tuning bandwidth of more than 6 GHz was observed by simply adjusting the position of the back short.[58]

In contrast to the GaAs/AlGaAs material system, lattice-matched In-GaAsP has a smaller bandgap than InP, as illustrated in Fig. 35. Therefore, electrons cannot be launched ballistically into the active region. However, a wide composition range in the InGaAsP material system can be lattice matched to InP. This wide composition range allows a wide range in the bandgap and conduction-band offset to be implemented. As a consequence, the proper current-limiting injection at the cathode can be designed. Theoretical investigations predict a significant improvement in DC-to-RF conversion efficiencies at W-band and D-band frequencies[61,69] while preserving the higher frequency limit of InP compared to GaAs.

We now have seen different ways for electrons to be "accelerated" and to be transferred faster into the upper valleys. Shorter transfer times reduce the dead space, increase the efficiency, and allow for shorter active regions to achieve higher operating frequencies. However, a reduced dead space increases the transit time and actually lowers the optimum operating frequency for an active region of the same length. The finite transfer times still

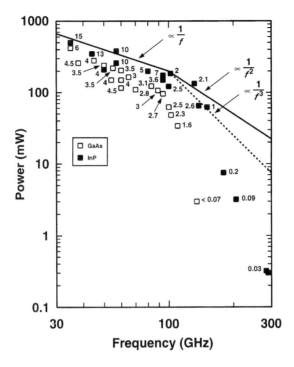

Fig. 36 State-of-the-art RF power levels from transferred-electron devices under CW operation in the frequency range 30–300 GHz. Numbers next to the symbols denote DC-to-RF conversion efficiencies (%).

impose a physical upper frequency limit. Additionally, if electrons gain too much energy and/or the active region is too short, the long energy relaxation time prevents the electrons from losing enough energy to transfer back to the lower valley. No domains can form in such a structure, and the dynamic resistance between the two terminals remains positive for all frequencies.

Figure 36 summarizes the best experimental results from GaAs and InP Gunn devices in the frequency range 30–300 GHz.

6.4.6 Noise

Thermal noise of electrons is the dominant effect in transferred-electron devices. If a TED is designed, using subcritical $N_D l < 1 \times 10^{12}/\text{cm}^2$, for and operated in the amplifier mode,[1] the small-signal noise measure M approaches the asymptotic limit M_0,

$$M_0 = \frac{qD}{k|\bar{\mu}_d| T_0} \tag{69}$$

where the ratio of the diffusion coefficient D and the differential mobility $|\bar{\mu}_d|$ is the crucial factor. InP shows an advantage of 142/72 (approx. 3 dB, see Table 9) over GaAs. If we introduce an equivalent differential mobility μ_{eff} for the oscillator mode (critical $N_D l > 1 \times 10^{12}/\text{cm}^2$) with the conductance per unit area G_D,

$$\mu_{eff} = \frac{|G_D|}{qN_D} \qquad (70)$$

we can define the large-signal noise measure M as

$$M = \frac{qD}{k\mu_{eff}T_0} \qquad (71)$$

The noise performance of Gunn devices near the carrier is dominated by flicker noise components with typical corner frequencies in the range 100 kHz to 1 MHz. As predicted by Eq. 10, the phase noise decreases -20 dB per decade at higher off-carrier frequencies. Table 10 summarizes typical results from GaAs and InP Gunn devices.

Although Eq. 71 predicts lower M for oscillators with InP Gunn devices, experimental results indicate little difference between the two. Figure 16 compares the FM noise measure M of various two-terminal devices in the frequency range 75–155 GHz. As illustrated in that figure, the large-signal FM noise measure of both GaAs and InP Gunn devices typically remains below 25 dB.[31,54,67,70] Some InP devices with current-limiting contacts show excess flicker-noise components near the carrier frequency. Oscillators with Gunn devices in the second-harmonic mode yield lower values for the phase noise, but similar values for the noise measure, because much higher Q values are achieved in a circuit without a resistive load at the fundamental frequency.

6.5   SUMMARY AND FUTURE TRENDS

6.5.1   Silicon and GaAs Devices

We have reviewed properties, performance, and fundamental power generation capabilities of two-terminal microwave devices. Advances in growth techniques, refined device-fabrication technologies, and improved circuits have contributed to the significant accomplishments that were achieved in two-terminal devices over the past two decades. If single devices cannot meet the power requirement in a system application, several devices are combined in a circuit. As one example, 16 Si IMPATT diodes in the output stage of a power-combiner circuit deliver an RF power of 20 W to the load in CW operation at 44 GHz.[71] As another example, eight GaAs IMPATT diodes

TABLE 10   Phase Noise of Free-Running Oscillators Using GaAs or InP Millimeter-Wave Gunn Devices

Material System	Phase Noise (dBc/Hz)	Off-Carrier Frequency (kHz)	Oscillation Frequency (GHz)	RF Output Power (mW)	Reference
GaAs	$<-80^*$	100	77	$>40^*$	58
GaAs	-70	100	80	55	70
GaAs	-100	1000	80	55	70
GaAs	-120	10000	80	55	70
GaAs	-80	100	94	10	31
GaAs	-105	1000	94	10	31
InP	-75	100	94	20	31
InP	-100	1000	94	20	31
InP	<-108	500	132	120	67
InP	<-103	500	151	58	67

*Reported as typical value, corresponding RF output power not mentioned.

in the output stage of a power-combiner circuit deliver an RF power of 5.9 to 6.5 W to the load in CW operation at frequencies of 60.5 to 62.5 GHz.[72] Corresponding overall DC-to-RF conversion efficiencies range from 3.8 to 4.2%. Reviews of various power-combining techniques can be found in Refs. 2 and 73. The roll-off in the performance of two-terminal devices for frequencies up to approximately 100 GHz (or up to approximately 60 GHz for GaAs Gunn devices and IMPATT diodes), as shown in Figs. 15 and 36, follows the rule $Pf = $ constant (see Ref. 1, pp. 589–592 and p. 671, for an explanation), which is typical of thermal limitation. This characteristic clearly indicates that the technology for Gunn devices and IMPATT diodes has reached a high level of maturity in this frequency range. Above 100 GHz, a roll-off can be observed that in most devices is much steeper than the rule $Pf^2 = $ constant (see Ref. 1 as above) predicts for electronic limitation. This type of roll-off indicates both fundamental physical limitations as well as a potential for improved performance. Significant improvements can be expected in GaAs IMPATT diodes between 60 GHz and 100 GHz as well as in Si IMPATT diodes above 160 GHz.

6.5.2   InP and Other Materials

Fundamental-mode operation up to 163 GHz was demonstrated with InP Gunn devices. Therefore, InP or InP with heterojunction barriers are promising material systems not only for device structures that exhibit improved performance in the fundamental mode at frequencies around or above 100 GHz, but also for other device structures that generate significant

RF power levels up to 320 GHz in the second-harmonic mode. InP is also emerging as a promising material for high-power millimeter-wave IMPATT diodes.[74] In all these devices, proper heat management is one of the important factors, and the employment of the appropriate diamond heatsink technologies is mandatory for maximum RF output power as well as reliable long-term operation. Again, power-combining techniques can be employed to increase the available RF power levels. As an example, four InP Gunn devices in a power-combiner circuit deliver an RF power of 260 mW to the load in CW operation at 98.6 GHz.[54,75]

Other material systems that hold great promise for high-power two-terminal devices include wide-bandgap materials such as diamond, SiC and GaN. Favorable material parameters, for example, are higher critical electric fields for breakdown, higher thermal conductivities, higher permissible operating temperatures, and expected higher carrier drift velocities. However, major improvements in material quality or availability and in fabrication technologies are needed before devices for system applications can be developed.

6.5.3   Planar Circuits and Monolithic Integration

Waveguide circuits are, in general, rugged, can dissipate heat easily, and offer high Q values. Therefore, they are the preferred circuits to obtain the maximum RF output power. However, they are bulky and in most cases must be machined, assembled, and tested individually, which prohibits low-cost mass production. Therefore, high-volume production for system applications such as wireless communication (high-speed data transmission) or collision avoidance in automobiles must be based on quite different approaches. Both hybrid and monolithic integration were demonstrated with various IMPATT diodes and Gunn devices. Examples are: (1) Hybrid microstrip-line oscillators with InP Gunn devices exhibited excellent performance. RF power levels (with corresponding DC-to-RF conversion efficiencies) of 52 mW (3.5%) at 94 GHz,[76] but also 40 mW (1.4%) 81 GHz[54,70] were measured and are considered comparable to values from similar Gunn devices in waveguide circuits. (2) GaAs double-drift flat-profile IMPATT diodes were successfully integrated with microstrip lines on polyimide and have high reproducibility. RF power levels (and corresponding DC-to-RF conversion efficiencies) of 1.1 W (10%) 58.5 GHz and 1.2 W (8.5%) were obtained in a two-diode push–pull configuration.[77] The thermal conductivity of the semiconductor materials silicon, InP, and GaAs is rather low (1.45, 0.68, and 0.46 W/cm-K, respectively) when compared to metals or diamond. Therefore, fully monolithic integration at millimeter-wave frequencies encounters severe thermal limitations, and high-efficiency IMPATT diodes are more suitable candidates for that than medium-efficiency Gunn devices. A summary of reports on integration of Gunn devices at frequencies up to approximately 68 GHz can be found elsewhere.[2]

A monolithically integrated oscillator with GaAs double-drift flat-profile IMPATT diodes was demonstrated at V-band frequencies: RF power levels (and corresponding DC-to-RF conversion efficiencies) of 110 mW (14.5%) are 58.1 GHz and 100 mW (12.5%) at 64.3 GHz were reported.[77] Monolithic integration of oscillators using Si IMPATT diodes was achieved in various configurations. Since its inception in 1986, the fundamental technological principle behind these circuits, that is, the use of high-resistivity silicon substrates,[78] has sparked tremendous progress in the so-called SIMMWIC (silicon monolithic millimeter-wave integrated circuit) technology. Examples for IMPATT diode oscillators monolithically integrated with antennas are: A slot antenna with a double-drift Si IMPATT diode radiated an RF power of 1 mW at 79 GHz,[79] and a dipole antenna with a double-drift Si IMPATT diode radiated an RF power (at a corresponding total DC-to-RF conversion efficiency) of 6.6 mW (0.58%) at 75.66 GHz.[80] Further significant improvements are expected to obtain higher RF power levels and higher operating frequencies. Quasi-optical power combining of several antenna modules can be employed to increase the available radiated RF power. A review of these combining techniques can be found in Ref. 81.

PROBLEMS

1. Design a single-drift GaAs TUNNETT diode for operation at 180 GHz.
   a) What is its estimated RF output power and DC-to-RF conversion efficiency assuming the same parameters as for Tables 2–8? (Note: $\epsilon_s/\epsilon_0 = 12.87$.)
   b) How are output power and conversion efficiency of the diodes affected if one device fabrication run results in poorer ohmic contacts with total specific contact (series) resistance of $\rho_s = 2.0 \times 10^{-6}$ Ωcm^2?

2. Design an InP single-drift IMPATT diode for 77 GHz using $v_D = 5.5 \times 10^6$ cm/s, $\epsilon_s/\epsilon_0 = 12.61$, $\mathscr{E}_a = 1000$ kV/cm, $\mathscr{E}_s = 5$ kV/cm, $A_i = 5.29 \times 10^6$/cm, $b = 3.20 \times 10^6$ V/cm, and $m = 1$. What RF output power is delivered into a load of 1 Ω, if the specific contact resistance is 2×10^{-6} Ωcm^2?

3. A phase noise of -106 dBc/Hz (in a bandwidth of 1 Hz) at 500 kHz off the carrier was measured with a spectrum analyzer on a free-running oscillator with an InP Gunn device. An RF output power of 120 mW at 120 GHz was measured with the power meter. What is the equivalent frequency-noise modulation of the device? A total locking range of 30 MHz was found in a self-injection locking experiment using a coupler with a coupling value of -20 dB and a back short with negligible loss. How does the noise measure M of this device compare to other results published for this frequency range (see Fig. 16)?

4. The current–voltage characteristics of the AlAs/GaAs/AlAs-RTD in Ref. 43 can be approximated using $V_p = 0.6\,\text{V}$, $V_v = 0.8\,\text{V}$, $J_p = 1.5 \times 10^5\,\text{A/cm}^2$, and $J_v = 1.09 \times 10^5\,\text{A/cm}^2$.

   a) Estimate the unstable RF output power and corresponding DC-to-RF conversion efficiency for a diode with a diameter of $4\,\mu\text{m}$, a depletion width of $75.7\,\text{nm}$, and a total contact (series) resistance of $\rho_s = 5.0 \times 10^{-7}\,\Omega\text{cm}^2$ at frequencies of $1\,\text{GHz}$, $100\,\text{GHz}$, and $500\,\text{GHz}$.

   b) What is the stable RF output power and the corresponding diode area for this RTD structure at $200\,\text{GHz}$ if the same total contact (series) resistance, the same depletion width, and a series inductance in the bias circuit of $0.15\,\text{nH}$ are assumed?

REFERENCES

1. S. M. Sze, *Physics of Semiconductor Devices*, 2nd Ed., Wiley, New York, 1981.

2. K. Chang, *Handbook of Microwave and Optical Components*, Vol. 2, Wiley, New York, 1990.

3. E. Kasper and J.-F. Luy, "State of the art and future trends in silicon IMPATT diodes for mm-wave seeker requirements," in *Proceedings of the Military Microwaves 90*, London, United Kingdom, 1990, p. 293.

4. B. Bayraktaroglu and H. D. Shih, "Integral packaging for millimeter-wave GaAs IMPATT diodes prepared by molecular beam epitaxy," *Electron. Lett.* **19**, 327 (1983).

5. H. Eisele, "Selective etching technology for 94-GHz GaAs IMPATT diodes on diamond heat sinks," *Solid-State Electron.* **32**, 253 (1989).

6. R. Kamoua, H. Eisele, and G. I. Haddad, "D-band (110 GHz–170 GHz) InP Gunn devices," *Solid-State Electron.* **36**, 1547 (1993).

7. H. Eisele and G. I. Haddad, "D-band InP Gunn devices with second-harmonic power extraction up to 290 GHz," *Electron. Lett.* **30**, 1950 (1994).

8. E. E. Palik, V. M. Bermudez, and O. J. Glembocki, "Ellipsometric study of the etch-stop mechanism in heavily doped silicon," *J. Electrochem. Soc.* **132**, 135 (1985).

9. H. Eisele, C. Kidner, and G. I. Haddad, "A CW GaAs TUNNETT diode source for 100 GHz and above," in *Proceedings of the 22nd European Microwave Conference*, August 24–27, 1992, Helsinki, Finland, p. 467.

10. M. Ino, T. Ishibashi, and M. Ohmori, "CW oscillation with p^+pn^+ silicon IMPATT Diodes in 200-GHz and 300-GHz bands," *Electron. Lett.* **12**, 148 (1976).

11. H. J. Kuno, "IMPATT devices for generation of millimeter waves," in *Infrared and Millimeter Waves*, Vol. 1, K. Button, ed., Academic Press, New York, 1979, Ch. 2.

12. K. Chang, W. F. Thrower, and G. M. Hayashibara, "Millimeter-wave silicon IMPATT sources and combiners for the 110-260-GHz range," *IEEE Trans. Microwave Theory Techn.* **MTT-29**, 1278 (1981).

13. Y. E. Ma, "Millimeter-wave active solid-state devices," in *Millimeter Wave Technol. III*, SPIE 544, 1985, p. 95.

14. K. Kurokawa, "Noise in synchronized oscillators," *IEEE Trans. Microwave Theory Techn.* **MTT-16**, 234 (1968).

15. S. M. Sze, *High-Speed Semiconductor Devices*, Wiley, New York, 1990.

16. H. Eisele, "GaAs W-band IMPATT diodes: the first step to higher frequencies," *Microwave J.* **34**(5), 275 (1991).

17. L. W. Holway and M. G. Adlerstein, "Approximate formulas for the thermal resistance of IMPATT diodes compared with computer calculations," *IEEE Trans. Electron. Dev.* **ED-24**, 156 (1977).

18. H. Eisele and G. I. Haddad, "Enhanced performance in GaAs TUNNETT diode oscillators above 100 GHz through diamond heat sinking and power combining," *IEEE Trans. Microwave Theory Techn.* **MTT-42**, 2498 (1994).

19. M. E. Elta and G. I. Haddad, "Mixed tunneling and avalanche mechanisms in p–n junctions and their effects on microwave transit-time devices," *IEEE Trans. Electron. Dev.* **ED-25**, 694 (1978).

20. C-C. Chen, R. K. Mains, G. I. Haddad, and H. Eisele, "Structure and simulation of GaAs TUNNETT and MITATT devices for frequencies above 100 GHz," in *Proceedings of the Fourteenth Biennial Cornell Conference*, August 2–4, 1993, Ithaca, New York, p. 194.

21. H. Eisele, C-C., Chen, R. K. Mains, and G. I. Haddad, "Performance of GaAs TUNNETT diodes as local oscillator sources," in *Proceedings of the Fifth International Symposium of Space Terahertz Technology*, May 10–May 13, 1994, Ann Arbor, Michigan, p. 622.

22. E. C. Niehenke, "GaAs: key to defense electronics," *Microwave J.* **28**(9), 24 (1985).

23. M. Tschernitz and J. Freyer, "140 GHz GaAs double-read IMPATT diodes," *Electron. Lett.* **31**, 582 (1995).

24. P. A. Rolland, M. R. Friscourt, D. Lippens, C. Dalle, and J. L. Nieruchalski, "Millimeter wave solid-state power sources," in *Proceedings of the International Workshop on Millimeter Waves*, Rome, April 2–4, 1986, p. 125.

25. M. Wollitzer, J. Büchler, F. Schäffler, and J.-F. Luy, "D-band Si IMPATT diodes with 300 mW CW output power at 140 GHz," *Electron. Lett.* **32**, 122 (1996).

26. U. Güttich, "BARITT-Dioden für das V-Band," *Mikrowellen Mag.* **13**, 37 (1987).

27. H. Presting, J.-F. Luy, F. Schäffler, and J. Puchinger, "Silicon Ka band low-noise BARITT diodes for radar system applications grown by MBE," *Solid-State Electron.* **37**, 1599 (1994).

28. J.-F. Luy, H. Jorke, H. Kibbel, A. Casel, and E. Kasper, "Si/SiGe heterostructure MITATT diode," *Electron. Lett.* **24**, 1386 (1988).

29. M. Pöbl, W. Bogner, and L. Gaul "CW GaAs MITATT source on copper heat sink up to 160 GHz," *Electron. Lett.* **30**, 1316 (1994).

30. W. Harth, W. Bogner, L. Gaul, and M. Claassen, "A comparative study on the noise measure of millimeter-wave GaAs IMPATT diodes," *Solid-State Electron.* **37**, 427 (1994).

31. C. Dalle, P. A. Rolland, and G. Lleti, "Flat doping profile double-drift silicon IMPATT for reliable CW high-power high efficiency generation in the 94-GHz-window," *IEEE Trans. Electron. Dev.* **ED-37**, 235 (1990).

32. D. M. Brookbanks, A. M. Howard, and M. R. B. Jones, "Si IMPATTs exhibit low noise at mm-waves," *Microwaves RF* **22**(2), 68 (1983).

33. S. J. J. Teng and R. E. Goldwasser, "High-performance second-harmonic operation W-band Gunn devices," *IEEE Electron. Dev. Lett.* **EDL-10**, 412 (1989).

34. M. G. Adlerstein and S. L. G. Chu "GaAs IMPATT diodes pulsed at 40 GHz," *Digest of the 1984 IEEE MTT-S International Microwave Symposium*, May 30–June 1, 1984, San Francisco, California, p. 481.

35. L. Esaki, "New phenomenon in narrow germanium *p–n* junctions," *Phys. Rev.* **109**, 603 (1958).

36. L. L. Chang, L. Esaki, and R. Tsu, "Resonant tunneling in semiconductor double-barriers", *Appl. Phys. Lett.* **24**(12), 593 (1974).

37. E. R. Brown, W. D. Goodhue, T. C. L. G. Sollner, and C. D. Parker, "Fundamental oscillations up to 200 GHz in resonant tunneling diodes and new estimates of their maximum oscillation frequency from stationary-state tunneling theory," *J. Appl. Phys.* **64**(3), 1519 (1988).

38. J. S. Scott, J. P. Kaminski, M. Wanke, S. J. Allen, D. H. Chow, M. Lui, and T. Y. Liu, "Terahertz frequency response of an $In_{0.53}Ga_{0.47}As/AlAs$ resonant-tunneling diode," *Appl. Phys. Lett.* **64**(15), 1995 (1996).

39. R. K. Mains and G. I. Haddad, "Time-dependent modeling of resonant-tunneling diodes from direct solution of the Schrödinger equation," *J. Appl. Phys.* **64**(7), 3564 (1988).

40. M. E. Hines, "High frequency negative-resistance circuit principles for Esaki diode applications," *Bell Syst. Tech. J.* **39**, 477 (1960).

41. A. Rydberg, H. Grönquist, and E. Kollberg, "A theoretical and experimental investigation on millimeter-wave quantum well oscillators," *Microwave Opt. Technol. Lett.* **1**, 333 (1988).

42. E. R. Brown, T. C. L. G. Sollner, C. D. Parker, W. D. Goodhue, and C. L. Chen, "Oscillations up to 420 GHz in GaAs/AlAs resonant tunneling diodes," *Appl. Phys. Lett.* **55**(17), 1777 (1989).

43. E. R. Brown, J. R. Söderström, C. D. Parker, L. J. Mahoney, K. M. Molvar, and T. C. McGill, "Oscillations up to 712 GHz in InAs/AlSb resonant tunneling diodes," *Appl. Phys. Lett.* **58**(20), 2291 (1991).

44. R. Blundell, D. C. Papa, E. R. Brown, and C. D. Parker, "Resonant tunneling diode as an alternative LO for SIS receiver applications," *Electron. Lett.* **29**, 288 (1993).

45. M. Reddy, R. Y. Yu, H. Kroemer, M. J. W. Rodwell, S. C. Martin, R. E. Muller, and R. P. Smith, "Bias stabilization for resonant tunnel diode oscillators," *IEEE Microwave Guided Wave Lett.* **MGWL-5**, 219 (1995).

46. K. D. Stephan, S.-C. Wong, E. R. Brown, K. M. Molvar, A. R. Calawa, and

M. J. Manfra, "5-mW parallel-connected resonant tunneling diode oscillator," *Electron. Lett.* **28**, 1411 (1992).

47. F. Capasso, S. Sen, F. Beltram, L. M. Lunardi, A. S. Vengurlekar, P. R. Smith, N. J. Shah, R. J. Malik, and A. Y. Cho, "Quantum functional devices: resonant-tunneling transistors, circuits with reduced complexity and multiple-valued logic," *IEEE Trans. Electron Dev.* **ED-36**, 2065 (1989).

48. B. K. Ridley and T. B. Watkins, "The possibility of negative resistance effects in semiconductors," *Proc. Phys. Soc. Lond.* **78**, 293 (1961).

49. C. Hilsum, "Transferred electron amplifiers and oscillators," *Proc. Inst. Radio Eng.* **50**, 185 (1962).

50. J. B. Gunn, "Microwave oscillation of current in III–V semiconductors," *Solid-State Commun.* **1**, 88 (1963).

51. J. B. Gunn, "Instabilities of current in III–V semiconductors," *IBM J. Res. Develop.* **8**, 141 (1964).

52. L. Wandinger, "mm-Wave InP Gunn devices: status and trends," *Microwave J.* **24**(3), 71 (1981).

53. B. Fank, J. Crowley, D. Tringali, and L. Wandinger, "Basics and recent applications of high-efficiency millimeter wave InP Gunn diodes," in *Proceedings of the First International Conference on Indium Phosphide and Related Materials for Advanced Electronic and Optical Devices*, Norman, Oklahoma, March 20–23, 1989, SPIE 1144, 1989, p. 534.

54. I. G. Eddison, "Indium phosphide and gallium arsenide transferred-electron devices," *Infrared and Millimeter Waves*, Vol. 11, *Millimeter Components and Techniques*, Part III, Academic Press, New York, 1984, p. 1.

55. J. A. Copeland, "LSA oscillator-diode theory," *J. Appl. Phys.* **38**, 3096 (1967).

56. P. N. Butcher, "Theory of stable domain propagation in the Gunn effect," *Phys. Lett.* **19**, 546 (1965).

57. J. F. Caldwell and F. E. Roxtoczy, "Gallium arsenide Gunn diodes for millimeter-wave and microwave frequencies," in *Proceedings of the 4th International Symposium on GaAs and Related Compounds*, Denver, Colorado, 1972.

58. I. Dale, J. R. P. Stephens, and J. Bird, "Fundamental-mode graded-gap Gunn diode operation at 77 and 84 GHz," in *Proceedings of the Microwaves 94 Conference*, London, United Kingdom, October 25-27, 1994, p. 248.

59. M. A. di Forte-Poisson, C. Brylinski, N. Proust, D. Pons, M. Secoué, P. Arsène Henry, M. Calligaro and J. Lacombe, "LP-MOCVD InP Gunn devices developed for 94 GHz millimeter range operation," in *Proceedings of the First International Conference on Indium Phosphide and Related Materials for Advanced Electronic and Optical Devices*, Norman, Oklahoma, March 20–23, 1989, SPIE 1144, 1989, p. 551.

60. A. Rydberg, "High efficiency and output power from second- and third-harmonic millimeter-wave InP-TED oscillators at frequencies above 170 GHz," *IEEE Electron. Dev. Lett.* **EDL-11**, 439 (1990).

61. M-R. Friscourt, P-A. Rolland, and M. Pernisek, "Heterojunction cathode contact transferred-electron oscillators," *IEEE Electron Dev. Lett.* **EDL-6**, 497 (1985).

62. B. Fank, J. Crowley, and C. Hang, "InP Gunn diode sources," *Millimeter Wave*

Technology III, SPIE 544, 1985, p. 22.

63. J. D. Crowley, C. Hang, R. E. Dalrymple, C. Hang, D. R. Tringali, F. B. Fank, and L. Wandinger, "InP Gunn diodes serve millimeter-wave applications," *Microwaves RF* **33**(3), 143 (1994).

64. J. D. Crowley, C. Hang, R. E. Dalrymple, D. R. Tringali, F. B. Fank, L. Wandinger, and H. B. Wallace, "140 GHz indium phosphide Gunn diode," *Electron. Lett.* **30**, 499 (1994).

65. J. Ondria and R. L. Ross, "Improved performance of fundamental and second-harmonic MMW oscillators through active doping concentration contouring," in *1987 IEEE MTT-S Digest*, p. 977.

66. K. Akamatsu, A. Yokohata, S. Kato, N. Ohkuba, and M. Ohmori, "High-efficiency millimeter-wave GaAs Gunn diodes operating in the fundamental mode," in *Digest of the Nineteenth International Conference on Infrared and Millimeter Waves*, Sendai, Japan, October 17–20, 1994, p. 89.

67. H. Eisele and G. I. Haddad, "High-performance InP Gunn devices for fundamental-mode operation in D-band (110–170 GHz)," *IEEE Microwave Guided Wave Lett.* **MGWL-5**, 385 (1995).

68. N. R. Couch, H. Spooner, P. H. Beton, M. J. Kelly, M. E. Lee, P. K. Rees, and T. M. Kerr, "High-performance, graded AlGaAs injector, GaAs gunn diodes at 94 GHz," *IEEE Electron. Dev. Lett.* **EDL-10**, 288 (1989).

69. R. Kamoua, "Heterojunction D-band (110 GHz–170 GHz) InP Gunn devices," *Solid-State Electron.* **37**, 269 (1994).

70. D. C. Smith, T. J. Simmons, and M. R. B. Jones, "A comparison of the performance of millimeter-wave semiconductor oscillator devices and circuits," *Digest of the 1983 IEEE MTT-S International Microwave Symposium*, May 31–June 3, Boston, Massachusetts, p. 127.

71. D. F. Peterson and D. P. Klemer, "Multiwatt IMPATT power amplification for EHF applications," *Microwave J.* **32**(4), 107 (1989).

72. M. K. Powers, J. McClymonds, D. Vye, and T. Arthur, "Solid-state power amplifier for 61.5 GHz," *NASA Tech. Briefs* **16**(8), 31 (1992).

73. K. Chang and C. Sun, "Millimeter-wave power-combining techniques," *IEEE Trans. Microwave Theory Techn.* **MTT-31**, 91 (1983).

74. H. Eisele, C-C. Chen, G. O. Munns, and G. I. Haddad, "The potential of InP IMPATT diodes as high-power millimeter-wave sources: first experimental results," *Digest of the 1996 IEEE MTT-S International Microwave Symposium*, June 17–21, San Francisco, California, p. 529.

75. J. J. Sowers, J. D. Crowley, and F. B. Fank, "CW InP Gunn diode power combining at 90 GHz," *Digest of the 1982 IEEE MTT-S International Microwave Symposium*, June 15–17, Dallas, Texas, p. 503.

76. C. Kim, C. Dunnrowicz, J. Crowley, B. Fank, and L. Wandinger, "Millimeter-wave tunable microstrip InP Gunn oscillators," *Microwave J.* **32**(4), 91 (1989).

77. B. Bayraktaroglu, "Monolithic IMPATT technology," *Microwave J.* **32**(4), 73 (1989).

78. J. Büchler, E. Kasper, P. Russer, and K. M. Strohm, "Silicon high-resistivity-substrate millimeter-wave technology," *IEEE Trans. Microwave Theory Techn.* **MTT-34**, 1516 (1986).

79. A. Stiller, E. M. Biebl, J.-F. Luy, K. M. Strohm, and J. Buechler, "A Monolithic integrated millimeter wave transmitter for automotive applications," *IEEE Trans. Microwave Theory Techn*. **MTT-43**, 1654 (1995).

80. M. Singer, A. Stiller, K. M. Strohm, J.-F. Luy, and E. M. Biebl, "A SIMMWIC 76-GHz front end with high polarizing purity," in *Digest of the 1996 IEEE MTT-S International Microwave Symposium*, June 17–21, San Francisco, California, p. 1079.

81. J. C. Wiltse and J. W. Mink, "Quasi-optical power combining of solid-state sources," *Microwave J*. **35**(2), 144 (1992).

7 High-Speed Photonic Devices

TIEN PEI LEE
Bell Communications Research, Red Bank, New Jersey

S. CHANDRASEKHAR
Lucent Technologies, Holmdel, New Jersey

7.1 INTRODUCTION

The rapid deployment of both terrestrial and undersea fiber-optic telecommunication systems in the last decade has paved the way for a modernized
photonic network at the turn of the next century. As the lower costs of fiber
cables and the higher bit rates of the photonic devices continue to drive down
the relative cost per bit, we anticipate that in a few years new transmission
and distribution facilities will be all fiber based. There is also every indication
that as the number of internet users increases the demands for broadband
services will grow. With an estimated bandwidth of 30 THz (1 THz = 10^{12} Hz)
available in single-mode fibers, a fiber-optic-based photonic network can
offer the broadband services envisioned for the "information age." Such
services include education, telecommuting, video-conferencing, interactive
database services (e.g., banking, shopping, library), computer networking,
high-definition television (HDTV) broadcast entertainment services, and
others that are not yet conceived.

The bit rate of early optical transmission systems in the late 1970s was
45 Mbit/s. By the mid-1980s, the transmission speed increased to rates
ranging from 400 Mbit/s to 1.7 Gbit/s. More recently, SONET (*synchronous
optical networks*) systems use rates of 155 Mbit/s, 622 Mbit/s, and 2.5 Gbit/s
for new interoffice and long-distance installations. SONET systmes with a
speed of 10 Gbit/s will become available in the near future.

Beyond 10 Gbit/s, it is believed that multiple-wavelength techniques by

Modern Semiconductor Device Physics, Edited by S. M. Sze.
ISBN 0-471-15237-4 © 1998 John Wiley & Sons, Inc.

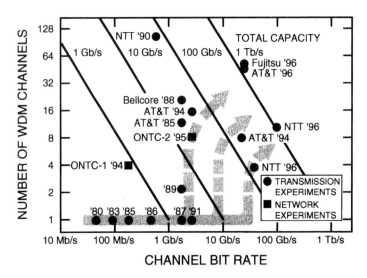

Fig. 1 The trend in lightwave systems capacity.

means of wavelength-division multiplexing (WDM) is a viable technology to exploit the extremely broad optical bandwidth (30 THz) in the low-loss transmission window of the optical fiber. The aggregated transmission capacity in a WDM system is the product of the bit rates and the number of WDM channels. Figure 1 shows this relationship. The solid diagonal lines are the aggregated total capacity of 1 Gbit/s, 10 Gbit/s, 100 Gbit/s, and 1 Tb/s respectively (1 Gbit/s = 10^9 bit/s, 1 Tb/s = 10^{12} bit/s). Included in Fig. 1 are both commercial single-channel systems deployed (marked by the year of deployment) and experimental laboratory WDM transmission and network experiments reported (marked by the year of publication). As can be seen in this figure, the speed of single-channel systems doubled every two years up to 2.5 Gbits in 1991. Further increase in capacity has been accomplished by adding more wavelength channels. Early laboratory WDM experiments were conducted between 1985 and 1990 with as many as 100 wavelength channels. More recently the trend is toward higher bit rates per wavelength channel to increase the total capacity, as indicated by the arrow bars in Fig. 1. In 1996, WDM transmission experiments with the aggregated capacity up to 1.1 Tbit/s have been reported.[1-3] Moreover, the wavelengths can be used for dynamically routing signals in an optical network. Several reconfigurable optical network testbeds have been demonstrated recently,[4] and their system capacities are shown by the solid squares in Fig. 1.

The rapid progress made in the lightwave systems in the last decade can be attributed in part to the technological advances in new semiconductor materials and high-speed photonic devices. Quantum-well and strained-layer quantum-well lasers have shown superior performance over conventional double-heterostructure lasers. The reduction of the threshold currents of semiconductor lasers by a factor of 1000 over the last three decades can be attributed to the new material growth techniques such as metalorganic chemical vapor deposition (MOCVD), molecular-beam epitaxy (MBE), and variations of these two techniques. These techniques permit the growth of very thin layers down to a single atomic layer. A stack of these thin layers with alternating compositions can be used to form quantum-wells and superlattices that fundamentally change the band structure of the materials. The lattice constants of the alternating layers can be slightly mismatched to allow unrelaxed strain that further improves the optical property of the materials.[5] Using such materials, lasers with low threshold currents, high output powers, very high modulation speeds, low frequency chirp, and narrow laser linewidths have been demonstrated.

For photoreceivers, the efforts in the last decade have been devoted to the development of very high-speed receivers, above 10 Gbit/s, and to optoelectronic integrated circuits (OEICs).[6] Waveguide detectors and metal–semiconductor–metal (MSM) detectors are favored structures for their low capacitance and ease of integration. However, p–i–n structures remain the simplest structures and best in performance. The advancement in epitaxial growth has helped to make better transistors, including heterojunction bipolar transistors (HBTs) and modulation-doped FET (MODFET) based on long-wavelength materials that are compatible with the p–i–n photodiode for optoelectronic integration. Using the high-performance transistors as preamplifiers, the sensitivity of OEIC receivers becomes comparable to that of the p–i–n-FET hybrids, as shown in Fig. 2. Further improvement of the receiver sensitivity can be obtained with avalanche photodiodes (APDs), coherent heterodyne receivers, and erbium-doped fiber amplifiers (EDFAs).

For data communication applications, low-cost light-emitting diodes (LEDs) are often used. The LED can operate up to 200 Mbit/s for short links using multimode fibers. The basic structures and operations of LEDs are much simpler than lasers and have been described elsewhere.[7,8] In this chapter, we review the present status of the high-speed photonic devices, namely, lasers, photonic integrated circuits (PICs) for optical transmitters, and optoelectronic integrated circuits (OEICs) for photoreceivers. The chapter is organized as follows. A brief tutorial on the basic laser design, operating principles, and characteristics is given in Section 7.2. The fundamental principles of quantum-well and strained quantum-well lasers and their advantages are discussed in Section 7.3. Various advanced laser structures, including submilliampere threshold-current lasers, uncooled lasers, high-speed DFB lasers, and photonic integrated circuits (PICs), such as external modulator integrated lasers, tunable lasers and multiwavelength laser arrays are presented in Section 7.4. Discussions on the photoreceiver OEICs are presented in Section 7.5, and a brief summary and discussion of future trends are presented in Section 7.6.

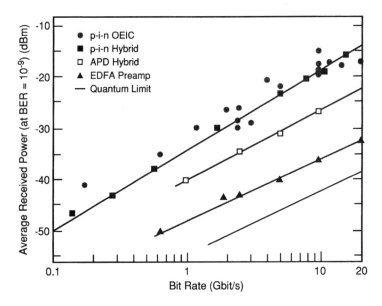

Fig. 2   Photoreceiver sensitivities as a function of the signal bit rates.

7.2   LASER DESIGN AND BASIC PRINCIPLES OF OPERATION

7.2.1   Semiconductor Laser Materials

The most important factor for making a double-heterostructure laser, as described in Section 7.2.2, is to select semiconductor materials that emit at the desired wavelength, and that can be grown epitaxially with a lattice constant matched to the substrate material. The most commonly used substrate materials are GaAs and InP for the 0.8–0.9-μm wavelength region and the 1.1–1.65-μm wavelength region respectively. The ternary AlGaAs alloy can be grown nearly lattice matched to GaAs. Because it has a larger bandgap than GaAs, the ternary was used for the carrier confinement layers in the double heterostructure laser that will be described in Section 7.2.2. With a 30% aluminum mole fraction, the bandgap is large enough to confine the injected electrons and holes in the GaAs active layer. AlGaAs alloys can also be used in the active layer for shorter wavelength devices such as compact-disc lasers. However, at aluminum mole fractions greater than 0.45 the electron–hole recombination becomes indirect, making photon generation inefficient.

In the long-wavelength region, which is primarily used for optical-fiber communications, InP is a favored substrate material. To match the lattice constant of InP ($a = 5.87$ Å) the quaternary alloy $In_{1-x}Ga_xAs_yP_{1-y}$ is used. For lattice match to InP ($y = 2.2x$), the bandgap E_g (in eV) varies as[9]

$$E_g = 1.35 - 0.72y + 0.12y^2 \tag{1}$$

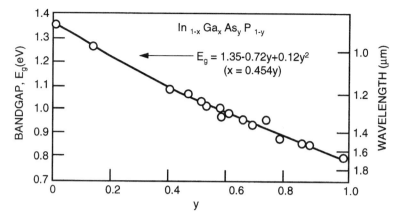

Fig. 3   Energy gap (and emission wavelength) vs composition of InGaAsP quaternary alloy. (After Nahory et al., Ref. 9.)

This relationship is shown in Fig. 3. The entire wavelength region from 0.92 to 1.65 μm is covered with this material system. Since the photon energy E is approximately equal to the bandgap energy, the lasing wavelength, λ, can be obtained using $E_g = hc/\lambda$, where h is the Planck constant and c is the speed of light in vacuum. If E_g is expressed in eV, the lasing wavelength in microns is given by

$$\lambda = \frac{1.24}{E_g} \qquad (2)$$

For the confinement layers, the composition is chosen so that the bandgap is larger than that of the active layer by approximately 200 meV.

The growth of these thin layers was mostly accomplished by liquid-phase epitaxy (LPE). In recent years, however, materials grown by metalorganic chemical vapor deposition (MOCVD) have produced thin layers that are much more uniform than layers made by LPE. In addition, multiple-quantum-well structures (MQW) with layers thinner than 100 Å were also grown by MOCVD. Multiple-quantum-well lasers have been fabricated and their performance is superior to those of the double-heterostructure (DH) lasers. See Section 7.3 for further discussion of quantum-well devices.

7.2.2   The Double Heterostructure

A typical semiconductor laser contains a double heterostructure (DH)[10] in the lasing region, as shown in Fig. 4a. In such a structure, an active layer of one semiconductor material of lower bandgap energy (such as GaAs), with a thickness less than the diffusion lengths of electrons and holes, is

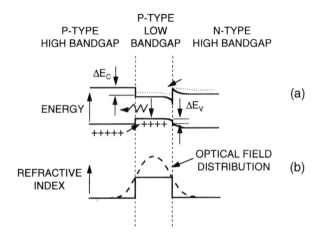

Fig. 4 The effect of (a) carrier confinement and (b) optical confinement in a double-heterostructure laser.

sandwiched between two confinement layers of another semiconductor material of higher bandgap energy (such as AlGaAs). The large bandgap of the confinement layers confines the injected electrons to the active layer, where efficient radiative recombination takes place. In addition, the higher index of refraction in the active layer of the DH structure also forms a planar waveguide, which confines the optical field close to the active layer. The optical confinement significantly reduces the internal loss that would otherwise occur in the absence of waveguiding resulting from spreading of the optical field in the lossy medium. The effects of both carrier and optical confinements are illustrated in Fig. 4b. The threshold current density reduces proportionally to the thickness of the active layer, until it is too thin to confine the optical field. The normalized threshold current density is about 5 kA/cm^2-μm for AlGaAs/GaAs laser diodes emitting in the 0.85-μm wavelength region and about 3.5 kA/cm^2-μm for InGaAsP/InP lasers emitting in the 1.3–1.55-μm wavelength region. The lowest thresholds were found in devices with an active layer thickness between 0.1 and 0.2 μm.

7.2.3 Gain- and Index-Guided Lasers

To achieve CW (continuous wave) room-temperature operation, stripe geometry structures were used for carrier confinement in the lateral dimension. These structures restrict current flow only to a narrow stripe contact approximately 10 μm wide using an SiO$_2$ film for the contact mask as shown in Fig. 5a. The stripe-geometry structure is also termed the "gain-guided" structure because the waveguiding is provided by the lateral variation of the optical gain. This structure was widely used in conjunction

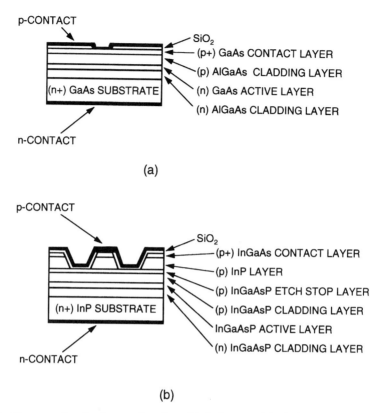

Fig. 5 Cross-sectional views of gain-guided lasers. (a) Stripe-geometry laser structure. (b) Ridge-waveguide laser structure.

with the double-heterostructure AlGaAs/GaAs lasers. A variation of this structure is the ridge-waveguide structure, as shown in Fig. 5b, which provides weak index guiding.

Since the widely available silica-based optical fiber has a loss minimum at the 1.55-μm wavelength region and zero dispersion occurs at a wavelength of 1.3 μm, the major development of semiconductor lasers for optical-fiber communication applications has focused on the InGaAsP/InP material system. The first demonstration of CW room-temperature operation at wavelengths beyond 1 μm was in 1976. The quaternary lasers now cover the entire wavelength range, 1.1–1.65 μm. Because of the slow oxidation rate in InP-based materials, it is possible to perform etching and epitaxial regrowth to construct so-called index-guided lasers. In such a structure, the InGaAsP active region is surrounded by the higher-bandgap InP in both the vertical and the lateral dimensions. Various index-guided structures, known as the buried heterostructures (BH), have been introduced with transverse

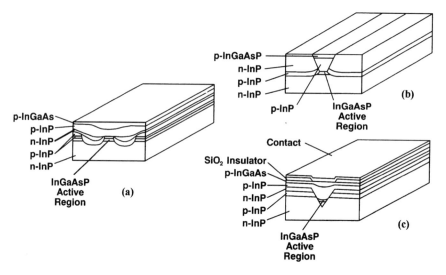

Fig. 6 Commonly used index-guided, double heterostructure lasers. (a) Double-channel planar buried heterostructure (DCPBH). (b) Planar buried heterostructure (PBH). (c) Channel substrate buried crescent (CSBC) structure.

modal stability superior to the gain-guided structures. Several of the most commonly used BH laser diodes are shown schematically in Fig. 6. Detailed discussions on the long-wavelength lasers can be found in Ref. 11.

7.2.4 Longitudinal Modes

Fabry–Perot Lasers.   The laser cavities shown in Figs. 5 and 6 are Fabry–Perot cavities, where the mirrors are formed by cleaving the semiconductor wafer. The mirror loss is independent of wavelength. All optical fields inside the cavity that experience a round-trip phase (equal to a multiple of 2π) will oscillate when the lasing threshold is reached. This gives rise to the longitudinal modes of the Fabry–Perot cavity laser. These modes are at wavelengths that satisfy the equation

$$\lambda_m = \frac{2n_g L}{m} \qquad (3)$$

where n_g is the group index of the medium, L the cavity length, and m the mode number (an integer). The spacing between adjacent modes is

$$\Delta\lambda = \lambda_m - \lambda_{m-1} = \frac{\lambda_m^2}{2n_g L} \qquad (4)$$

Each mode, starting from the spontaneous emission into the mode, grows as the optical gain increases by an increase in the injection current. Because of the parabolic density of states, the gain for each mode, g_m, can be approximated by[12]

$$g_m = g_p - \frac{(\lambda_p - \lambda_m)^2}{\lambda_G^2} \tag{5}$$

where λ_p is the wavelength at the gain peak, g_p, and λ_G characterizes the width of the gain spectrum. The steady-state photon density of mode m, S_m, can be expressed by[12]

$$S_m = \frac{\gamma N / \tau}{(c/n_g)[\Gamma(\alpha_c + \alpha_m - g_m)]} \tag{6}$$

In Eq. 6, γ is the spontaneous emission factor,[12] N the carrier density, τ the carrier lifetime, c the velocity of light in vacuum, n_g the group refractive index, Γ the optical confinement factor, α_c the free carrier loss, and $\alpha_m = (1/2L)\ln(1/R_1 R_2)$ the mirror loss, where L is the cavity length, and R_1 and R_2 are the mirror reflectivities. Equation 6 shows that the power of each mode depends inversely on the difference of the loss and the modal gain. Thus, the mode that is closest to the peak of the gain curve has the highest power (the main mode). As the current increases above threshold, the main mode power increases faster than the side modes. Figure 7a shows a computed result based on a multimode rate equation model.[12] Figure 7b shows the experimentally measured mode spectra[13] of a 1.3-μm InGaAsP laser at three different current levels, which is in good agreement with the computed results.

Distributed-Feedback Lasers.   The multimode Fabry–Perot lasers are only useful for systems at rates below 1 Gbit/s. For advanced lightwave systems, such as multi-gigabit/s, wavelength-division-multiplexed systems, and coherent lightwave systems, single-frequency lasers are necessary. Single-frequency oscillation can be achieved by integrating a frequency-selective grating element inside the laser cavity. A major achievement in the 1980s was the realization of high-performance single-frequency lasers. The distributed-feedback (DFB) laser concept[14] was applied to semiconductor lasers. Distributed-feedback lasers[15–16] and distributed-Bragg-reflector (DBR) lasers[17] were first demonstrated in AlGaAs/GaAs lasers, and the same concept was applied to InGaAsP/InP lasers later.[11]

In the DFB laser, the grating region is built in a waveguide layer adjacent to the active layer, as shown in Fig. 8a, whereas in the DBR laser the grating is outside the pumped active region along the length of the cavity, as shown in Fig. 8b. Detailed analysis of the DFB and DBR lasers can be found in Ref. 11. The basic operation of the DBR laser is conceptually understand-

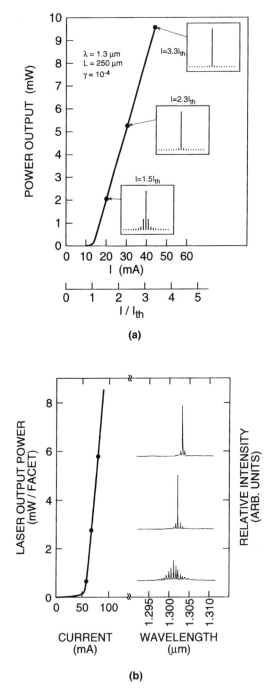

(a)

(b)

Fig. 7 Power output and mode spectrum for an InGaAsP laser with a Fabry–Perot cavity. (After Lee et al., Ref. 12.) (b) Measured. (After Nelson et al., Ref. 13.)

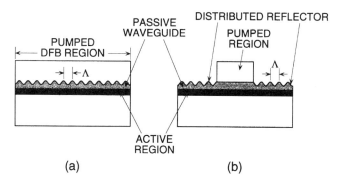

Fig. 8 Schematic diagram of (a) a distributed feedback (DFB) laser and (b) a distributed Bragg-reflector (DBR) laser.

able by considering that the reflection is enhanced at the wavelength λ_B, known as the Bragg wavelength, which is related to the period of the grating Λ by

$$\lambda_B = \frac{2n_{\text{eff}}\Lambda}{m} \tag{7}$$

where n_{eff} is the effective refractive index of the mode and m is the integer order of the grating. The mode at the Bragg wavelength that has the lowest loss and thus lowest threshold gain will lase predominantly. The operation of the DFB laser is more complex but can be understood qualitatively as follows. The grating region has a periodically varying index of refraction that couples two counter propagating traveling waves. The coupling is maximum for wavelength close to the Bragg wavelength. In the ideal case where there are no facet reflections, longitudinal modes are spaced symmetrically around λ_B at a wavelength given by

$$\lambda = \lambda_B \pm \left(\frac{(m + \frac{1}{2})\lambda_B^2}{2n_g L_{\text{eff}}} \right) \tag{8}$$

where m is the mode number and L_{eff} is the effective grating length. Because of the periodic index variation due to the grating structure, there is a stop band defined as the wavelength band within which the transmission through this periodic structure is zero and, consequently, the reflection is the largest. Ordinarily, only the two lower-order modes in Eq. 8 ($m = 0$) are found within the stop band, and oscillations of these two wavelengths can occur simultaneously because of the symmetry. In practice, however, the randomness in the cleaving process results in different end phases, removing the degeneracy of the modal gain, and giving rise to a single-mode operation. Facet asymmetry can be further enhanced by coating one facet with either high-reflection or anti-reflection films to improve the single-frequency yield.

An alternative approach to obtain a high single-frequency yield is to incorporate a quarter-wave phase shift in the grating of the DFB laser.[18–20] The schematic of the quarter-wave phase-shifted DFB laser is shown in Fig. 9. In this structure, a λ/4-phase shift is incorporated in the grating corrugation at the center of the laser cavity with both facets anti-reflection coated. In this case, the mode at the Bragg wavelength has the lowest threshold gain, and hence lases predominantly. Figure 10a shows the output spectrum of a 1.5-μm GaInAsP λ/4-shifted DFB laser above threshold. The side-mode suppression ratio is better than 40 dB. Figure 10b shows the spectrum at a current just below threshold, and the stop band can be noticed.

7.2.5 Modulation Characteristics

The intensity of semiconductor lasers can be conveniently modulated by modulating the injection current to the laser. The speed of modulation can be as high as several tens of gigahertz because stimulated emission shortens the carrier lifetime considerably. In this section we discuss the modulation characteristics of semiconductor lasers with particular emphasis on the high-speed capabilities and the limitations of long wavelength InGaAsP laser diodes.

The dynamic characteristics of semiconductor lasers can be understood best by analysis of the rate equations. To simplify the analysis, we make the following assumptions:

1. Only a single mode is present.
2. The carrier diffusion can be ignored.

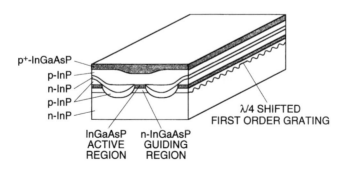

Fig. 9 Schematic of a λ/4-shifted DFB DCPBH laser.

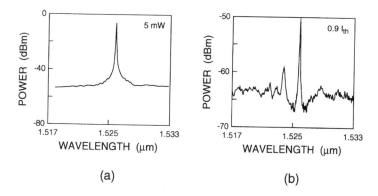

Fig. 10 Output spectrum of a $\lambda/4$-shifted DFB laser at 5 mW (a) and at $0.9I_{th}$ (b).

3. The carrier injection is uniform and the photon density fluctuations along the cavity is negligible.
4. The gain saturates according to $g = g_0/(1 + \varepsilon S)$, where S is the photon density and ε is a parameter characterizing the nonlinear gain *versus* photon density.

With these assumptions the rate equations take the simple form

$$\frac{dN}{dt} = \frac{I}{qV} - \frac{N}{\tau_n} - \frac{v_g \Gamma A (N - N_{tr}) S}{1 + \varepsilon S} \tag{9}$$

$$\frac{dS}{dt} = \gamma \frac{\Gamma N}{\tau_n} - \frac{S}{\tau_p} + \frac{v_g \Gamma A (N - N_{tr}) S}{1 + \varepsilon S} \tag{10}$$

In Eqs. 9 and 10, $v_g = c/n_g$ is the group velocity, A is the differential gain ($= dg/dN$), N_{tr} the transparent carrier density defined as the carrier density at which the optical gain equals the loss in the laser medium, V the volume of the active region, τ_n and τ_p the carrier recombination lifetime and the photon lifetime respectively. Assuming a small-signal modulation, that is, $I = I_0 + ie^{j\omega t}$, $N = N_0 + ne^{j\omega t}$, and $S = S_0 + se^{j\omega t}$, and ignoring the small product terms ns, s^2, and other terms small compared to $1/\tau_p$, Eqs. 9 and 10 can be readily solved to yield the photon transfer function in the form

$$\frac{s(\omega)}{s(0)} = \frac{\omega_0^2}{\omega_0^2 - \omega^2 + j\omega\gamma_d} \tag{11}$$

where ω is the angular frequency, ω_0 is given approximately by[21]

$$\omega_0 = \sqrt{\frac{v_g \Gamma A S_0}{\tau_p(1 + \varepsilon S_0)}} \tag{12}$$

and γ_d is given approximately by

$$\gamma_d = \frac{\varepsilon S}{\tau_p} \tag{13}$$

Equation 11 indicates that the transfer function has a pole at

$$\omega = j\frac{\gamma_d}{2} \pm \sqrt{\omega_0^2 - \frac{\gamma_d^2}{4}} \tag{14}$$

The first term in Eq. 14 gives the damping term, and the second term is the frequency at the peak of the response, defined as the relaxation oscillation frequency.

The small-signal response curves calculated using Eq. 11, with the nonlinear gain parameter $\varepsilon = 0$ and $\varepsilon = 1.3 \times 10^{-23}\, m^3$ are shown in Fig. 11. Note that the magnitude of the peak response reduces as ε increases, indicating gain saturation resulting from the nonlinear effect. Equation 12 indicates that the relaxation resonant frequency is proportional to the square root of the output power up to a maximum, beyond which the resonant frequency decreases because of the nonlinear gain. Figure 12 is a theoretical calculation[22] of the peak response frequency *versus* the output optical power for various values of ε. It is important to note that the peak response frequency can be increased in several ways.

1. By having a large differential gain, A, in the gain medium, such as in the quantum-well active regions.
2. By shortening the photon lifetime, such as in a short-cavity laser.
3. By decreasing the nonlinear gain coefficient.[23-25]

   The practical limitations, however, are due to the parasitic capacitance and inductance associated with the particular laser structure. The parasitic capacitance comes from the laser *p–n* junction and the current-blocking junctions in buried-heterostructure lasers. The inductance stems from the bonding wires of the laser chip inside the package. Thus, to reach the intrinsic high-frequency response of the laser, these parasitics must be kept to a minimum. Various high-speed-laser structures are discussed in Section 7.4.3.

7.2.6 Frequency Chirp

The gain variation during intensity modulation by the injection current also produces a change in the refractive index in the gain medium, which leads to a frequency modulation. This effect is known as frequency chirp by direct-current modulation. The refractive index consists of a real part and

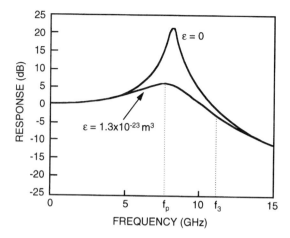

Fig. 11 Small-signal laser modulation response showing the effect of damping due to gain saturation. f_p is the frequency at the peak of the response and f_3 is the 3-dB modulation bandwidth. (After Bowers and Pollack, Ref. 22.)

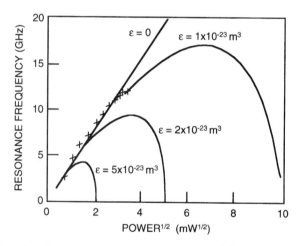

Fig. 12 Relaxation frequency as a function of (power)$^{1/2}$, showing the effect of gain saturation. The solid curves are theoretical calculations and the crosses are measured values. (After Bowers and Pollack, Ref. 22.)

an imaginary part, that is, $n = n' + jn''$. A change in the gain $\Delta g = -4\pi\Delta n''/\lambda$ is linked to a change in the frequency $\Delta\nu/\nu = \Delta n'/n'$, by a frequency-dependent material constant α defined as

$$\alpha = -\left(\frac{4\pi}{\lambda}\right)\left(\frac{dn'/dN}{dg/dN}\right) \tag{15}$$

where dN is the incremental change in carrier density, and α is commonly known as the linewidth enhancement factor.[26]

Using the photon rate equation (Eq. 10) and expanding about a DC operating point, the frequency chirp, $\Delta\nu$, can be related to the variation in the output power under direct-current modulation:[27]

$$\Delta\nu(t) = -\left(\frac{\alpha}{4\pi}\right)\left(\frac{d}{dt}\ln P(t) + \kappa P(t)\right) \tag{16}$$

where $P(t)$ is the time variation of the optical power, and κ is a parameter given by

$$\kappa = \frac{2\Gamma\varepsilon}{\eta_d h\nu V} \tag{17}$$

where V is the volume of the laser active region. For sinusoidal modulation, the RF component of Eq. 16 becomes

$$\frac{\Delta\nu}{\Delta P} = -\frac{j\alpha}{2}\frac{f}{P} + \kappa \tag{18}$$

Equation 18 indicates that at high modulation frequencies the frequency chirp increases with α and becomes worse as the frequency increases. It also gets worse when there is a large optical power variation, such as during relaxation oscillation. Strong damping of the relaxation oscillation would reduce the amount of frequency chirp. The factor α is inversely proportional to the differential gain and is strongly material and frequency dependent, which will be discussed further in the next section. The second term in Eq. 18, dominating at low frequencies, is an adiabatic chirp, which can be minimized by increasing the volume of the active region in the laser structure. Typical chirp width of commercial DFB lasers is 0.4–0.6 nm at 1–2-Gbit/s modulation rates. Although this amount of chirp can allow transmission distances up to several hundred kilometers at 2.5 Gbit/s with single-mode fibers at a wavelength of 1.55 μm, the transmission distance is limited to only few kilometers at 10 Gbit/s because of the chromatic dispersion in single-mode fibers. The chromatic dispersion stems from the difference in the group velocities at different wavelengths of the lightwaves traversing the optical fiber. The different wavelengths within a light pulse resulting from the frequency chirp of the direct-current modulation of the laser will introduce pulse spread, which, in turn will cause interference in adjacent pulses (intersymbol interference) and degrade the transmitted signal.

7.2.7 Laser Linewidth

Under CW operation, a typical semiconductor laser exhibits a Lorentzian lineshape with a linewidth of about 100 MHz at 1 mW output power. The measured linewidth is about a factor of 50 larger than that predicted by the classical theory.[26] The linewidth broadening is due to the change in lasing frequency with gain, as described in Section 7.2.5. The fluctuation in gain of a laser under CW operation is due to spontaneous emission in the semiconductor laser. This spontaneous emission is the source of the frequency noise that gives rise to the linewidth broadening.

The modified linewidth formula,[26] which is derived in Appendix 7.A, is given by

$$\partial v = v_g^2 \left(\frac{\eta_{sp} h v}{8 \pi P_0} \right) \alpha_m (\alpha_m + \alpha_c)(1 + \alpha^2) \qquad (19)$$

where v_g is the group velocity and η_{sp} (= 2–3) is the spontaneous emission factor. The line enhancement factor, α, is defined in Eq. 15. The value of α ranges from 2 to 6 depending on the material system and the device structure. It should be noted that the linewidth is inversely proportional to the output power, because spontaneous emission becomes relatively small as the laser power increases. The linewidth also decreases as the laser cavity increases because the mirror loss α_m decreases.

For DFB lasers the linewidth can be narrowed by detuning the lasing wavelength toward the shorter wavelength side of the gain peak because of the larger differential gain and, thus, a smaller α. For the same reason, quantum-well lasers tend to have narrower linewidths because they have higher differential gain than do DH lasers. The best quantum-well lasers exhibit linewidths in the range 100 kHz to 1 MHz, whereas conventional DH lasers show linewidths at least an order of magnitude larger.

7.2.8 Wavelength Tunability

Wavelength tuning can be achieved by (1) using an external cavity containing a tunable wavelength filter and (2) monolithically integrated multiple-electrode DFB and DBR lasers. External cavity tunable lasers employing a diffraction grating filter, an electro-optic filter, and an acousto-optic filter, have been demonstrated. The advantage of using an external cavity is the large tuning range that can be achieved because of the broad gain width of the semiconductor. The disadvantages are that the wavelength tuning is discrete, and the external cavity arrangement requires extremely good mechanical stability.

Continuous wavelength tuning has been achieved with monolithic integrated multiple electrode DFB and DBR lasers. It is based on the principle that the injected free carriers shift the absorption edge, which in turn changes the index of refraction through the Kramers–Kronig relation, which describes the relationship between the refractive index and the absorption near the band edge. The wavelength is tuned by the change in the Bragg wavelength through the change in the effective index of refraction.

The carrier-induced refractive index change is approximated by

$$\Delta n_{\text{eff}} = \Gamma \left(\frac{d\bar{n}'}{dN} \right) \Delta N \tag{20}$$

where $\Gamma (= 0.3–0.5$ for a DH laser) is the optical mode confinement factor, $(d\bar{n}'/dN) = -(2.8 \pm 0.6) \times 10^{-20}\,\text{cm}^3$ for $1.3\,\mu\text{m}$ InGaAsP lasers, $d\bar{n}'$ is the change in the real part of the refractive index, and ΔN is the incremental change in the injected carrier density. The wavelength tuning range can be estimated by

$$\frac{\Delta\lambda}{\lambda} = \frac{\Delta\bar{n}_{\text{eff}}}{\bar{n}_{\text{eff}}} \tag{21}$$

Practically, the maximum index change is about 1%, which results in an estimated tuning range of 10–15 nm. Detailed discussions on tunable lasers are found in Section 7.4.5.

7.3 QUANTUM-WELL AND STRAINED-LAYER QUANTUM-WELL LASERS

A significant improvement of semiconductor lasers was made with quantum-well structures[28-30] fabricated by means of MBE and MOCVD. Using these growth techniques, very thin active layers that have a thickness comparable to the de Broglie wavelength ($\lambda = h/p$ or 200–300 Å) of the confined carriers are grown while maintaining high material quality. Quantum confinement results in significant improvement in laser performance, such as reduction in the threshold currents, high output powers, and very high speed, compared to the conventional double heterostructure lasers.

The quantization of energy states in the quantum-well also made possible the tailoring of the emission wavelength. In the mid 1980s, there was an interest in extending the emission wavelength beyond $0.88\,\mu\text{m}$, which corresponds to the bandgap of GaAs. By adding a small amount of indium to a GaAs quantum well, the emission wavelength can be increased to $0.98\,\mu\text{m}$ for use as the pump laser wavelength for the erbium-doped fiber amplifiers. The small amount of strain due to the mismatch of InGaAs to GaAs does not degrade the device reliability as long as the strained layer is within a critical thickness. Rather, the strained-layer quantum-well devices have better performance than the unstrained and bulk devices.

Conventional semiconductor lasers are made of double heterostructures (DH) and are available commercially today. Although the active layer in a DH laser is thin (1000–3000 Å) in order to confine electrons and the optical field, the electronic and the optical properties remain the same as in the bulk material. In a quantum-well laser, on the other hand, the active layer is made very thin ($\sim$100 Å). The kinetic energy of the electrons in the direction of the layer growth becomes quantized into discrete energy levels. Assuming an infinitely deep well, the confined energy levels are given by

$$E_n = \frac{\hbar^2}{2m^*}\left(\frac{n\pi}{L_z}\right)^2 \tag{22}$$

where $\hbar$ is the reduced Planck constant, m^* is electron effective mass, L_z is the thickness of the quantum-well, and n is an integer. A similar expression holds for the valence band. Figure 13 illustrates the two lowest energy levels in the conduction band in a one-dimensional quantum-well. The density of states changes from the parabolic dependence of energy to a step-like function, as depicted in Fig. 14. Since the density of states is constant, rather than gradually increasing from zero, there is a group of electrons of nearly the same energy available to recombine with a group of holes of nearly the same energy. This results in a differential gain that is larger than that in conventional double heterostructures. The implication for laser characteristics is significant. First, the reduced density of states in the step-like structure requires fewer electrons to reach optical transparency (a small value of N_{tr}), leading to a low threshold current density. Second, the differential gain, A ($= dg/dN$), is much larger than that of bulk materials,[29] giving rise to high speed (Eq., 14) and narrow linewidth (Eq. 19). Third, increased gain spectral width can be achieved by optimizing the laser cavity design to extend the lasing wavelength range from the lowest energy state to higher energy states.

Quantum-well lasers have been successfully made in GaAs/AlGaAs material systems for some time. A comprehensive review of earlier work can be found in Refs. 28 and 30. Threshold current densities as low as 65 A/cm^2 and submilliampere threshold currents have been reported. Recently, significant progress has also been made in InGaAs/InGaAsP multiple-quantum-well (MQW) systems for the 1.3 μm and 1.5 μm wavelength regions.[31,32] Examples of MQW laser structures are shown in Fig. 15. Figure 15a shows a schematic diagram of the separate-confinement-heterostructure (SCH) multiple-quantum-well laser where four quantum wells of InGaAs with InGaAsP barrier layers are sandwiched between the InP cladding layers to form a waveguide with a step index change.[31] The active region is composed of four 80-Å-thick InGaAs quantum-wells (undoped) separated by 300-Å-thick InGaAsP (1.3 μm wavelength, undoped) barriers. A bandgap diagram of the active region is shown in Fig. 15b. The n- and p-cladding InP layers are doped with sulfur (10^{18}/cm^3) and zinc (10^{17}/cm^3), respectively. Figure 15(c) shows a graded-index separate confinement-heterostructure (GRIN-SCH) where a graded index of the waveguide is accomplished by several small stepwise increases of the bandgap energies of multiple cladding layers.[32] The GRIN-SCH structure confines both the carriers and the optical

ONE DIMENSIONAL CASE

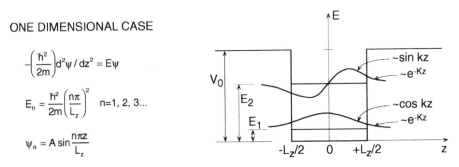

$$-\left(\frac{\hbar^2}{2m}\right)d^2\psi/dz^2 = E\psi$$

$$E_n = \frac{\hbar^2}{2m}\left(\frac{n\pi}{L_z}\right)^2 \quad n=1, 2, 3...$$

$$\psi_n = A\sin\frac{n\pi z}{L_z}$$

Fig. 13   Wave functions and energy subbands of an infinitely deep quantum well.

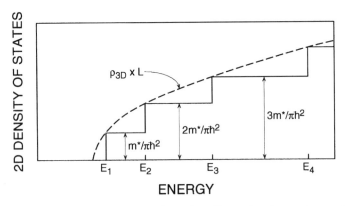

Fig. 14   Density of states of a two-dimensionally confined quantum well compared to that of a bulk semiconductor.

field more effectively than the SCH structure, and, consequently, leads to a lower threshold current density. These structures also exhibited low waveguide losses, permitting the fabrication of long cavities for both high output power and narrow laser linewidth.

Although $In_xGa_{1-x}As$ is lattice matched to InP at $x = 0.53$, biaxial strain can result from tensile strain at $x < 0.53$ to compressive strain at $x > 0.53$, as illustrated in Fig. 16. When $x < 0.53$, the lattice is stretched in the direction of the growth plane (x, y), and consequently the lattice constant in the z direction is reduced. Conversely, when $x > 0.53$, the lattice is compressed in the growth plane and the lattice constant is elongated in the z direction. It has been proposed[5] that under biaxial strain the valence band structure can be modified and the conduction-band discontinuity can be increased.[33] Figure 17 shows the modified band structures of $In_xGa_{1-x}As$ resulting from the compressive and tensile strain while maintaining the emission wavelength at 1.5 μm. In the unstrained case (middle), the heavy-hole and the light-hole

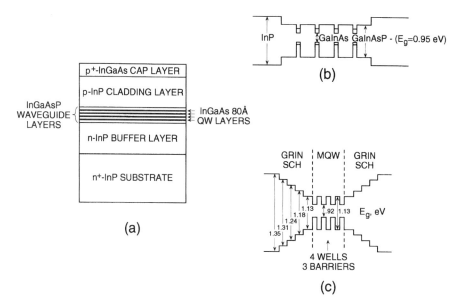

Fig. 15 (a) Schematic of the cross-section of an InGaAs/InGaAsP multiple-quantum-well laser structure. (b) Detail schematic of the bandgaps of the SCH-MQW layers shown in (a). (After Koren et al., Ref. 31.) (c) GRINSCH-MQW structure with several thin cladding layers of increasing bandgaps to approximate the graded-index change. (After Kasukawa et al., Ref. 32.)

bands are degenerate at the zone center. The effective mass of the heavy hole is much larger than that of the electrons in the conduction band, whereas the effective mass of the light hole is similar to that of the electron. The optical transition occurs between the conduction band and the heavy-hole band. Under compressive strain (right), the heavy-hole band level moves up. The heavy-hole effective mass is reduced in the in-plane (x–y) direction. The optical transition is transverse-electric (TE) polarized. Under biaxial tensile strain (left), the light-hole band level moves up. The in-plane effective mass remains small. The optical transition is transverse-magnetic (TM) polarized. The low effective hole mass reduces the carrier density needed for population inversion. Furthermore, the splitting of the hole subbands reduces the nonradiative Auger recombination and intervalence band absorption. These effects, together with the increased conduction-band discontinuity in the compressively strained layer reduces the lasing threshold current and improves the quantum efficiency of strained-layer MQW lasers.

To estimate the threshold-current-density reduction, we consider the best case for a symmetric conduction and valence band structure, that is, $m_v = m_c = m$, (m_v and m_c are effective masses of holes and electrons respectively). The transparency carrier area-density per well for a quantum-

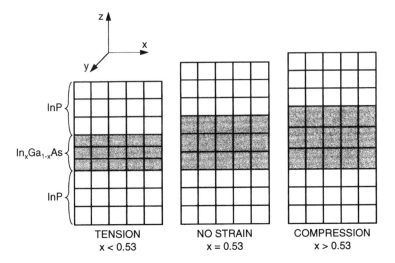

Fig. 16 Schematic diagram of the lattice structures of unstrained and strained layers.

well laser with a symmetric band structure, N_{tr}^s, which is derived in Appendix 7.B, is given by[34]

$$N_{tr}^s = \frac{4\pi mkT}{h^2}\ln(2) \tag{23}$$

By curve fitting to numerical solutions for a quantum-well laser with an asymmetric band structure (i.e., $m_v > m_c$), the transparency carrier area-density per well, N_{tr}^a, can be approximated by a heuristic formula,

$$N_{tr}^a = \left(\sqrt{\frac{m_v}{m_c}} + \frac{2m_v}{m_c + m_v}\right)\frac{N_{tr}^s}{2} \tag{24}$$

Assuming $m_c = 0.041m_0$ and $m_v = 0.70m_0$, where m_0 is the free electron mass, the transparency carrier area density of the unstrained-quantum-well laser is about a factor of 3 larger than that of a strained quantum-well laser.

Also as shown in Appendix 7.B, because of the step-like density of states, the gain per quantum-well from the lowest energy subband transition is finite and given by[34]

$$G_{max} = \frac{8\pi^3 h\nu|M_T|^2}{\varepsilon_s v_g h^3}\frac{m_c m_v}{m_c + m_v}\frac{1}{L_z} \tag{25}$$

where $|M_T|^2$ is the transition matrix element and ε_s is the dielectric

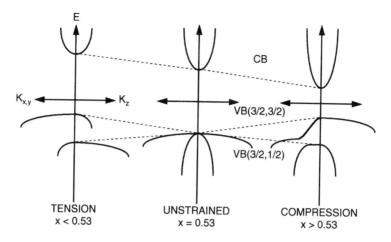

Fig. 17 Modified band structures of $In_xGa_{1-x}As$ under tensile and compressive strain.

permittivity of the material. Thus, the maximum gain of an asymmetric conduction and valence band structure is twice that of the symmetric band structure, that is $G_{max}^a = 2G_{max}^s$. The maximum differential gain at transparency is given by

$$\frac{dG_{max}}{dN} = \frac{\pi^2}{\ln(2)} \frac{\nu |M_T|^2}{\varepsilon_s v_g kT} \tag{26}$$

Equation 26 is derived in Appendix 7.B. The differential gain increases with reduced asymmetry between the electron and hole effective masses and becomes optimal for $m_c = m_v$. Because of the increased differential gain in strained-layer quantum-well lasers, the modulation bandwidth will be enhanced, and the linewidth enhancement factor, hence, the linewidth and frequency chirp, will be reduced. Furthermore, the threshold current is proportional to $T^{3/2}$ in bulk lasers but proportional to T in strained-layer quantum-well lasers (Eq. 23). Thus, the temperature sensitivity of the threshold current for strained-layer quantum-well lasers is reduced.

Figure 18 shows a summary[35] of reported threshold-current densities per quantum-well of both compressive and tensile-strained single-quantum-well lasers emitting at a 1.5 μm wavelength. Significantly lower threshold current densities are obtained compared to unstrained quantum-well lasers. With increasing compressive strain, the in-plane hole effective mass monotonically decreases, contributing to the observed decrease in the threshold current densities. However, for large compressive strains, the InGaAs well thickness becomes very thin to maintain the 1.5 μm wavelength emission. As a result, the variations in the well thickness become larger and reduce the gain.

Consequently, a larger threshold current density is required. This effect can be remedied by using the larger bandgap InGaAsP for the quantum-well with thicker wells.

Note that in Fig. 18 there is no low threshold region for the tensile-strained branch between 0 and 1% strain. This is due to the fact that for the particular well thickness there is a bandmixing effect that increases the heavy hole effective mass. As the strain increases more than 1% with a simultaneous increase in the well thickness, the separation of the light-hole and the heavy hole bands is large enough to reduce the bandmixing and, consequently, the hole effective mass. As a result, the threshold current is significantly reduced. For larger strains, the well thickness becomes wider than the critical thickness and in some cases results in misfit dislocations. Thus, the increased threshold current densities can be attributed to the defects introduced in the crystal by the large amount of strain.

In addition to the reduction in the threshold current densities, a high quantum efficiency of 82% and a small increase of threshold current with temperature were reported for strained-layer MQW lasers.[36] Furthermore, lasers with powers as high as 380 mW have been obtained for 1.3-μm strained-layer MQW lasers, and over 300 mW was achieved in 1.5-μm strained-layer MQW lasers.[36,37] With the MQW and strained-layer MQW structures, a variety of advanced lasers and photonic integrated circuit (PIC) structures are made possible for future system applications, which are described in the following section.

7.4   ADVANCED LASER STRUCTURES AND PHOTONIC INTEGRATED CIRCUITS (PICs)

7.4.1   Submilliampere Threshold Current Lasers

To achieve a low threshold current using the strained-layer quantum-well, the laser cavity volume must be minimized. Thus, a narrow active stripe and short cavity length are necessary. Since the mirror loss is dominant in the short cavity, high-reflection mirrors (made by dielectric coating to minimize the mirror loss) are used. In addition, the leakage current is minimized by using either the buried heterostructure (BH) or the semi-insulating planar buried heterostructure (SI-PBH). For both compression-strained and tensile-strained quantum-well lasers, threshold currents below 1 mA have been achieved.[38–41] Table 1 summarizes the results published in the literature. The threshold current remained less than 10 mA at 100°C.

7.4.2   Uncooled Lasers

The low threshold current achieved in strained-layer quantum-well lasers results in a small increase in the junction temperature caused by Joule

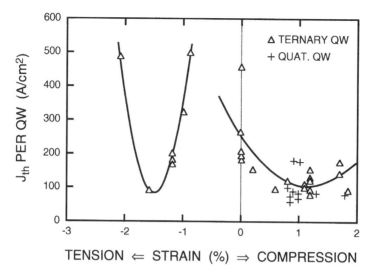

Fig. 18 Summary of threshold current densities vs the strain in 1.5 μm wavelength InGaAsP QW lasers. (After Thijs, Ref. 35.)

heating. Combined with the reduced Auger recombination, the threshold current is expected to increase slowly with temperature. This leads to laser devices that do not need a Peltier (thermoelectric) cooler to keep the junction temperature low at high ambient temperatures. Thus, they allow reduced cost and improved reliability.

The conventional quaternary InGaAsP quantum-well lasers have been optimized in terms of the number of wells, barrier composition, amount of strain and well thickness to improve performance at temperatures between -40 to $+85°C$, that is, the operating temperature range of telecommunications equipment in an uncontrolled environment. The difficulty for InGaAsP lasers to operate at high temperatures is due to the fact that electrons tend to leak through the barrier layer at elevated temperatures, since the conduction-band offset of an InGaAsP quantum-well is small ($\Delta E_c = 0.4\Delta E_g$). Recently, it has been found that AlGaInAs strained-layer quantum wells have a larger conduction-band offset ($\Delta E_c = 0.7\Delta E_g$) than that of the InGaAsP quantum wells.[42] Therefore, the electrons in the wells are confined even at high temperatures. The AlGaInAs material system also allows the growth of a GRIN-SCH structure because of the similar ion radii of aluminum and gallium, and thereby, provides a better optical-field confinement. Ridge-waveguide-structure lasers, similar to the structure shown in Fig. 5b, using 1.3-μm AlGaInAs compressively strained quantum-well materials were fabricated with ridge waveguide 3 μm wide, and cavity 300 μm long.[42] The rear end facet was coated to provide 70% reflectivity. Figure 19 shows the light-output vs injection-current characteristics at various

TABLE 1 Summary of Submilliampere Threshold Lasers

Structure	I_{th} (mA)	Strain	Reference
DCPBH	0.9	Compression	38
SI-BH	0.9	Compression	39
BH	0.88	Compression	41
SI-BH	0.8	Compression	40
SI-BH	0.62	Tensile	40

temperatures. The threshold current at room temperature was 14 mA, and increased to 35 mA at 100°C. Note that the differential quantum efficiency dropped less than 1 dB over the temperature range 25–100°C. The maximum intrinsic 3-dB bandwidths are measured as 19.5, 15, and 13.9 GHz at 25, 65, and 85°C. The improvement of the high-temperature performance is significant when compared with that of the conventional InGaAsP lasers.

7.4.3 High-Speed DFB Lasers and Narrow-Linewidth Lasers

The design of high-speed lasers with narrow linewidth and low chirp requires optimization of various materials and device parameters, as discussed in Sections 7.2.5 and 7.2.6. Typically the optimization process involves:

1. Increasing the differential gain by using quantum-well structures and by applying wavelength detuning in DFB lasers.
2. Reducing the nonlinear gain coefficient by minimizing spatial hole burning.
3. Shortening the photon lifetime by using a short cavity.

Quantum-well structures can be further enhanced by including compressive or tensile strain to increase the differential gain. However, for quantum-well lasers the effects of carrier transport time across the optical confinement layers and the carrier escape time from the active wells must be considered.[43] These lead to the optimization of the well and barrier thickness for high-speed lasers. Finally, the RC time constant of the finished laser must be minimized. The RC time constant is a function of the series resistance and the chip capacitance. Since the series resistance of a forward-biased junction is less than a few ohms, the chip capacitance is the major contribution of the RC time constant.

Various examples of DFB laser structures with low capacitance for high-speed applications are shown in Fig. 20. Figure 20a shows an etched mesa buried heterostructure (EM-BH) laser with semi-insulating Fe–InP

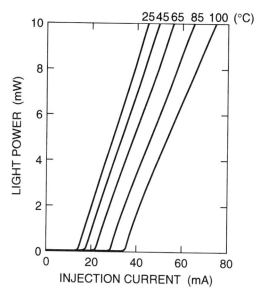

Fig. 19 Light–current characteristics of 1.3-μm AlGaInAs strained-QW ridge waveguide lasers at various temperatures. (After Zah et al., Ref. 42.)

regrown inside the etched grooves.[44] The contact metalization is restricted to the top of the mesa stripe by applying a dielectric film over the entire chip except for the contact stripe to minimize the chip capacitance. Figure 20b shows another etched-mesa flat buried heterostructure (EM-FBH) without regrowth.[45] A very low capacitance structure is achieved by using a self-aligned constricted mesa structure[46] (SA-CM) as shown in Fig. 20c. The use of a polyimide layer to minimize parasitic capacitance has become fairly common because the process is simpler than the regrowth shown in Fig. 20a and c. An etched mesa BH structure DFB laser (PIQ-BH) planarized with a Hitachi Chemical PIQ6200 polyimide layer is shown in Fig. 20d.[47] The best bandwidth was 17 GHz for a DFB laser with a bulk active layer and 22.5 GHz for lasers with an MQW active layer, both obtained by optimizing the doping in the active layer and the waveguide width, and by detuning the wavelength.[48] In the short wavelength region, a relaxation frequency as high as 30 GHz has been demonstrated in GaAs/AlGaAs MQW lasers.[49]

Theoretically, QW lasers have higher differential gain that leads to a small linewidth enhancement factor and lower waveguide loss, which permits a longer laser cavity. Both of these features help to reduce the linewidth. In practice, the laser linewidth ranged from 3 MHz for a DH DFB laser with a standard length of 300 μm to as narrow as 70 kHz for a compressively strained multiple-quantum-well DFB laser.

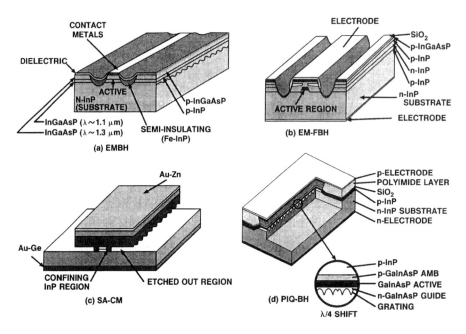

Fig. 20   Various high-speed DFB laser structures.

7.4.4   DFB Lasers Integrated with External Modulators

Since the laser chirp width is directly proportional to the modulation frequency (see Eq. 18), the transmission distance in conventional fibers using direct-current-modulated DFB lasers is limited at high bit rates (2.5 Gbit/s and above) by the chromatic dispersion in the fiber at the 1.55 μm wavelength region, as described in Section 7.2.6. The frequency chirp of the laser can be reduced by the use of an external modulator. The most attractive modulator is the electro-absorption (EA) modulator that can be monolithically integrated with a DFB laser.

The absorption modulator is based on the quantum-confined Stark effect (QCSE), which stems from a large field-induced variation of the absorption coefficient near the band edge. The MQW material produces a large QCSE at small bias, therefore, it is attractive for modulator applications.[50] The EA modulator can be operated at high speed with low chirp ($\alpha \approx 1$) at a small voltage drive. Another advantage of the EA modulator is that it is possible to integrate it with the DFB laser. This not only reduces the size and cost, but also improves performance. The light output, for example, increases substantially because the insertion loss is very small compared with discrete modulators. The long-term reliability is also improved by the robustness of the packaging.

Monolithic integration of the laser and the EA modulator has been achieved by etch and regrowth, and by the selective-area growth techniques.[51] For MQW structures, the desired bandgap of the EA modulator can be obtained by designing the quantum-well thickness and composition. In selective-area growth by MOCVD, where the substrate wafer is partially covered by a patterned oxide mask, the growth rate varies with the size of the oxide pattern. Therefore, different quantum-well thickness can be obtained with a single epitaxial growth by varying the size of oxide patterns. This method is very attractive because the well thickness varies gradually in the transition region and produces nearly perfect coupling between the laser and the modulator.

Figure 21 shows the photoluminescence (PL) peak wavelength of a wafer sample by selective-area growth with MOCVD.[51] The measurement was made along the optical axis in the masked and unmasked areas. The sample contained masked regions with $W_g = 15\,\mu m$ and three oxide mask widths, $W_m = 15$, 20, and 25 μm. As shown in the figure, the PL peak wavelength in the unmasked area was 1.48 μm, and that in the masked areas varied from 1.53 μm to 1.55 μm depending on W_m. DFB lasers were fabricated in the masked areas, and modulators were fabricated in the unmasked areas. Figure 22 shows the structure of an MQW electro-absorption modulator integrated with a DFB laser. The device was based on a separate confined-heterostructure (SCH) InGaAs/InGaAsP MQW, and a semi-insulating buried heterostructure, grown by a three-step, low-pressure MOCVD process. The modulator absorption layer and the laser active layer form one continuous SCH-MQW structure with slightly different thicknesses and compositions. A partial cross-sectional view of the modulator–laser section is shown in the inset. Notice that a window section exists near the modulator end facet. The window section is necessary to reduce the end-facet reflections, which may increase the frequency chirp under modulation.

The static characteristics of a modulator-integrated laser are shown in Fig. 23 for various laser bias currents. The integrated laser has a total length of 600 μm (laser 400 μm, modulator 150 μm, and separation 50 μm). The threshold current and the slope efficiency at zero modulator bias were 17 mA and 0.10 W/A respectively. Extinction (on–off) ratios of 15 dB at −1 V and 28 dB at −2 V were observed. The dependence on the laser optical power was sufficiently low to permit high-power operation of this device using an EA modulator. A 500-km transmission experiment using a dispersion-shifted fiber was conducted. Two erbium-doped fiber amplifiers were employed as the booster and preamplifier, and another five were used as the in-line amplifiers. Error-free transmission at 10 Gbit/s was achieved with negligible dispersion penalties.

7.4.5 Tunable Lasers

Grating-Tuned External-Cavity Lasers. The most conventional method to construct a tunable laser is to use a semiconductor laser chip as the gain

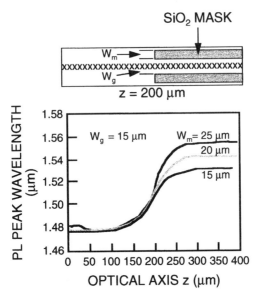

Fig. 21 Distribution of photoluminescent peak wavelength measured along the optical axis in the masked and unmasked growth regions in selective-area growth by MOCVD. (After Aoki et al., Ref. 51.)

medium in an external cavity with a diffraction grating serving both as a mirror and as a narrow-band filter. This arrangement is shown in Fig. 24. One facet of the laser diode is anti-reflection (AR) coated, and the light from this end is collimated by a lens. The other cleaved facet and the diffraction grating form the external cavity. The lasing frequency is tuned by rotating the grating. Fine tuning can be achieved by axial displacement of the grating or by adding an adjustable phase plate. In principle, a tuning range over the entire width of the gain spectrum is possible. In practice, however, the maximum range obtained was 55 nm, centered at a wavelength of 1.5 μm, limited by the gain available in the semiconductor chip to overcome the total loss of the assembly. The light-coupling efficiency of the lens imposed a major limitation. Because of the long cavity, such a grating-tuned external-cavity laser has exhibited a very narrow linewidth of ~10 kHz. An extremely wide tuning range of 105 nm centered at an wavelength of 0.8 μm has been obtained using an optimized GaAs/AlGaAs single-quantum-well laser.

Multiple-Electrode DFB Lasers.   The high injected-carrier density (10^{18}/cm^3) in the semiconductor laser reduces the effective index of refraction in the corrugation region (the Bragg region), thereby decreasing the lasing wavelength. However, in a single-electrode DFB laser operated above threshold, most injected carriers recombine to produce photons, resulting in a very small increase in carrier density, which leads to a small change in

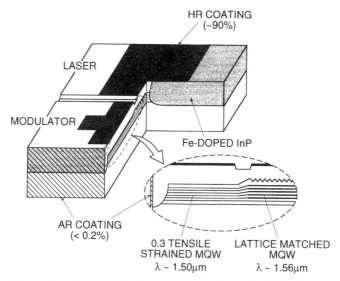

Fig. 22 Schematic diagram of the structure of an MQW electro-absorption-modulator-integrated DFB laser by selective-area growth. (After Aoki et al., Ref. 51.)

lasing wavelength. The range of wavelength tuning can be improved by using a two- or three-electrode DFB laser, with a large current applied to one electrode and a small current to the other.[52-54] Schematic diagrams are shown in Fig. 25.

The operational principle of wavelength tuning can be qualitatively understood as follows. In the asymmetric structure of the DFB laser the optical field is higher in the region near the output port where the facet is not reflecting (e.g., where AR coated), and the wavelength is primarily determined by the effective index of refraction in this region. With this section biased at current densities at or slightly below the threshold current density, it serves as a Bragg reflector. In addition, because of the low pumping level, the injected carriers do not contribute significantly to photon generation, resulting in a large change of the refractive index, and, thereby, substantial wavelength tuning. The gain is provided by the other section pumped substantially above threshold. A continuous wavelength tuning range of 1–2 nm was reported.[52]

A three-electrode, $\lambda/4$-shifted, DFB laser[53] also showed good tunability. The two outer electrodes were electrically connected to a common current supply, whereas the central electrode was supplied with a different current. As shown in Fig. 26, a continuous tuning range of 1.9 nm was obtained by varying the ratio of these two currents. Because of the long cavity length (1.2 mm) used in this tunable laser, a linewidth as low as 500 kHz was obtained.

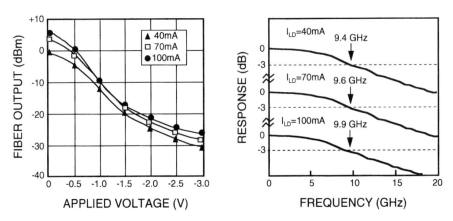

Fig. 23 Static characteristics of an EA modulator-integrated DFB laser diode. The insert shows the laser bias currents. (After Aoki et al., Ref. 51.)

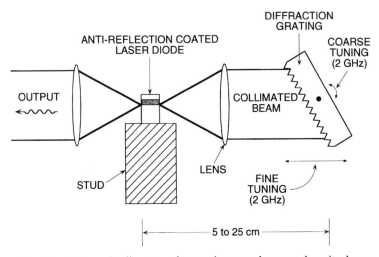

Fig. 24 Schematic diagram of a grating-tuned external-cavity laser.

The advantage of the tunable DFB laser is its ease of fabrication. The disadvantage is the limited tuning range, because the tuning section must be biased below threshold. When the tuning section is biased above threshold, both sections can oscillate independently, resulting in mode competition, which causes the laser output to switch between different modes. This behavior is known as mode hopping.

Another approach is to use a twin-guide DFB laser structure,[55] as shown in Fig. 27. Above the active layer in the waveguide is a tuning layer, which

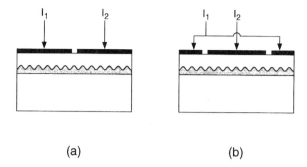

Fig. 25 Schematic diagrams of two-electrode (a) and three-electrode (b) wavelength tunable DFB lasers.

is electrically separated from the active layer, and independently biased. Since the carriers in the tuning layer only change the refractive index of the waveguide, continuous tuning can be achieved without mode hopping. A continuous tuning range of 7.1 nm is achieved under a constant current in the active layer and a varying current in the tuning layer. A 5.4-nm tuning range has been obtained when both currents are tuned to maintain a constant output power of 1 mW.

Multiple-Electrode DBR Lasers. The wavelength tuning range can be improved by separating the Bragg region in the passive waveguide (a large bandgap material) from the active region (a small bandgap material) inside the laser cavity.[56–58] Figure 28 shows a schematic of a wavelength tunable DBR laser. The active layer extends a short length above the waveguide layer, made by selective etching and regrowth processes. The corrugated region serves as a tunable DBR. The wave length is electronically tuned by the current injection into the DBR section. The injected carrier density in the passive Bragg region can be high to increase the refractive index changes because carriers in the passive region do not contribute to the gain.

The Bragg reflector exhibits a high reflection within a certain wavelength band (the stop band), which is nominally between 2 nm and 4 nm wide. The mode that is nearest to the center of the band and simultaneously satisfies the $2m\pi$ round-trip phase condition will lase. Thus, by introducing a phase region in the waveguide, independently controlled by the injected current, the lasing wavelength can be tuned to each Bragg wavelength. A theoretical treatment is found in Ref. 59. With proper design and independent adjustment of the three currents in the active, the Bragg, and the phase sections, quasi-continuous tuning ranges from 8 nm to 10 nm have been achieved.

Finally, a two-section DBR laser with accurate channel spacing of 50 GHz has been demonstrated.[60] Figure 29 shows the optical frequency tuning characteristics vs the tuning current.

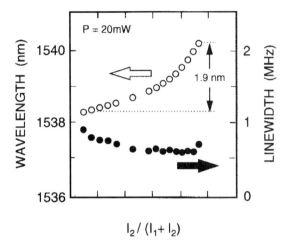

Fig. 26 Wavelength tuning and linewidth characteristics of a three-electrode DFB laser. (After Kotaki et al., Ref. 53.)

The tuning range of ~10 nm is nearly equal to the maximum estimated range according to Eqs. 20 and 21. To further increase the tuning range, three independently tunable DBR lasers with an output combiner and an amplifier were monolithically integrated on one chip,[61] as shown in the schematic diagram of Fig. 30. The passive waveguide combines the output of three lasers, and the amplifier boosts the output power. The overall tuning range was extended to 21 nm.

Tunable DBR lasers have been used as the local oscillator in a balanced coherent receiver made by photonic integrated circuit (PIC) design. Detail discussion of the PIC coherent receiver is given in Section 7.5.4.

7.4.6 Multiwavelength DFB Laser Arrays

Single-mode optical fiber provides an extremely broad transmission bandwidth, over 30 THz. The traditional time-division-multiplexed (TDM) systems employing electronic multiplexers are limited to a bandwidth of about 40 GHz, beyond which optics must be used for multiplexing and demultiplexing the signal. The high cost of electronics at this speed may prohibit the use of ultrahigh-speed TDM systems. On the other hand, increasing the transmission bandwidth in fibers by wavelength-division multiplexing (WDM) is rather simple. It is also cost-effective if it is used in conjunction with erbium-doped fiber amplifiers (EDFAs) to eliminate the electronic regenerators. In addition, multiple wavelength operation can switch and route signals to permit reconfiguration of an optical network.

A multiwavelength DFB laser array is attractive as the light source in

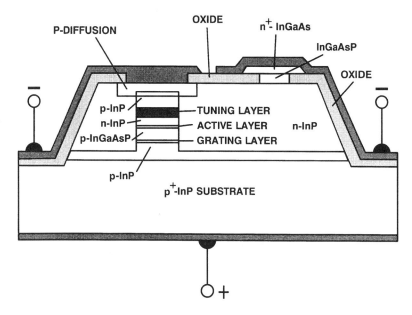

Fig. 27 Schematic diagram of a tunable twin-guide laser structure. (After Illek et al., Ref. 55.)

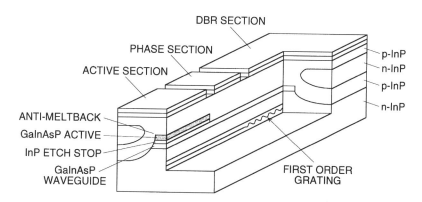

Fig. 28 Schematic diagram of a wavelength-tunable three-electrode DBR laser.

multiwavelength optical networks.[62,63] For the multiwavelength signals to propagate in the network with minimum degradation through many narrow-band optical channel filters, the accuracy of channel wavelengths and channel spacing is extremely important. It is difficult to meet the accuracy requirement with the tunable lasers discussed previously. However, multiwavelength DFB laser arrays can indeed meet the challenge.[64]

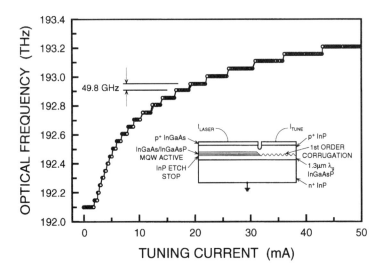

Fig. 29 Frequency tuning characteristics of a DBR laser with a constant 50 GHz per channel. (After Shankaranarayanan et al., Ref. 60.)

To make DFB laser arrays easier to package, photonic-integrated-circuit (PIC) techniques can integrate the laser arrays with a power combiner (star coupler) based on the array waveguide design on an InP substrate.[65–67] Thus, any one of the output waveguides contains all wavelengths of the DFB laser array. To overcome the inherent splitting loss of the star coupler, a booster amplifier is built in series with the output waveguide. Only a single-mode fiber pigtail is needed to couple the multiwavelength light signals into the transmission fiber. This leads to a simple and robust package.

Integrated laser arrays with as many as 21 wavelengths have been fabricated.[65] Figure 31 shows an improved design of a 20-DFB laser array.[66] Figure 31a shows a schematic top view of the laser, combiner, and amplifier layout, and Fig. 31b is a micrograph of a completed chip showing the top contact pads and the output waveguides. For the 20-DFB lasers, every two lasers have the same wavelength. This wavelength redundancy is helpful for improving chip yield.[64,66] The wavelengths are from 1544 nm to 1562 nm and are within the gain bandwidth of the EDFA. Wavelength channel spacing is 2 nm. The star coupler combines the output from the DFB laser array through the passive waveguide array, then couples to four output waveguides. Two of the output waveguides contain an optical amplifier to compensate for the splitting and diffraction losses of the star coupler and the waveguide losses. Two passive output waveguides are used for redundancy.

The passive waveguide structure is a 0.2-μm-thick InGaAsP layer with a composition corresponding to a bandgap wavelength of 1.25 μm. An array of 3-μm-wide waveguides are etched and then buried with a regrown

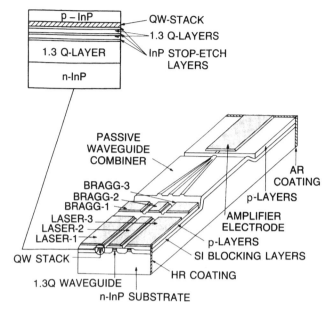

Fig. 30 Schematic diagram of an integrated wavelength-tunable laser with an extended tuning range of 21 nm. (After Koren et al., Ref. 61.)

semi-insulating-InP layer. In the DFB laser region, the active layer consists of six strain-compensated quantum-wells grown on top of the waveguide layer. The gratings with ten different pitches for the DFB lasers are patterned by e-beam direct writing. Each DFB laser is 370 μm long. The spacing between the lasers is 50 μm.

The two output booster amplifiers are 590 μm and 880 μm long, respectively, to provide different maximum gain and saturation power levels. A window structure, similar to that of the modulator described previously, is incorporated to reduce the end-facet reflection that may degrade the laser performance. An anti-reflection coating is applied to the facets to further reduce the facet reflections. The entire chip size is 4.3 mm $\times$ 1.4 mm.

Figure 32 shows the wavelength and the threshold current distribution of the integrated laser array.[67] The slope of the wavelength line indicates the wavelength spacing of 1.6 nm (or 200 GHz in frequency) between channels, which have a channel spacing accuracy of 0.2 nm. The absolute wavelength variation is $\pm$1 nm. Precise wavelength registration can be obtained by additional temperature tuning. Figure 33 shows the output spectrum of 10 wavelengths.[67] Note that in Fig. 32, the threshold current distribution is quite uniform. The nominal threshold current is around 25 mA. Driven at a bias current of 60 mA, the output power per channel into the single mode fiber is -13 ± 1.5 dBm. The output power can be boosted by the on-chip optical

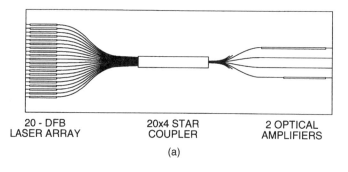

20 - DFB   20x4 STAR   2 OPTICAL
LASER ARRAY   COUPLER   AMPLIFIERS

(a)

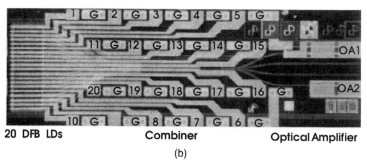

20 DFB LDs   Combiner   Optical Amplifier

(b)

Fig. 31 Design of a 20-DFB laser array. (a) Schematic top view of the laser, combiner, and amplifier layout. (b) Micrograph of a completed chip showing the top contact pads and the output waveguides. (After Zah et al., Ref. 67.)

amplifier. An output power of 0 dBm (1 mW) per wavelength channel is achieved with the 590-μm-long amplifier biased at 200 mA. The accuracy of the channel spacing and the uniformity of the threshold currents in the laser array can be attributed to the proximity effect in that the material is more uniform within a local region than in the entire wafer.

7.5 PHOTORECEIVERS AND OPTOELECTRONIC INTEGRATED CIRCUITS (OEICs)

7.5.1 Receiver OEICs

In the last decade, OEIC research focused on photoreceivers. In its simplest form, the photoreceiver incorporates a photodetector to convert the optical signal into an electrical signal, followed by a low-noise electronic preamplifier IC to raise the electrical signal level to a value that can be used for further signal processing. In compound semiconductors, there is a wide variety of photodetector and transistor device technologies available for both the circuit designer and the process engineer to realize OEIC photoreceivers. We shall

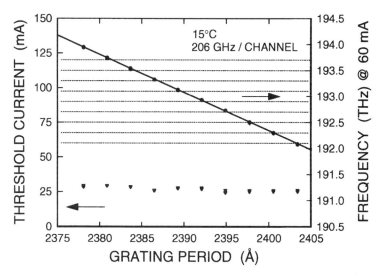

Fig. 32 Wavelength and the threshold current distributions of the integrated laser array. (After Zah et al., Ref. 67.)

briefly consider the choices for photodetectors and transistors suitable for monolithic integration.

Photodetectors.   The role of the photodetector is to convert the incoming optical signal into an electrical signal efficiently. There are several types of detectors,[7] including the Schottky-barrier (SB) diode, metal–semiconductor–metal (MSM) photodiode, *p–i–n* photodiode, avalanche photodiode (APD), and heterojunction phototransistor (HPT).

The SB diode and the MSM detector[68] are probably the simplest from the aspect of implementation. The device uses a Schottky barrier between the metal and the semiconductor to form a depletion region where photoabsorption can take place. The MSM is a planar device with inter-digitated electrode and therefore has very low capacitance. Both the SB and MSM devices have high-speed performance capability, with the MSM detector trading some quantum efficiency for bandwidth. Both detectors have been used in receiver OEICs as they are very amenable to integration with planar electronic devices. The devices do, however, exhibit certain nonlinear characteristics with respect to bias voltages and optical intensity, and these are related to the charge-trapping effects at the Schottky barrier as well as the zero-electric-field regions under the electrodes.

The *p–i–n* photodetector is the most commonly used device because of its high quantum efficiency, low-voltage operation, high speed, and ease of fabrication. Both planar (diffused junction) and mesa-structure *p–i–n*

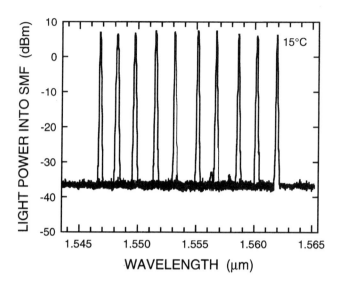

Fig. 33 Output spectrum of a 10-wavelength integrated laser array. (After Zah et al., Ref. 67.)

photodiodes have been used for monolithic integration. The avalanche photodetector (APD) provides gain through the avalanche mechanism, but requires very critical epitaxial layers and high operating voltage. It has not been used, so far, for OEICs. The heterojunction phototransistor[69] (HPT) is basically a bipolar transistor with the base–collector junction acting as a p–i–n photodetector. The transistor mechanism provides gain, but the speed is limited by the charging times of the two junction capacitances and the base-storage times. Nevertheless, it is a very attractive device for monolithic integration.

Transistors.   The electronic functions following photodetection are generally performed by a low-noise preamplifier. Here again, several types of device structures are available for monolithic integration. They can be categorized into field-effect and bipolar devices.

In the field-effect transistors (FETs), the transport of majority carriers from the source to drain is controlled by a field applied to a gate terminal. There are different ways of achieving the field effect, which produce JFETs, MESFETs, and modulation-doped FETs (MODFETs).[7] The FETs are fundamentally planar devices with the speed of the device dependent on the gate length, which is defined lithographically. For a 1-μm gate length, these devices can have unity-current gain cutoff frequencies f_T in the range 15–35 GHz. The small gate-to-source capacitance results in a large f_{max} (the maximum frequency of oscillation), typically in the range 20–50 GHz.

Submicron electron-beam lithography would be required to achieve higher speeds.

The bipolar device in compound semiconductors is the heterojunction bipolar transistor (HBT). The HBTs incorporate an emitter with a bandgap wider than that of the base, and so can have higher base doping and lower emitter doping than their homojunction counterparts. This results in low base resistance and low emitter–junction capacitance, together with high injection efficiency. In addition, when compared to compound semiconductor FETs, the HBTs have higher transconductance and drive capability. More importantly, HBTs with high-speed performance can be fabricated with modest optical lithographic design rules. For emitter feature size between $2\,\mu$m to $5\,\mu$m, f_T values in the range 50–200 GHz and f_{max} values in the range 35–120 GHz have been reported. These frequencies are possible because the transport of the minority carriers is in the direction of epitaxial growth, and these distances can easily be controlled to hundreds of angstroms by epitaxy.

Integration Technology.   There are several different approaches to integration of the photodetector and the transistor on a single chip. The requirements on the epitaxial layers for the photodetector and the transistors are in general not the same, and therefore one must exploit crystal growth technology to implement the OEIC receiver efficiently. In one approach, all the required epitaxial layers for the photodetector and the transistor are grown sequentially on a planar substrate with one device layer physically above or below the other device layers. Using photolithography and selective delineation, the two devices are realized and interconnected on a somewhat terraced surface topography. The advantage of this approach is that the integration is independent of the epitaxial growth technique and one can individually optimize the two devices. This method has been used to integrate *p–i–n* and MSM photodetectors with JFETs, MODFETs, and HBTs.

In the second approach, trenches are made in the substrate in the region where the photodetector is to be realized and, then, the epitaxial layers are grown everywhere. The trenching allows one to place the photodetector at about the same level as the transistor, resulting in a semiplanar surface topography. This is particularly helpful when implementing FETs, which are planar by nature, and which require fine-line lithography for high-speed operation. The disadvantage of this approach is that artifacts of the epitaxial growth on a patterned substrate have to be considered, which limits the transportability of the OEIC technology. This implementation method has been used in integrating *p–i–n* photodetectors with MODFETs and JFETs.

Finally, in a third approach, the photodetector and the transistor are realized from the same basic epitaxial layers with the two devices sharing some of the layers. The simplicity of this approach is only partially influenced by the resulting compromise on the performance of each of the devices. As an example of this implementation, the base–collector p–n junction of the HBT has been used to realize the p–i–n both in the GaAs and the InP material systems. In another version, the phototransistor (HPT) doubles as an HBT in an all-bipolar OEIC photoreceiver.

7.5.2 FET-Based Receiver OEICs

The very first monolithic photoreceiver was fabricated in the GaInAs/InP material system in 1980.[70] It used a GaInAs JFET whose gate was extended to form the p–i–n photodiode. The simplicity of fabricating a FET spurred world-wide effort to integrate it with a photodetector and to date, FET-based OEIC receivers far outnumber those made using HBTs. Early research work was concentrated in the short-wavelength regime (750–900 nm) with GaAs-based devices. Both p–i–n and MSM photodiodes in the AlGaAs/GaAs material system were integrated with GaAs MESFETs to realize high-performance single-channel receivers[71] as well as multichannel arrays of receivers.[72]

More recently, a recessed InP substrate was used to fabricate a p–i–n/ MODFET OEIC receiver,[73] as shown in Fig. 34. Because the p–i–n was fabricated in the recess, the top surface of the detector was at the same level as the transistor and the problems related to pattern delineation in photolithography of the transistor gate were minimized. Using a single epitaxial growth on such a recessed substrate, monolithic photoreceivers were demonstrated with a 3-dB bandwidth at frequencies as high as 6 GHz and with a high yield of working circuits.

Finally, a single epitaxial growth on a patterned InP substrate was applied to realize integration of MSM photodetectors and MODFETs,[74] as shown in Fig. 35. The MSM detector and the MODFET both require fine lithographic definition for high-speed operation and, therefore, need a planar surface topography. This surface was made by first patterning the substrate so that at the end of the epitaxial growth, the MODFET and the MSM would be at the same level. A 3-GHz transimpedance OEIC receiver was demonstrated that incorporated 1.3-μm-gate-length MODFETs and two pairs of interdigitated fingers, $2\ \mu$m $\times$ $30\ \mu$m, for the MSM photodetector.

7.5.3 HBT-Based Receiver OEICs

The first HBT-based OEIC receiver reported was in the GaAs material system.[75] A phototransistor was used as the detector in an all-bipolar receiver. It showed modest performance at 140 Mbit/s at a wavelength of 840 nm. It was followed soon after with the first all-bipolar OEIC receiver in the InP material system, using a similar implementation method and operated at 100 Mbit/s in the 1300–1600-nm wavelength range.[76] Since then there has been excellent progress in this field with speeds reaching 20 Gbit/s and performance comparable to the best p–i–n/FET and p–i–n/bipolar hybrid

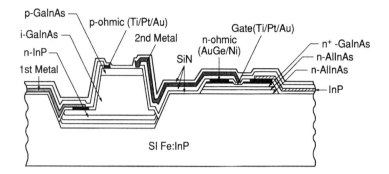

Fig. 34 Cross-sectional view of p–i–n and MODFET used in the monolithic p–i–n/MODFET photoreceiver. (After Yano et al., Ref 73.)

receivers. The most successful approach has been the one shown schematically in Fig. 36.[77] The p–i–n epitaxial layers are grown first, followed by the HBT layers, in a single growth run on a planar, nonpatterned, semi-insulating InP substrate. Selective wet chemical etching is used to delineate the HBTs and the p–i–n. The p–i–n layers below the HBTs are electrically and optically inactive and serve only as a pedestal.

This integration approach has several advantages. The individual devices can be independently optimized for best performance, with the HBTs fabricated using any advanced self-aligned technology. In addition, the HBT structure could be changed from a single-heterojunction device (SHBT) to that of a double-heterojunction device (DHBT) with its superior breakdown characteristics. The planar substrate allows any available epitaxial growth technique to be used without having to consider growth artifacts, making the technology highly transportable. The feature size increases as one proceeds from the top to the bottom epilayer. This allows for relaxed lithographic tolerances with the smallest feature size at the top. The p–i–n can be tailored for top or bottom illumination by an appropriate design of the metalization and it also exploits a double-pass scheme for light absorption that minimizes the requirement on the thickness of the absorbing layer. Nevertheless, the skyscraper-like topography necessitates the use of a planarizing medium (polyimide in this case) and would be the limiting factor in achieving very small feature sizes.

The complete photoreceiver fabricated with this implementation methodology involves a total of 16 lithographic levels, including levels for thin-film "NiCr" metal resistors and anti-reflection coating of the p–i–n. Using this approach, monolithic receivers with speeds ranging from 1 Gbit/s to 20 Gbit/s were realized using different epitaxial growth techniques and circuit designs, and increasing optimization of the device epilayers.[78] Figure 37 shows the circuit schematic used in one such realization, and its

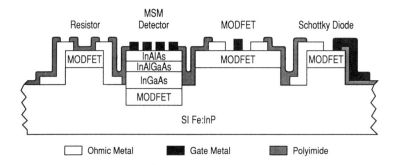

Fig. 35 Cross-sectional view of MSM, MODFET, and resistor used in the monolithic MSM/MODFET photoreceiver. (After Chang et al., Ref. 74.)

small-signal response curve is shown in Fig. 38. The preamplifier design is a double-feedback input stage, which has the advantages of increased bandwidth with high gain. The OEIC receiver operated up to 20 Gbit/s with a sensitivity of -17 dBm for a bit-error rate of 10^{-9}. The bit-error rate as a function of the optical incident power at 10, 15, and 20 Gbit/s is shown in Fig. 39.

In the second implementation, shown schematically in Fig. 40, the base–collector p–n junction of the HBT doubles as the p–i–n photodetector and, therefore, the two devices can be fabricated from a single-device epilayer structure. Sharing the epitaxial layers necessarily implies some compromise on individual performance, but results in a simpler fabrication process that requires only 10 lithographic levels for the entire photoreceiver. The thickness of the GaInAs collector has to be judiciously chosen so as to obtain good quantum efficiency for the p–i–n photodetector and, at the same time, not unduly influence the transit time of the electron across the collector depletion layer. Nevertheless, this approach restricts one to the use of the single-heterojunction transistor structure (SHBT), which limits operation to lower values of base–collector breakdown voltage. This approach was first demonstrated in the GaAs material system.[79] It was subsequently implemented in the InP material system[80] and demonstrated both high speed (5 Gbit/s) and high sensivity (-22.5 dBm). In addition, since the epitaxial structure is identical to that required for a phototransistor, the same p–i–n/HBT OEIC receiver circuit was used to demonstrate operation as an all-bipolar receiver by moving the input optical signal from the p–i–n location to the first transistor of the preamplifier. There is only a 1-dB penality in operating under such conditions.

Similar p–i–n/HBT OEIC photoreceivers using the base–collector junction for the p–i–n photodiode have been reported recently,[81–83] and it is evident that as HBTs attain higher speeds, HBT-based OEICs will continue to be attractive components for high-speed applications.

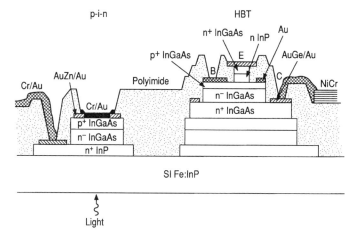

Fig. 36 Cross-sectional view of $p-i-n$ and HBT used in the monolithic $p-i-n$/HBT photoreceiver. (After Chandrasekhar et al., Ref. 76.)

7.5.4   Other Functional Receiver OEICs

Multichannel Photoreceiver Arrays.   The strongest feature of monolithic integration is the high yield in realizing replicas of the same circuit. This is particularly attractive for optical communication systems employing the wavelength-division-multiplexing architecture, where multichannel photoreceiver arrays are required. Recent advances in OEICs have enabled arrays of photoreceivers with either $p-i-n$ or MSM photodetectors integrated with a variety of FET and HBT structures. In Table 2, we have listed several reported long-wavelength OEIC receiver arrays[84] and have used a figure-of-merit (defined as the product of the bandwidth, the transimpedance and the number of channels) to compare the different results. It is desirable to have a large number of channels on a single monolithic chip, each with high bandwidth and high gain (or high transimpedance), resulting in a large figure-of-merit.

An eight-channel $p-i-n$/HBT monolithic OEIC receiver array has been fully packaged. The $p-i-n$ photodetectors, centered on a 250-μm pitch, are edge illuminated. This scheme of coupling light into the photodetectors facilitates easy mating with either ribbon fiber connectors or guided-wave multiplexer–demultiplexer devices. The array chip size was 1 mm × 2.5 mm and has 80 elements per chip, comprising eight detectors, 40 HBTs, and 32 "NiCr" resistors. There are a total of 24 electrical pinouts. The small-signal frequency response of the packaged array showed a 3-dB bandwidth of 1.7 GHz and a crosstalk better than −20 dB. Each channel was operated up to 2.5 Gbit/s, giving an overall throughput of 20 Gbit/s.

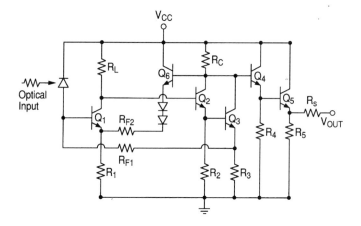

Fig. 37 Schematic circuit diagram of the monolithic *p–i–n*/HBT photoreceiver incorporating a double-feedback input stage preamplifier. (After Lunardi et al., Ref. 78.)

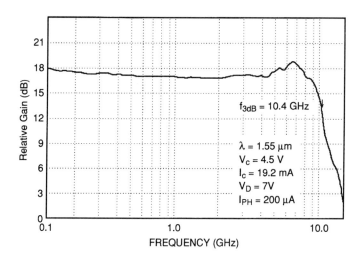

Fig. 38 Optical-to-electrical small-signal response of the fiber-pigtailed package OEIC module with a 1.55-μm light source. The bias conditions for the amplifier (V_C and I_C) and the photodetector (V_D and I_{ph}) are indicated. (After Lunardi et al., Ref. 78.)

Coherent Heterodyne Receivers.   One of the potential advantages of OEICs is the ability to realize new functionality through the application of photonic circuit integration. Once we know how to put together a photodetector and a transistor on a single chip, it becomes simple to realize different functions for different applications. One such application is in coherent heterodyne

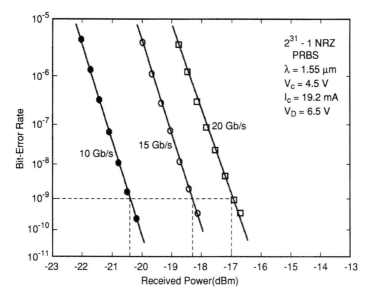

Fig. 39 Measured bit-error rate as a function of received optical power at 10, 15, and 20 Gbit/s for the *p–i–n*/HBT OEIC photoreceiver. (After Lunardi et al., Ref. 78.)

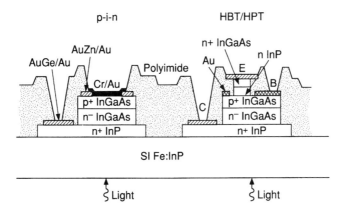

Fig. 40 Cross-sectional view of HBT with a base–collector/*p–i–n* used in the monolithic photoreceiver. (After Chandrasekhar et al., Ref. 80.)

communications, where a dual-detector balanced optical receiver[85–87] has become the preferred method of obtaining high-sensitivity reception. This receiver has two *p–i–n* photodetectors in a balanced configuration followed by a low-noise preamplifier.

In a coherent optical receiver, the incoming optical frequency signal is mixed with a local oscillator signal to produce an intermediate-frequency (IF)

TABLE 2   Long-Wavelength OEIC Receiver Arrays[84]

No. of Channels (N)	Technology	Bandwidth BW (GHz)	Transimpedance $Z(\Omega)$	Figure-of-Merit (BW × Z × N)
2	p–i–n/JFET	0.6	790	950
4	p–i–n/JFET	0.8	1,000	3,200
4	p–i–n/JFET	0.49	1,700	3,330
4	MSM/MODFET	1.3	650	3,380
4	p–i–n/MODFET	4.0	227	3,630
8	p–i–n/MODFET	1.2	550	5,280
8	p–i–n/JFET	0.3	2,500	7,000
4	p–i–n/JFET	0.2	10,000	8,000
5	p–i–n/MODFET	4.4	450	9,900
8	p–i–n/HBT	1.7	750	10,200

beat signal, usually in the radio frequency range. Coherent detection of an optical signal is completely analogous to the detection method universally used for the reception of radio-frequency signals. Coherent optical-transmission systems and networks allow the use of very dense WDM channels and also offer enhanced sensitivity compared with direct-detection systems. Although the greater complexity of optical coherent receivers and the recent introduction of low-noise fiber-optical amplifiers has diminished the immediate need for coherent transmission, the introduction of a high-performance, cost-effective photonic integrated circuit (PIC) chip to perform all of the functions of a coherent receiver could rekindle interest in this technology.

A coherent optical receiver requires a stable, tunable local-oscillator laser, a coupler to coherently combine the signal and local-oscillator fields, and a single or balanced photodetector/mixer to produce the beat frequency. Electronics to drive the local-oscillator laser, to stabilize its frequency or phase, and to amplify and process the IF signal are also required. Tunable laser chips were discussed in Section 7.4.5. Polarization control is also required so that signal and local oscillator signals can maintain the proper polarization through the couplers and at the photodetectors.

Some progress toward the complete integration of a coherent receiver PIC based on GaInAsP/InP technology for 1550-nm operation has been reported in the past few years.[88–90] A continuously tunable MQW-DBR laser was integrated with a directional coupler/switch and zero-bias waveguide photodetectors,[88] as shown schematically in Fig. 41. Error-free reception of FSK (frequency-shift-keying)-modulated digital signals was achieved at 105 Mbit/s, reportedly limited in this experiment by the transmitter FM bandwidth. A similar chip utilizing a multi-electrode DFB laser as the tunable local oscillator was also reported.[89] Preliminary results on a polarization-diversity heterodyne receiver PIC with a tunable four-section DBR laser as a local oscillator have been reported recently.[90] This PIC chip, shown schematically in Fig. 42, consists of 16 elements: a tunable laser, a passive polarization rotator, a signal input port, a polarization-diversity waveguide network (two TE/TM mode splitters, two TE/TM filters and two 3-dB couplers), and a balanced receiver with four photodetectors, two field-effect transistors, and a load resistor. This certainly represents one of the most complex PIC chips fabricated and reported to date.

7.6 SUMMARY AND FUTURE TRENDS

The advanced lasers, small-scale photonic integrated circuits (PICs), and receiver OEICs, discussed in this chapter have been designed primarily for specialized high-performance, high-speed applications such as telecommunications. This trend should continue, with emphasis in OEIC research extending to higher-speed circuits, receivers with greater sensitivity and more extensive electronics, and greater use of array-based circuits. The trend in PICs towards higher levels of integration also should continue. Examples are circuits for advanced WDM systems, and even heterodyne receivers such as those described above.

Beyond long-distance telecommunications needs, future efforts should soon become focused on higher volume applications. These new systems will cover much shorter distances and will support both telephone networks, data communication, and cable television distribution, for which low cost is an overriding consideration. The availability of functional blocks that are truly lower-cost alternatives to discrete optoelectronic circuits should trigger the widespread deployment of OEICs and PICs. As costs are reduced, applications to optical interconnections between various components also will become feasible. Photonic switching, routing, and signal processing, and perhaps even computing, are technologies which eventually may become practical with OEICs and PICs.

Many challenges must be overcome before OEICs and PICs can fully achieve their anticipated potential. Their performance levels now are often comparable to those of optoelectronic or photonic circuits with discrete components and predicted higher performance has been realized only in certain special cases. So far, most of the research on OEICs and PICs has been concerned with the integration of existing discrete devices. A major thrust of future integration research should be in the direction of increased functionality, with new ways of incorporating novel device elements.

APPENDIX 7.A   DERIVATION OF THE LINEWIDTH FORMULA

Following Ref. 26, the linewidth formula of a semiconductor laser is derived as follows.

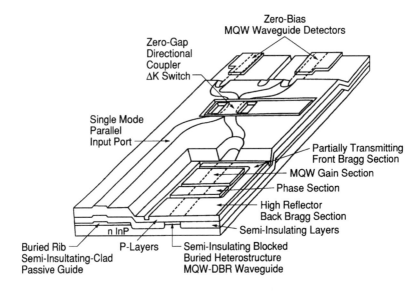

Fig. 41 Schematic of a heterodyne receiver PIC, including a tunable MQW-DBR laser, coupler, and waveguide detectors. (After Koch et al., Ref. 88.)

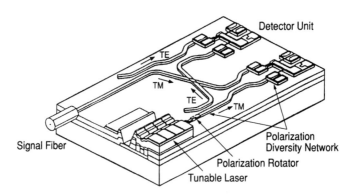

Fig. 42 Schematic of a polarization-diversity heterodyne receiver PIC. (After Kaiser et al., Ref. 90.)

The width of the laser line can be considered as resulting from fluctuations in the phase of the optical field. These fluctuations arise from spontaneous emission events which randomly alter the phase and intensity of the lasing field, as shown in Fig. 43.

Referring to Fig. 43, the instantaneous phase change is denoted by $\Delta\phi_i'$, where the subscript i denotes the ith spontaneous emission event. Each spontaneous emission on the average causes a field-intensity change equivalent to adding one photon to the optical field. To restore the steady-state field intensity, the laser will undergo relaxation oscillations, which last about 1 ns. During this time, there will be a net gain change $\Delta g(t) = (-2\omega/c)\Delta n''(t)$, where $\Delta n''(t)$ is the derivation of the imaginary part of the refractive index from its steady-state value. The change in n'' is caused by a change in the carrier density, which will also alter the real part of the refractive index n'. The ratio of these changes is

$$\alpha = \Delta n'/\Delta n'' \tag{A1}$$

A change in $\Delta n'$ during a limited period of time results in an additional phase shift of the laser field and additional line broadening. By relating $\Delta g = (-2\omega/c)\Delta n''$, Eq. A1 can be written as

$$\alpha \equiv -\left(\frac{4\pi}{\lambda}\right)\left(\frac{(\Delta n'/\Delta N)}{(\Delta g/\Delta N)}\right) \tag{A2}$$

where the quantity α is defined as the linewidth enhancement factor.

In Fig. 43, the optical field is represented by a complex amplitude β normalized so that the average intensity $I = \beta^*\beta$ is also equal to the average number of photons in the cavity. $I(t)$ and $\phi(t)$ represent the intensity and phase of the laser field. The ith spontaneous emission event alters β by $\Delta\beta_i$, where $\Delta\beta_i$ has unit magnitude and a random phase

$$\Delta\beta_i = \exp(i\phi + i\theta_i) \tag{A3}$$

where θ_i is random. This is illustrated in Fig. 43.

We now solve for the phase change $\Delta\phi_i$ due to a single spontaneous

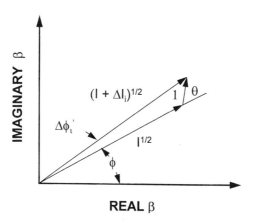

Fig. 43 Instantaneous changes of the phase ϕ and intensity I of the optical field caused by the ith spontaneous-emission event.

emission event that alters I from the steady-state value. There are two contributions to the phase change, and these are denoted by $\Delta\phi_i'$ and $\Delta\phi_i''$. The change $\Delta\phi_i'$ is due to the out-of phase component of $\Delta\beta_i$. It is clear from Fig. 43 that this change is

$$\Delta\phi_i' = I^{1/2}\sin(\theta_i) \tag{A4}$$

The second contribution $\Delta\phi_i''$ is due to the intensity change and the fact that the intensity and phase changes are coupled, as we have described previously. From Fig. 43, using the law of cosines, we see that the amplitude changes from $I^{1/2}$ to $(I+\Delta I_i)^{1/2}$, where

$$\Delta I_i = 1 + 2I^{1/2}\cos(\theta_i) \tag{A5}$$

To relate the intensity change ΔI_i to the second component of the phase change $\Delta\phi_i''$, we use the rate equations for I and ϕ,

$$\frac{d\phi}{dt} = \frac{\alpha}{2}(G - \gamma) \tag{A6}$$

$$\frac{dI}{dt} = (G - \gamma)I \tag{A7}$$

where G is the net rate of stimulated emission and γ is the rate of cavity loss caused by facets and waveguide losses. Combining Eqs. A6 and A7, we obtain

$$\frac{d\phi}{dt} = \frac{\alpha}{2I}\frac{dI}{dt} \tag{A8}$$

Upon integration of Eq. A8 and observing that $I(0) = I + \Delta I_i$ and $I(\infty) = I$, we obtain

$$\Delta\phi_i'' = \left(-\frac{\alpha}{2I}\right)\Delta I_i = \left(-\frac{\alpha}{2I}\right)[1 + 2I^{1/2}\cos(\theta_i)] \tag{A9}$$

The total phase change is the sum of Eqs. A4 and A9:

$$\Delta\phi_l = \Delta\phi_l' + \Delta\phi_l'' = \frac{\alpha}{2I} + I^{-1/2}[\sin(\theta_i) - \alpha\cos(\theta_i)] \tag{A10}$$

The first term is a small constant phase change that causes a frequency shift due to the spontaneous emission events. Ignoring this constant term, the total phase fluctuation for $N = Rt$ spontaneous emission events, where R is the spontaneous emission rate, will be

$$\Delta\phi = \sum_i I^{-1/2}[\sin(\theta_i) - \alpha\cos(\theta_i)] \tag{A11}$$

The value of $\langle\Delta\phi^2\rangle$ can be calculated from (A11), by letting the cross terms vanish for random angles, to yield

$$\langle\Delta\phi^2\rangle = \left(\frac{Rt}{2I}\right)(1 + \alpha^2) \tag{A12}$$

It is well known that the power spectrum of the laser is Lorentzian with a full width at half maximum of

$$\delta\nu = \frac{1}{\pi t_{\text{coh}}} \tag{A13}$$

where the coherence time t_{coh} is given by

$$\frac{1}{t_{\text{coh}}} = \frac{\langle\Delta\phi^2\rangle}{2t} \tag{A14}$$

Combining Eqs A12, A13, and A14, we have

$$\delta\nu = \frac{R}{4\pi I}(1 + \alpha^2) \tag{A15}$$

The intensity I can be expressed in terms of the output power of the laser. For a laser with equal output power from the front and the back facet, the intensity I is related to the power per facet P_0 by

$$I = \frac{2P_0}{h\nu v_g \alpha_m} \tag{A16}$$

where v_g is the group velocity and α_m is the mirror loss defined previously.

The rate of spontaneous emission R and the gain g are related by

$$R = gn_{\text{sp}}v_g = (\alpha_m + \alpha_c)n_{\text{sp}}v_g \tag{A17}$$

where n_{sp} is the spontaneous emission factor, and $g = \alpha_m + \alpha_c$. Substituting I and R in Eq. A15 by Eqs. A16 and A17 respectively, we obtain

$$\delta\nu = v_g^2\left(\frac{n_{\text{sp}}h\nu}{8\pi P_0}\right)\alpha_m(\alpha_m + \alpha_c)(1 + \alpha^2) \tag{A18}$$

APPENDIX 7.B APPROXIMATE EXPRESSION FOR THE TRANSPARENCY CARRIER AREA–DENSITY AND THE DIFFERENTIAL GAIN FOR STRAINED-LAYER SINGLE-QUANTUM-WELL LASERS

Long-wavelength quantum-well lasers employing biaxial compressively strained active layers fabricated to date have exhibited significantly lower lasing threshold current densities and higher differential gain. The biaxial compressively strained systems have an unusual band-structure. In particular, the highest energy valence band, a heavy-hole band, is well separated in energy from the the next highest valence bands. Furthermore, this band has light-hole characteristics for crystal wavevectors in the plane of the quantum-well. The in-plane effective mass is predicted to be 1.5 times that of the light-hole effective mass for typical well thicknesses. This leads to a nearly symmetric bandstructure for the conduction and valence bands involved in the optical transitions. Simple expressions can be developed for transparency carrier density, and the differential gain for strained-layer QW lasers having a nearly symmetric band structure.

Let us assume that all subbands are parabolic and that optical transitions obey k-selection rules. Furthermore, transition broadening is neglected. We focus on gain resulting from the lowest energy subband in the conduction and valence bands. As conceptually illustrated in Fig. 44, the point of maximum gain in this case always occurs for the band-edge transitions. (In lasers with a bulk gain layer, the gain peak lies above the band edge and depends strongly on excitation.) Under these assumptions the maximum gain G_{max} for a single quantum-well is given by

$$G_{max} = \frac{8\pi^3 h\nu |M_T|^2 m_r}{\varepsilon_s v_g h^3 L_z} [f_c(n) - f_v(n)] \tag{B1}$$

where m_r is the reduced effective mass, and $f_c(n)$ and $f_v(n)$ are Fermi electronics occupancies for states at the lowest energy conduction and valence

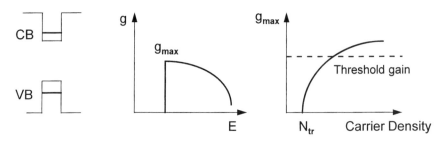

Fig. 44 Schematics of band filling, gain vs E, and the maximum gain vs current characteristics of a single-quantum-well active layer.

subband edges, respectively. The charge neutrality is assumed in Eq. B1. It can be shown that for a quantum-well the Fermi functions are related to the electron density by a simple approximation[34]

$$f_c = 1 - \exp\left(-\frac{n}{N}\right) \tag{B2}$$

$$f_v = \exp\left(-\frac{n}{P}\right) \tag{B3}$$

where

$$N = \sum_{l=0} n_l \exp(-\varepsilon_{cl}) \tag{B4}$$

In Eq. B4, ε_{cl} is the lth subband edge location (in units of kT) higher than the lowest subband edge. Thus, for the lowest subband under consideration, $\varepsilon_c = 0$, and we obtain

$$N = n_0 = \frac{4\pi m_c kT}{h^2 L_z} \tag{B5}$$

Combining Eqs. B1, B2, and B3, we obtain the following simple expressions:

$$G_{max} = \frac{8\pi^3 h\nu |M_T|^2 m_r}{\varepsilon_s v_g h^3 L_z}\left[1 - \exp\left(-\frac{n}{N}\right) - \exp\left(-\frac{n}{P}\right)\right] \tag{B6}$$

$$\frac{dG_{max}}{dn} = \frac{8\pi^3 h\nu |M_T|^2 m_r}{\varepsilon_s v_g h^3 L_z}\left[\frac{1}{N}\exp\left(-\frac{n}{N}\right) + \frac{1}{P}\exp\left(-\frac{n}{P}\right)\right] \tag{B7}$$

It is obvious that for a symmetric band structure, as in the case of a biaxial-strained single-quantum-well layer, $N = P$, and Eq. B6 reduces to

$$G_{max} = \frac{8\pi^3 h\nu |M_T|^2 m_r}{\varepsilon_s v_g h^3 L_z}\left[1 - 2\exp\left(-\frac{n}{N}\right)\right] \tag{B8}$$

The transparency carrier density is defined as the carrier density required at which $G_{max} = 0$. This leads to

$$n = n_{tr}^s = N\ln(2) \tag{B9}$$

The transparency area carrier density N^s_{tr} for a symmetric-band-structure single-quantum-well laser is

$$N^s_{tr} = n^s_{tr} L_z \tag{B10}$$

Combining Eqs. B5, B9, and B10, we obtain

$$N_{tr}^s = \frac{4\pi m_c kT}{h^2} \ln(2)$$ (B11)

Now, we proceed to calculate the differential gain expressed in Eq. B7. Using Eq. B7, and assuming that $N = P$ and $2\exp(-n/N) = 1$ at transparency, we obtain the differential gain at transparency, given by

$$\frac{dG_{max}}{dN} = \frac{\pi^2 v |M_T|^2}{\ln(2)\,\varepsilon_s v_g kT}$$ (B12)

which is Eq. 26 in the text.

PROBLEMS

1. Assuming that the optical gain of the semiconductor laser medium is g (cm^{-1}), the free-carrier absorption loss is α (cm^{-1}), and the mirror reflectivities are R_1 and R_2 for a Fabry–Perot cavity, show that the lasing threshold gain is expressed by

$$g_{th} = \alpha + \frac{1}{2L}\ln\left(\frac{1}{R_1 R_2}\right)$$

2. An InGaAsP Fabry–Perot laser operating at a wavelength of 1.3 μm has a cavity length of 300 μm. The index of refraction of InGaAsP is 3.39.
   a) What is the mirror loss expressed in cm^{-1}?
   b) If one of the laser facets is coated to produce 90% reflectivity, how much threshold current reduction (as a percentage) can be expected, assuming $\alpha = 10\,cm^{-1}$?

3. a) Derive Eq. 4 for the mode spacing of a Fabry–Perot laser.
   b) For an InGaAsP laser operating at a wavelength of 1.3 μm, calculate the mode spacing in nanometer for a cavity of 300 μm, assuming the group index of refraction is 3.4.
   c) Express the mode spacing obtained above in GHz.

4. How does one maximize the 3-dB modulation bandwidth of a semiconductor laser?

5. What is the physics behind the frequency chirping when a semiconductor laser is intensity modulated?

6. The limitation of the (bit rate) × (transmission distance) product for 1-dB

power penalty resulting from chromatic dispersion in a single-mode fiber for a chirp-free light source is given by

$$B^2 L \leq \frac{c}{2D\lambda^2}$$

where B is the bit rate, L is the fiber transmission distance, c is the velocity of light in vacuum, D is the fiber dispersion coefficient, and λ is the wavelength. The conventional single-mode fiber has a dispersion coefficient $D = 15$ ps/km-nm at $\lambda = 1550$ nm. Calculate the value of $B^2 L$ in $(GHz)^2$km.

7. Using the result in Problem 6, calculate the chromatic-dispersion-limited fiber transmission distance for (a) 2.5 Gbit/s and (b) 10 Gbit/s.

8. The dispersion-limited (bit rate) × (distance) product for a 1-dB power penalty in a single-mode fiber with a chirped light source is given by

$$BL \leq \frac{1}{4D\sigma_\lambda}$$

where σ_λ is the laser-chirped spectral width. If an optical fiber has a chromatic dispersion of 15 ps/km-nm, and the laser under pulsed modulation has a frequency chirp of 0.4 nm, what is the maximum distance that such a system can transmit without regeneration, neglecting other system degradation effects, at (a) 2.5 Gbit/s and (b) 10 Gbit/s?

9. A dispersion-shifted single-mode fiber usually has zero dispersion near 1550 nm.

   a) If the dispersion coefficient of a dispersion-shifted fiber is 1 ps/km-nm, what is the maximum transmission distance for a system operated at 10 Gbit/s using a chirp-free light source?

   b) If a low-chirp electro-absorption modulator with an equivalent chirp width of 0.1 nm is used, what is the expected transmission distance?

   c) How do the above results, including Problems 7 and 8, compare with a system that uses a direct-current-modulated DFB laser which has a chirp width of 0.4 nm?

10. Show that the quantum efficiency η of a p–i–n photodetector is related to the responsivity R at a wavelength λ (μm) by the equation

$$R = \frac{\eta\lambda}{1.24}$$

11. A p–i–n photodetector has a 1-μm InGaAs absorbing layer. There is an

anti-reflection coating (reflectivity = 0%) on the side where the light enters the photodetector.

a) What is the external quantum efficiency of the photodiode at a wavelength of 1.55 μm?

b) What would be the external quantum efficiency if light were to travel through the absorbing layer twice?

12. An electronic preamplifier has a small-signal 3-dB bandwidth of 10 GHz and a noise spectral density of 10 pA/$\sqrt{\text{Hz}}$. What would be the calculated sensitivity (received optical power) of a photoreceiver built from such a preamplifier at a bit rate of 10 Gbit/s for a bit-error rate of 10^{-9}? Assume a responsivity of 1 A/W for the p–i–n photodetector.

13. A photoreceiver has a transimpedance of 1000 Ω. What should be the incident peak-to-peak optical power for obtaining an output voltage of 10 mV peak-to-peak? Assume a quantum efficiency of 1 A/W for the p–i–n photodetector.

REFERENCES

1. H. Onaka, H. Miyata, G. Ishikawa, K. Otsuka, H. Ooi, Y. Kai, S. Kinoshita, M. Seino, H. Nishimoto, and T. Chikama, "1.1 Tbit/s WDM transmission over 150 km 1.3 μm zero-dispersion single-mode fiber," in *OFC'96*, San Jose, CA, PD19, February 1996.

2. A. H. Gnauck, R. W. Tkach, F. Forghieri, R. M. Derosier, A. R. McCormick, A. R. Chraplyvy, J. L. Zyskind, J. W. Sulhoff, A. J. Lucero, Y. Sun, R. M. Jobson, and C. Wolf, "One terabit/s transmission experiment," in *OFC'96*, San Jose, CA, PD20, February 1996.

3. T. Morioka, H. Takara, S. Kawanishi, O. Kamatani, K. Takiguchi, K. Uchiyama, M. Saruwatari, H. Takahashi, M. Yamada, T. Kanamori, and H. Ono, "100 Gbit/s × 10 channel OTDM/WDM transmission using a single supercontinuum WDM source," in *OFC'96*, San Jose, CA, PD21, February 1996.

4. A. Brackett, A. S. Acampora, J. Sweitzer, G. Tangonan, M. T. Smith, W. Lennon, K. C. Wang, and R. H. Hobbs, "A scalable multiwavelength multihop optical network: a proposal for research on all-optical networks," *IEEE J. Lightwave Technol.* **11**, 736 (1993). Also, for general reading on the multiwavelength technology, see the Special Issue on Multiwavelength Optical Technology and Networks, *IEEE J. Lightwave Technol.* **14**, 932 (1996).

5. E. Yablonovitch and E. O. Kane, "Reduction of lasing threshold current density by the lowering of valence band effective mass," *IEEE J. Lightwave Technol.* **LT-4**, 504 (1986).

6. J. Mun, "Photodetectors and OEIC receivers," in *Optoelectronic Integration: Physics, Technology and Application*, O. Wada, Ed., Kluwer, Boston, 1994, Ch. 6.

7. S. M. Sze, Physics of Semiconductor Devices, 2nd Ed., Wiley, New York, 1981.

8. T. P. Lee, C. A. Burrus, and R.H. Saul, "Light emitting diodes for telecommunication," in *Optical Fiber Telecommunications II*, S. E. Miller and I. P. Kaminow, Eds., Academic Press, New York, 1988, Ch. 12.

9. R. E. Nahory, M. A. Pollack, W. D. Johnston, and R. L. Barnes, "Bandgap versus composition and demonstration of Vegard's law for $In_{1-x}Ga_xAs_{1-y}P_y$ lattice matched to InP," *Appl. Phys. Lett.* **33**, 659 (1978).

10. H. C. Casey and M. B. Panish, *Heterostructure Lasers*, Parts A and B, Academic Press, New York, 1978.

11. G. P. Agrawal and N. K. Dutta, *Long-wavelength Semiconductor Lasers*, Van Nostrand–Reinhold, Princeton, N.J., 1986.

12. T. P. Lee, C. A. Burrus, J. A. Copeland, A. G. Dentai, and D. Marcuse, "Short cavity InGaAsP injection lasers: dependence of mode spectra and single-longitudinal-mode power on cavity length," *IEEE J. Quant. Electron.* **QE-18**, 1101 (1982).

13. R. J. Nelson, R. B. Wilson, P. D. Wright, P. A. Barnes, and N. K. Dutta, "CW electrooptical properties of InGaAsP buried heterostructure laser diodes," *IEEE J. Quant. Electron.* **QE-17**, 202 (1981).

14. H. Kogelnik and C. V. Shank, "Stimulated emission in a periodic structure, " *Appl. Phys. Lett.* **18**, 152 (1971).

15. C. V. Shank, R. V. Schmidt, and B. I. Miller, "Double-heterostructure GaAs distributed feedback laser," *Appl. Phys. Lett.* **25**, 200 (1974).

16. M. Nakamura, K. Aiki, J. Umeda, A. Yariv, H. W. Yen, and T. Morikawa, "GaAs-GaAlAs double-heterostructure distributed feedback diode lasers," *Appl. Phys. Lett.* **25**, 487 (1974).

17. F. K. Reinhart, R. A. Logan, and C. V. Shank, "GaAs/Al_xGa_{1-x}As injection lasers with distributed reflectors," *Appl. Phys. Lett.* **27**, 45 (1975).

18. S. L. McCall and P. M. Platzman, "An optimized $\pi/2$ distributed feedback laser," *IEEE J. Quant. Electron.* **QE-21**, 1899 (1985).

19. K. Utaka, S. Akiba, K. Sakai, and Y. Matsushima, "$\lambda/4$-shifted InGaAsP/InP DFB lasers," *IEEE J. Quant. Electron.* **QE-22**, 1042 (1986).

20. M. Okai, S. Tsuji, and N. Chinone, "Stability of the longitudinal mode in $\lambda/4$-shifted InGaAsP/InP DFB lasers," *IEEE J. Quant. Electron.* **QE-25**, 1314 (1989).

21. K. Y. Lau and A. Yariv, "High frequency current modulation of semiconductor injection lasers," in *Semiconductors and Semimetals: Lightwave Communication Technology*, R. K. Willardson and A C. Beer, Eds., Vol. 22B, W. T. Tsang, Vol. Ed., Academic Press, New York/London, 1985.

22. J. E. Bowers and M. A. Pollack, "Semiconductor lasers for telecommunications," in *Optical Fiber Telecommunications, II*, S. E. Miller and I. P. Kaminow, Eds., Academic Press, New York, 1988.

23. Y. Arakawa and A. Yariv, "Theory of gain, modulation response, and spectral linewidth in AlGaAs quantum well lasers," *IEEE J. Quant. Electron.* **QE-21**, 1666 (1985).

24. Y. Arakawa, and T. Takahashi, "Effect of nonlinear gain on modulation dynamics in quantum well lasers," *Electron. Lett.* **25**, 169 (1989).

25. K. Uomi, H. Nakano, and N. Chinone, "Intrinsic modulation bandwidth in ultra-high-speed 1.3 μm and 1.5 μm GaInAsP DFB lasers," *Electron. Lett.* **25**, 1689 (1989).

26. C. H. Henry, "Theory of linewidth of semiconductor lasers," *IEEE J. Quant. Electron.* **QE-18**, 259 (1982).

27. T. L. Koch and R. A. Linke, "Effect of nonlinear gain reduction on semiconductor laser wavelength chirping," *Appl. Phys. Lett.* **48**, 613 (1986).

28. W. T. Tsang, "Quantum confinement heterostructure semiconductor lasers," in *Semiconductor and Semimetals*, Vol. 24, R. K. Willardson and A. C. Beer, Eds., R. Dingle, Vol. Ed., Academic Press, New York, 1987, Ch. 7.

29. Y. Arakawa and A. Yariv, "Theory of gain, modulation response, and spectral linewidth in AlGaAs quantum well lasers," *IEEE J. Quant. Electron.* **QE-21**, 1666 (1985).

30. For a review on quantum wells see C. Weisbuch, "Fundamental properties of III-V semiconductor two-dimensional quantized structures: the basis for optical and electronic device applications," in *"Semiconductors and Semimetals,"* Vol. 24, Raymond Dingle, Vol. Ed., Academic Press, New York, 1987, Ch. 1.

31. U. Koren, B. I. Miller, Y. K. Su, T. L. Koch, and J. E. Bower, "Low internal loss separate confinement heterostructure InGaAs/InGaAsP quantum well laser," *Appl. Phys. Lett.* **51**, 1744 (1987).

32. K. Kasukawa, Y. Imajo, and T. Makino, "1.3 μm GaInAsP/InP buried heterostructure graded index separate confinement multiple quantum well (BH-GRIN-SC-MQW) lasers entirely grown by metalorganic chemical vapor deposition," *Electron. Lett.* **25**, 104 (1989).

33. R. E Cavicchi, D. V. Lang, D. Gershoni, A. M. Sergent, J. M. Vanderberg, S. N. G. Chu, and M. B. Panish, "Admittance spectroscopy measurement of band offsets in strained-layers of $In_xGa_{1-x}As$ grown on InP," *Appl. Phys. Lett.* **54**, 739 (1989).

34. K. J. Vahala and C. E. Zah, "Effect of doping on the optical gain and the spontaneous noise enhancement factor in quamtum well amplifiers and lasers studied by simple analytical expressions," *Appl. Phys. Lett.* **52**, 1945 (1988).

35. P. J. A. Thijs, *Strained-layer InGaAs(P)/InP quantum well semiconductor lasers grown by organometallic vapour phase epitaxy*, Ph.D Thesis, Delft University, Delft, The Netherlands, (1993).

36. P. J. A Thijs and T. Van Dongen, "High quantum efficiency, high power, modulation doped GaInAs strained-layer quantum well laser diodes emitting at 1.5 μm," *Electron. Lett.* **25**, 1735 (1989).

37. P. J. A. Thijs, L. F. Tiemeijer, T. van Dongen, and J. J. M. Binsma, "High performance $\lambda = 1.3$ μm strained-layer InGaAsP/InP quantum well lasers," *IEEE J. Lightwave Technol.* **12**, 28 (1994).

38. C. E. Zah, F. J. Favire, R. Bhat, S. G. Menocal, N. C. Andreadakis, D. M Hwang, M. Koza, and T. P. Lee, "Submilliampere threshold 1.5-μm strained-layer multiple-quantum-well lasers," *IEEE Photon. Technol. Lett.* **2**, 852 (1990).

39. H. Temkin, N. K. Dutta, T. Tanbun-Ek, R. A. Logan, and A. M. Sergent, "InGaAs/InP quantum well lasers with sub-mA threshold current," *Appl. Phys. Lett.* **57**, 1610 (1990).

40. L. F. Tiemeijer, P. J. A. Thijs, J. M. Binsma, and T. van Dongen, "Direct measurement of the transparency current and valence band effective messes in tensile and compressively strained InGaAs/InP multiple quantum-well laser amplifiers," *Appl. Phys. Lett.* **60**, 554 (1992).

41. J. S. Osinski, P. Grodzinski, Y. Zou, P. D. Dapkus, Z. Karim, and A. R. Tanguay, Jr., "Low threshold current 1.5 μm buried heterostructure lasers using strained quaternary quantum wells," *IEEE Photon. Technol. Lett.* **4**, 1313 (1992).

42. Zah, R. Bhat, B. Pathak, F. Favire, W. Lin, M. C. Wang, N. C. Andreadakis, D. M. Hwang, M. A. Koza, T. P. Lee, Z. Wang, D. Darby, D. Flanders, and J. J. Hsieh, "High performance uncooled 1.3 μm $Al_x Ga_y In_{1-x-y} As$/InP strained-layer quantum well lasers for subscriber loop applications," *IEEE J. Quant. Electron.* **QE-30**, 511 (1994).

43. R. Nagarajan, D. Tauber, and J. E. Bowers, "High-speed semiconductor lasers," in *Current Trends in Integrated Optoelectronics*, T. P. Lee, Ed., World Scientific, Singapore, 1994, Ch. 1.

44. R. L. Cella, Brown, Y. Twu, J. L. Ziko, N. K. Dutta, "High speed 1.3 micron InGaAsP distributed feedback lasers," in *11th IEEE International Semiconductor Laser Conference, Technical Digest*, 50, Boston, MA, 1988.

45. K. Kamite, H. Sudo, M. Yano, H. Ishkawa, and H. Imai, "Ultra-high speed InGaAsP/InP DFB lasers emitting at 1.3 microns wavelength," *IEEE J. Quant. Electron.* **QE-23**, 1054 (1987).

46. Y. Hirayama, H. Furuyama, M. Morinagaa, N. Suzuki, Y. Uematsu, K. Eguchi, and M. Nakamura, "High speed (13 GHz) 1.5-μm self-aligned constricted-mesa DFB lasers grown entirely by MOCVD," in *11th IEEE International Semiconductor Laser Conference, Technical Digest*, 46, Boston, MA, 1988.

47. K. Uomi, H. Nakano, N. Chinone, "Ultra-high-speed 1.55 μm λ/4-shifted DFB lasers with bandwidth of 17 GHz," *Electron. Lett.* **25**, 668 (1989).

48. P. A. Morton, T. Tanbumn-Ek, R. A. Logan, P. f. Sciortino Jr., A. M. Sergent, and K. W. Wecht, "Superfast 1.55 μm DFB lasers," *Electron. Lett.* **29**, 1429 (1993).

49. K. Uomi, T. Mishima, and N. Chinone, "Ultra-high relaxation oscillation frequency (up to 30 GHz) of highly p-doped GaAs/GaAlAs multiple quantum-well lasers," *Appl. Phys. Lett.* **51**, 78 (1987).

50. T. H. Wood, "Multiple quantum-well waveguide modulators," *IEEE J. Lightwave Technol.* **6**, 743 (1988).

51. M. Aoki, M. Suzuki, H. Sano, S. Sasaki, T. Kawano, and H. Kodera, "Monolithic integration of DFB lasers and electroabpsorption modulators using in-plane quantum energy control of MQW structures," in *Current Trends in Integrated Optoelectronics*, T. P. Lee, Ed., World Scientific, Singapore, 1994, Ch. 3.

52. M. Okai, S. Sakano, and N. Chinone, "Wide range continuous tunable double-sectioned distributed feedback lasers," in *15th European Conference on*

Optical Communication, Gothenburg, Sweden, 1989.

53. Kotaki, S. Ogita, M. Mstauda, Y. Kuwahara, and H. Ishkawa, "Tunable, narrow-linewidth and high-power λ/4-shifted DFB laser," *Electron. Lett.* **25**, 990 (1989).

54. M. Kuznetsov, "Theory of wavelength tuning in two-segment distributed feedback lasers," *IEEE J. Quant. Electron.* **QE-24**, 1837 (1988).

55. S. Illek, W. Thulke, C. Schanen, H. Lang, and M.-C. Amann, "Over 7 nm (875 GHz) continuous wavelength tuning by tunable twin-guide (TTG) laser diode," *Electron. Lett.* **26**, 46 (1990).

56. Kobayashi and I. Mito, "Single frequency and tunable laser diodes," *IEEE J. Lightwave Technol.* **6**, 1623, (1988).

57. Kotaki, M. Matsuda, H. Ishikawa, H. Imai, "Tunable DBR laser with wide tuning range," *Electron. Lett.* **24**, 503 (1988).

58. T. L. Koch, U. Koren, R. P. Gnall, C. A. Burrus, and B. I. Miller, "Continuously tunable 1.5 μm multiple-quantum-well GaInAs/GaInAsP distributed-Bragg-reflector lasers," *Electron. Lett.* **24**, 1431 (1988).

59. X. Pan, H. Olesen, and B. Tromborg, "A theoretical model of multielectrode DBR lasers," *IEEE J. Quant. Electron.* **QE-24**, 2423 (1988).

60. N. K. Shankaranarayanan, U. Koren, B. Glance, and G. Wright, "Two-section DBR laser transmitters with accurate channel spacing and fast arbitrary-sequence tuning for optical FDMA networks," *Tech. Digest, Optical Fiber Communication*, TuI2, 36 (1994).

61. U. Koren, T. L. Koch, B. I. Miller, G. Eisenstein, G. Raybon, "An integrated tunable light source with extended tunability range," in *Integrated Optics and Optical Communication Conference (IOOC)*, Kobe, Japan, 19A2-3, August 1989.

62. C. E. Zah, B. Pathak, F. Favire, R. Bhat, C. Caneau, P. S. D. Lin, A. S. Gozdz, N. C. Andreadakis, M. A. Koza, and T. P. Lee "1.5-μm tensile-strained single-quantum-well 20-wavelength distributed feedback laser arrays," *Electron. Lett.* **28**, 1585 (1992).

63. C. E. Zah, P. S. D. Lin, F. Favire, B. Pathak, R. Bhat, C. Caneau, A. S. Gozdz, N. C. Andreadakis, M. A. Koza, T. P. Lee, T. C. Wu, and K. Y. Lau, "1.5-μm compressive-strained multiple-quantum-well 20-wavelength distributed-feedback laser arrays," *Electron. Lett.* **28**, 824 (1992).

64. T. P. Lee, C. E. Zah, R. Bhat, W. C. Young, B. Pathak, F. Favire, P. S. D. Lin, N. C. Andreadakis, C. Caneau, A. Rahjel, M. Koza, J. Gamelin, L. Curtis, D. D. Mohoney, and A. Lepore, "Multiwavelength DFB laser array transmitters for ONTC reconfigurable optical network testbed," *IEEE J. Lightwave Technol.* **14**, 967 (1996).

65. C. E. Zah, F. J. Favire, B. Pathak, R. Bhat, C. Caneau, P. S. D. Lin, A. S. Gozdz, N. C. Andreadakis, M. A. Koza, and T. P. Lee, "Monolithic integration of a multi-wavelength compressive-strained multi-quantum-well distributed-feedback laser array with a star coupler and optical amplifiers," *Electron. Lett.* **28**, 824 (1992).

66. C. E. Zah, M. R. Amersfoort, B. Pathak, F. Favire, P. S. D. Lin, A. Rajhel, N. C. Andreadakis, R. Bhat, C. Caneau, and M. A. Koza, "Wavelength accuracy

and output power of multiwavelength DFB laser array with integrated star coupler and optical amplifiers," *IEEE Photonic Technol. Lett.* **8**, 864 (1996).

67. C. E. Zah, B. Pathak, M. R. Amersfoort, F. Favire, P. S. D. Lin, N. C. Andreadakis, A. Rajhel, R. Bhat, C. Caneau, M. A. Koza, and L. Curtis, "High power 10-wavelength DFB laser arrays with integrated combiner and optical amplifier," in *15th IEEE International Semiconductor Laser Conference*, Haifa, Israel, October 13–18, 1996.

68. J. B. D. Soole and H. Schumacher, "InGaAs metal-semiconductor-metal photodetectors for long wavelength communications," *IEEE J. Quant. Electron.* **QE-27**, 737 (1991).

69. J. C. Campbell, "Phototransistors for lightwave communications," in *Semiconductors and Semimetals, Lightwave Communications Technology*, Vol. 22, Part D, Academic Press, New York, 1985.

70. R. F. Leheny, R. E. Nahory, M. A. Pollack, A. A. Ballman, E. D. Beebe, J. C. Dewinter, and R. J. Martin, "Integrated $In_{0.53}Ga_{0.47}As$ p–i–n/FET. photoreceiver," *Electron. Lett.* **16**, (1980).

71. O. Wada, H. Hamaguchi, S. Miura, M. Makiuchi, K. Nakai, T. Horimatsu, and T. Sakurai, "AlGaAs/GaAs p–i–n photodiode/preamplifier monolithic photoreceiver integrated on a semi-insulating GaAs substrate," *Appl. Phys. Lett.* **46**, 1031 (1985).

72. O. Wada, H. Hamaguchi, M. Makiuchi, T. Kumai, M. Ito, K. Nakai, and T. Sakurai, "Monolithic four channel photodiode/amplifier array integrated on a GaAs substrate," *Technical Digest, IOOC-ECOC'85*, Venice, Italy, 1985, p. 303.

73. H. Yano, G. Sasaki, M. Murata, and H. Hayashi, "An ultra-high-speed optoelectronic integrated receiver for fiber-optic communications," *IEEE Trans. Electron Dev.* **39**, 2254, (1992).

74. G.-K. Chang, W. P. Hong, L. Gimlett, R. Bhat, C. K. Nguyen, G. Sasaki, and J. C. Young, "A 3 GHz transimpedence OEIC receiver for 1.3–1.55 μm fiber-optic systems," *IEEE Photon. Technol. Lett.* **2**, 197 (1990).

75. H. Wang and D. Ankri, "Monolithic integrated photoreceiver implemented with GaAs/AlGaAs heterojunction bipolar phototransistor and transistors," *Electron. Lett.* **22**, 391 (1986).

76. S. Chandrasekhar, J. C. Campbell, A. G. Dentai, C. H. Joyner, G. J. Qua, A. H. Gnauck, and M. D. Feuer, "Integrated InP/GaInAs heterojunction bipolar photoreceiver," *Electron. Lett.* **24**, 1443 (1988).

77. S. Chandrasekhar, L. M. Lunardi, A. H. Gnauck, D. Ritter, R. A. Hamm, M. B. Panish, and G. J. Qua, "10-Gbit/s OEIC photoreceiver using InP/InGaAs heterojunction bipolar transistors," *Electron. Lett.* **28**, 466 (1992).

78. L. M. Lunardi, S. Chandrasekher, A. H. Gnauck, C. A. Burrus, and R. A. Hamm, "20-Gb/s monolithic p–i–n/HBT photoreceiver module for 1.55-μm applications," *IEEE Photonic Technol. Lett.* **10**, 1201 (1995).

79. K. D. Pedrotti, R. L. Pierson, Jr., R. B. Nubling, C. W. Farley, E. A. Sovero, and M. F. Chang, "Ultra-high speed PIN/HBT monolithic OEIC photoreceiver," *Technical Digest, 49th Annual Dev. Res. Conf.*, Boulder, CO, June 1991.

80. S. Chandrasekhar, L. M. Lunardi, A. H. Gnauck, R. A. Hamm, and G. J. Qua,

"High speed monolithic *p–i–n*/HBT and HPT/HBT photoreceivers implemented with simple phototransistor structure," *IEEE Photon. Technol. Lett.* **5**, 1316 (1993).

81. R. H. Walden, W. E. Stanchina, R. A. Metzger, R. Y. Loo, J. Schaffner, M. W. Pierce, Y. K. Brown, F. Williams, V. Jones, J. Pikulski, M. Rodwell, K. Giboney, R. A. Mullen, and J. F. Jensen, "An InP-based HBT 1x8 OEIC array for a WDM network," in *Technical Digest*, LEOS *Summer Topical Meeting on ICs for New Age Lightwave Communications*, FC1, Keystone, CO, August 1995.

82. A. L. Gutierrez-Aitken, J. Cowles, P. Bhattacharya, and G. I. Haddad, "High bandwidth InAlAs/InGaAs PIN-HBT monolithically integrated photoreceiver" in *Technical Digest, Sixth International Conference on Indium Phosphide and Related Materials*, TuB3, Santa Barbara, CA, 1994.

83. E. Sano, M. Yoneyama, H. Nakajima, and Y. Matsuoka, "A monolithically integrated photoreceiver compatible with InP/InGaAs HBT fabrication process," *IEEE J. Lightwave Technol.* **12**, 638 (1994).

84. S Chandrasekhar and M. A. Pollack, "Optoelectronic and photonic integrated circuits," in *Perspectives in Optoelectronics*, S. S. Jha, Ed., World Scientific, Singapore, 1995, Ch. 4.

85. B. M. Oliver, "Signal-to-noise ratio in photoelectric mixing," *Proc. IRE* **49**, 1960 (1961).

86. G. L. Abbas, V. W. S. Chan, and T. K. Lee, "Local oscillator excess noise suppression for homodyne and heterodyne detection," *Opt. Lett.* **8**, 419 (1983).

87. B. L. Kasper, C. A. Burrus, J. R. Talman, and K. L. Hall, "Balanced dual-detector receiver for optical heterodyne communicatons at Gbit/s rates," *Electron. Lett.* **22**, 413 (1986).

88. T. L. Koch, U. Koren, R. P. Gnall, F. S. Choa, F. Hernandez-Gil, C. A. Burrus, M. G. Young, M. Oron, and B. I. Miller, "GaInAs/GaInAsP multiple-quantum-well integrated heterodyne receiver," *Electron. Lett.* **25**, 1623 (1989).

89. H. Takeuchi, K. Kasaya, Y. Hondo, H. Yasaka, K. Oe, and Y. Imamura, "Monolithic integrated coherent receiver on InP substrate," *IEEE Photon. Technol. Lett.* **1**, 398 (1989).

90. P. Kaiser, D. Trommer, H. Heidrich, F. Fidorra, S. Malchow, D. Franke, W. Passenberg, W. Rehbein, H. Schroeter-Janßen, R. Stenzel, and G. Unterbörsch, "Polarization diversity heterodyne receiver OEIC on InP:Fe substrate" in *Technical Digest, Fifth Optoelectron. Conf. (OEC '94)*, PD II-1, Japan, 1994.

8 Solar Cells

MARTIN A. GREEN

Photovoltaics Special Research Centre, University of New South Wales, Sydney, Australia

8.1 INTRODUCTION

In their most common form, solar photovoltaic cells are large-area p–n junction diodes designed to convert light, usually sunlight, into electricity. Figure 1a shows a typical p–n junction solar cell. Photons within the incoming light that have energy greater than the semiconductor bandgap give up their energy by creating electron–hole pairs in the cell. The asymmetrical electronic properties of the p–n junction impart a directionality to the flow of these photogenerated carriers. As shown in Fig. 1b, the current resulting from this directional flow is superimposed upon the normal rectifying current–voltage characteristics of the junction, displacing them downwards by an amount that depends on the light intensity. A portion of the curve is forced into the fourth quadrant where power can be extracted from the solar cell terminals, as from a normal electrochemical battery. The most important parameters of a cell are generally cost and the efficiency of the conversion process from light to electricity.

Some of the earliest functional semiconductor devices were solar cells, with the first thin-film selenium cells dating from the 1880s.[1] However, these early devices, and the cuprous oxide devices that became popular around 1930,[1] were not sufficiently efficient for power generation but found use as large-area photodetectors. The evolution of crystalline silicon technology in the 1950s made possible the first practical energy-conversion applications.[2]

Even with this leap forward, the cells were too expensive for general use.

Modern Semiconductor Device Physics, Edited by S. M. Sze.
ISBN 0-471-15237-4 © 1998 John Wiley & Sons, Inc.

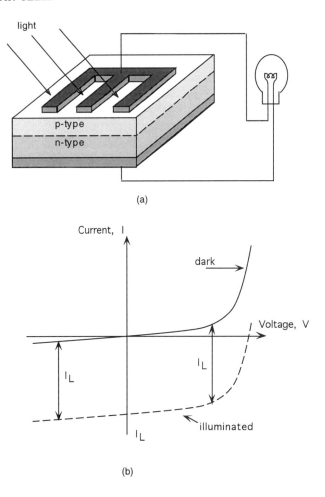

Fig. 1 (a) Illuminated *p–n* junction diode converts light into electrical power that is supplied to an electrical load placed across the diode terminals. (b) Current–voltage characteristics of the diode in the dark (solid line) and when illuminated (dashed line).

However, the use of silicon cells in spacecraft from the late 1950s onwards established a small industry supplying space cells.[3] The oil embargoes of the 1970s reawakened interest in renewable energy sources and, in particular, in the large-scale use of photovoltaics (PV) terrestrially. Several new companies began specializing in supplying solar cells for terrestrial use, primarily for telecommunications and similar applications involving small "remote" electrical loads such as navigational aids. The cost of these terrestrial cells has decreased rapidly over the last two decades as a result

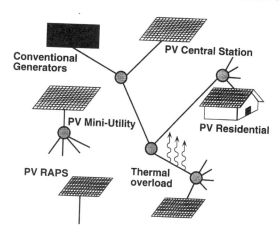

Fig. 2 Actual and potential solar-cell applications ranging from photovoltaic (PV) remote-area power supplies (RAPS) to central generating stations.

of increased manufacturing volumes and improved processing. With the promise of even lower costs offered by a number of thin-film cell approaches, far more widespread application seems likely in the future.

Figure 2 shows a range of possible applications of solar cells. As well as being used in the small remote-area power supplies (RAPS) previously mentioned, cells are becoming increasingly competitive for some of the larger-scale applications shown in Fig. 2. Particularly promising are applications where the modularity of the cells can be used to advantage, such as when solar cells are distributed throughout the electricity supply grid and contribute to other operational requirements of the grid, apart from just the need for electrical power.[4] Residential use of photovoltaics is an attractive future option given the public interest in this technology and the large numbers indicating a willingness to pay above present market prices for an environmentally sound source of electricity.[5] Increasingly, electricity utilities around the world have responded to this interest by introducing incentive schemes for the use of photovoltaics and other renewable energy sources. Solar cells are likely to be too expensive in the foreseeable future for the larger-scale central power station applications shown uppermost in Fig. 2, although several large systems demonstrating this use are already operating around the world.[6]

Current information on photovoltaic devices can be found in the Conference Records of the IEEE Photovoltaic Specialists Conferences,[7] held every 18 months, and also in specialist journals such as *Progress in Photovoltaics*[8] and *Solar Energy Materials and Solar Cells.*[9] More detailed treatments of photovoltaic-device principles,[10,11] technology,[12,13] and applications[12,14] are given elsewhere.

8.2 SOLAR RADIATION AND IDEAL ENERGY-CONVERSION EFFICIENCY

8.2.1 Outline

Solar cells respond to individual photons in sunlight. Provided a photon has sufficient energy to excite an electron from the valence band to the conduction band to create an electron–hole pair, it is of secondary importance whether or not the photon has an energy corresponding to long-wavelength red light or shorter-wavelength blue light. This feature makes the cell power output dependent on the number of photons in the incident light rather than just the power content. To allow comparison of results, cell performance is generally specified under light of both a standard intensity and spectral content. Since solar-cell performance also varies with temperature, resulting mainly from the temperature dependence of the diode dark current–voltage characteristics, cell temperature also needs to be specified.

A variety of approaches have been developed for calculating upper limits of the energy conversion efficiency possible by the photovoltaic process. Figure 3 shows the fourth quadrant region of Fig. 1b inverted so that it lies in the first quadrant, as is normal practice. From Fig. 3, the efficiency is obviously very dependent on the "knee" voltage of the p–n junction diode which, in turn, is determined by the diode's dark-saturation-current density.[10] The key question in calculating energy conversion efficiency limits is how small this dark-saturation-current density can become.

Early approaches to answering this question were empirical, based on assigning suitably optimistic values to the parameters of standard p–n

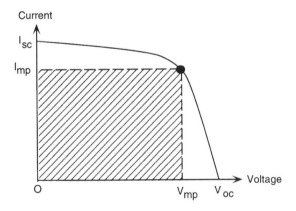

Fig. 3 Cell output characteristics reflected in the voltage axis to bring it into first quadrant. The area of the shaded rectangle is proportional to the power output of the cell.

junction diode theory.[15] The most sophisticated approaches developed to date are based on "detailed balance" between light absorption by the solar cell and the inverse process of light emission by a forward-biased *p–n* junction.[16] However, even this treatment requires assumptions that may be circumvented by creative device design in the future. The only assured bound is that imposed by the Carnot limit, which predicts a maximum energy conversion efficiency of 96% for a system operating between a source temperature corresponding to the sun's photosphere ($\sim$6000 K) and a sink temperature corresponding to the earth's surface ($\sim$300 K).[16]

8.2.2 Standard Sunlight Spectra and Performance Measurement

In the early 1980s, standard sunlight spectra for measuring solar cell performance were defined and have been accepted internationally.[17]

Sunlight outside the earth's atmosphere, at the mean distance of its orbit around the sun, is reasonably constant in intensity, varying by only a few percent. A standard value of 1367 W/m^2 perpendicular to the sun's rays is now commonly accepted.[18] Terrestrially, apart from total obscuration of the sun at night, there is a variable attenuation of sunlight by clouds and by atmospheric scattering and absorption, even in clear weather. The extent of the latter attenuation depends primarily on the length of the light's path through the atmosphere, or the mass of air through which it passes. The "Air Mass" is defined as 1/cos ϕ, where ϕ is the angle between the vertical and the sun's position (Air Mass can most easily be estimated from the length of the shadow, s, of a vertical structure of height, h, as $\sqrt{1 + (s/h)^2}$). The sunlight intensity is also affected by other factors, such as the amount of other atmospheric constituents such as water vapor, which acts as an efficient absorber of sunlight, and dust, which can both absorb and scatter the light.

In clear skies, the maximum sunlight intensity occurs when the sun is straight overhead (i.e., Air Mass 1 conditions) with a corresponding peak intensity of about 1 kW/m^2. Terrestrial solar-cell performance is specified with reference to the more representative Air Mass 1.5 (AM1.5) spectrum[17] of Fig. 4. This spectrum was calculated for a given set of atmospheric conditions and is now accepted as the international standard. There would be few, if any, solar cells that have had their performance measured under precisely this spectrum. In general, solar-cell and solar-cell-system perfor- mance is measured under natural or artificial sunlight of a different spectral content from the standard and "spectral mismatch" corrections applied to calculate performance referred to the standard. Since manufacturers prefer to specify the "peak" or full "one-sun" output of their devices and systems, the performance is also normalized to an intensity of 1 kW/m^2 to give the "peak watt" (W_p) rating of the device or system. (There is a certain inconsistency in this procedure, since the reference illumination has a spectral distribution corresponding to the AM1.5 spectrum, but a scaled intensity higher than would be obtained by integrating the AM1.5 spectrum.)

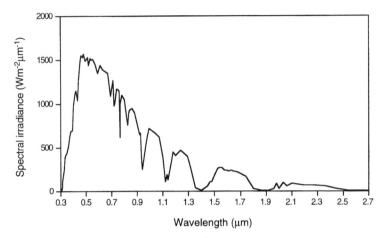

Fig. 4 Spectral content of the standard global Air Mass 1.5 spectrum used as a reference in terrestrial photovoltaic cell and system testing. (After Hulstrom et al., Ref. 17.)

The measurement temperature must also be known, since, as noted, cell performance is sensitive to temperature. This sensitivity results mainly from the same effect which causes an approximately −2 mV/°C shift in the silicon diode voltage required to maintain fixed current. One standard measurement temperature is 25°C, the standard for other semiconductor devices. Solar-cell components are usually designed to operate at as low a temperature as possible to maximize cell output and to enhance the longevity of the cell packaging. Even so, terrestrial operating temperatures as high as 80°C have been measured under extreme conditions, where the cell and system power output might be 30% lower than at 25°C. To address this issue, a nominal operating cell temperature (NOCT) is also specified for cell packages, better known as modules. This is the temperature that solar cells within a module will reach under a standard set of operating conditions, as discussed further in Section 8.5.

Because light scatters during its passage through the atmosphere, sunlight contains both a "direct" component from the direction of the sun and a "diffuse" component from other directions in the sky, originating from multiple scattering events. Even under very clear skies, the diffuse component accounts for over 10% of the total incident energy and can be larger, on average, than the direct energy in cloudy climates. Since short-wavelength blue light is scattered more effectively by atmospheric molecules than longer wavelengths, not only is the sky blue in color, but the diffuse spectrum has a higher blue content than the combined or "global" spectrum.

The distinction between "direct" and "diffuse" sunlight becomes particularly important for solar-cell systems based on focused or concentrated sunlight. It is not possible to concentrate diffuse sunlight to any significant extent, although direct sunlight can be focused to up to 46,000 "suns' concentration," in principle. This difference arises from the different acceptance angles required for the optical concentrating system to focus the two types of sunlight. Diffuse light is scattered over the whole sky, whereas direct sunlight is constrained within the 0.267° angle subtended by the sun's disk at the earth's surface. Concentrating systems must move to "track" the sun to achieve any reasonable level of sunlight concentration (any concentration greater than about 4 in practice).

As a consequence, a standard "direct normal AM1.5 spectrum" is also defined[17] for the measurement of the performance of solar-cell devices and systems designed for concentrated sunlight. The performance of these concentrating systems is generally specified only on the basis of the amount of direct sunlight intercepting the system's aperture. This concession makes the rating of such systems not directly comparable to those of the more prevalent nonconcentrating option. The seemingly higher performance offered by concentrating systems is often largely an illusion arising from this different measurement convention.

8.2.3 Empirical Efficiency Limits

Figure 3 suggests that it is relatively straightforward to calculate the power output of a p–n junction diode under illumination, and hence to calculate the cell energy-conversion efficiency. An upper limit to the current at short-circuit (I_{sc}) is determined by the number of photons in the incident sunlight with sufficient energy to create an electron–hole pair.[10] A certain fraction of these photons will create electron–hole pairs in the active collection region of the cell and so contribute to I_{sc}. The open-circuit voltage, V_{oc}, can be calculated from the diode equation as modified under illumination:[10,11]

$$I = I_0(e^{qV/nkT} - 1) - I_L \qquad (1)$$

where I_0 is the dark saturation current of the diode, kT/q is the "thermal voltage", n is the ideality factor (equal to unity for an ideal diode), and I_L is the light-generated current causing the displacement of the I–V curves into the fourth quadrant (Fig. 1), equal to I_{sc} in most cases. I_0 is determined by the diode geometry and other design and material parameters such as doping levels and surface recombination velocities.[10,11] From Eq. 1, V_{oc} is given by

$$V_{oc} = \frac{nkT}{q} \ln\left(\frac{I_L}{I_0} + 1\right) \qquad (2)$$

Although this equation implies that a high ideality factor, n, and a high temperature, T, gives high open-circuit voltage, I_0 proves to be a more important term, and the opposite is the case! We see from Fig. 3 that the power output of the cell is always less than the product of $V_{oc}I_{sc}$. This feature is quantified by the introduction of a third parameter, the fill factor, FF, that has a value less than unity, with the power output given by $V_{oc}I_{sc}(FF)$. The value of FF is deduced by finding where the product of V and I from Eq. 1 is largest and then dividing by $V_{oc}I_{sc}$. Although no explicit analytical expression exists, an empirical expression for FF is:[11]

$$FF = \frac{v_{oc} - \ln(v_{oc} + 0.72)}{v_{oc} + 1} \tag{3}$$

where v_{oc} is the normalized open-circuit voltage given by $V_{oc}/(nkT/q)$. This equation is accurate to four significant digits for $v_{oc} > 15$. In practice, both series and shunt resistances associated with the cell will reduce the fill factor below the ideal value given by Eq. 3.

The empirical approach to determining efficiency limits relies on making optimistic assumptions about I_L and I_0. For I_L, all photons of energy above the bandgap energy are assumed to create electron–hole pairs that contribute to cell output current. For I_0, optimistic material values are inserted into the expressions for I_0 for different diode designs for the more developed semiconductor materials. Such work suggests that the most important parameter determining I_0 is the semiconductor bandgap E_g with the following empirical expression giving a lower bound on I_0:[11]

$$I_0 \geq 1.5 \times 10^5 \exp\left(-E_g/kT\right) \text{ A/cm}^2 \tag{4}$$

Since I_0 decreases strongly with increasing E_g, V_{oc} increases as E_g increases. Because I_{sc} shows the opposite trend (since fewer photons in sunlight have energy above E_g as E_g increases), this formulation shows there is an optimum E_g for maximum energy conversion efficiency. Using Eq. 4, this optimum is predicted to lie at $E_g = 1.4$ eV, close to the bandgap of GaAs.[15] The highest-efficiency single-junction cells fabricated to date have indeed been based on GaAs, with efficiencies above 25% demonstrated,[19] although the best silicon cells are now quite close to this performance level.

The disadvantage of the empirical approach to calculating efficiency limits is that it is firmly rooted in concepts of cell design that presently prevail. More fundamental approaches ideally would also give limits for solar cell designs that are still awaiting conception.

8.2.4 Radiative Recombination

In 1960, a more fundamental approach to calculating solar cell efficiency limits was published by Shockley and Quiesser.[16] This approach recognized that, regardless of how material quality might be improved to reduce carrier recombination associated with impurities and other defects, the radiative recombination of carriers would provide a lower bound on recombination throughout the device. Since I_0 depends on the recombination rate integrated over the whole volume of the device, radiative recombination, therefore, also places a lower bound on attainable values of I_0.

Radiative recombination introduces the concept of "photon recycling."[20] In this process, a photon creates an electron–hole pair, which recombines radiatively. This radiative recombination produces a photon that has an energy just above the bandgap. This photon can be "recycled" by creating an electron–hole pair by reabsorption elsewhere in the device. Generation of this carrier pair reduces the net recombination rate.

Rather than keeping track of events internal to the cell, Shockley and Quiesser[16] analyzed the light emitted externally by an idealized cell. A "black body" is a perfect absorber of incoming light, clearly a desirable property for a cell, as well as a perfect emitter of light (which is not so desirable, since we will see that this increases I_0!). In thermal equilibrium, the number of photons of frequency ν emitted from one surface of a solar-cell per-unit-area per second is given by Planck's black-body distribution function:[16]

$$N_{ph}(\nu) = (2\pi\nu^2/c^2)\,[\exp{(h\nu/kT)} - 1]^{-1} \qquad (5)$$

Each net radiative recombination event in the semiconductor produces an emitted photon with an energy above the bandgap. Therefore, a lower bound on the recombination rate throughout the cell volume, including photon recycling effects, can be deduced from the number of photons emitted from the cell surfaces that have an energy above the bandgap. By considering the optimal case, where the rear of the cell is a perfect reflector (and neglecting the area of the sides of the cells), emission from only the front surface of the cell needs to be considered.

To proceed further, Shockley and Quiesser[16] noted that the integrated radiative recombination rate would increase exponentially with increasing forward voltage applied across the cell. This can be proved if very large carrier mobilities are assumed, which is a condition consistent with estimating upper bounds on the efficiency. Under these conditions, large currents flow under small carrier and quasi–Fermi-level gradients. In the case of infinite mobilities, both electron and hole quasi–Fermi levels are constant across the cell with a separation equal to the applied cell voltage, in the absence of "contact resistance" effects. It follows that radiative recombination must increase uniformly and exponentially with voltage throughout the cell volume. Hence, for an ideal black-body cell, where unavoidable radiative recombination is the only recombination mechanism, an equation for cell characteristics identical to Eq. 3 is deduced but with I_0 equal to q times the photon emission rate given by Eq. 5, integrated over all photon energies higher than the bandgap:

$$I_0 = qA \int_{E_g}^{\infty} \left(\frac{2\pi E^2}{h^2 c^2}\right) \left[\exp\left(\frac{E}{kT}\right) - 1\right]^{-1} dE \qquad (6)$$

that is,

$$I_0 \approx qA \left(\frac{2\pi kT}{h^3 c^2} \right) E_g^2 e^{-E_g/kT} \tag{12}$$

We can simplify the integral by neglecting the -1 term in the square brackets, since $E \gg kT$ for the energies involved and by noting that $\exp(-E/kT)$ is strongly peaked near the lower limit of the integration. The rest of the integrand can be regarded as constant compared to this rapidly varying quantity over the portion of the integration which gives the major contribution to the integral. The exact solution with the -1 term neglected has the E_g^2 term above replaced by $E_g^2 + 2E_g(kT) + 2(kT).^2$ The -1 term can be included by expanding the term in the square brackets in Eq. 6 as a power series[21] in $\exp(-E/kT)$. (The first simplification is a handy technique to remember in the semiconductor field where integrals involving exponentials frequently occur!)

Two points emerge by comparing Eqs. 4 and 7. Both predict a strong exponential dependence on the bandgap, but the more fundamental approach shows a slightly stronger preference for lower-bandgap material because of the E_g^2 term. This is due, fundamentally, to the lower density of photon states near the bandgap edge in low-bandgap material, and pushes the optimum bandgap to slightly lower values than the earlier empirical approach. Also, the calculated values of I_0 are much lower. The prefactor of the exponential in Eq. 7 is about 5.8×10^2 A/cm^2 for $E_g = 1.4$ eV, compared to 1.5×10^5 A/cm^2 for Eq. 4. This is not unexpected, given that one value is based on the best experimental data, whereas the other is the best possible value.

The radiative efficiency limit for a single cell is close to 33% for cells of any bandgap intermediate between that of silicon (1.12 eV) and GaAs (1.42 eV) under the standard AM1.5 spectrum.[22] A material may not be capable of reaching this limit even if it is of perfect quality, since intrinsic processes other than radiative recombination may come into play before the radiative limit is reached. For example, intrinsic Auger recombination processes in silicon limit silicon-cell efficiency to around 29%, rather than 33%.[22,23]

The simplest way, conceptually, to exceed these efficiency limits is by "spectrum splitting." By splitting sunlight into narrow wavelength bands and directing each band to a cell that has a bandgap optimally chosen to convert just this band, as shown in Fig. 5a, much higher performance (efficiency above 60%) is possible in principle.[24,25] Fortunately, simply stacking cells on top of one another with the highest bandgap cell uppermost, as in Fig. 5b, automatically achieves an identical spectral-splitting effect, making this "tandem" cell approach a reasonably practical way of increasing cell efficiency.

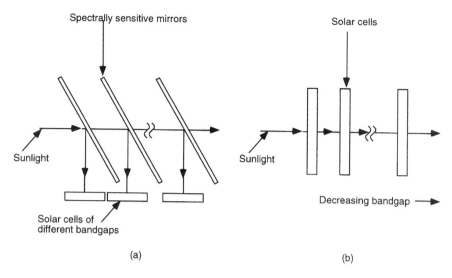

Fig. 5 Multigap-cell concepts. (a) Spectrum-splitting approach. (b) Tandem-cell approach. (After Green, Ref. 11.)

8.3 SILICON SOLAR CELLS: CRYSTALLINE, MULTICRYSTALLINE, AND AMORPHOUS

8.3.1 Advantages of Silicon

Silicon is an ideal material for use in solar-cells. It is nontoxic and is the second most abundant element in the earth's crust, posing minimal environmental or resource depletion risks if used on a large scale. It also has a well established technological base because of its use in microelectronics, as detailed in earlier chapters.

Given these advantages, particularly the latter, it is, perhaps, not too surprising that the overwhelming majority of commercial solar-cells fabricated to date have been based on silicon in one of three forms; crystalline, multicrystalline (large-grain polycrystalline), and amorphous. To produce adequate electronic quality in the amorphous form, silicon is alloyed with hydrogen at about the 10% atomic level. This results in material with properties distinctly different from the elemental crystalline counterpart.

The following sections describe the evolution of each silicon cell type, present cell designs, and topics of current research interest.

8.3.2 Crystalline Silicon Cells

Evolution of Cell Design.[23]   The first silicon cells, described in 1941,[26] were multicrystalline, with the first crystalline cells reported in 1954.[27] The evolution of laboratory cell performance from these early beginnings is shown

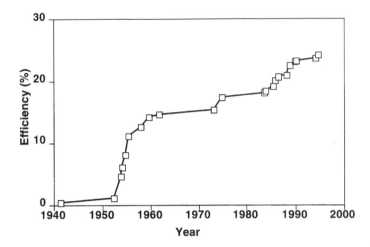

Fig. 6 Evolution of silicon laboratory cell efficiency.

in Fig. 6. As seen in this figure, from the mid-1950s until the early 1960s, there was a very rapid evolution in cell performance stimulated by the growing demand for cells for spacecraft. Performance stabilized in the 1960s, with a new burst of improvements in the early 1970s. Since the early 1980s, there has been a period of sustained improvement in the performance of research devices, although commercial silicon solar technology remains based largely upon technology of the 1970s.

The structure of a standard commercial cell for use in space is shown in Fig. 7. Standard microelectronics processing is used for these space cells. The crystalline substrates are prepared from cylindrical ingots grown by the Czochralski (CZ) method and sliced into wafers by inner-diameter sawing.[11,28] The ingots are doped with boron during growth, with target resistivities of either 2 Ω-cm or 10 Ω-cm. The p–n junction is formed by lightly diffusing phosphorus into the top cell surface to about 100 $\Omega/\square$ sheet resistivity. Conditions are chosen to produce a doped surface region that is heavily doped with phosphorus, just below solid solubility limits, but which is very shallow (~0.25 μm junction depth). By producing a shallow junction from this diffusion, traditionally referred to as the "emitter" diffusion, carriers photogenerated between the junction and the surface have a better chance of being collected by the junction, that is, are more likely to travel from the point of generation to the junction than to recombine at the surface or in transit. Thin emitter regions with properties as above are often described as being "transparent" to minority carriers. The high doping near the surface improves tolerance to poor surface properties by suppressing surface minority-carrier concentrations near the surface and, hence, their surface recombination rates.

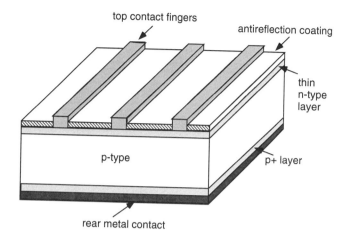

Fig. 7   Conventional silicon space cell.

The "high–low" p^+–p junction at the back of the cell is usually formed by alloying a layer of evaporated aluminum. This high-low junction feature is often known as a "back surface field" (BSF) because of the electrostatic field present at this junction. Although this nomenclature is universally accepted, it is not entirely appropriate, since it is fundamentally the suppression of the minority-carrier concentration near the rear contact by the p^+ region that makes this region effective, rather than the previous field. The effect of the p^+ region is usually characterized by assigning an effective surface recombination velocity along the p–p^+ interface. This velocity is a lot lower than along a silicon–metal interface. The reduction in surface recombination velocity has beneficial effects upon both the I_0 of the cell and the prospects for collecting carriers generated in the p-type region near the rear of the cell. The introduction of a BSF, therefore, results in both improved V_{oc} and I_{sc}.

A Ti/Pd/Ag multilayer is universally used as the top- and rear-surface metalization in silicon space cells. The titanium is deposited first by vacuum evaporation to ensure good adhesion of the multilayer to the underlying silicon. The silver is deposited last, either by vacuum evaporation or by electrolytic plating, and imparts high conductivity to the trilayer. The intermediate palladium layer acts as a barrier that eliminates a detrimental reaction between the titanium and silver in the presence of moisture. For the rear contact, an aluminum layer is often evaporated before the titanium to improve the rear surface reflectance. Such an enhanced reflectance rear contact is referred to as a back-surface reflector (BSR).

To reduce reflection from the top surface of the cell, either a single-layer antireflection (SLAR) coating of TiO_2 or Ta_2O_5 is evaporated onto this surface or a higher performance, double-layer (DLAR) coating of TiO_2 followed by Al_2O_3 is used.

The performance of space cells fabricated in this way is quite modest by the standards of the best cells of Fig. 6. A DLAR/BSF/BSR cell has an energy conversion efficiency of about 14.5–15% under zero Air Mass (AM0) space radiation, and a higher efficiency of about 16–17% under AM1.5 radiation because of the more favourable spectrum of the latter for silicon. However, initial or beginning-of-life (BOL) performance is not the only important parameter for a space cell. In space, cells are usually exposed to damaging high-energy particles (electrons, protons, etc.) present in the earth's radiation fields,[29] from which ground-based cells (and, fortunately, ourselves!) are protected by the earth's atmosphere. The design of the cell also determines the end-of-life (EOL) performance after such damaging exposure during the life of the space mission.[29] More modern designs can give much higher BOL performance but only marginally higher, or even lower, EOL outputs.

The other important cell operational feature is the thermal performance in space. Since thermal radiation is the main heat-loss process in the vacuum of space, thermal performance is determined by how much light the cell absorbs as well as the cell's emissivity at the infrared wavelengths at which it reradiates absorbed energy. BSR cells have excellent thermal performance, since infrared light in the solar spectrum is reflected out of the cell, rather than being absorbed.

The space cell design of Fig. 7 was first developed in the early 1970s. There have been two important directions in which cell design has subsequently evolved. One is towards lower-cost fabrication for terrestrial use and the other is towards higher-efficiency devices for research and specialized high-performance applications. Although apparently completely different directions, there has been considerable interchange between developments in the two areas.

PERL Cell.   The structure of a 24% efficient PERL (Passivated Emitter Rear Locally-diffused) cell[23] is shown in Fig. 8. These and other cells with almost identical structure have been made in reasonable quantities for specialized applications where performance is at a premium, such as solar-car racing. Compared to the standard space cell in Fig. 7, several performance enhancing features can be discerned. Perhaps the most apparent is the use of a texture on the top surface of the cell, in the form of "inverted pyramids." These are formed by using anisotropic etches to expose slowly etching (111) crystallographic planes. This technique has been used in other areas of microelectronics such as to form the V-groove in the gate region of VMOS field-effect transistors.[10] For solar-cells, all four (111) equivalent planes are exposed by etching the surface, originally chosen to be of (100) orientation. The surface of PERL cells is selectively masked by a patterned oxide to define the pyramid location. The bases of the pyramids are about 10 μm wide.

The pyramids serve several optical functions. They reduce reflection of

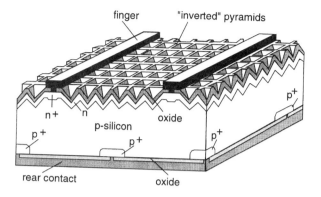

Fig. 8 High-performance PERL cell.

light incident on the top surface of the cell. Light incident perpendicularly to the cell will strike one of the inclined (111) equivalent planes obliquely. Most will be refracted obliquely into the cell at this point. Since the (111) planes are at an angle of 54.7° to the original surface of the cell, the reflected light is reflected downwards, giving at least one more chance of being coupled into the cell.[23] Even without antireflection coatings, only about 30% reflection occurs at each point of incidence on a silicon surface. Hence, after two reflections, only about 9% of the light would remain unabsorbed (30% × 30%). With antireflection coatings, reflection at the first point of incidence can be kept below 10% over a reasonable wavelength range, reducing reflection to around 1% after two "bounces".

Since the pyramids cause light refracted into the cell to travel obliquely across the cell, a second function of the surface texture is to cause light to be absorbed closer to the active top junction of the cell. A third benefit arises for light that is so weakly absorbed that it is not absorbed on a single pass across the entire cell thickness. The rear contact of the PERL cell is separated from the silicon by an intervening oxide layer. This gives much better rear reflectance than if it was in intimate contact (97% vs about 85% for the lowermost aluminum layer in the rear contact multilayer). Hence, most of the weakly absorbed light reaching the rear is reflected by this efficient reflector and heads back internally towards the front surface. It is possible for this light to hit any one of three faces of the pyramids internally. If this light hits the face that is oppositely inclined to the face that coupled the light in, the light is coupled straight out of the cell at this stage. About half the light has this fate.[23] For light striking the other two faces, light strikes at angles where it is totally internally reflected, at least once, and again travels towards the rear of the cell. Since the light must strike the silicon surface at an angle within 16° or so of the perpendicular to escape ($\sin^{-1}(1/\bar{n})$, where $\bar{n}$ is the refractive index of silicon), only a small fraction of light escapes after each subsequent double pass as the direction of light is increasingly diversified ($1/\bar{n}^2$ or 9% for completely randomized light directions[30]).

By increasing the optical pathlength for weakly absorbed infrared wavelengths, the current output of the cell can be substantially increased. A pathlength enhancement of weakly absorbed light of $4\bar{n}^2$ (about 50) is possible by such "light trapping" approaches.[30] This factor, in principle, can be increased even further for cells, such as those in concentrator systems, which only have to accept light from well-defined directions.[25,31] Since light trapping increases the absorbance of a cell, it may not always be a desirable feature for a space cell, where it may increase operating temperatures. The resulting loss in output voltage may offset the increased current output.

Another important feature is the almost complete enshroudment of the PERL cell by a thin layer of thermally grown oxide. The ability of such oxides to reduce interface trap densities and so reduce the electronic activity of (i.e., passivate) silicon surfaces is well known from other branches of microelectronics.[10] High levels of surface passivation produce low surface recombination rates along the cell surfaces without the need to have surfaces heavily doped, increasing design options. Contact to the n-type region is made through slots in this passivating oxide, whereas contact to the p-type region is made through small holes. The regions immediately under the contacts are heavily doped to suppress minority-carrier concentrations and, hence, recombination rates.

The cell is finished by adding a DLAR coating. For the DLAR coating to be optically effective, the thickness of the underlying passivating oxide has to be less than about 20 nm. This thin top-surface passivating oxide is grown separately from the rear passivating oxide, since the latter is most effective when grown 10–20 times thicker.

Highest performance is obtained if these devices are fabricated on either float-zoned (FZ) or magnetically confined Czochralski grown (MCZ) silicon with a resistivity of 1–2 Ω-cm. This is because the high minority-carrier lifetimes possible with such material can be retained during cell processing. Internal gettering using oxide precipitates, as commonly used in other areas of microelectronics, is not a feasible option for cells, since the entire bulk must be of uniformly good quality. Carrier lifetimes in the range 1–10 ms are routinely obtained in such devices, quite close to the value of 30 ms that appears to be the highest ever measured for silicon.[32] Good, but slightly poorer results have also been reported for cells fabricated with the PERL structure using conventional CZ substrates.

Although PERL cells were developed primarily for research purposes, they have found applications as advanced space cells and in solar-car racing, as previously mentioned. Features such as oxide passivation and the use of heavily doped regions to passivate contact areas have found their way into low-cost cell production via the buried-contact cell, discussed later. The PERL cells are also very efficient at converting narrow-band or monochromatic light with efficiencies approaching 50%.[33] This makes such cells of interest for laser-beamed energy-transmission concepts or for thermophotovoltaic energy conversion.[34] PERL cells also can convert concentrated sunlight with high efficiency up to about 50 suns' concentration. Above this concentration level, a different silicon cell design, the rear contact cell, has given best results.

Rear-Contact Cells.   The rear-contact solar-cell[12,23,35] of Fig. 9 introduces a number of interesting features into cell design, including the location of both *p*- and *n*-type contacts on the rear surface of the cell in the form of interspersed stripes. Since carriers are photogenerated mostly within a few microns of the top surface in sunlight, they have to diffuse to the rear of the cell to be collected. The cell therefore has to be very thin (usually thinner than 150 μm) and to have very long carrier lifetimes, to reduce carrier loss by recombination during the journey to the rear. Extremely good-quality surface passivation is also essential along the front illuminated surface of the cell to reduce carrier recombination there. Although very high levels of passivation can be obtained by growing oxide directly onto the lightly doped 100-Ω-cm *n*-type substrates normally used in these devices, the resulting interface is unstable under ultraviolet (UV) radiation, presumably due to the injection into this oxide of hot carriers generated in the silicon.[36] Doping the surface with a phosphorus diffusion greatly improves this stability, as does adding a UV-absorbing layer on top of the cell, such as provided by a TiO_2 SLAR coating.[36]

As opposed to the PERL cells, which operate at relatively low levels of minority-carrier injection, these rear-contact cells are designed to operate under high-level injection.[35] At high sunlight intensities, the bulk *n*-type regions are flooded by approximately equal quantities of electrons and holes,

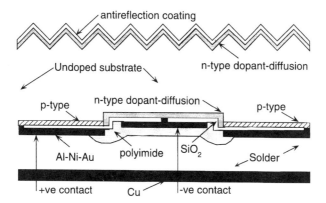

Fig. 9   Rear-contact solar cell. Both *p*-type and *n*-type contacts are made to the rear surface of the cell, insulated by a layer of polyimide. (After Verlinden et al., Ref. 37.)

overriding the carrier contribution from the low concentration of dopants. This largely eliminates the resistive losses which might be expected from a lightly doped region in low-level injection.[35] Placing both contacts on the rear not only eliminates the shading losses that occur in conventional designs, but also allows a much larger fraction of the cell area to be dedicated to extracting current from each polarity contact. As can be seen from Fig. 9, considerable attention has been given to the design of the rear metalization to allow low resistance electrical and thermal connections to the cell while providing electrical isolation between the two contact polarities.

Cells of this and similar types have been used in a number of concentrated sunlight demonstration systems. Over 7000 were fabricated for nonconcentrating use for a solar car, the "Honda Dream," which won the 1993 World Solar Challenge across Australia.[37] These cells were used in an 8-m^2 solar array, which was the first of this size to exceed 20% energy-conversion efficiency.

Screen-Printed Commercial Cells.[23]   For more general terrestrial use, the interest is in reducing cell costs as much as possible, usually at the expense of performance. The most common fabrication approach is based on technology developed for thick-film hybrid microelectronics, to produce the cell structure shown in Fig. 10.

The starting material is again a silicon wafer, usually boron doped with a resistivity in the range 0.5-10 Ω-cm. These wafers are either "off-specification" for microelectronics, recycled from the microelectronics industry, or grown from a polysilicon source that is "off-specification." The cylindrical CZ grown ingot is often squared off by sawing prior to wafering to give "quasi-square" wafers. The wafers are cut using inner diameter or continuous wire saws to a thickness of 200–400 μm. After etching in NaOH solution to remove saw damage, the wafers are generally "textured" in a weaker anisotropically etching NaOH solution to give the random layout of upright pyramids shown in Fig. 10, ideally with a pyramid-base dimension of 5–10 μm.

Next the wafers are diffused, either in a conventional tube furnace, or in a belt furnace commonly used in hybrid microelectronics. The top and rear contacts are then applied by screen-printing silver paste, another hybrid microelectronics approach. For the top contact, the construction of the wire screening mesh limits the width of the lines that can be produced economically to about 150 μm. The low aspect ratio (height–width ratio), the relatively broad linewidth and the relatively poor conductivity of the fired silver fundamentally restricts the performance of these devices.[23] Another restriction is that the diffused top layer has to be quite thick and heavily doped, to a sheet resistivity of about 40 Ω/□, to ensure sufficiently low contact resistance. This is not an ideal situation as apparent from the earlier discussion of silicon space cells. Carriers generated near the surface are not collected, causing a poor response to blue wavelengths (these wavelengths

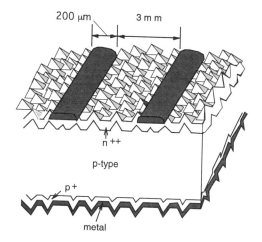

Fig. 10 Screen-printed silicon solar cell (not to scale).

are absorbed quickly in silicon). Phosphorus is often added to the top surface paste to help reduce contact resistance.

A variety of approaches are used for the rear contact. In the simplest case, some manufacturers screen print silver paste directly, as they do for the top contact. Others add aluminum to the silver paste in an attempt to simultaneously form a BSF by aluminum alloying during the paste firing. Yet others first screen and fire an aluminum paste to produce the BSF layer, and then screen and fire a silver paste. However, the silver and aluminum can interact, necessitating additional processing variants.

An inexpensive antireflection coating, generally TiO_2, can then be added. This improves cell performance by about 8% in air, but only by about 4% after encapsulation under glass or material of a similar refractive index.

Cells are generally either round, with a diameter of 10–12 cm, or quasi-square, with a length of 10–12 cm along each side. Energy-conversion efficiency in the range 12–14% is generally produced by cells made with this screen-printing approach. When encapsulated into modules, module efficiency is in the range 10–13%.

This technology for making cells was first demonstrated in the early 1970s, and became the commercial standard by the early 1980s. As apparent from Fig. 6, there have been quite substantial improvements in silicon laboratory cell performance since this time. The buried-contact cell, discussed below, has been developed to incorporate many of these improvements into commercial cells.

Buried-Contact Solar Cells.[23,38] It is difficult to modify the screen-printing approach to increase cell efficiency because of the requirements imposed by the very simple metalization approach. Most success has come by pursuing

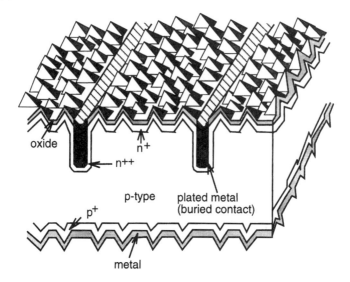

Fig. 11 Buried-contact solar cell.

a completely different route. The most successful commercial alternative has been the buried-contact cell of Fig. 11. This device was first commercialized by the solar-cell manufacturer BP Solar in the early 1990s as the company's "Saturn" product line, and has been produced in increasing volume since that time.

As previously noted, performance from the screen-printed cell approach is sacrificed to processing simplicity. The key to overcoming the resulting limitations proved to be finding an equally simple way of metalizing cells that did not demand a similar performance compromise. In the buried contact approach, metalization is achieved by the electrodeless deposition of an Ni/Cu/Ag multilayer. This is inherently a much less expensive metalization step than screen printing, which also gives much better electrical performance in terms of metal conductivity and contact resistance.

The disadvantage of this metalization approach was that no simple technique was available to define the pattern of the deposited metal. The buried-contact approach solves this difficulty in a particularly elegant way.

The buried-contact cell processing starts identically to that for a screen-printed cell. Incoming wafers are etched to remove saw damage and then chemically textured. The top junction of the cell is then formed by a diffusion that is much lighter than that in the screen-printed cell, and an oxide is thermally grown along the diffused surface or, alternatively, silicon nitride may be deposited.

A laser is used to cut deep grooves through the resulting dielectric layer and part of the underlying silicon. After cleaning, the cells undergo a second diffusion, much heavier than the first. Since all but the grooved regions are covered by a dielectric that effectively blocks diffusion, the diffusion is automatically localized to the grooved area. Aluminum is then deposited on the rear surface and sintered to give a rear BSF. The cell is then placed in the electrodeless plating solutions, which selectively deposit metal on the conducting areas. Because the rest of the top surface is protected by dielectric, only the grooves and the entire rear surface are plated.

The buried-contact approach provides a very effective way of incorporating a number of high-efficiency features into the cell. The key to this effectiveness is the multiple ways in which the top surface dielectric is used. It is not only used as both a diffusion and a plating mask during processing, but serves as both top-surface passivation and antireflection coating in the completed device.

High-efficiency features, in approximate order of importance, are the nearly ideal top-surface diffusion and surface passivation, which can be optimized independently of contact-shading and contact-resistance constraints. Because of this, buried-contact cells can respond almost ideally to the rapidly absorbed blue wavelengths, giving an estimated 15–20% contribution to the performance advantage over a screen-printed cell. A large performance-improvement component also comes from the reduced shading loss by the top-contact metal, since the metal lines can be much narrower than screen-printed lines. This accounts for another 5–10% advantage, depending upon cell design. An additional advantage arises from superior metal conductance. Not only is the conductivity of the plated metal (mainly copper) about three times that of screen-printed silver but the grooves also can be very deep, resulting in a large cross-sectional area. Another advantage is much lower contact resistance. These resistance improvements combined yield a further 5% performance advantage. Additional advantages that are more difficult to quantify arise from the ability of the laser grooving, the heavy phosphorus groove diffusion, and the rear aluminum BSF formation steps to provide very effective gettering of impurities from the bulk regions of the cells.[23]

On the standard CZ silicon used in production, the relative cell efficiency advantage is in the 25–30% range.[38] The buried-contact sequence produces cells with efficiency in the range 16–18% on the same substrates used for screen-printed cells that have an efficiency in the range 12–14%. Data based on several years of production experience suggest that the costs of making these improved cells is the same per unit area as screen-printed cells,[38] giving a pronounced advantage in cost-per-unit power output and additional advantages in installation costs.

Further developmental work is concentrating on replacing the rear aluminum BSF step by a grooving approach similar to that used on the top surface. This gives a structure conceptually very similar to the PERL cells described earlier, the highest efficiency silicon cells yet demonstrated. Combining this approach with thinner wafers, a cell efficiency of 20% on CZ substrates may eventually be reached in commercial production.

8.3.3 Multicrystalline and Ribbon Wafers

Poorer material quality can be used for photovoltaics than for integrated circuits, without major performance loss. This has prompted the development of alternatives to CZ ingot growth and wafering for preparing silicon solar-cell substrates.

Rather than using CZ growth to produce crystalline ingots, large-grain polycrystalline (multicrystalline) ingots can be produced by technologically less-demanding processes, such as casting, involving the solidification of a melt in a crucible as shown in Fig. 12a. These more robust processes reduce ingot costs, since much larger ingots can be produced, with much higher throughput per unit investment in growth equipment. The ingots can be sawn to smaller units, as in Fig. 12b, before they are sliced into wafers. Offsetting these advantages has been the much larger size and the maturity of the integrated-circuits industry. This means there is a supply of "off-specification" wafers and secondhand CZ crystal growers, for example, which cost much less than mainstream product.

The most desirable process would be one that produced silicon directly in the form of large-area sheets. The closest to achieving this ideal are two ribbon-growth processes:[39] the edge-defined film-fed growth (EFG) method shown in Fig. 13 and the dendritic-web approach. The first approach is based on the use of a graphite die to define the shape of the growing ribbon, whereas the second relies on close control of temperature in the solidification zone.

These customized approaches produce substrates with many features differing from CZ substrates. The substrate quality is invariably poorer and more variable following from both higher levels of chemical contamination and the presence of crystallographic defects, the most severe being grain boundaries. The surface generally is not of the (100) orientation required for crystallographic texturing, to reduce reflection loss. Because of these differences, changes in processing are required to extract good performance from these materials.

Since crystallographic texturing is not generally possible, some other form of reflection control is required. In production, quarter-wavelength antireflection coatings are used, usually either TiO_2 or Si_3N_4. The former is used because of its almost ideal refractive index for this application and its ease of deposition, being suitable for simple spraying and other atmospheric-pressure deposition approaches. Si_3N_4 is more difficult to deposit, with silane (SiH_4) generally used as the silicon source material. The hydrogen produced by the dissociation of silane is incorporated into the film and substrate and reduces the electronic activity of many defects.

Grain boundaries in silicon tend to be very active electronically, acting as undesirable recombination centers, particularly when decorated by impurities. For reasonable cell performance, grains in silicon wafers have to be at least several millimeters in width. The variability in grain properties between wafers introduces variability in performance. If hydrogen, in atomic

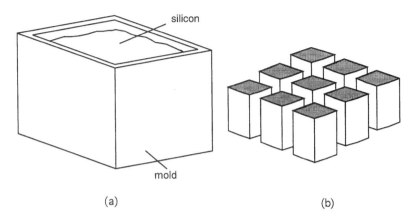

silicon

mold

(a)   (b)

Fig. 12   (a) Solidification of multicrystalline ingot in a crucible. (b) Ingot sawn into multiple blocks.

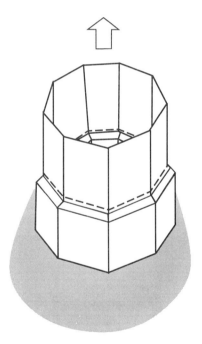

Fig. 13   Growth of a nonogonal, tubular ribbon of silicon using the edge-defined, film-fed growth (EFG) method. (After Green, Ref. 23.)

form, is introduced into the wafer, it can neutralize some of the electronic activity at the grain boundary and at other crystallographic and chemical defects by attaching itself to dangling chemical bonds or otherwise complexing with defected regions. This improves the cell performance, particularly the performance of the poorest cells, tightening the distribution of cell characteristics.

One way of introducing the hydrogen used commercially is to use an Si_3N_4 antireflection coating as previously noted, although quite a complex processing sequence is involved. Another way, used in pilot production, is by bombardment with hydrogen ions near the end of cell processing.[40]

The chemical contamination of the material also produces different processing requirements from those for CZ substrates. The material often benefits from lower processing temperatures than the CZ counterpart and also benefits more from gettering treatments. Heavy phosphorus diffusion and aluminum BSF treatments are particularly effective gettering steps which appear at least partly complementary in their effect.[41] Sometimes, adding additional impurities can improve performance. Copper is reputed to improve cell performance in a way similar to hydrogen, whereas a high oxygen content is known to improve the quality of EFG ribbon, possibly by complexing with carbon incorporated from the graphite die.

Most commercial sequences for processing multicrystalline and ribbon material tend to be variations of the screen-printing approach described for CZ silicon. The chemical texturing step is usually excluded. Texturing can at least partially reduce reflection loss for multicrystalline material by texture etching those grains not too far from (100) orientation. However, texturing can roughen the surface by etching some grains more rapidly than others, jeopardizing the continuity of the screen-printed metalization fingers. The temperatures used during processing are often different from those used for CZ material, for reasons previously noted. An antireflection coating of TiO_2 is generally added near the end of processing as an essential feature, rather than as an option as in the case of CZ material. Overall, performance is lower than from CZ material due to the poorer material quality. However, since wafers are generally manufactured with a square or rectangular shape, active cell material can be packed more densely in the final module package than those based on circular or even quasi-square CZ wafers. This geometrical advantage at the module level largely offsets the efficiency disadvantage at the cell level.

Improved cell-processing sequences, such as the buried-contact sequence, can be applied to those substrates with good effect. Mechanical texturing of the cell surface has been demonstrated at the laboratory level and can give an additional boost to cell performance. However, such material is not as responsive to high-efficiency processing sequences as the better-quality crystalline material. The best result demonstrated by using the most advanced laboratory processing with such substrates is, at the time of writing, 18.6% for a $1\,cm^2$ cell,[42] appreciably lower than the 24% for crystalline cells.[43]

8.3.4 Polycrystalline Silicon Thin-Film Cells

From the early 1970s, there has been a strong interest in trying to reduce the costs of silicon solar-cells by depositing the cells in thin-film form onto a foreign supporting substrate. One approach involved applying silicon in molten form onto a substrate, such as ceramic, that can withstand high temperatures; another investigated lower-temperature deposition such as physical vapor deposition onto a low-temperature substrate such as an aluminum sheet.[44]

Exciting developments in the mid-1970s with hydrogenated amorphous silicon (Section 8.3.5) dampened enthusiasm for these approaches. It was not until the mid-1980s that interest in polycrystalline silicon thin films resurged.[45] Several developments during the intervening period were responsible for this reawakened interest.

One was the development of light-trapping theory[23,30] and the appreciation of how effective this could be in improving the performance of thin films made from a weakly absorbing semiconductor, such as silicon.[22,45] A second was the progress with improving bulk-silicon cell efficiency, which led to a re-evaluation of limiting silicon cell efficiency, and highlighted the good performance possible, in principle, from very thin silicon cells.[22,23] The third was the growing interest in silicon-on-insulator technology and in the use of thin polycrystalline films of silicon for active-matrix liquid-crystal displays.[46]

Cells formed on films about $100\,\mu m$ thick are expected to be the first commercial product of this type.[47] The situation with much thinner cells is evolving very rapidly. Some recent highlights have been the fabrication of simple $10\text{-}\mu m$-thick cells with a claimed 9% energy conversion efficiency.[48] Initially, amorphous silicon material was deposited on a metal substrate and crystallized at a temperature of only 600°C. This was also the maximum temperature reached during processing.

Another development was the invention of an integrated parallel multi-layer-cell concept[49] that is more tolerant of low-quality material than conventional cell structures. The completed cell and module structure is shown in Fig. 14. Several junctions connected in parallel in the cell structure reduce the distance for photogenerated carriers to travel before being collected by any one junction, relaxing material quality requirements. An approach similar to that used in the buried-contact cell is used to contact the layers of the multilayer stack and to connect adjacent cell regions together in series. It appears feasible to fabricate these devices directly on a glass superstrate with efficiencies up 15% projected for completed modules.

8.3.5 Amorphous Silicon Solar Cells

Historical Development. Pure amorphous silicon has poor electronic properties. However, in the late 1960s, it was found that amorphous silicon

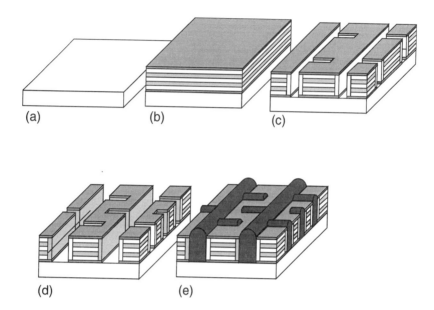

Fig. 14 Fabrication of multilayer cells. (a) Glass substrate (becomes superstrate in final module). (b) Multilayer deposition. (c) First polarity groove. (d) Second polarity groove. (e) Metalization.

formed by the decomposition of silane in a glow discharge showed interesting properties. This was ultimately linked to the incorporation of substantial quantities of hydrogen (about 10% atomic). The resulting alloy (a-Si:H) exhibited significant photoconductivity and could be doped n-type or p-type, by incorporating the appropriate dopant-containing gases (PH_3 and B_2H_6) into the discharge.[50]

By the mid-1970s, the first operational cells had been reported.[50] By the early 1980s, cell performance had been improved to about a 10% conversion efficiency by improvements such as a wider bandgap a-SiC:H "window" layer on the top surface of the cell and the incorporation of light trapping. The first commercial product on the market was an instant success. This product, targeted at the consumer market, used small arrays of cells to power pocket calculators and digital watches. Such consumer-product applications have dominated commercial a-Si:H solar-cell sales to date, although other nonsolar applications have proved even more significant. The material is used in large volumes as a photoreceptor coated on photocopier drums and in active-matrix liquid-crystal displays.[51]

For larger outdoor "power" modules, progress has been less rapid than originally expected because of stability problems. The beneficial effect of the hydrogen upon the amorphous-silicon properties deteriorates under illumination. Such "Staebler–Wronski" degradation typically causes a steady drop in solar module output over the first few months in the field, after which the output stabilizes.[50] Amorphous-silicon-based modules are generally rated by manufacturers in terms of such "stabilized" output.

Cell Design.   The glow-discharge approach remains the preferred deposition method for a-Si:H and its alloys. This deposition approach produces material of acceptable performance, although of quite poor quality compared to that of the more established crystalline semiconductors. In particular, carrier mobilities are low ($\sim 1\,cm^2/V$-s) as are minority-carrier diffusion lengths ($\sim 0.1\,\mu m$). Moreover, there is a continuous distribution of electron states lying at energies within the normally "forbidden" gap between the valence and conduction bands of the semiconductor. Although the material can be doped like the material's crystalline counterpart, not all dopants are electronically active, and the material quality decreases significantly as the doping level increases.

The combination of hydrogen and disorder in the material increases the material's bandgap significantly above that of crystalline silicon, to about 1.7 eV. The disorder within the amorphous material also relaxes the quantum-mechanical selection rules that make crystalline silicon such a weak absorber of light. However, the a-Si:H material is still not as strongly absorbing as some of the direct-bandgap materials to be discussed later. Layers about 0.5 μm thick are required to capture a reasonable fraction of the available sunlight.

The cell design that has proved most suited to take advantage of a-Si:H material properties is the *p–i–n* cell structure of Fig. 15. The doped layers are generally kept very thin (<50 nm), since very few of the carriers generated in these layers contribute to the collected photocurrent. These doped layers, however, do establish an electric field in the better quality *i* layer, which aids the collection of carriers generated in this region. This field allows carriers to be collected over distances longer than a minority-carrier diffusion length.

The important parameters determining carrier collection from this region is the material quality as reflected by the "mobility–lifetime" product multiplied by the strength of the electric field in the *i* layer. The latter field depends on the "built-in" voltage across the *i* layer, established by the doped layers that terminate this region, and the *i*-layer thickness. The strength of this field reduces as the output voltage across the cell increases. This gives rise to an effect not seen to any extent in crystalline cells. The light-generated current component, I_L, is not constant as suggested by Fig. 1b, but reduces with increasing voltage across the cell. This results in an inherently lower fill factor for these devices, decreasing the power output from the cell.

The material degradation under light exposure also reduces carrier collection from the *i* layer. Therefore, the *i*-layer thickness has to be chosen for stabilized material quality rather than for that of as-deposited material.

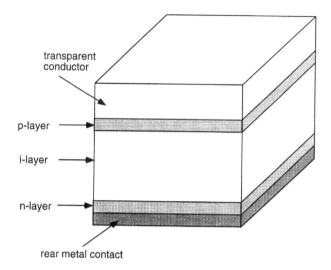

Fig. 15   Amorphous silicon *p–i–n* solar-cell structure.

The lateral conductivity of the doped layers in the cell is small, since these layers are thin and have low carrier mobility. To conduct the generated current laterally, conducting contacts are required on both cell surfaces. Generally, the contact on the surface exposed to sunlight is a transparent conductor, usually the degenerately doped semiconductor SnO_2. The contact on the opposite surface is generally metallic and also acts as a rear reflector for the cell.

Commercial Cells.   The first large amorphous-silicon installation was a 400-kW system installed at Davis, California.[52] The strategy adopted in this installation was to produce the amorphous silicon power modules as inexpensively as possible per unit area, with quite modest stabilized efficiency (only 4–5%). Such low efficiencies would normally have a negative impact upon area-related "balance-of-systems" costs such as mounting structures and wiring costs. The use of innovative support structures and installation techniques[52] neutralized this impact with this installation setting new standards for such "balance-of-system" costs.

The basic cell structure used in these modules is shown in Fig. 16.[53] A layer of SiO_2 followed by a transparent conducting layer of the large bandgap, degenerately doped semiconductor, SnO_2, is deposited onto a glass sheet 1.55 m × 0.78 m and patterned using a split-beam laser. The SnO_2 is deposited under conditions that create a surface texture that improves "light-trapping" in the final cell. Forty-eight of these substrates are then simultaneously coated by a *p–i–n* junction stack of amorphous silicon by the decomposition of silane in a radio-frequency (RF) plasma-discharge system.

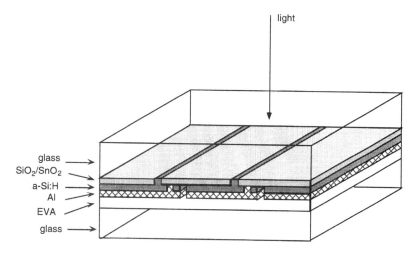

Fig. 16   Series-interconnected a-Si solar cells deposited onto a glass sheet with a rear glass cover bonded using ethylene vinyl acetate (EVA).

For good material quality, the deposition rate has to be slow (< 5 Å/s). Fortunately, only a thin, 0.3-μm layer of a-Si:H is required. After deposition, the a-Si:H layers are patterned in a second laser system. A layer of aluminum is then sputtered onto the rest of the silicon and this layer is also patterned by laser. This technique forms a series of interconnected cells, as shown in Fig. 16. Modules are completed by laminating a backing layer of glass onto the rear of the cells as shown.

These large-area modules show an energy conversion efficiency in the range 4–5% after being stabilized by outdoor exposure. Minor performance improvements would be expected rather simply by improving light trapping using a transparent conductor such as ZnO or SnO_2 to contact the rear of the device and then applying the aluminum layer over the top. This increases the reflectance for light reaching the rear of the cell. Replacing the aluminum by silver can further boost this reflectance.

Further improvements can be obtained by stacking two cells on top of each other to produce a tandem-cell configuration, as indicated in Fig. 17. The top cell responds to the strongly absorbed blue wavelengths, whereas the bottom cell responds to the longer-wavelength red light. Because of the marginal quality of the amorphous silicon after stabilization, this tandem approach can improve the performance of stabilized devices, even if the bottom cell is made from the same material as the top cell.

One reason the tandem-cell approach works so well for amorphous silicon is that it is very easy to produce the necessary shunting of the reverse-polarity p–n junction interposed between the top and rear cell. This shunting is required so that this junction does not generate a photovoltage opposing that

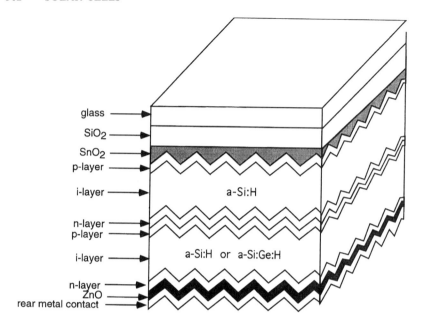

Fig. 17   Tandem a-Si: H solar cell.

generated by the other two junctions of Fig. 17. This reverse-polarity junction is between two heavily doped regions in which it is difficult to produce the normal rectifying *p–n* junction characteristics, due to the poor electronic quality of these regions.

Best performance from this tandem approach can be obtained if the semiconductor materials used in the top and rear cell are different, with the wider bandgap material used as the top cell. With this arrangement, the tandem cell can respond to a wider range of wavelengths than a single cell of intermediate bandgap.

Alloying amorphous silicon with carbon increases the material's bandgap, whereas alloying it with germanium reduces the gap. Also, microcrystalline material, consisting of a mixed phase intermediate between amorphous and polycrystalline, adds additional flexibility in the design of amorphous-silicon-based tandem cells.

A commercial process has been developed to produce a cell structure similar to that of Fig. 17 with standard a-Si:H used as the material for the top cell and a-Si:Ge:H alloy as the bottom cell. This process has produced a stabilized efficiency around 8% for large-area modules.[54]

By stacking three cells on top of one another, higher performance is possible, in principle. The first a-Si:H research module with a stabilized efficiency above 10% was based upon such a triple-junction approach, with the top cell consisting of a layer of a-Si:H, and with the bottom two cells of increased thicknesses and containing increasing percentages of germanium.[55]

Areas of present activity address production issues such as improving material deposition rates while maintaining acceptable cell performance and increasing the utilization of the silicon content of the source gases. Other work addresses incremental improvements to cell performance such as an improved tradeoff between the transmittance and sheet resistance of the transparent conducting oxide used in these cells, or improved rear-contact reflectance. Yet other areas of activity are directed at improved understanding of material properties or ways to overcome deficiencies of present materials, such as by new deposition approaches. For example, "hot-wire" deposition is being widely investigated, where the silane source material is dissociated by the high temperature generated by a resistively heated wire.

8.4 COMPOUND-SEMICONDUCTOR CELLS

8.4.1 III–V Crystalline Cells

Historical Outline.[56] The earliest semi-empirical analyses of cell-efficiency limits predicted an optimum semiconductor bandgap of around 1.4 eV,[15] somewhat higher than silicon's 1.1 eV. This stimulated the fabrication of the first GaAs cells in 1961, which produced devices with about 9% efficiency. The development of GaAs technology through the 1960s for semiconductor lasers and light-emitting diodes (Chapter 7) produced rapid improvement in device performance. The first solar cells that had an efficiency around 20% were produced in the early 1970s.[56] One feature was the use of a largely transparent $Al_xGa_{1-x}As$ window layer on the top surface of the cell to exploit the formidable properties of the III–V alloy system. Since that time, the best GaAs-based cells have maintained a slight efficiency advantage over the best silicon cells, with this lead fluctuating in the (relative) range 3–10%. The application of metalorganic chemical vapor deposition (MOCVD) techniques in the early 1980s instead of the earlier liquid-phase epitaxial (LPE) techniques reportedly produced a marked jump in efficiency.[56] The use of $Ga_{0.5}In_{0.5}P$ as the window layer produced the first 25%-efficient devices in the late 1980s.

About the same time, progress was made in implementing III–V tandem cells. By choosing a suitable III–V alloy, a lattice-matched crystalline tandem cell can be deposited on a substrate by first depositing a low-bandgap cell, followed by a tunneling junction and a higher-bandgap top cell, connected in series with the lower cell by the tunneling junction. Monolithic tandem cells of this type first exceeded the performance of the simpler single junction towards the end of the 1980s and produced a 30%-efficient device in 1996.[19]

Because of the higher cost of III–V technology compared to silicon, these high-performance crystalline III–V devices have not been used in low-cost terrestrial applications. Attempts to produce thin-film polycrystalline III–V cells have been notably disappointing, attributed to the high electronic activity of grain boundaries in these materials. Some interest has been shown in thin-film crystalline devices, produced by chemically removing the thin active layer from a potentially reusable substrate.[57] Good experimental devices have been demonstrated, but producing 100,000 m^2 per year of such cells with this approach, the output of some terrestrial-cell manufacturers, would pose serious challenges.

However, III–V cells, have found a market niche in providing electrical power for spacecraft. Here the higher cost of the cells is offset by a lower launch weight, resulting from their higher efficiency. The performance advantages in space applications are increased over terrestrial applications. This is because these cells have a higher tolerance to the damaging high-energy radiation found in space, particularly the cells with higher-efficiency designs. A feature of this space application has been the fabrication of the GaAs-based cells on germanium substrates, rather than on the GaAs substrates used for the best laboratory devices. This allows the better mechanical and thermal properties of germanium substrates to be used to advantage, and sacrifices little in electrical performance. Although these are single-junction devices, the lower-bandgap germanium also offers the allure of a performance increase by utilizing a GaAs/Ge tandem device.

Another III–V compound, InP, also appears promising for space applications because of its exceptionally high tolerance to the damaging space environment.[56] Great interest has also been shown in the use of III–V cells for terrestrial systems based on concentrated sunlight. The high cost of these cells is offset by the smaller area requirement for a given power output. However, only the tandem devices have opened up any appreciable performance advantage over inherently much less expensive silicon concentrator cells, making the use of single-junction III–V concentrator cells very unlikely.

A newer potential application for III–V cells is in thermophotovoltaic energy conversion.[34,58] In this application, conventional fuels are used to heat an emitter material to a high temperature, with the emitter chosen to emit energy by luminescence in a small wavelength range. This energy can be converted to electricity very efficiently by a solar-cell of appropriate bandgap. An overall conversion efficiency of over 40% has been projected even for the small generators that might be of interest in spacecraft.

GaAs Cells.   The structure of a typical GaAs space cell is shown in Fig. 18. Starting with an *n*-type germanium substrate, a heavily doped layer of GaAs is grown, which acts as a BSF in the completed device. This is followed by the growth of a more lightly doped GaAs region, which forms the main body of the cell. Since GaAs is a direct-bandgap semiconductor that absorbs

metal grid

p-GaAs (0.5 μm)

DLAR
window

emitter

base

buffer

inactive
substrate

contact

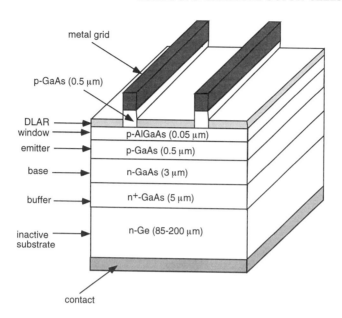

p-AlGaAs (0.05 μm)

p-GaAs (0.5 μm)

n-GaAs (3 μm)

n+-GaAs (5 μm)

n-Ge (85-200 μm)

Fig. 18   GaAs space cell.

sunlight strongly, this region need be only a few microns thick to capture most of the incident sunlight. The lightly doped GaAs is followed by the growth of a *p*-type GaAs emitter layer and then a thin, *p*-type AlGaAs window layer. Since contact to AlGaAs is difficult, a patterned layer of heavily doped *p*-type GaAs is grown on top of this layer to improve the contact resistance to, and reliability of, the overlying top metal contact. Alternatively, the *p*-type AlGaAs layer can be patterned to allow contact directly to the underlying *p*-type GaAs layer. Finally a double layer antireflection coating of TiO_2/Al_2O_3 is applied. Such cells typically display an 18% AM0 conversion efficiency, which corresponds to efficiencies of about 20% under terrestrial sunlight.

Major differences from silicon devices arise in several areas. The AlGaAs window, rather than thermal oxide, is used for passivation of the top surface. It is not as effective in producing low surface-recombination velocity and it is absorbing, reducing the blue response of cells. Because of the strong light absorption in GaAs, light-trapping schemes are not as beneficial as in silicon cells. In addition, the narrower bandwidth of GaAs cell response and the low refractive index of AlGaAs reduces reflection loss from the top surface of a planar cell, which reduces the benefits of surface texturing.

InP Cells.[56]   InP cells appear promising for space applications since they are very resistant to the damaging high-energy radiation encountered in space. However, one disadvantage is that the lattice constant of InP is not well matched to that of the semiconductors produced in high volume that might be used as substrates (Si, Ge, GaAs). Another is that although there are higher bandgap semiconductors, which might appear suitable for window layers on the grounds of lattice match, the window approach has performed poorly in practice.[59] The best performance has been demonstrated by simple homojunction devices grown by MOCVD on expensive and mechanically fragile InP substrates.

Highest performance is obtained when the top n^+ layer is thin (0.1 μm). There is no deliberate passivation of the surface apart from that provided by the deposited antireflection coating. Terrestrial energy-conversion efficiency above 20% has been demonstrated,[19] despite a relatively poor blue response of the cells resulting from the less than ideal surface passivation. Cells have also been fabricated with a deposited ZnO layer forming the top-surface emitter. Rather than providing a high-quality lattice-matched window layer, this is a fine-grained polycrystalline layer deposited at low temperature. Performance demonstrated by this approach has been lower than for homojunction cells.

Tandem Cells.   The wide choice of material with the same lattice constant within the III–V compound-semiconductor and alloy system has made this system ideal for producing lattice-matched crystalline tandem cells. Building upon developments in optoelectronics, early work in this area concentrated on the AlAs/GaAs alloy system. Subsequent work has been more diversified with good performance demonstrated for monolithic tandem cells based on GaInP/GaAs and InP/GaInAs.

A major challenge for crystalline devices has been the development of the tunneling junctions required to connect the cells. By having both sides of this junction heavily doped, the resulting tunnel currents effectively short-out the normally rectifying junction characteristics. This prevents an opposing photovoltage from being developed across this junction. Since this tunneling region acts as a plane of very high effective surface-recombination velocity, thin layers of the heavily doped tunneling regions are required. These can serve as a BSF region for the upper cell and a front-surface field (FSF) for the lower cell.

Alternatively, separate contacts could be made to this tunneling junction region to make a monolithic three-terminal device or two different cells could be mechanically stacked on top of each other to make a four-terminal device. The advantage in both cases is that the current output of the top and bottom cells does not have to be matched, increasing both design flexibility and performance. The disadvantage is that a separate circuit has to be set up for each cell type, adding to circuit complexity. Schemes that rely on local parallel connection of the two circuits appear the most promising way to deal with this complexity,[60] for example, connecting four of the lower-bandgap cells in parallel with three of the higher-bandgap cells.

Monolithic two-terminal tandem devices have performance levels about

15–20% above the best single-junction devices, but are estimated to cost only a similar amount more than a III–V single-junction device. They would, therefore, appear a better choice in specialized high-performance systems, such as for high-concentration terrestrial systems or in space.

Production quantities of tandem cells have been available since 1996 for use on spacecraft. Figure 19 shows the structure of one of the devices under development. Note the evolution of this technology from the well-established GaAs/Ge single-junction production technology.

8.4.2 Polycrystalline Thin-Film Compound-Semiconductor Cells[61]

Introduction. Although the III–V compound semiconductors can be strong absorbers of light and are optically ideal for fabricating very thin devices, the performance from polycrystalline thin-film III–V devices has been disappointing. This is attributed to the high level of electronic activity at the grain boundaries in III–V material, as previously noted.

There is, however, an enormous variety of other compound semiconductors and their alloys with a bandgap suitable for efficient energy conversion. Although bandgaps of such materials can be predicted with reasonable reliability, other properties that determine the suitability of the material for use in photovoltaics, such as grain-boundary activity and the ease of producing reasonable grain size, are not so readily forecast. Consequently, extended experimental evaluation is the only reliable method for determining the suitability of any given photovoltaic-material candidate.

From the limited evaluations of candidate materials that have been conducted to date, two compound semiconductors have shown good potential in thin-film form. One is the II–VI compound semiconductor CdTe, whereas the other is based on the I–III–VI$_2$ compound CuInSe$_2$ (CIS) and its related alloy system, $CuIn_xGa_{1-x}(Se_yS_{1-y})_2$.

The strengths of CdTe arise from its robustness in the range of preparation techniques that have given good cell results, including very crude techniques such as electroplating. A weakness is an uncertainty as to the environmental acceptability of deploying large areas of such material, because special precautions are likely to be required for its final disposal because of the toxicity of its constituents.

For CIS and its alloys, the major strengths are the following: the ease with which large grains can be grown during deposition and with which grain boundaries can be rendered inactive electronically; the material's tolerance to deviations from perfect stoichiometry (ratio of I, III, and VI constituents); and the design flexibility imparted by varying the alloy composition. A possible weakness relates to challenges in manufacturing, with a longer-term issue concerning the limited supplies of materials such as indium, gallium and selenium.

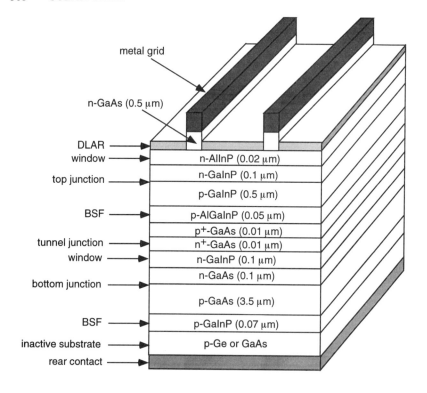

metal grid

n-GaAs (0.5 μm)

DLAR
window

top junction

BSF

tunnel junction
window

bottom junction

BSF

inactive substrate

rear contact

n-AlInP (0.02 μm)

n-GaInP (0.1 μm)

p-GaInP (0.5 μm)

p-AlGaInP (0.05 μm)

p+-GaAs (0.01 μm)

n+-GaAs (0.01 μm)

n-GaInP (0.1 μm)

n-GaAs (0.1 μm)

p-GaAs (3.5 μm)

p-GaInP (0.07 μm)

p-Ge or GaAs

Fig. 19 Monolithic tandem space cell.

CdTe Cells.[61,62,63] These cells were originally investigated because CdTe has a nearly ideal bandgap and, unlike some other compound semiconductors, this material can be doped both *n*-type and *p*-type. Subsequently, it was found that the material could be prepared by a range of simple techniques and still have good properties. This is attributed to the ability of post-deposition treatments to increase the grain size and to reduce the activity of grain boundaries in this material.

A variety of device structures have been investigated with this material, including the *p–n* homojunction and a variety of heterojunctions, including junctions with Cu_2Te, SnO_2, and various II–IV compounds.[62] However, best results have been obtained with the $CdTe/CdS/SnO_2$ heterojunction structure of Fig. 20.

The deposition of the transparent conducting oxide (TCO) layer of SnO_2 onto the glass substrate is followed by the deposition of a thin layer of CdS. This layer is usually heat treated in a reducing atmosphere or in $CdCl_2$ to increase grain size and reduce defect density.[62,63] Next, the CdTe layer is deposited by one of a variety of techniques, followed by a heat treatment in $CdCl_2$ or another chlorine-containing compound. The heat treatment not

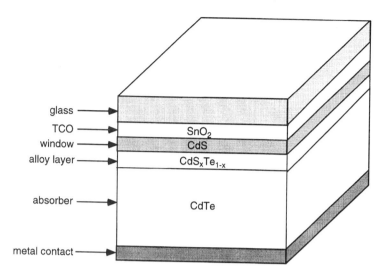

Fig. 20   CdTe cell structure.

only increases grain size and reduces defect density, but encourages the interdiffusion of the CdS and CdTe layers. The junction moves into the CdTe, rather than remaining at the original metallurgical interface, improving the junction quality. A variety of approaches have been used to make rear contact to the CdTe.[62,63] These generally are based on a two-layer approach, with the first layer being a heavily doped layer that makes good electrical contact vertically with the CdTe, whereas the second layer is metallic and provides good lateral conductivity. Most fabrication procedures include:[52]

1. An etch or surface-preparation step, which may produce a tellurium-rich surface layer.
2. Creation of the primary layer, either by deposition of a p^+ layer of SnTe-Cu, HgTe, or PbTe, or by modification of the CdTe surface by supplying a p-type dopant (e.g., Cu, Hg, Pb, or Au).
3. A subsequent heat treatment above 150°C.
4. Application of the secondary contact by sputtering, vacuum evaporation, or screen printing.

As for a-Si cells, the first CdTe cells produced in any volume were for pocket calculators. These cells were fabricated using the simple screen-printing sequence, whereby the CdS window layer is screen printed onto a borosilicate glass substrate as a paste consisting of CdS, $CdCl_2$, and a binder (propylene glycol).[62] The paste is fired at 120°C and then sintered in a nitrogen ambient at about 700°C in a belt furnace. The $CdCl_2$ flux encourages grain growth within the CdS films, resulting in grain sizes in the range

20–30 μm in films that are 20–30 μm thick. The CdTe paste consists of a mixture of cadmium and tellurium, together with the same flux and binder. The paste is fired, then sintered at around 600°C. The cadmium and tellurium react during this sintering step, with subsequent grain growth encouraged by the CdCl$_2$ flux. Considerable alloying occurs in the CdS/CdTe interfacial region, which improves the properties of this interface. Finally, the back contact is screen printed in the form of a carbon paste containing a small amount of copper, and the entire device annealed.

The performance of the screen-printed CdTe cells is quite modest, mainly because of the thick CdS window layer. Absorption in this layer and in the alloyed interfacial region results in a very poor response to any wavelengths shorter than 500 nm. The thickness of the layers used in these devices would exacerbate concerns over the toxicity and sustainability of this technology, if this approach were used in any volume.

The first CdTe cells to reach 15% energy conversion used more sophisticated processing, although they were still relatively simple compared to other photovoltaic technologies. A very thin CdS window layer was deposited by a solution-growth technique. This approach gives thin films of good stoichiometry, but the films are porous and the surface is contaminated by absorbed impurities such as hydroxides.[62] Prior to the deposition of the CdTe, the CdS film is heated in hydrogen to remove oxygen-containing species and to densify the film. The CdTe film is then deposited by close-spaced sublimation (CSS).

In this process,[62] a heated CdTe source dissociates into its cadmium and tellurium constituents in gaseous form. These recombine on the cooler substrate surface to reform CdTe. The deposition rate can be controlled in the range 0.1–10 μm/min by varying parameters such as the temperatures of the source and substrate, their separation, the deposition chamber ambient, and its pressure. The microstructure of the deposited films is determined by the substrate temperature and the source–substrate temperature gradient. Generally, the grain size increases with increasing source temperature and increasing film thickness, with an average grain size typically in the range 2–5 μm and, generally, of random orientation. The electrical resistivity of the p-type CdTe films can be controlled by using cadmium-deficient or antimony-doped source materials.

For the first 15% efficient CdTe cells, the CdS layer was about 70 nm thick and the CdTe layer was 4–5 μm thick. A 100-nm-thick MgF$_2$ antireflection coating layer was deposited on the illuminated glass surface.

The CSS approach has also been the focus of a commercial sequence for fabricating large area modules.[62] Both CdS and CdTe are sequentially deposited onto a SnO$_2$-coated glass substrate by a modified CSS technique. After a post-deposition heat treatment, electrical contact is made to the CdTe by depositing of a Ni/Al bilayer contact. Laser scribing at various stages during processing patterns the SnO$_2$ layer, the CdS/CdTe active layers, and the rear contact layer to provide automatic series interconnection of the cells within a module. This process is identical to that previously described for a-Si (Fig. 16). Efficiencies up to 8% were demonstrated in the early 1990s for large-area modules.

Environmental Issues.[63]   Environmental issues stemming from the toxicity of cadmium and its compounds have slowed the introduction of this CdTe-based technology. Issues arise during manufacture, during deployment in the field, and during disposal at "end of life."

Manufacturing hazards can undoubtedly be controlled. Hazards during deployment stem mainly from incidents such as fire, which could cause the release of toxic vapors. Because of the potential for leaching of cadmium into groundwater, special attention may have to be given to the final disposal of these modules. Some manufacturers believe the cadmium-based materials can be recycled, although the collection of a product dispersed into widely different cultural and geographical regions would pose significant challenges.

Copper Indium Diselenide and Its Alloys.[62,64]   Copper indium diselenide (CIS) is a direct-bandgap semiconductor with a bandgap of 1.04 eV at room temperature. A small cell with a reported efficiency of 12% was made by the evaporation of CdS onto a $CuInSe_2$ single crystal in 1975.[65] In the early 1980s, the first efficient thin-film cells were made with this material using coevaporation of the copper, indium and selenium elemental constituents. Thin-film cell efficiency close to 18% has been demonstrated by incorporating $CuGaSe_2$ into $CuInSe_2$ to increase the bandgap of the material.[19] $CuInS_2$ is another candidate for increasing the bandgap and has also given good results.

Figure 21 shows the generic structure of such a thin-film alloy cell. A molybdenum back contact is deposited on a glass substrate by sputtering or electron-beam evaporation. After deposition of the main $Cu(In,Ga)(S,Se)_2$ absorber layer by techniques to be described, a thin CdS or $Cd_{1-x}Zn_xS$ window layer is deposited by evaporation or, for best results, by solution growth. This is followed by the deposition of ZnO, by RF sputtering or by chemical vapor deposition. The best devices are made with a two-step process in which about 50 nm of lightly doped ZnO is deposited followed by 300 nm of aluminum-doped material, to reduce lateral resistance. Ni/Al contacts are applied to contact the ZnO.

Three techniques had been used to deposit the $Cu(In,Ga)(S,Ge)_2$ absorber layer for those cells which display over about 16% efficiency:[64]

1. Coevaporation of the elements onto a heated substrate
2. Selenization of sputtered or evaporated Cu/In precursors in an H_2Se or selenium atmosphere
3. Diffusion of copper and selenium into $(In,Ga)_2Se_3$ precursors.

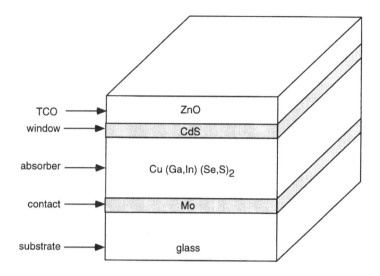

TCO ——→ ZnO
window ——→ CdS
absorber ——→ Cu (Ga,In) (Se,S)$_2$
contact ——→ Mo
substrate ——→ glass

Fig. 21   Solar cell based on CuInSe$_2$ and related alloys.

For the highest-performance devices, attempts are made to control composition and, hence, the alloy bandgap across the thickness of the absorber layer. Not all aspects are well understood. With excess copper concentration, large crystallites are grown, an effect believed to be promoted by the segregation of a copper chacogenide phase to the surface of the films. This phase can be removed by subsequent chemical treatment with KCN or converted to a more desirable compound. The surface composition of the grains also depends on the material composition. In indium-rich material, a stable CuIn$_3$Se$_5$ phase has been observed on film surfaces that have a larger bandgap than the bulk regions.[64] It has been postulated that layers of this type are located at the interface with the CdS window layer. Furthermore, sodium outdiffusion from soda–lime glass substrates has beneficial effects upon cell performance, by increasing carrier concentrations in the absorber layer. Such material complexities may be responsible for reported difficulties in attempts to commercialize this material in the late 1980s.[61] Module efficiencies above 10% have been reported in associated pilot-production activities.

The completed devices use the glass layer as a substrate rather than a superstrate as is the case for the a-Si and CdTe technologies described earlier (Fig. 16). The same laser-patterning steps can be used, however, for the CIS-based cells as for these other thin-film technologies, although conducted in the reverse order from that discussed earlier in connection with Fig. 16. The very high efficiencies already obtained with this material in thin-film polycrystalline make it a very strong candidate for a future, low-cost photovoltaic product.

8.5 MODULES

8.5.1 Terrestrial Modules

With the rapid growth of interest in the terrestrial use of photovoltaics in the mid-1970s, a variety of approaches were explored for packaging cells into weatherproof modules.[11,66] Since there is no inherent "wearout" mechanism involved in cell operation, the quality of the packaging, in terms of the mechanical and chemical protection it provides for the cells and their metallic contacts and interconnects, determines the effective operational life of the cells. Also important is the ability of the packaging to provide electrical insulation of the cells, for the protection of both the cell and the equipment. The thermal attributes of the module package are also important. Not only are the cells more efficient at low temperatures, but processes that degrade module integrity typically slow down by a factor of 2 for every 10°C reduction in module temperature. Another important consideration is keeping the costs of modules as low as possible.

By the early 1980s, module design had stabilized with the laminated-module design of Fig. 22 becoming the industry standard for bulk silicon cells.[66] The glass superstrate provides a low-cost transparent cover as well as structural support for the module. The cells, after soldering together, are attached to this cover by a laminating process. In this process, the layers shown in Fig. 22 are stacked with the glass layer lowermost, and inserted into a laminating machine. This machine heats the layers and applies hydrostatic pressure while the laminate is held under vacuum. The ethylene vinyl acetate (EVA) layers soften during this treatment and flow around the other solid components, filling the gaps between the layers. On cooling, the EVA binds the layers to the glass and provides a relatively soft encapsulating ambient to accommodate the differential thermal expansion of the stiffer layers within the laminate. The optional fiberglass-cloth layer provides resistance against tearing of the laminated layers and, reportedly, helps prevent air from being trapped in the module.

The glass superstrate layer generally consists of a 3-mm-thick sheet of low-iron tempered glass. Generally, iron is deliberately added to the melt in window glass production to improve manufacturability. However, iron causes absorption at infrared wavelengths, reducing the achievable output of silicon cells by about 5%. With the present cell manufacturing cost structure, economics favor the use of tempered glass with reduced iron content. Tempering is a heat treatment involving the rapid cooling of the glass surfaces, causing a compressive stress within a thin surface layer. This reduces the prospects for the initiation of cracks, greatly strengthening the glass. The 3 mm thickness of glass allows the module to pass standard wind-loading and hail-impact tests. In the latter case, the modules have to be able to withstand a 32-mm-diameter hailstone traveling at its terminal velocity.[67]

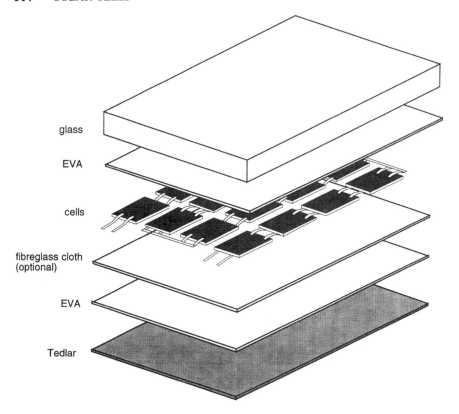

Fig. 22 Standard lamination approach to module fabrication.

The rear layer of the laminate is generally a film of one or more prelaminated layers formed from tedlar or mylar. This layer plays a role in keeping moisture out of the module, provides electrical insulation for the cells, and contributes to the optical and thermal environment seen by the cells.

If circular cells are used, the module will have a low packing factor with unused space between the cells. If the rear layer is white as viewed from the glass superstrate, light is reflected from these unused areas. If the reflection has a significant diffuse component, some of this reflected light will be trapped within the module by total internal reflection from the top glass surface. Some of this trapped light will reach active cell material. The output of a cell can be increased by up to 10% by this "zero-depth concentrator" effect.[68] Alternatively, the rear layer is chosen to have a dark color similar to that of the cells as seen from this direction. This improves module aesthetics because of the more uniform appearance but reduces the concentrator effect. The associated loss in output is tolerable if the spaces between the cells are small, such as when the cells are fabricated on square wafers.

The rear surface of this backing layer is generally white. This reduces absorption of light striking the rear of the module. Darker layers would, generally, have higher thermal emissivity, increasing heat radiation from the rear. However, this advantage is offset by the higher absorption of stray light. Ideally, this surface would have a spectrally selective emissivity—reflective at visible and near infrared wavelengths, absorptive at longer wavelengths.

If a transparent back cover is required, a transparent plastic layer (tradename Tefzel) has proved suitable. Such transparency may be required where cells are designed to respond to light incident on both top and rear cell surfaces. Light scattered onto the rear of the module by clouds or ground features can give a 20% boost in module output in such cases. Larger boosts are possible if special effort is made to further capitalize on this bifacial response, such as by strategically placing reflective material in the module background.

Prior to lamination, the positive and negative connections to the modules are brought out from the rear of the module through slits in the rear plastic layer or through holes drilled in the rear glass plate if a glass sheet is used at the rear of the module. After lamination, a plastic junction box is usually attached to the rear of the modules. Housed in this box are screw connections to the modules and other module circuit components, such as series and bypass diodes, discussed later. An aluminum frame is generally attached to the module to provide mounting points. Unframed modules, known as "laminates," are also used in some installations and are attached to supports by adhesives.

Cells in modules of this standard design generally reach temperatures of about 30°C above ambient in bright sunlight, when air can circulate freely over both the front and rear module surfaces, since the main heat loss is by convection from these surfaces.[66] The temperature rise above ambient will be lower in windy conditions or at lower illumination levels. It will be higher if air flow around the module is restricted, for example, when the module is mounted too close to a supporting structure, such as a wall or roof. The thermal performance of a module is generally specified by a parameter known as the nominal operating cell temperature (NOCT). This is the temperature that the module would reach under a standard set of conditions (intensity 800 W/m^2, ambient temperature 20°C, wind velocity 1 m/s). The cell operating temperature, T_c, under different ambient temperatures, T_a, and intensities, I_a (W/m^2), is given by

$$T_c = T_a + (\text{NOCT} - 20°C)\, I_a/800 \qquad (8)$$

The standard module encapsulation approach has proved very reliable in the field. One problem observed in the late 1980s after several years field experience was a browning of the EVA layer, particularly when modules were deployed in hot climates. The most severe case was that of a large installation in California where mirrors were used to augment the light intensity on the module, simultaneously increasing the module operating temperature. Improved formulations of EVA and the specification of glass containing cerium oxide to block some of the otherwise-transmitted high-energy ultraviolet photons are reported to have solved this problem.[69] The experience demonstrates the care that has to be taken if modules are used in nonstandard ways that are likely to increase operating temperatures.

Thin-film cells offer additional module design options, particularly for a-Si films which can be deposited on flexible plastic substrates. However, since many of these thin-film cells can be deposited on a glass substrate, glass is also a popular choice for encapsulating these cells. For the large 400-kW a-Si installation mentioned in Section 8.3.5, the cells were encapsulated between two sheets of standard window glass, with the structure again laminated using EVA as the binder. In this case, the use of low-iron glass was not justified because a-Si cells do not respond strongly past wavelengths of about 800 nm, reducing the effect of iron-induced absorption on these cells. The lower cell efficiency would also favor cheaper encapsulation options per unit area. The strengthening provided by the lamination of two glass sheets also was more than adequate for the modules to pass standard qualification testing.[67]

For spacecraft, moisture is not an important issue, although launch weight and tolerance to high-energy space-radiation are important. A minimalist approach to encapsulation is used, whereby a thin ($150 \mu m$) sheet of glass is attached to the top surface of each cell, leaving some areas of the cell busbar exposed to allow subsequent cell interconnection. The cells with this glass coverslide attached are then bonded to a honeycombed substrate, which provides lightweight structural support.

Similar techniques have been used to produce modules for other specialized applications such as solar-car racing. In such applications, there may be less severe requirements upon durability and effects such as dust retention may be of reduced importance, since modules can be cleaned several times each day. Some innovative approaches have been developed for reducing weight and reducing the surface reflection from the top layer of the encapsulant by appropriate texturing.[70]

8.5.2 Module Circuit Design

Mismatch between the characteristics of cells within a module can have surprisingly serious consequences in the field. Mismatch can arise from the normal production spread in cell parameters, by the blocking of sunlight falling on the cell surface (e.g., by leaves, bird droppings, or shadowing), by changes in cell and module characteristics caused by the discoloration of module encapsulation, or by cell cracking or corrosion.

A simple example of mismatch is the idealized case where a number of

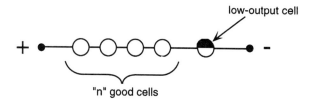

Fig. 23   Series-connected cells.

identical cells are connected in series, but one has reduced output, for example, because it has cracked or has been shaded (Fig. 23).

Figure 24 shows both the output current–voltage characteristic of the series connection of the good cells and that of the reduced-output cell. The output of the combination (dashed line) can be readily deduced since, at each value of current, the voltage of the good cells can be added to that of the cell with reduced output to determine the total voltage.

Note that the short-circuit current of the combination is determined by that of the poor cell. One low-output cell in a series combination can, therefore, cause a disproportionately large loss in the output of a series connection of cells. It is therefore essential to sort cells into different current-output categories (generally 5% tolerance bands) prior to encapsulation, to prevent excessive "mismatch loss" when the cells are series connected into a module. Similarly, when connecting modules into series, it is important to have the module current outputs closely matched.

More serious consequences from mismatch can occur when modules are operated close to short-circuit. Individual cells can be short-circuited with no damage. However, when a number of cells are connected in series and short-circuited, problems can arise.

Figure 25 shows the construction of Fig. 24 that was used to deduce the output of a combination of good cells and one lower-output cell. The short-circuit current of the combination is determined by the condition when the voltage across the low-output cell is equal and opposite to that of the combined voltage of the good cells. This operating point can be found graphically by reflecting the curve for the good cells about the current axis as shown. The point where this crosses the low-output cell characteristic gives the short-circuit current of the combination, as well as the voltage across the low output cell. The product of these quantities, shown by the shaded rectangle in Fig. 25, equals the power dissipation in the low-output cell.

To summarize, at short-circuit, a low-output cell becomes reverse biased as the high-output cells combine to attempt to force more current through it. It can be seen from Fig. 25 that it is possible for the full power output of the good cells to be dissipated in the poor cell. This will cause the poor cell to heat up above the rest of the module. If the heat were uniformly distributed across the cell, the temperature rise would be similar to the 30°C

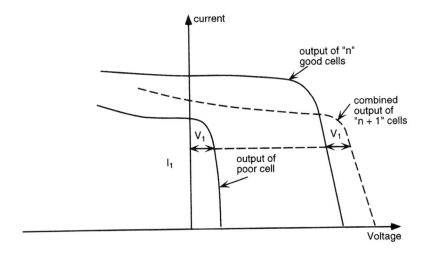

Fig. 24   Current–voltage curves for cells of Fig. 23.

increase for every 1 kW/m^2 dissipated, as for energy supplied from sunlight. However, the actual situation is much worse that this. The heat dissipation in the reverse-biased low-output cell can be confined to small localized breakdown areas, creating much higher local temperatures. These high local temperatures can discolor or damage the encapsulation layers or crack the glass in the module, causing module failure.

Experimentally, such destructive effects are possible if more than 10–15 cells are connected in series and the combination is operated near short-circuit. The solution is to connect a small discrete bypass diode, in opposite polarity to the cells, across each group of 10–15 cells. This restricts the reverse bias voltage that can be built up across the group to the forward voltage drop of the diode and, therefore, limits the voltage across any member of the group. Bypass diodes are sometimes inserted in junction boxes, as mentioned previously.

For more demanding applications such as spacecraft and solar cars, a bypass diode across each individual cell will not only protect the cell against this effect but will also eliminate the disproportionate loss in output that occurs when individual cells in a series connection are shaded.

A "series blocking diode" is often included in the module junction box. This diode is connected in series with the cells in the module but in reverse polarity to that of the diodes forming the cells. Its purpose is to prevent current flowing into the module when the module is not generating a current output. An example is when the module is used to charge a storage battery. The current generated by the module will pass in the forward direction through the diode with little power loss. At night, the module output voltage

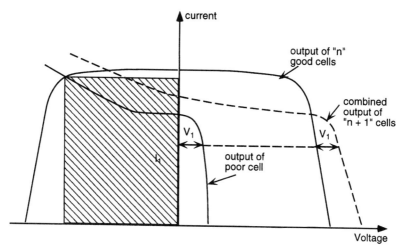

Fig. 25 Power dissipation in the low-output cell when the connection of Fig. 23 is short-circuited.

would drop and the battery would discharge through the solar-cells if this series or "blocking" diode were not present.

In larger systems, the blocking diode can also protect against catastrophes. For example, without such diodes, a ground short could make one branch of a parallel connection of modules an attractive path for the entire current-generating capacity of the system.

Small systems may consist of one or a small number of modules generally with series-connected cells. Large systems may have more complicated arrangements of series and parallel connections. Figure 26 shows the standard nomenclature used to describe the configuration of cells in modules and in solar arrays in such cases.

8.5.3 AC Modules

One strong feature of photovoltaics is its modularity. Cells can be used in systems generating from microwatts to megawatts. Extra capacity can be added to a system by adding modules.

For "stand-alone" systems, there is a DC section of the system that includes the modules and battery storage. There may be an AC section of the system if AC electrical loads are to be supplied. A DC-to-AC inverter is incorporated in such systems to provide the required interface.

In a grid-connected system, however, there may be no need for dedicated energy storage, and hence less rationale for collecting energy in DC form. Electricity could be converted to AC form at the module level and collected in this form. Small efficient inverters would be attached at the rear of each module, as shown in Fig. 27. This would take full advantage of the modularity

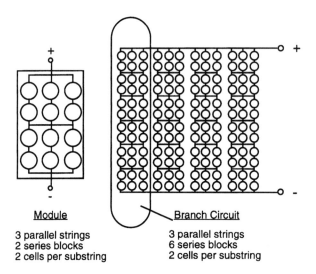

Module

3 parallel strings
2 series blocks
2 cells per substring

Branch Circuit

3 parallel strings
6 series blocks
2 cells per substring

Fig. 26   Module circuit nomenclature.

of photovoltaics and provide some safety advantages and savings in installation and wiring costs, as well as minimizing mismatch losses between modules due to effects such as partial shading. By choosing an inverter design such as a line-commutated design, the module would not deliver output unless it was connected to an AC source. The inverter could be supplied with a "male" outlet plug and simply plugged into a normal electrical outlet. The inverter would also have design features that would prevent "islanding," the condition whereby a number of these modules could sustain one another's output,[71] which is a potential safety hazard when sections of a system have apparently been switched off.

8.6   SUMMARY AND FUTURE TRENDS

The photovoltaics field has a history of steadily reducing costs and, consequently, expanding areas of application. Since the early 1970s, solar-cell applications have grown from use in small remote "stand-alone" systems, such as on spacecraft, to a wide range of potentially large applications including building integrated power generation and residential electricity generation. A number of utilities around the world have introduced schemes to encourage the use of this environmentally benign form of electricity generation, although the total installed power will take many years to grow to any significant fraction of total power generated by conventional means. Countries such as Indonesia are exploring the use of photovoltaic power in its remote villages, where it is difficult to distribute centrally generated

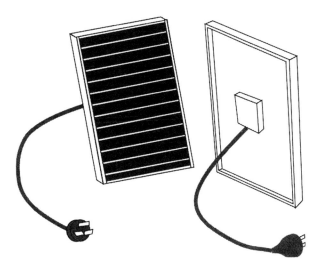

Fig. 27   AC photovoltaic module concept.

electricity. More imaginatively, a proposal has been made that international aid agencies focus on equipping families in less-developed regions of the world with a 40-W photovoltaic module to help redress global inequities in access to energy resources.[72]

Technology based on silicon wafers, as used in the broader microelectronics field, has been used in the bulk of solar-cell product supplied in the past. A number of newer thin-film technologies have been developed to the stage of commercial readiness, including those based on hydrogenated amorphous silicon, cadmium telluride, copper indium diselenide and polycrystalline silicon. These thin-film technologies involve lower material costs and a change of the unit of production from individual cells, as in the silicon wafer case, to a much larger module of interconnected cells. These newer technologies lead to the expectation that such thin-film technologies can be produced more cheaply in large production volumes. The energy-conversion efficiency that can be obtained in a production setting is a key issue with each of these thin-film alternatives. It is often argued that an efficiency above 15% is essential for the widespread development of photovoltaics, just to be able to keep costs proportional to the total area of the installation under control, including the cost of mounting and wiring the modules. For technologies involving toxic heavy metals, such as cadmium, careful attention has to be given to the environmental consequences of encouraging their widespread deployment.

To reach 15% thin-film efficiency, high-efficiency concepts such as stacked "tandem" cells will be essential for some of these above technologies. The tandem-cell approach is already well developed for amorphous-silicon technology, although production efficiencies are only about half of the 15%-efficiency target.

In more specialized applications, for example, spacecraft, systems relying on concentrated sunlight, and thermophotovoltaic conversion, more expensive cell technology can be used. GaAs cells grown by MOCVD on germanium substrates became a well-established technology for space cells in the first half of the 1990s, with higher efficiency GaInP$_2$/GaAs tandem cells on germanium substrates produced in pilot production quantities during the second half. The flexibility offered by the III–V alloy system also makes these cells well suited for thermophotovoltaic conversion. Silicon remains a strong challenger for converting concentrated or focused sunlight, with 26–27%-efficient concentrator cells already in production.

Low-cost cells based on inexpensive plastics are the subject of exploratory research, and have appeal, although their performance to date has been modest. Other technology, such as that based on dye-sensitized nanocrystalline cells,[73] offers the attraction of very simple production sequences, but has the disadvantage of low efficiency and of concerns about durability.

As well as an expanding role in meeting the rapidly increasing demand for electricity in the developing parts of the world, photovoltaics has the potential to contribute to the growing demand for environmentally sustainable energy production in wealthier countries. One way this might come about is in the form of small, modular, residential units equipped with a sophisticated inverter for each module that could be directly plugged into the mains as a "demand-side" management contributor.

Whatever the direction in which photovoltaic technology develops, the next 10 years is expected to be an exciting period for this environmentally benign form of electricity generation with an enormous future.

PROBLEMS

1. a) What is the length of a horizontal shadow cast by a vertical pole of height 10 m under ambient conditions corresponding to the standard Air Mass 1.5 solar-cell test spectrum?

   b) What is the angle of the sun above the horizon under Air Mass 2 conditions?

   c) The following is an approximate expression giving the intensity of the direct component of sunlight on a plane perpendicular to the sun's rays as a function of the prevailing Air Mass (AM):

   $$I_D = 1.367 \times 0.7^{AM^{0.678}} \text{ kW/m}^2$$

   Compare the intensities on the plane this predicts under AM1 and AM2 conditions, assuming an additional 10% contribution from diffuse sunlight in both cases. (Expression based on that given by C. Hu and R. M. White, Ref. 74.)

d) Given that the AM2 spectrum has a higher relative component of red wavelengths than blue compared to the AM1 spectrum, under which spectrum would you expect a silicon cell at fixed temperature to have:

i) A higher power output?
ii) A higher energy-conversion efficiency?

2. The performance ratio of a photovoltaic system is defined as the ratio of the AC electrical energy produced by the system divided by the product of the total solar energy on the array plane and the efficiency of the modules in the array under standard test conditions (i.e., efficiency under 1 kW/m^2 of sunshine at a cell temperature of 25°C).

This factor takes into account a number of nonidealities, including DC-to-AC conversion efficiency, cell operation at temperatures above 25°C, dust accumulation, partial shading of the cells, and system downtime. Given that utilization factors in the 60–80% range are common, that there are typically between 1000–2000 hours of equivalent peak sunshine (1 kW/m^2) at different geographical locations worldwide for arrays of appropriate orientation, that residential electricity demands typically lie in the range 15–40 kW-h/day, and that commercial solar modules typically have an energy conversion efficiency in the 10–14% range under standard test conditions, calculate:

a) The peak rating of a 20-m^2 roof-mounted photovoltaic array for both the low and the high end of the module efficiency range given above.
b) The maximum and minimum percentage of the total AC household load which could be generated by a well-orientated, roof-mounted photovoltaic array of this area.

3. a) At 300 K, an ideal solar-cell has a short-circuit current of 3 A and an open-circuit voltage of 0.6 V. Calculate and sketch its power output as a function of operating voltage and find its fill-factor from this power output. Compare this fill-factor value with that given by Eq. 3.
b) A second ideal cell has a short-circuit current of 4 A and an open-circuit voltage of 0.55 V at 300 K . Using Eq. 3, calculate its fill factor and, using this value, its maximum output power.
c) The cell of part (b) is connected in parallel with the cell of part (a). Calculate the short-circuit current, open-circuit voltage and maximum output power of the parallel combination. What is the relative mismatch loss (sum of maximum power possible from individual cells in isolation minus that of the combination divided by the sum of that possible from individual cells)?
d) The two cells of part (b) are then connected in series. Calculate the

open-circuit voltage and short-circuit current of the series combination.

e) Calculate and sketch the power output of the series combination as a function of output voltage, and find the fill factor of the series-connected combination from these calculations. What is the relative mismatch loss in this case?

4. A hypothetical module consists of 36 identical cells, each with ideal current–voltage characteristics and each giving an open-circuit voltage of 0.6 V and a short-circuit current of 3 A at a cell operating temperature of 300 K. The module is short-circuited and one cell shaded so that half its area is obscured. Calculate the short-circuit current of the combination and the power dissipated in the shaded cell.

5. A standard commercial module consists of 36 series-connected cells. Given a digital multimeter, clear outdoor weather, a sheet of cardboard, and scissors, suggest how you could estimate:
   a) The temperature rise of the cells above ambient when illuminated.
   b) The cell limiting the module current output when the module is short-circuited.

6. a) Sketch the output characteristics of an ideal solar-cell in the form shown in Fig. 3. Noting that a parasitic resistance, R_s, in series with the cell will change the output voltage by an amount IR_s at any given current I, sketch how the form of the output characteristics are modified when the cell has a parasitic series resistance associated with it.
   b) Show that, if FF_o is the fill factor of the cell in the absence of series resistance, the fill factor, FF, in the presence of a moderate R_s is given by:

   $$FF_o < FF < FF_o(1 - R_s/R_{mp})$$

   where R_{mp} is defined as V_{mp}/I_{mp}, and has a value close to the characteristic resistance of the cell (V_{oc}/I_{sc}).
   c) Noting that a shunt resistance, R_{sh}, across an ideal cell will change the current output at any voltage V by an amount V/R_{sh}, sketch how the output characteristics of a cell with zero series resistance are modified when the cell has a parasitic shunt resistance associated with it.
   d) Show that the fill factor of the cell in the presence of R_{sh} is given by

   $$FF_o < FF < FF_o(1 - R_{mp}/R_{sh})$$

7. a) The spectral responsivity of a solar-cell is defined as the current output per unit incident power from a monochromatic illumination

source. Show that the upper limit upon the spectral responsivity of a cell is given by $q\lambda/hc$, where q is the electronic charge, h is Planck's constant and c is the velocity of light in a vacuum.

b) Calculate the value of the ideal spectral responsivity under moderate illumination levels at a wavelength of $0.8\,\mu$m for:

    i)   A homojunction cell of $1\,$eV bandgap.

    ii) A homojunction cell of $2\,$eV bandgap.

    iii) A heterojunction cell with the junction formed between material of $2\,$eV and $1\,$eV bandgaps.

    iv) A two-terminal, monolithic, series-connected tandem cell when the upper cell has a bandgap of $2\,$eV and the second lower cell has a bandgap of $1\,$eV.

8. a) For the silicon space cell of Fig. 7, consider a vertical line which passes through a finger of the top contact and the rear contact. Sketch the energy-band diagram as a function of position along this line at thermal equilibrium (no illumination or bias applied to cell), relative to the Fermi level of the system, assuming that the metal contacts form Schottky barriers to the underlying heavily doped silicon.

   b) Sketch the energy-band diagram of the cell of part (a) at open-circuit voltage under illumination, showing the carrier quasi–Fermi levels and clearly indicate changes from the sketch of part (a). Assume large carrier mobilities and carrier lifetimes, low contact resistance at both metal contacts (essentially zero in both cases) and low injection throughout the device.

   c) Repeat part (b) using the same assumptions except assume a more lightly doped p-type base region that is under high-level injection when open-circuited. Clearly show any differences from the sketch for part (b). (Assume the same open-circuit voltage in both cases.)

9. Repeat parts (a) and (b) of Problem 8 for the advanced space cell of Fig. 19. Assume bandgaps for the GaAs, GaInP, AlInP, and AlGaInP regions of $1.4\,$eV, $1.6\,$eV, $1.8\,$eV and $1.7\,$eV, respectively. Conduction-band offsets depend upon preparation conditions and the stress in the layers, but assume values of $0.3\,$eV, $0.2\,$eV, $0.5\,$eV, $0.1\,$eV, $0.4\,$eV, and $0.5\,$eV between GaAs and GaInP, GaInP and AlInP, GaAs and AlInP, GaInP and GaAlInP, GaAs and GaAlInP, and Ge and GaInP, respectively, the second bandgap edge being at higher electron energy than the first in all cases. For part (b), additionally, assume a perfect tunneling junction (infinite shunt conductance).

10. A polycrystalline solar-cell has a grain size comparable to the thickness of its base region. Consider the grain to be a cube, with the top surface corresponding to the edge of the junction collection region, the bottom surface corresponding to the rear cell contact, and the sides of the cube corresponding to the grain boundaries. Assume the semiconductor

material within the grain boundaries is of high quality (very high carrier mobilities and lifetimes) and is doped to a uniform, intermediate doping concentration. The cell is short-circuited, which reduces the excess minority-carrier concentration at the edge of the junction depletion region to zero. Consider an electron–hole pair generated right at the center of the cube. Estimate the probability of collection when:

a) A BSF is incorporated at the rear contact, giving a very low effective surface recombination velocity along this surface, and the grain boundaries are of low activity or are well passivated, also giving very low effective recombination velocities at the grain boundaries.

b) An ohmic rear contact is used rather than a BSF, giving a very high effective recombination velocity along the rear of the cell, but the grain boundaries remain of low activity or well passivated.

c) An ohmic rear contact is used but the grain boundaries are of high activity, corresponding to a very high effective surface-recombination velocity. If grain boundary recombination cannot be effectively passivated, what criterion does this suggest for the required grain size for good cell performance?

REFERENCES

1. M. A. Green, "Photovoltaics: coming of age," in *Conference Record, 21st IEEE Photovoltaic Specialists Conference*, Orlando, May 1990, *IEEE Publ. No. 90CH 2838-1*, p. 1.

2. D. M. Chapin, C. S. Fuller, and G. L. Pearson, "A new silicon *p–n* junction photocell for converting solar radiation into electrical power," *J. Appl. Phys.* **8**, 676 (1954).

3. M. Wolf, "Historical development of solar cells," in *Solar Cells*, C. E. Backus, Ed., IEEE Press, New York 1976.

4. D.L. Travers and D. S. Shugar, "Value of grid-support photovoltaics to electric distribution lines," *Progr. Photovolt.* **9**, 293 (1994).

5. J. Oppenheim, "A program to demonstrate that consumers place value on environmentally benign electricity: residential rooftop PV," in *Conf. Proceedings, 13th European Photovoltaic Solar Energy Conference*, Nice, France, October 1995, p. 836.

6. See Special Sub-Issue, "Large photovoltaic systems," *Progr. Photovolt.* (1997).

7. *Conference Records, IEEE Photovoltaic Specialists Conference*. Available from IEEE Service Center, 445 Hoes Lane, PO Box 1331, Piscataway, N.J., 08855-1331, USA.

8. *Progress in Photovoltaics*, M. A. Green, E. Lorenzo, H. N. Post, H. W. Schock, K. Zweibel, P. A. Lynn, Eds., Wiley, New York.

9. *Solar Energy Materials and Solar Cells*, C. M. Lampert, Ed., Elsevier, Amsterdam/New York.

10. S. M. Sze, *Physics of Semiconductor Devices*, 2nd ed., Wiley, New York, 1981 (Chapter 14, in particular).

11. M. A. Green, *Solar Cells: Operating Principles, Technology and System Applications*, Prentice–Hall, Englewood Cliffs, N.J., 1982. (Available from Photovoltaics Special Research Centre, UNSW, Sydney, Australia, 2052.)

12. L. D. Partain, Ed., *Solar Cells and Their Applications*, Wiley, New York, 1995.

13. T. B. Johansson, H. Kelly, A. K. N. Reddy, R. H. Williams, and L. Burnham, Eds., *Renewable Energy: Sources for Fuels and Electricity*, Island Press, Washington, 1993.

14. T. Markvart, *Solar Electricity*, Wiley, New York, 1994.

15. J. J. Loferski, "Theoretical considerations governing the choice of the optimum semiconductor for photovoltaic solar energy conversion," *J. Appl. Phys.* **27**, 777 (1956).

16. W. Shockley and H. J. Queisser, "Detailed balance limit of efficiency of *p–n* junction solar cells," *J. Appl. Phys.* **32**, 510 (1961).

17. R. Hulstrom, R. Bird, and C. Riordan, "Spectral solar irradiance data sets for selected terrestrial conditions," *Solar Cells* **15**, 365 (1985).

18. O. Kawasaki, S. Matsuda, Y. Yamamoto, Y. Kiyota, and Y. Uchida, "Study of solar simulator calibration method and round robin calibration plan of primary standard solar cell for space use,"in *Conf. Record, IEEE First World Conference on Photovoltaic Energy Conversion*, Hawaii, December 1994, p. 2100.

19. M. A. Green, K. Emery, K. Bücher, D. L. King, and S. Igari, "Solar cell efficiency tables (Version 9)," *Progr. Photovolt.* **5**, 51 (1996).

20. J. E. Parrott, "Radiative recombination and photon recycling in photovoltaic solar cells," *Solar Energy Mater. Solar Cells* **30**, 221 (1993).

21. P. Würfel, *Physik der Solarzellen*, Spektrum Akademischer Verlag, Heidelberg, 1995 (in German).

22. T. Tiedje, E. Yablonovitch, G. D. Cody, and B. G. Brooks, "Limiting efficiency of silicon solar cells," *IEEE Trans. Electron Dev.* **ED-31**, 711 (1984).

23. M. A. Green, *Silicon Solar Cells: Advanced Principles and Practice*, Bridge Printery, Sydney, 1995. (Available from Photovoltaics Special Research Centre, UNSW, Sydney, Australia, 2052.)

24. A. L. Fahrenbuch and R. H. Bube, *Fundamentals of Solar Cells*, Academic Press, New York, 1983.

25. A. Luque, Ed., *Physical Limitations to Photovoltaic Energy Conversion*, IOP Publishing, 1990.

26. R. S. Ohl, "Light sensitive electric device," *US Patent 2402662*, filed 27 March 1941; "Light-sensitive electric device including silicon," *US Patent 2443542*, filed 27 May 1941.

27. D. M. Chapin, C. S. Fuller, and G. L. Pearson, "A new silicon *p–n* junction photocell for converting solar radiation into electrical power," *J. Appl. Phys.* **8**, 676 (1954).

28. G. E. McGuire Ed., *Semiconductor Materials and Process Technology Handbook*, Noyes Press, Park Ridge, N.J., 1988.

29. T. Markvart, "Radiation damage in solar cells," *J. Mater. Sci.: Mater. Electron.* **1**, 1 (1990).

30. E. Yablonovitch and G. D. Cody, "Intensity enhancement in textured optical sheets for solar cells," *IEEE Trans. Electron Dev.* **ED-29**, 300 (1982).

31. P. Campbell and M. A. Green, "The limiting efficiency of silicon solar cells under concentrated sunlight," *IEEE Trans. Electron Dev.* **ED-33**, 234 (1986).

32. E. Yablonovitch, D. L. Alara, C. C. Chang, T. Gmitter, and T. B. Bright, "Unusually low surface-recombination velocity on silicon and germanium surfaces," *Phys. Rev. Lett.* **57**, 249 (1986).

33. M. A. Green, J. Zhao, A. Wang, and S. R. Wenham, "45% efficient silicon photovoltaic cell under monochromatic light," *IEEE Electron Dev. Lett.* **13**, 317 (1992).

34. D. L. Chubb, "Reappraisal of solid selective emitters," in *Conf. Record, 21st IEEE Photovoltaic Specialists Conference*, Kissimimee, Florida, May 1990, p. 1326.

35. R. M. Swanson, "Point-contact solar cells: modelling and experiment," *Solar Cells* **17**, 85 (1986).

36. P. E. Gruenbaum, J. Y. Gan, R. R. King, and R. M. Swanson, "Stable passivations for high-efficiency silicon solar cells," in *Conf. Record, 21st IEEE Photovoltaic Specialists Conference*, Kissimimee, Florida, May 1990, p. 317.

37. P. J. Verlinden, R. M. Swanson, and R. A. Crane, "7000 high-efficiency cells for a dream," *Progr. Photovolt.* **2**, 143 (1994).

38. S. R. Wenham, C. B. Honsberg, and M. A. Green, "Buried contact silicon solar cells," *Solar Energy Mater. Solar Cells* **34**, 101 (1994).

39. A. Eyer, A. Rauber, and A. Goetzberger, "Silicon sheet materials for solar cells," *Optoelectronics* **5**, 239 (1990).

40. J. E. Johnson, J. I. Hanoka, and J. A. Gregory, "Continuous mode hydrogen passivation," in *Conf. Record, 18th IEEE Photovoltaic Specialists Conf.*, Las Vegas 1985, p. 1112.

41. M. Pasquinelli, S. Martinuzzi, J. Y. Natoli, and F. Floret, "Improvement of phosphorus gettered multicrystalline silicon wafers by aluminium treatment," in *Conf. Record, 22nd IEEE Photovoltaic Specialists Conference*, Las Vegas, October 1991, p. 1035.

42. A. Rohatgi, S. Narasimha, S. Kamra, P. Doshi, C. P. Khattak, K. Emery, and H. Field, "Record high 18.6% efficient solar cell on HEM multicrystalline material," paper presented at 25th IEEE Photovoltaic Specialists Conference, Washington, May 1996, p. 741.

43. J. Zhao, A. Wang, P. Altermatt, and M. A. Green, "24% efficient silicon solar cells with double layer antireflection coatings and reduced resistance loss," *Appl. Phys. Lett.* **66**, 3636 (1995).

44. A. K. Ghosh, C. Fishman, and T. Feng, "Theory of the electrical and photovoltaic properties of polycrystalline silicon," *J. Appl. Phys.* **51**, 446 (1980).

45. A. Barnett, M. G. Mauk, J. C. Zolper, I. W. Hall, W. A. Tiller, R. B. Hall, and J. B. McNeely, "Thin-film silicon and GaAs solar cells," in *Conf. Record, 17th IEEE Photovoltaic Specialists Conference*, Kissimmee, Florida, May 1984, p. 747.

46. J. S. Im and R. S. Sposile, "Crystalline Si films for integrated active-matrix liquid-crystal displays," *MRS Bull.* **21**(3), 39 (1996).

47. J. E. Cotter, A. M. Barnett, D. H. Ford, M. A. Goetz, R. B. Hall, A. E. Ingram, J. A. Rand, and C. J. Thomas, "Advanced Silicon-Film*M solar cell design and development," in *Conf. Record, 1st World Conference on Photovoltaic Energy Conversion*, Hawaii, December 1994, p. 1732.

48. T. Baba, M. Shima, T. Matsuyama, S. Tsuge, K. Wakisaka, and S. Tsuda, "9.2% efficiency thin-film polycrystalline silicon solar cell by a novel solid phase crystallization method," in *Conf. Proceedings, 13th European Photovoltaic Solar Energy Conference*, Nice, October 1995, p. 1708.

49. M. A. Green and S. R. Wenham, "Novel parallel multijunction solar cell," *Appl. Phys. Lett.* **65**, 2907 (1994).

50. D. E. Carlson and S. Wagner, "Amorphous silicon photovoltaic systems," in *Renewable Energy: Sources for Fuels and Electricity*, T. B. Johansson, H. Kelly, A. K. N. Reddy, R. H. Williams, and L. Burnham, Eds., Island Press, Washington, 1993.

51. P. G. LeComber, "Non-photovoltaic applications of amorphous silicon," in *Conf. Proceedings, 8th E.C. Photovoltaic Solar Energy Conference*, Florence, May 1988, p. 1229.

52. D. S. Shugar, "Applications of amorphous photovoltaics: myths and facts," in *Conf. Record, 1st World Conference on Photovoltaic Energy Conversion*, Hawaii, December 1994, p. 670.

53. J. Macneil, E. Eser, F. Kampas, J. Xi, A. Delahoy, F. Essis Jr., and C. H. Liu, "Recent improvements in very large area a-Si PV module manufacturing," in *Conf. Proceedings, 10th European Photovoltaic Solar Energy Conference*, Lisbon, April 1991, p. 1188.

54. R. R. Arya, R. S. Oswald, Y. M. Li, N. Maley, K. Jansen, L. Yang, L. F. Chen, F. Willing, M. S. Bennett, J. Morris, and D. E. Carlson, "Progress in amorphous silicon based multijunction modules," in *Conf. Record, 1st World Conference on Photovoltaic Energy Conversion,* Hawaii, December 1994, p. 394.

55. J. Yang, A. Banerjee, T. Glatfelter, K. Hoffman, X. Xu, and S. Guha, "Progress in triple-junction amorphous silicon-based alloy solar cells and modules using hydrogen dilution," in *Conf. Record, 1st World Conference on Photovoltaic Energy Conversion*, Hawaii, December 1994, p. 380.

56. P. Iles and Y. C. M. Yeh, "Silicon, gallium arsenide and indium phosphide cells," in *Solar Cells and Their Applications*, L. D. Partain, Ed., Wiley, New York, 1995.

57. C. Hardingham, A. Hayward, T. A. Cross, and C. Goodbody, "Direct glassed and ultrathin GaAs cells," in *Conf. Record, 1st World Conference of Photovoltaic Energy Conversion*, Hawaii, December 1994, p. 2217.

58. P. A. Iles and C. L. Chu, "Design and fabrication of thermophotovoltaic cells," in *Conf. Record, 1st World Conference on Photovoltaic Energy Conversion*, Hawaii, December 1994, p. 1740.

59. J. Lammasniemi, K. Tappura, and K. Smekalin, "Recombination mechanisms at window/emitter interface in InP and other III–V semiconductor based solar cells," in *Conf. Record, 1st World Conference on Photovoltaic Energy Conversion*,

Hawaii, December 1994, p. 1771.

60. L. M. Fraas, "Concentrator modules using multijunction cells," in *Solar Cells and Their Applications*, L. D. Partain, Ed., Wiley, New York 1995.

61. K. Zweibel, "Thin films: past, present, future," *Progr. Photovolt.* **3**, 279 (1995).

62. T. L. Chu and S. S. Chu, "Thin film II-VI photovoltaics," *Solid-State Electronics* **38**, 533 (1995).

63. P. V. Meyers and R. W. Birkmire, "The future of CdTe photovoltaics," *Progr. Photovolt.* **3**, 393 (1995).

64. W. H. Bloss, F. Pfisterer, M. Schubert, and T. Walter, "Thin-film solar cells," *Progr. Photovolt.* **3**, 3 (1995).

65. J. L. Shay, S. Wagner, and H. M. Kasper, "Efficient CuInSe$_2$/CdS solar cells," *Appl. Phys. Lett* **27**, 89 (1975).

66. E. Christensen, Ed., "Flat-plate solar array project," Jet Propulsion Laboratory, *Report JPL400-279*, October 1985.

67. J. Bishop and H. Ossenbrink, "Results of four years of module qualification testing to CEC specification 503," in *Conf. Proceedings, 13th European Photovoltaic Solar Energy Conference*, Nice, October 1995, p. 2104.

68. N. F. Shepard and L. E. Sanchez, "Development of shingle-type solar cell module," in *Conf. Record, 13th IEEE Photovoltaic Specialists Conference*, Washington, DC 1978, p. 160.

69. J. P. Calica, W. H. Holley, S. C. Agro, R. S. Yorgensen, M. Ezrin, P. Klemchuk, and G. Lavigne, "Advanced development of non-discoloring EVA-based PV encapsulants," in *Conf. Proceedings, 13th European Photovoltaic Solar Energy Conference*, Nice, October 1995, p. 2370.

70. D. M. Roche, A. E. T. Schinckel, J. W. V. Storey, C. P. Humphris, and M. R. Guelden, *"Speed of Light: The 1996 World Solar Challenge,"* UNSW Photovoltaics Special Research Centre, Sydney 2052, Australia, 1997.

71. S. Marte and P. Kremer, "Methods against islanding for small grid connected inverters," in *Conf. Proceedings, 13th European Photovoltaic Solar Energy Conference*, Nice, October 1995, p. 1839.

72. W. Palz, "Power for the World," *Int. J. Solar Energy* **14**, 231 (1994).

73. A. J. McEvoy, M. Grätzel, H. Wittkopf, D. Jestel, and J. Benemann, "Nanocrystalline Electrochemical Solar Cells," in *Conf. Record, 1st World Conference on Photovoltaic Energy Conversion*, Hawaii, 1994, p. 1779.

74. C. Hu and R. M. White, *Solar Cells*, McGraw–Hill, New York, 1983.

Appendix A: List of Symbols

Symbol	Description	Unit
a	Lattice constant	Å
$\mathscr{B}$	Magnetic induction	Wb/m^2
c	Speed of light in vacuum	cm/s
C	Capacitance	F
$\mathscr{D}$	Electric displacement	C/cm^2
D	Diffusion coefficient	cm^2/s
E	Energy	eV
E_c	Bottom of conduction band	eV
E_F	Fermi energy level	eV
E_g	Energy bandgap	eV
E_v	Top of valence band	eV
$\mathscr{E}$	Electric field	V/cm
$\mathscr{E}_c$	Critical field	V/cm
$\mathscr{E}_m$	Maximum field	V/cm
f	Frequency	Hz(cps)
$F(E)$	Fermi–Dirac distribution function	
h	Planck constant	J-s
$h\nu$	Photon energy	eV
I	Current	A
I_c	Collector current	A
J	Current density	A/cm^2
J_t	Threshold current density	A/cm^2
k	Boltzmann constant	J/K
kT	Thermal energy	eV
L	Length	cm or μm
m_0	Electron rest mass	kg
m^*	Effective mass	kg
n	*Density of free electrons*	cm^{-3}
n_i	*Intrinsic density*	cm^{-3}
N	*Doping concentration*	cm^{-3}

Symbol	Description	Unit
N_A	Acceptor impurity density	cm^{-3}
N_C	Effective density of states in conduction band	cm^{-3}
N_D	Donor impurity density	cm^{-3}
N_V	Effective density of states in valence band	cm^{-3}
p	Density of free holes	cm^{-3}
P	Pressure	Pa
q	Magnitude of electronic charge	C
Q_{it}	Interface trapped charge	charges/cm^2
R	Resistance	Ω
t	Time	s
T	Absolute Temperature	K
v	Carrier velocity	cm/s
v_s	Saturation velocity	cm/s
v_{th}	Thermal velocity	cm/s
V	Voltage	V
V_{bi}	Built-in potential	V
V_{EB}	Emitter–base voltage	V
V_B	Breakdown voltage	V
W	Thickness	cm or μm
W_B	Base thickness	cm or μm
x	x direction	
∇	Differential operator	
∇T	Temperature gradient	K/cm
ε_0	Permittivity in vacuum	F/cm
ε_s	Semiconductor permittivity	F/cm
ε_i	Insulator permittivity	F/cm
$\varepsilon_s/\varepsilon_0$ or $\varepsilon_i/\varepsilon_0$	Dielectric constant	
τ	Lifetime or decay time	s
θ	Angle	rad
λ	Wavelength	μm or A
ν	Frequency of light	Hz
μ_o	Permeability in vacuum	H/cm
μ_n	Electron mobility	cm^2/V-s
μ_p	Hole mobility	cm^2/V-s
ρ	Resistivity	Ω-cm
ω	Angular frequency ($2\pi f$ or $2\pi\nu$)	Hz
Ω	Ohm	Ω

Appendix B: International System of Units (SI Units)

Quantity	Unit	Symbol	Dimensions
Length[a]	meter	m	
Mass	kilogram	kg	
Time	second	s	
Temperature	kelvin	K	
Current	ampere	A	
Light intensity	candela	Cd	
Angle	radian	rad	
Frequency	hertz	Hz	$1/s$
Force	newton	N	$kg\text{-}m/s^2$
Pressure	pascal	Pa	N/m^2
Energy[a]	joule	J	N-m
Power	watt	W	J/s
Electric charge	coulomb	C	A-s
Potential	volt	V	J/C
Conductance	siemens	S	A/V
Resistance	ohm	Ω	V/A
Capacitance	farad	F	C/V
Magnetic flux	weber	Wb	V-s
Magnetic induction	tesla	T	Wb/m^2
Inductance	henry	H	Wb/A
Light flux	lumen	Lm	Cd-rad

[a]It is more common in the semiconductor field to use cm for length and eV for energy $(1 \text{ cm} = 10^{-2} \text{ m}, 1 \text{ eV} = 1.6 \times 10^{-19} \text{ J})$

Appendix C: Unit Prefixes[*]

Multiple	Prefix	Symbol
10^{18}	exa	E
10^{15}	peta	P
10^{12}	tera	T
10^{9}	giga	G
10^{6}	mega	M
10^{3}	kilo	k
10^{2}	hecto	h
10	deka	da
10^{-1}	deci	d
10^{-2}	centi	c
10^{-3}	milli	m
10^{-6}	micro	μ
10^{-9}	nano	n
10^{-12}	pico	p
10^{-15}	femto	f
10^{-18}	atto	a

[*]Adopted by International Committee on Weights and Measures. (Compound prefixes should not be used, e.g., not $\mu\mu$ but p.)

Appendix D: Greek Alphabet

Letter	Lowercase	Uppercase
Alpha	α	A
Beta	β	B
Gamma	γ	Γ
Delta	δ	Δ
Epsilon	ε	E
Zeta	ζ	Z
Eta	η	H
Theta	θ	Θ
Iota	ι	I
Kappa	κ	K
Lambda	λ	Λ
Mu	μ	M
Nu	ν	N
Xi	ξ	Ξ
Omicron	o	O
Pi	π	Π
Rho	ρ	P
Sigma	σ	Σ
Tau	τ	T
Upsilon	υ	Y
Phi	ϕ	Φ
Chi	χ	X
Psi	ψ	Ψ
Omega	ω	Ω

Appendix E: Physical Constants

Quantity	Symbol	Value
Angstrom unit	Å	$1\ \text{Å} = 10^{-4}\ \mu\text{m} = 10^{-8}\ \text{cm} = 10^{-10}\ \text{m}$
Avogadro constant	N_{av}	6.02214×10^{23}
Bohr radius	a_B	$0.52917\ \text{Å}$
Boltzmann constant	k	$1.38066 \times 10^{-23}\ \text{J/K}\ (R/N_{av})$
Elementary charge	q	$1.60218 \times 10^{-19}\ \text{C}$
Electron rest mass	m_0	$0.91094 \times 10^{-30}\ \text{kg}$
Electron volt	eV	$1\ \text{eV} = 1.60218 \times 10^{-19}\ \text{J}$
		$= 23.053\ \text{kcal/mol}$
Gas constant	R	$1.98719\ \text{cal/mol} - \text{K}$
Permeability in vacuum	μ_0	$1.25664 \times 10^{-8}\ \text{H/cm}\ (4\pi \times 10^{-9})$
Permittivity in vacuum	ε_0	$8.85418 \times 10^{-14}\ \text{F/cm}\ (1/\mu_0 c^2)$
Planck constant	h	$6.62607 \times 10^{-34}\ \text{J-s}$
Reduced Planck constant	$\hbar$	$1.05457 \times 10^{-34}\ \text{J-s}\ (h/2\pi)$
Proton rest mass	M_p	$1.67262 \times 10^{-27}\ \text{kg}$
Speed of light in vacuum	c	$2.99792 \times 10^{10}\ \text{cm/s}$
Standard atmosphere		$1.01325 \times 10^5\ \text{Pa}$
Thermal voltage at 300 K	kT/q	$0.025852\ \text{V}$
Wavelength of 1-eV quantum	λ	$1.23984\ \mu\text{m}$

Appendix F: Lattice Constants at 300 K

Element or Compound	Name	Crystal Structure[a]	Lattice Constant at 300 K (Å)
Element			
C	Carbon (diamond)	D	3.56683
Ge	Germanium	D	5.65790
Si	Silicon	D	5.43102
Sn	Gray tin	D	6.4892
IV–IV			
6H-SiC	Silicon carbide	W	$a = 3.08,\ c = 15.117$
III–V			
AlAs	Aluminum arsenide	Z	5.660
AlN	Aluminum nitride	W	$a = 3.11,\ c = 4.98$
AlP	Aluminum phosphide	Z	5.4635
AlSb	Aluminum antimonide	Z	6.1355
BN	Boron nitride	Z	3.6157
BP	Boron phosphide	Z	4.5383
GaAs	Gallium arsenide	Z	5.65325
GaN	Gallium nitride	W	$a = 3.16,\ c = 5.12$
GaP	Gallium phosphide	Z	5.4505
GaSb	Gallium antimonide	Z	6.09593
InAs	Indium arsenide	Z	6.0583
InN	Indium nitride	W	$a = 3.5446,\ c = 5.7034$
InP	Indium phosphide	Z	5.8687
InSb	Indium antimonide	Z	6.47937
II–VI			
CdS	Cadmium sulfide	Z	5.8320
CdS	Cadmium sulfide	W	$a = 4.16,\ c = 6.756$
CdSe	Cadmium selenide	Z	6.050
CdTe	Cadmium telluride	Z	6.482
ZnO	Zinc oxide	R	4.580

Element or Compound	Name	Crystal Structure[a]	Lattice Constant at 300 K (Å)
ZnS	Zinc sulfide	Z	5.420
ZnS	Zinc sulfide	W	$a = 3.82$, $c = 6.26$
ZnSe	Zinc selenide	Z	5.6676
ZnTe	Zinc telluride	Z	6.1037
IV–VI			
PbS	Lead sulfide	R	5.9362
PbTe	Lead telluride	R	6.4620

[a]D, diamond; W, wurtzite; Z, zincblende; R, rock salt.

Appendix G: Properties of Important Element and Binary Semiconductors

Semiconductor	Bandgap (eV)		Mobility at 300 K (cm^2/V-s)[a]		Energy Band[b]	Effective Mass m^*/m_0		Dielectric Constant $\varepsilon_s/\varepsilon_0$
	300 K	0 K	Electrons	Holes		Electrons[c]	Holes[d]	
Element								
C	5.47	5.48	2000	2100	I	1.4/0.36	1.08/0.36	5.7
Ge	0.66	0.78	3900	1800	I	1.57/0.082	0.28/0.04	16.2
Si	1.124	1.17	1450	505	I	0.92/0.19	0.54/0.15	11.9
Sn	0	0.94	10^5 @ 100 K	10^4 @ 100 K	D	0.023	0.195	24
IV–IV								
6H-SiC	2.86	2.92	300	40	I	1.5/0.25	1.0	9.66
III–V								
AlAs	2.15	2.23	294	—	I	1.1/0.19	0.41/0.15	10
AlN	6.2	—	—	14	D	—	—	9.14
AlP	2.41	2.49	60	450	I	3.61/0.21	0.51/0.21	9.8
AlSb	1.61	1.68	200	400	I	1.8/0.26	0.33/0.12	12
BN	6.4	—	4	—	I	0.752	0.37/0.15	7.1
BP	2.4	—	120	500	I	—	—	11

Semiconductor	Bandgap (eV)		Mobility at 300 K (cm^2/V-s)[a]		Energy Band[b]	Effective Mass m^*/m_0		Dielectric Constant
	300 K	0 K	Electrons	Holes		Electrons[c]	Holes[d]	$\varepsilon_s/\varepsilon_0$
GaAs	1.424	1.519	9200	320	D	0.063	0.50/0.076	12.4
GaN	3.44	3.50	440	130	D	0.22	0.96	10.4
GaP	2.27	2.35	160	135	I	4.8/0.25	0.67/0.17	11.1
GaSb	0.75	0.82	3750	680	D	0.0412	0.28/0.05	15.7
InAs	0.353	0.42	33000	450	D	0.021	0.35/0.026	15.1
InN	1.89	2.15	250	—	D	0.12	0.5/0.17	9.3
InP	1.34	1.42	5900	150	D	0.079	0.56/0.12	12.6
InSb	0.17	0.23	77000	850	D	0.0136	0.34/0.0158	16.8
II–VI								
CdS	2.42	2.56	340	50	D	0.21	0.80	5.4
CdSe	1.70	1.85	800	—	D	0.13	0.45	10.0
CdTe	1.56	—	1050	100	D	0.1	0.37	10.2
ZnO	3.35	3.42	200	180	D	0.27	1.8	9.0
ZnS	3.68	3.84	180	10	D	0.40	—	8.9
ZnSe	2.82	—	600	300	D	0.14	0.6	9.2
ZnTe	2.4	—	530	100	D	0.18	0.65	10.4
IV–VI								
PbS	0.41	0.286	800	1000	I	0.22	0.29	17.0
PbTe	0.31	0.19	6000	4000	I	0.17	0.20	30.0

[a]The values are for drift mobilities obtained in the purest and most perfect materials available to date.
[b]I, indirect, D, direct.
[c]Longitudinal/transverse effective mass for ellipsoid constant energy surfaces.
[d]Heavy-hole/light-hole effective mass for degenerate valence band.

Appendix H: Properties of Si and GaAs at 300 K

Properties	Si	GaAs
Atoms/cm^3	5.02×10^{22}	4.42×10^{22}
Atomic weight	28.09	144.63
Breakdown field (V/cm)	$\sim 3 \times 10^5$	$\sim 4 \times 10^5$
Crystal structure	Diamond	Zincblende
Density (g/cm^3)	2.329	5.317
Dielectric constant	11.9	12.4
Effective density of states in conduction band, N_c (cm^{-3})	2.8×10^{19}	4.7×10^{17}
Effective density of states in valence band, N_v (cm^{-3})	1.04×10^{19}	7.0×10^{18}
Effective mass, m^*/m_0		
Electrons	$m_l^* = 0.92$ $m_t^* = 0.19$	0.063
Holes	$m_{lh}^* = 0.15$ $m_{hh}^* = 0.54$	$m_{lh}^* = 0.076$ $m^*_{hh} = 0.50$
Electron affinity, χ(V)	4.05	4.07
Energy gap (eV) at 300 K	1.124	1.424
Index of refraction	3.42	3.3
Intrinsic carrier concentration (cm^{-3})	1.02×10^{10}	2.1×10^6
Intrinsic Debye length (μm)	41	2900
Intrinsic resistivity (Ω-cm)	3.16×10^5	3.1×10^8
Lattice constant (A)	5.43102	5.65325
Linear coefficient of thermal expansion, $\Delta L/L\Delta T$ (°C^{-1})	2.59×10^{-6}	5.75×10^{-6}

Properties	Si	GaAs
Melting point (°C)	1412	1240
Minority-carrier lifetime (s)	3×10^{-2}	$\sim 10^{-8}$
Mobility (drift) (cm²/V-s):		
μ_n (electrons)	1450	9200
μ_p (holes)	505	320
Optical-phonon energy (eV)	0.063	0.035
Phonon mean free path $\lambda_0(\text{Å})$	76 (electrons) 55 (holes)	58
Specific heat (J/g-°c)	0.7	0.35
Thermal conductivity at 300 K (W/cm-°K)	1.31	0.46
Thermal diffusivity (cm²/s)	0.9	0.44
Vapor pressure (Pa)	1 at 1650°C 10^{-6} at 900°C	100 at 1050°C 1 at 900°C

Appendix I: Properties of Selected Ternary III–V Compound Semiconductors

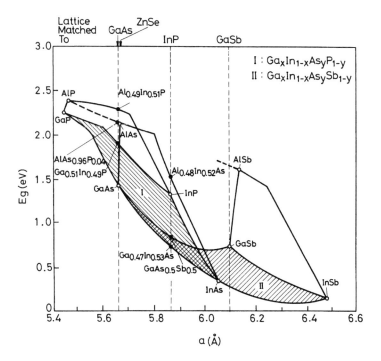

Energy bandgap vs lattice constant for various III–V compounds and their alloys. (After Madelung, Ref. 34 in the Introduction.)

Aluminum gallium arsenide ($Al_xGa_{1-x}As$)

Crystal Structure	Zincblende
Energy bandgap (eV)	$1.424 + 1.087x + 0.438x^2$ ($x < 0.43$) $1.905 + 0.10x + 0.16x^2$ ($x > 0.43$)
Lattice constant (Å)	$5.6533 + 0.0078x$
Electron mobility (cm^2/V-s)	$9200 - 22,000x + 10,000x^2$ ($x < 0.43$) $-255 + 1160x - 720x^2$ ($x > 0.43$)
Hole mobility (cm^2/V-s)	$320 - 970x + 740x^2$
Electron effective mass (m_0)	$0.063 + 0.083x$ (Γ minimum, density of states) $0.85 - 0.14x$ (X minimum, density of states) $0.56 + 0.10x$ (L minimum, density of states)
Hole effective mass (m_0)	Heavy hole: $0.50 + 0.14x$ (density of states) Light hole: $0.076 + 0.063x$ (density of states)
Dielectric constant	$12.4 - 3.12x$

Aluminum indium arsenide ($Al_xIn_{1-x}As$)

Crystal Structure	Zincblende
Energy bandgap, Γ(eV)	$0.37 + 1.91x + 0.74x^2$
Energy bandgap, X(eV)	$1.8 + 0.4x$
Bandgap crossover	$x = 0.68$, $E_g = 2.05$ eV
$Al_{0.48}In_{0.52}As$	$E_g = 1.45$ eV, lattice matched to InP

Gallium indium arsenide with $x = 0.47$ ($Ga_{0.47}In_{0.53}As$)

Crystal Structure	Zincblende
Energy bandgap (eV)	0.75
Lattice constant (Å)	5.8687, lattice matched to InP
Electron mobility (cm^2/V-s)	13800
Electron effective mass (m_0)	0.041
Hole effective mass (m_0)	Heavy hole: 0.465 Light hole: 0.05

Appendix J: Properties of SiO$_2$ and Si$_3$N$_4$ at 300 K

	SiO$_2$	Si$_3$N$_4$
Structure	Amorphous	Amorphous
Melting point (°C)	~1600	—
Density (g/cm^3)	2.2	3.1
Refractive index	1.46	2.05
Dielectric constant	3.9	7.5
Dielectric strength (V/cm)	10^7	10^7
Infrared absorption band (μm)	9.3	11.5–12.0
Energy gap (eV)	9	~5.0
Thermal-expansion coefficient (°C^{-1})	5×10^{-7}	
Thermal conductivity (W/cm-K)	0.14	
DC resistivity (Ω-cm):		
at 25°C	10^{14}–10^{16}	~10^{14}
at 500°C		~2 × 10^{13}
Etch rate in buffered HF^a (Å/min)	1000	5–10

^{a}Buffered HF: 34.6% (wt.) NH$_4$F, 6.8% (wt.) HF, 58.6% H$_2$O.

INDEX